Condensed Matter Theories

Volume 23

Condensed Matter Theories

Volume 23

Virulh Sa-yakanit
Chulalongkorn University, Thailand

Editor

World Scientific

NEW JERSEY • LONDON • SINGAPORE • BEIJING • SHANGHAI • HONG KONG • TAIPEI • CHENNAI

Published by

World Scientific Publishing Co. Pte. Ltd.
5 Toh Tuck Link, Singapore 596224
USA office: 27 Warren Street, Suite 401-402, Hackensack, NJ 07601
UK office: 57 Shelton Street, Covent Garden, London WC2H 9HE

British Library Cataloguing-in-Publication Data
A catalogue record for this book is available from the British Library.

CONDENSED MATTER THEORIES
Volume 23
Proceedings of the 31st International Workshop

ISBN-13 978-981-283-661-8
ISBN-10 981-283-661-6

Printed in Singapore by B & JO Enterprise

PREFACE

The International Workshop on Condensed Matter Theories is an annual Scientific meeting which has been hosted by prominent research institutions and universities in the Americas, Europe, and Asia. The Thirty-first Workshop of the series (CMT31) took place during 3–8 December 2007, in the **Chumpot-Pantip Conference Room, 4th floor, Prajadhipok-Ramphai Barni Building, Chulalongkorn University**, Bangkok, Thailand. The meeting was held under the joint sponsorship of Chulalongkorn University, The National Research Council of Thailand (NRCT), The Asia Pacific Center for Theoretical Physics (APCT), The Schwinger Foundation and U.S. Army Research Office.

The International Workshops on Condensed Matter Theories have a strong interdisciplinary tradition, in recognition of the commonality of problems faced by theorists and computational scientists when they seek fundamental and practical understanding of many-body systems in diverse areas of physics. Researchers working in the subfields of Solid-State, Soft-Matter, Low Temperature, Materials, Atomic, Nuclear, Particle, Statistical, Astrophysical, Chemical, and Biological Physics have gathered to share new concepts and strategies, as well as to present novel developments in analysis and computation. Over the years, the CMT Workshop Series has provided extraordinary opportunities for physicists from the full spectrum of nations to interact and learn from each other in a stimulating atmosphere of collegiality, often in exotic settings. The Series has been the origin of much fruitful international collaboration.

Fifty-six invited papers were presented, of which forty-eight appear as chapters in this volume. Reports of recent results generated lively debate on two-dimensional electron systems, the metal-insulator transition, dilute magnetic semiconductors, effects of disorder, magnetoresistence phenomena, ferromagnetic stripes, quantum Hall systems, strongly correlated Fermi systems, superconductivity, dilute fermionic and bosonic gases, nanostructured materials, plasma instabilities, quantum fluid mixtures, and helium in reduced geometries.

During the opening main session of the meeting, we took the opportunity to honor Charles Campbell's 65th Birthday and his outstanding research achievements and leadership within condensed matter physics, which received amply deserved recognition in his election to fellowship of the American Physical Society. Charles Campbell's knowledge and expertise extend over many areas of condensed matter physics and theoretical physics. He is a master of the synthesis of ideas and techniques drawn from diverse sources, and an articulate spokesman for the unity of physics. Blessed with extraordinarily acute physical intuition, he has developed into one of the intellectual leaders of the subdiscipline of microscopic many-body theory. Complementing the breadth of his knowledge, there is clarity, depth, and solidity about his thinking that is a rare find. Running against the current trend, he is not one who simply grinds out results using some fashionable theoretical recipe. Rather, he is a true pioneer and innovator. The impact of his work on the present shape of *ab initio* many-body theory has been substantial; in fact it is abundantly clear that his contributions to correlated wave-function theories have been critical for the success that this approach has enjoyed since the 1970s.

The Workshop and its participants were the beneficiaries of generous financial support from the National Research Council of Thailand (NRCT), Asia Pacific Center for Theoretical Physics (APCT), Schwinger Foundation and U.S. Army Research Office.

V. Sa-yakanit

XXXI International Workshop on Condensed Matter Theories (CMT31)

VIRULH SA-YAKANIT
(Editor)

International Advisory Committee

List of Participants

Charles Edwin Campbell	Department of Physics University of Minnesota, Minneapolis, USA Email: campbell@umn.edu
Eckard Krotscheck	Institute for Theoretical Physics, Altenberger Strasse 60 JKU Linz, Austria Email: eckhard.krotscheck@jku.at
Manfred Louis Ristig	Department of Physics, University of Cologne Germany Email: ristig@thp.uni-koeln.de
Mikko Juhani Saarela	Physical Sciences, P.O. Box 3000, University of Oulu, Finland Email: Mikko.Saarela@oulu.fi
Susana Hernandez	Department of Physics – University of Buenos Aires Ciudad Universitaria – Pab. I Argentina Email: shernand@df.uba.ar
Mukunda Prasad Das	Department of Theoretical Physics Institute of Advanced Studies, ANU, Canberra, The Australian National University, ACT, Australia Email: mukunda.das@anu.edu.au
Khandker Fazlul Quader	Kent State University Dept of Physics, Kent State University, Kent, OH 44242, USA Email: quader@scorpio.kent.edu
Manuel de Llano	Instituto de Investigaciones en Materiales, UNAM, Mexico City Email: dellano@servidor.unam.mx
Gertrud Elisabeth Zwicknagl	Inst. f. Mathemat. Physik, Mendelssohnstr. 3, Braunschweig, Germany Email: g.zwicknagl@tu-bs.de
Julius Heinz Ranninger	CNRS, Institut Neel – MCBT 25 Avenue de Martyrs, GRENOBLE, France Email: julius.ranninger@grenoble.cnrs.fr
T. A. Mamedov	Faculty of Engineering, Baskent University, 06530 Ankara, Turkey and Institute of Physics, Academy of Sciences of Azerbaijan, 370143 Baku, Azerbaijan Email: tmamedov@baskent.edu.tr

Virulh Sa-yakanit	Center of Excellence in Forum for Theoretical Science Department of Physics, Faculty of Science Chulalongkorn University, Bangkok, 10330 Thailand Email: vsayakanit@gmail.com
F. Bary Malik	Southern Illinois University, Carbondale, Illinois 62901, USA Email: fbmalik@physics.siu.edu
Johannes Richter	ITP, University Magdeburg P.O. BOX 4122, D-39016 Magdeburg Email: Johannes.Richter@Physik.Uni-Magdeburg.DE
Shigeji Fujita	UB, Physics Buffalo-Amherst Email: fujita@buffalo.edu
Erich K. R. Rung	University of Ilmenau, Weimarer Str. 25, 98693 Ilmenau Germany Email: Erich.Runge@tu-ilmenau.de
Andrey Varlamov	COHERENTIA CNR-INFM via Politecnico, 1, Italy Email: varlamov@ing.uniroma2.it
John W. Clark	Department of Physics Washington University in St. Louis, Missouri, USA Email: jwc@wuphys.wustl.edu
James Frederick Annett	Physics Department, University of Bristol, Bristol, UK Email: james.annett@bristol.ac.uk
Victor Galitski	Physics Department and Joint Quantum Institute University of Maryland, College Park USA Email: galitski@umd.edu
S.-R. Eric Yang	Korea University South Korea Email: eyang@venus.korea.ac.kr
Feodor V. Kusmartsev	Department of Physics, Loughborough University, LE11 3TU, UK Email: F.Kusmartsev@lboro.ac.uk
Rajeev Ahuja	Physics Department, Uppsala University, Uppsals, Sweden Email: rajeev@fysik.uu.se
Bilal Tanatar	Physics Department, Bilkent University, Ankara, Turkey Email: tanatar@fen.bilkent.edu.tr

Ali Zaoui — L.M.L. (UMR 8107)
Polytech'Lille.Université des Sciences et Technologies de Lille. Cite scientifique, Avenue Paul Langevin,
59655 Villeneuve D'Asq Cedex, France
Email: azaoui@polytech-lille.fr

Gerasimos Spyridon Anagnostatos — Institute of Nuclear Physics
National Center for Scientific Research "Demokritos"
Aghia Paraskevi, Greece
Email: anagnos4@otenet.gr

Prafulla Kumar Panda — Department of Physics, Chemistry
Indian Association for the Cultivation of Sciences
Jadavpur, Kolkata, India
Email: pcpkp@iacs.res.in

Gerd Roepke — Institut Physic,
University of Rostock, Germany
Email: gerd.roepke@uni-rostock.de

Abdullah Shams Bin Tariq — Rajshahi University
Department of Physics,
Rajshahi University, Bangladesh
Email: asbtariq@ru.ac.bd, asbtariq@yahoo.com

Taiichi Yamada — Laboratory of Physics,
Kanto Gakuin University,
Yokohama 236-8501, Japan
Email: yamada@kanto-gakuin.ac.jp

Raymond F. Bishop — School of Physics and Astronomy,
The University of Manchester,
UK
Email: raymond.bishop@manchester.ac.uk

John Stuart Briggs — University of Freiburg
H-Herder Str. 3, 79104 Freiburg, Germany
Email: briggs@physik.uni-freiburg.de

Supitch Khemmani — Department of Physics,
Faculty of Science, Srinakharinwirot University,
Bangkok 10110, Thailand
Email: supitch@swu.ac.th

Siu-Tat Chui — University of Delaware
Sharp Lab, Newark, DE, USA
Email: chui@udel.edu

David Neilson — Dipartimento di Fisica
Univer Pisa, Italy
Email: david.neilson@unicam.it

Hisazumi Akai — Department of Phyics,
Osaka University, Toyonaka, Osaka, Japan
Email: akai@phys.sci.osaka-u.ac.jp

M Howard Lee	University of Georgia Physics Department, Athens, GA, USA Email: mhlee@uga.edu
Eva Y. Andrei	Department of Physics Rutgers University, USA Email: eandrei@phyics.rutgers.edu
Angelo Plastino	National University La Plata La Plata – Argentina Email: plastino@fisica.unlp.edu.ar
R. N. Bhatt	Princeton Center for Theoretical Physics, Princeton University, Princeton, New Jersey, USA Email: ravin@princeton.edu
Ou-Yang Zhong-can	Institute of Theoretical Physics Chinese Academy of Sciences, China Email: oy@itp.ac.cn
Henrik G. Bohr	Quantum Physics Centre Department of Physics Technical University of Denmark, Denmark Email: hbohr@fysik.dtu.dk
Heidi Ella Mechthild Reinholz	Rostock University Universitätsplatz 3, Germany Email: heidi.reinholz@uni-rostock.de
Eduardo Vicente Ludena	Instituto venezolano de Investigaciones Cientificas, Chemistry Center, IVIC, Apartado 21827, Venezuela Email: popluabe@yahoo.es
Stewart John Clark	Dept of Physics, Science Laboratories, University of Durham, Durham, UK Email: s.j.clark@durham.ac.uk
Helga M. Boehm	Inst. f. Theoretical Physics Johannes Kepler University Linz, Austria Email: helga.boehm@jku.at
Alexander Juergen Eisfield	University of Freiburg H-Herder Str. 3, 79104 Freiburg, Germany Email: alexander.eisfield@physik.uni-freiburg.de
Eric Surauad	LPT-IRSAMC University P. Sabatier, France Email: suraud@irsamc.ups-tlse.fr
Karl E. Kurten	Faculty of Physics, Vienna University 5 boltzmanngasse, Austria Email: karl.kuerten@univie.ac.at

Dao Tien Khoa	Institute for Nuclear Science and Technique, VAEC P.O. Box 5T-160, Nghia Do, Hanoi, Vietnam Email: khoa@mail.vaec.gov.vn
Arun Kumar Basak	Department of Physics Rajshahi University, Rajshahi 6205, Bangladesh Email: akbasak2001@yahoo.com
F. F. Karpeshin	Fock Institute of Physics, St. Petersburg State University, 198504 St. Petersburg, Russia Email: karpesh@nuclpc1.phys.spbu.ru
Tawee Tunkasiri	Department of Physics, Faculty of Science, Chiang Mai University, Chiang Mai, 50200 Thailand Email: tawee@chiangmai.ac.th
Sukum Eitssayeam	Department of Physics, Faculty of Science, Chiang Mai University, Chiangmai, 50200 Thailand Email: sukum99@yahoo.com
Treedej–Kittiauchawal	Department of Physics, Faculty of Science, King Mongkut's University of Technology, Thonburi, 10140 Thailand Email: opticslaser@hotmail.com
Chittra–Kedkaew	Department of Physics, Faculty of Science, King Mongkut's University of Technology, Thonburi, 10140 Thailand Email: opticslaser@hotmail.com
Kamonpan Pengpat	Department of Physics, Faculty of Science, Chiang Mai University, Chiangmai, 50200 Thailand Email: kpengpat@gmail.com
Vudhichai Parasuk	Department of Chemistry, Faculty of Science Chulalongkorn University, Bangkok 10330, Thailand Email: pvudhich@pioneer.netserv.chula.ac.th
S. Boonchui	Department of Physics, Faculty of Science Kasetsart University Bangkok 10110, Thailand Email: sooty_th@yahoo.com

Group photo.

Charles Campbell's 65th Birthday.

Charles Campbell at the Chumpot-Pantip Conference Room.

CONTENTS

Main Session

Condensed Matter

Main Session

CHARLES CAMPBELL AT SIXTY-FIVE: A TRIBUTE TO INNOVATION AND ENDURING DEDICATION

JOHN W. CLARK

Department of Physics, Washington University
St. Louis, MO 63130 USA

jwc@wustl.edu

Received 31 July 2008

A retrospective of the career of Charles E. Campbell in condensed matter physics is presented as a tribute to his pathbreaking contributions to quantum many-body theory and his selfless dedication to the advancement of the research community associated with the Condensed Matter Workshop series.

1. Career History

A native of Ohio, Charles E. Campbell attended Ohio State University and received his B.S. in 1964, having been elected to Phi Beta Kappa. I first came to know him in the fall of the same year, when he entered our graduate program at Washington University in St. Louis. He was a student in several of my courses, developed rapidly as one of the top theory students of an outstanding class, and joined our many-body theory group, working under the direction of Eugene Feenberg. His promise for exceptional achievement was already visible at the time, and indeed was exemplified in a brilliant thesis on the correlation structure of quantum fluids, completed in 1969. Chuck (as he is known to most of us) then spent two years as a postdoc with Michael Schick at the University of Washington in Seattle, followed by two more postdoctoral years with Alexander Fetter at Stanford. During this period his research led to significant advances in the theory of helium monolayers.

In 1973, Campbell joined the physics faculty of the University of Minnesota (Twin Cities) and rose through the ranks to become full professor in 1981. He has served as Head of the School of Physics and Astronomy and played an important role in the establishment of the Theoretical Physics Institute. His distinguished service to the University of Minnesota has been interrupted only by research leaves spent at the University of Cologne (as a Humboldt Fellow), Los Alamos National Laboratory (Associated Western Universities Sabbatical Fellow), Hong Kong Institute of Science & Technology (Visiting Professor), and the University of Linz (Fulbright Senior Lecturer and Guest Professor).

2. Benchmarks of Innovation in Many-Body Theory

In 1996, Charles Campbell's outstanding research achievements and leadership within condensed matter physics received amply deserved recognition in his election to Fellow of the American Physical Society. Chuck's knowledge and expertise extend over many areas of condensed matter physics and theoretical physics. He is a master at the synthesis of ideas and techniques drawn from diverse sources, and an articulate spokesman for the unity of physics. Blessed with extraordinarily acute physical intuition, he has developed into one of the intellectual leaders of the subdiscipline of microscopic many-body theory. Complementing the breadth of his knowledge, there is a clarity, depth, and solidity about his thinking that is rare to find. Running against the current trend, he is not one who simply grinds out results using some fashionable theoretical recipe. Rather, he is a true pioneer and innovator. The impact of his work on the present shape of *ab initio* many-body theory has been substantial; in fact it abundantly clear that his contributions to correlated wave-function theories have been critical for the success that this approach has enjoyed since the 1970's.

Over the years, Chuck Campbell's primary research efforts have been focused on the theory of quantum fluids. He has played a pioneering role in bringing correlated-basis (or CBF) approaches to a high level of quantitative accuracy, through the introduction of procedures for optimal determination of the essential correlation functions. Early in his career, he developed the *paired-phonon analysis* into a practical method for microscopic description of the ground states and elementary excitations of strongly-interacting Bose systems. Paired-phonon analysis (PPA), which he formulated together with Eugene Feenberg,[1] is an approach within CBF theory in which the many-particle Hamiltonian is diagonalized in a subspace spanned by paired-phonon states. Importantly, the two of them recognized that the paired-phonon subspace contains the *optimal* Jastrow trial ground state for the system. They went on to show how to find this state consistently within the hypernetted-chain (HNC) approximation, in a paper that preceded any similar work by nearly a decade. At Minnesota, Campbell extended the optimization of ground-state structure beyond the Jastrow trial function to include triplet (or intrinsic three-body) correlations[2] and (with C. C. Chang, one of his first doctoral students) applied this extension[3] to study the density dependence of the roton spectrum in liquid ^{4}He. An important calculational breakthrough within PPA came in 1977, when Campbell and Chang[4] transformed the approach into an efficient and powerful iterative scheme for optimal determination of the static structure function and the two-body pseudopotential.

This body of work has been of fundamental importance for the development of modern many-body theory,[5] allowing us to move beyond model problems and asymptotic behaviors to quantitative prediction of structure, excitations, and dynamics from first principles. Applications of PPA optimization theory have expanded far outside the original context of bulk liquid ^{4}He, with successes in such diverse problems as helium films and surfaces, helium clusters, ^{3}He–^{4}He mixtures, the electron gas, metal surfaces, atoms, nuclei, spin systems, and even lattice gauge models. One of the most valuable properties of variational-CBF/HNC theory, as formulated by Campbell and collaborators, is that it fails when it should fail – e.g. if one is seeking the optimal translationally invariant Jastrow ground state, the equations have no solutions at densities below the spinodal point, or in the high-density region corresponding to a solid phase. Such behavior stands in strong contrast with that of more traditional (and sometimes more famous) approaches rooted in perturbation theory – which continue to yield predictions (and sometimes continue to be taken seriously) in regions where they are known to be incorrect on qualitative physical grounds.

Charles Campbell's productivity has been sustained over many years in a series of contributions of exceptional quality. In a not-quite-random sampling, these include his research on the stability of quantum fluid mixtures with Karl Kürten,[6] on liquid metallic hydrogen with John Zabolitzky,[7] on the quantized Hall effect with Tao Pang,[8] and on the static pair-pair correlation function in classical fluids.[9] One may point especially to the superb work on electron correlations in atomic systems carried out with Pang and Krotscheck[10] within the optimal variational theory of inhomogeneous Fermi systems. In this last-cited effort there is the promise (yet to be fully exploited) of fundamental advances in the realistic calculation of electronic structure and correlations. These studies are indicative of the breadth of Campbell's command of central issues in condensed matter physics.

Another profoundly important development spearheaded by Campbell is the extension of variational-CBF theory to finite temperatures, an advance with impact comparable to that of PPA optimization theory. The essential steps were taken while Chuck was visiting and working with Fred Ristig in Cologne on a Humboldt Fellowship (1981–82), the first papers[11,12] being coauthored with Fred, Karl Kürten, and Fred's student Gerd Singer. This effort epitomizes the path-breaking nature of much of Campbell's research. Trial density matrices are constructed that build in strong correlations along with thermal excitations; the Gibbs-Delbrück-Molière variational principle is applied to the corresponding free energy, to yield Euler-Lagrange equations for the temperature-dependent correlations and excitations. Implementation of this methodological breakthrough continues to spread across a broad range of strongly-correlated many-body problems, notably in studies of the thermodynamic properties of lattice gauge models and lattice-spin systems as well as helium surfaces and films. Further work along this line by Chuck includes (i) seminal conceptual papers[13,14] with Clements, Krotscheck, and Smith, (ii) the landmark calculations on the thermodynamics of boson quantum films with Clements, Krotscheck, and Saarela,[15] and (iii) steps toward a true microscopic understanding of the λ transition in liquid ^{4}He and its relation to Bose-Einstein condensation.[16]

For the last five years or so, Chuck has been active in a quite separate area of condensed matter physics, exploring new ideas and empirical findings[17,18,19] on micromagnetic domain evolution in magnetic thin films, stripes, and nanoscale dots, interacting strongly with experimental colleagues and students in the Magnetic Microscopy Center at Minnesota. However, he comes to CMT31 fresh from a lengthy stay at Johannes Kepler University in Linz, where he returned with new vigor to revisit the dynamical properties of boson quantum fluids with Eckhard Krotscheck's group.

3. A Life of Service and Dedication

Charles Campbell's service to the profession, to his university, and to his students has been prodigious in its generosity – inspiring admiration (and not a small glint of guilt) in all who know him. He is universally recognized as a leading figure and contributor to two sizable communities within physics, namely quantum many-body theory and quantum fluids and solids. His has played indispensable roles in the promotion, organization, and guidance of three prominent conference series in these subfields: the International Conferences on Recent Progress in Quantum Many-Body Theories (the main series of the field, the seventh meeting being held at the University of Minnesota in 1991 with Campbell as chief local organizer), the International Symposia on Quantum Fluids and Solids (QFS2000 having been held in Minneapolis, again with Chuck as a key organizer), and of course our annual International Workshops on Condensed Matter Theories, which is distinguished by strong involvement of scientists from developing countries. As a major voice in these communities, Campbell has gained respect from his colleagues for his integrity and wisdom, his dedication, and his clear-headed approach to problems. He has given freely of his time and energy, in spite of heavy obligations to teaching, service, and research at his home institution. We could not do without him.

Although less less familiar to his professional colleagues at other institutions, Campbell's dedication to his own school and university has been phenomenal in its scope and impact. His appointments and committee assignments, far too numerous to list, some reaching to the highest levels of policy formation, reflect an intense commitment to the welfare of his university and its people. In 2001, this important aspect of Chuck's life and career was recognized in part by the University of Minnesota's Institute of Technology with the George W. Taylor Award for Distinguished Service.

The third dimension of Campbell's service, at least as important as the others, is expressed in his extraordinary dedication to teaching. He has a brilliant record as a master classroom teacher, admired for his excellent rapport with both undergraduate and graduate students. As a research mentor, he is a splendid example, a wise and thoughtful counselor to his doctoral students. He is not only a teacher to the young: although I was once his teacher, I continue to learn from him, year by year.

Chuck Campbell's not-so-secret secret is that in everything he does — research, service, teaching — *He cares, and cares deeply!*

4. A Personal Note: Campbell and Feenberg

Eugene Feenberg regarded Chuck Campbell as one of his very best students (probably the deepest among us). His great promise was apparent to Eugene at a very early stage. It was with great pride that he followed Chuck's development as a physicist of unusual breadth and originality — recognized today as among the true pioneers and scholars of microscopic many-body theory. I know of no one who has been more successful, in his life work, in re-expressing the Feenberg pattern of insight, clarity, and integrity.

References

1. C. E. Campbell and E. Feenberg, *Phys. Rev.* **188**, 396 (1969).
2. C. E. Campbell, *Phys. Lett.* **44A**, 471 (1973).
3. C. C. Chang and C. E. Campbell, *Phys. Rev. B* **13**, 3779 (1976).
4. C. C. Chang and C. E. Campbell, *Phys. Rev. B* **15**, 4238 (1977).
5. C. E. Campbell, in *Progress in Liquid Physics*, edited by C. A. Croxton (Wiley, NY, 1978), pp. 213-308.
6. K. E. Kürten and C. E. Campbell, *Phys. Rev. B* **26**, 124 (1982).
7. C. E. Campbell and J. G. Zabolitzky, *Phys. Rev. B* **29**, 123 (1984).
8. T. Pang and C. E. Campbell, *Phys. Rev. B* **35**, 1459 (1987).
9. B. E. Clements, C. E. Campbell, P. J. Samsel and F. J. Pinski, *Phys. Rev. A* **44**, 1139 (1991).
10. T. Pang, C. E. Campbell and E. Krotscheck, *Phys. Rep.* **223**, 1 (1992).
11. C. E. Campbell, K. E. Kürten, M. L. Ristig and G. Senger, *Phys. Rev. B* **30**, 3728 (1984).
12. G. Senger, M. L. Ristig, K. E. Kürten and C. E. Campbell, *Phys. Rev. B* **33**, 7562 (1986).
13. B. E. Clements and C. E. Campbell, *Phys. Rev. B* **46**, 10957 (1992).
14. J. A. Smith, B. E. Clements, E. Krotscheck and C. E. Campbell, *Phys. Rev. B* **47**, 5239 (1993).
15. C. E. Campbell, B. E. Clements, E. Krotscheck and M. Saarela, *Phys. Rev. B* **55**, 3769 (1997).
16. G. Senger, M. L. Ristig, C. E. Campbell and J. W. Clark, *Ann. Phys.* **218**, 160 (1992).
17. C. Bayer, J. P. Park, H. Wang, M. Yan, C. E. Campbell and P. A. Crowell, *Phys. Rev. B* **69**, 134401 (2004).
18. G. D. Skidmore, A. Kunz, C. E. Campbell and E. D. Dahlberg, *Phys. Rev. B* **70**, 012410 (2004).
19. H. Wang and C. E. Campbell, Phys. Rev. B **76**, 220407 (2007).

MANY-BOSON DYNAMIC CORRELATIONS

C. E. CAMPBELL[†‡] and E. KROTSCHECK[†]
[†]*Institut für Theoretische Physik, Johannes-Kepler-Universität, A-4040 Linz, Austria*
[‡]*School of Physics and Astronomy, University of Minnesota, Minneapolis, Minnesota 55455*

In this paper we present an overview of a systematic development of the linear equations of motion for a dynamically correlated wave function that moves beyond the previous theories that include time-dependent pair correlations at most. We argue that these time-dependent pair correlations are insufficient to describe important physical effects in the energy/momentum regime of the ^{4}He roton; minimally, time-dependent three-body correlations are necessary to capture the relevant physics. For simplicity we illustrate this on the problem of atomic impurities in ^{4}He.

1. Introduction

The high level of success of the semi-analytic Jastrow-Feenberg method for describing the ground state of strongly correlated boson fluids such as liquid ^{4}He has not yet been matched in the theory of the dynamics of such systems. However, there has been significant progress in the theory of dynamics using an action extremum principle which parallels the Jastrow-Feenberg functional optimization procedure. Extremization in that case produces equations of motion for reduced densities, leading, most importantly, to a theory of the dynamic structure function $S(k,\omega)$ that is experimentally measurable using inelastic neutron scattering. The first steps in this procedure have been successfully completed, giving e.g. a semi-quantitative theory of the phonon-roton spectrum in liquid ^{4}He.[1] Moreover there is formal agreement between this theory — at a well-understood level of approximation — with the alternative approach of using the method of Correlated Basis Functions (CBF) together with Brillouin-Wigner (BW) perturbation theory for the phonon-roton spectrum.[2,3,4] In this paper, we briefly report on progress beyond this stage that raises hope for the accurate inclusion of pair *and* triplet dynamic fluctuations which are important for an understanding of the short wavelength structure of $S(k,\omega)$ and the phonon-roton spectrum in the short-wavelength plateau region. A more detailed report on this work may be found in Ref. 5.

2. Correlated Dynamics

The Jastrow-Feenberg functional optimization procedure makes use of the fact that the exact ground state wave function for an N-body Bose system may be written in the Jastrow-Feenberg form:[6]

$$\Psi_0(\mathbf{r}_1, \dots \mathbf{r}_N) = \exp \frac{1}{2} \left\{ \sum_i u^{(1)}(\mathbf{r}_i) + \sum_{i<j} u^{(2)}(\mathbf{r}_i, \mathbf{r}_j) + \dots \right\} \tag{1}$$

thereby exhibiting the fundamental n-body static correlation factors as the exponentials of the real functions $u^{(n)}(\mathbf{r}_{i_1}, \dots \mathbf{r}_{i_n})$. The n-body factors must satisfy the set of equations obtained by minimizing the energy expectation value

$$\frac{\delta}{\delta u^{(n)}} \frac{\langle \Psi_0 \mid H \mid \Psi_0 \rangle}{\langle \Psi_0 \mid \Psi_0 \rangle} = 0\,. \tag{2}$$

The one-body function $u^{(1)}(\mathbf{r}_i)$ determines the spatial geometry, and the two-body function $u^{(2)}(\mathbf{r}_i, \mathbf{r}_j)$ accounts for fundamental correlations between pairs of particles. The short-range behavior of $u^{(2)}(\mathbf{r}_i, \mathbf{r}_j)$ is determined by the short-range properties of the interaction, whereas its long-range behavior is determined by the physics of the long wavelength excitations.

The Jastrow-Feenberg functional optimization procedure provides a variational approximation to the ground state wavefunction by truncating the multi-particle correlation factors of Eq. (1) at a finite value of n and then solving the Euler-Lagrange equations (2). This must include triplet correlation factors ($n = 3$) to obtain quantitatively accurate results in the very dense helium liquids.[6] If the sum in the exponential is not truncated and the set of equations (2) is (in principle) solved for all n, this Jastrow-Feenberg representation of the ground state wave function is exact for bosons.

In order to describe dynamics, it is natural to generalize the Jastrow-Feenberg theory to time-dependent wave functions. In this paper, we describe how the theoretical description of the dynamics of a Bose fluid may be advanced to the next level by including *time dependent triplet correlations.*

For simplicity — and because of renewed interest — we focus here on the problem of an atomic impurity in the Bose fluid, which was first done by Owen[7] in the long-wavelength limit and later generalized to finite wavelengths, *cf.* Ref. 8. Restricting most of our attention to the impurity problem simplifies the theory while illustrating the relevant physics. The simplification comes about because there is only one impurity particle, from which it follows that the theory involves distribution functions of one less particle than the comparable theory of the bulk excitations. Moreover there is renewed experimental interest in the impurity problem in the form of a muonic helium impurity in ^{4}He. An interesting question of experimental relevance has been raised: to what extent would there be energy loss by a muonic helium impurity due to coupling to the background excitations?[9] Damping can occur if the impurity can lose energy by coupling to the phonon-roton spectrum. The relevancy of this question is illustrated by the fact that an estimate of the effective mass of muonic helium in liquid ^{4}He shows that its free-particle spectrum, modified by the effective mass, crosses the phonon-roton ^{4}He spectrum near the experimentally determined roton energy, where there is a high density of states, and thus one would expect strong contributions of the roton to the actual impurity excitation spectrum and damping of muonic helium in ^{4}He. To be able to determine whether damping can occur, one must first have a quantitative theory of both the phonon-roton spectrum and the impurity excitations. Then, to predict the coupling between the roton and the impurity, it is also important to know the strength of the roton excitation and the coupling matrix element.

The dynamic correlation theory begins with a time-dependent wave function in the form

$$\Phi(t) = \frac{1}{\sqrt{\langle \Psi^I \mid \Psi^I \rangle}} e^{-iE^I_{N+1}t/\hbar} \Psi^I(\mathbf{r}_0, \mathbf{r}_1, ...\mathbf{r}_N; t) \ , \tag{3}$$

where $\Psi^I(\mathbf{r}_0, \mathbf{r}_1, ...\mathbf{r}_N; t)$ contains the time-dependent correlations, written in a Jastrow-Feenberg form

$$\Psi^I(\mathbf{r}_0, \mathbf{r}_1, ...\mathbf{r}_N; t) = \exp\left[\frac{1}{2}\delta U(\mathbf{r}_0, \mathbf{r}_1, ...\mathbf{r}_N; t)\right] \Psi^I_0(\mathbf{r}_0, \mathbf{r}_1, ..., \mathbf{r}_N) \ , \tag{4}$$

$$\delta U(\mathbf{r}_0, \ldots, \mathbf{r}_N; t) = \delta u_1(\mathbf{r}_0; t) + \sum_i \delta u_2(\mathbf{r}_0, \mathbf{r}_i; t) + \sum_{i<j} \delta u_3(\mathbf{r}_0, \mathbf{r}_i, \mathbf{r}_j, ; t) + \cdots \tag{5}$$

where E^I_{N+1} is the ground state energy of N-body liquid ^{4}He with one impurity particle and Ψ^I_0 is the corresponding ground state wave function, which is the generalization of (1) to include an impurity particle, and is assumed to be the exact ground state. The impurity particle coordinates are labeled by $\mathbf{r}_0$. Note that, unlike the correlation factors in the ground state, where the functions $u^{(n)}$ must be real, the time dependent functions $\delta u_n(t)$ in the time dependent correlation factor $\exp\left[\frac{1}{2}\delta U(t)\right]$ are in general complex.

As in the Jastrow-Feenberg ground state theory, the sum in $\delta U(t)$ is approximated by truncating it at a small value of n. The step that takes the present work beyond previous dynamical theories is the inclusion of the $\delta u_3(\mathbf{r}_0, \mathbf{r}_i, \mathbf{r}_j, ; t)$ in the calculation. It is this *three-body* dynamical function that includes the effect of pair fluctuations on the impurity spectrum.

Stationarity of the action integral,

$$\mathcal{S} = \int dt \mathcal{L}(t) = \int dt \left\langle \Phi(t) \right| H^I_{N+1} - i\hbar \frac{\partial}{\partial t} \left| \Phi(t) \right\rangle \ , \tag{6}$$

with respect to the $\delta u_n(t)$ is used to generate the equations of the dynamics. H^I_{N+1} is the Hamiltonian of the system containing one impurity particle in the N particle host. Since we are concerned with linear response, we expand the Lagrangian to second order:

$$\begin{aligned} \mathcal{L}_{\rm int}(t) &= \left\langle \Psi^I(t) \right| H^I_{N+1} - E^I_{N+1} \left| \Psi^I(t) \right\rangle = \frac{1}{8} \left\langle \Psi^I_0 \right| [\delta U^*, [T, \delta U]] \left| \Psi^I_0 \right\rangle \\ &= \sum_i \frac{\hbar^2}{8m_i} \left\langle \Psi^I_0 \right| \left| \nabla_i U \right|^2 \left| \Psi^I_0 \right\rangle \end{aligned} \tag{7}$$

$$\mathcal{L}_t(t) = \left\langle \Psi^I(t) \right| i\hbar \frac{\partial}{\partial t} \left| \Psi^I(t) \right\rangle = \frac{i\hbar}{8} \left\langle \Psi^I_0 \right| \left[\delta U^* \delta \dot{U} - \delta U \delta \dot{U}^* \right] \left| \Psi^I_0 \right\rangle . \tag{8}$$

The linearized equations of motion are then

$$\frac{\delta\left[\mathcal{L}_{\rm int}(t)-\mathcal{L}_t(t)\right]}{\delta(\delta u_i^*(\mathbf{r}_0,..\mathbf{r}_{i-1};t))}=0\,, \tag{9}$$

where

$$\frac{\delta\mathcal{L}_t(t)}{\delta(\delta u_i^*(\mathbf{r}_0,..\mathbf{r}_{i-1};t))}=\frac{\mathrm{i}\hbar}{4(i-1)!}\dot\rho_i(\mathbf{r}_0,\ldots,\mathbf{r}_{i-1};t)\,. \tag{10}$$

The $\dot\rho_i(\mathbf{r}_0,\ldots,\mathbf{r}_{i-1};t)$ are complex functions, with the physical density fluctuation being their real part. The i-body functions $\rho_i(\mathbf{r}_0,\ldots,\mathbf{r}_{i-1};t)$ are the time-dependent i-body densities.

The one-body equation is simply the (complex) continuity equation

$$\nabla_0\mathbf{j}(\mathbf{r}_0,t)+\delta\dot\rho_1(\mathbf{r}_0,t)=0\,, \tag{11}$$

where $\delta\rho_1(\mathbf{r}_0,t)$ and $\mathbf{j}(\mathbf{r}_0,t)$ are functionals of δu_1, δu_2 and δu_3 given by

$$\delta\rho_1(\mathbf{r}_0;t)\equiv\rho_1(\mathbf{r}_0)\Bigg[\delta u_1(\mathbf{r}_0;t)+\int d^3\rho_1 g_2(\mathbf{r}_0,\mathbf{r}_1)\delta u_2(\mathbf{r}_0,\mathbf{r}_1;t)$$
$$+\frac{1}{2}\int d^3\rho_1 d^3\rho_2 g_3(\mathbf{r}_0,\mathbf{r}_1,\mathbf{r}_2)\delta u_3(\mathbf{r}_0,\mathbf{r}_1,\mathbf{r}_2;t)\Bigg] \tag{12}$$

and

$$\begin{aligned}\mathbf{j}_1(\mathbf{r}_0;t)&\equiv\frac{\left\langle\Psi^I_{N+1}\right|\hat{\mathbf{j}}^I(\mathbf{r}_0)\delta U\left|\Psi^I_{N+1}\right\rangle}{\left\langle\Psi^I_{N+1}\mid\Psi^I_{N+1}\right\rangle}\\&=\frac{\hbar}{2m_I\mathrm{i}}\rho_1(\mathbf{r}_0)\Big[\nabla_0\delta u_1(\mathbf{r}_0;t)+\int d^3\rho_1 g_2(\mathbf{r}_0,\mathbf{r}_1)\nabla_0\delta u_2(\mathbf{r}_0,\mathbf{r}_1;t)\\&\quad+\frac{1}{2}\int d^3\rho_1 d^3\rho_2 g_3(\mathbf{r}_0,\mathbf{r}_1,\mathbf{r}_2)\nabla_0\delta u_3(\mathbf{r}_0,\mathbf{r}_1,\mathbf{r}_2;t)\Big]\,.\end{aligned} \tag{13}$$

Here, $\hat{\mathbf{j}}^I(\mathbf{r}_0)$ is the particle current operator for the impurity, and the dimensionless integration notation $d^3\rho_i\equiv\rho_1(\mathbf{r}_i)d^3r_i$ is employed for brevity. As usual, the i-body *ground state* distribution functions which appear here are defined by

$$g_i(\mathbf{r}_0,\ldots,\mathbf{r}_{i-1};t)=\frac{\rho_i(\mathbf{r}_0,\ldots,\mathbf{r}_{i-1};t)}{\rho_1(\mathbf{r}_0)\ldots\rho_1(\mathbf{r}_{i-1})} \tag{14}$$

where the coordinate $\mathbf{r}_0$ signifies that one of the particles in these distribution functions is the impurity particle.

Similar but cumbersome equations of motion result from varying the action with respect to δu_2^* and δu_3^*. However a more tractable and transparent set of equations of motion is obtained by introducing an auxiliary one body function $\delta v_1(\mathbf{r}_0;t)\equiv\delta\rho_1(\mathbf{r}_0;t)/\rho_1(\mathbf{r}_0)$ instead of $\delta u_1(\mathbf{r}_0;t)$ as the fluctuating one-body variable suggested by the structure of Eq. (12), and after a bit more scrutiny, an auxiliary two-body fluctuation, δv_2 as the independent function replacing δu_2 is very useful:

$$\delta v_2(\mathbf{r}_0,\mathbf{r}_2;t)\equiv\delta u_2(\mathbf{r}_0,\mathbf{r}_2;t)+\int d^3\rho_3 d^3\rho_4\mathcal{G}(\mathbf{r}_0;\mathbf{r}_2,\mathbf{r}_3,\mathbf{r}_4)\delta u_3(\mathbf{r}_0,\mathbf{r}_3,\mathbf{r}_4;t), \tag{15}$$

where the auxiliary four-point function $\mathcal{G}$ is defined by

$$\mathcal{G}(\mathbf{r}_0,\mathbf{r}_1,\mathbf{r}_2,\mathbf{r}_3) \equiv \frac{1}{2\rho_1(\mathbf{r}_1)}\frac{\delta g_3(\mathbf{r}_0,\mathbf{r}_2,\mathbf{r}_3)}{\delta g_2(\mathbf{r}_0,\mathbf{r}_1)}\,. \tag{16}$$

The three equations of motion are then obtained by extremizing the action with respect to the three independent functions δv_1^*, δv_2^*, and δu_3^*. With this choice of independent dynamic functions, the equation of motion for δv_2 decouples from the three-body equation. Referring to Ref. 5 for details, including the definition of the operator $\mathcal{T}$, the equation for δv_2 becomes, after Fourier transforming with respect time:

$$\mathcal{T}(\omega)\delta v_2 = \mathrm{i}\hbar\mathbf{j}(\mathbf{r}_0,\omega)\cdot\nabla_0 g_2(\mathbf{r}_0,\mathbf{r}_1)\,. \tag{17}$$

One objective of this theory is to obtain an expression for the impurity self-energy $\Sigma^{(I)}(q,\omega)$. For simplicity, we restrict our attention to a homogeneous system. In that case, the one-body variable is a plane wave whose amplitude is a variational function, $\delta v_1(\mathbf{r}_0) = v_1 e^{\mathrm{i}\mathbf{q}\cdot\mathbf{r}_0}$, and we can also separate out the "external" momentum of the two- and three-body variational functions:

$$\delta v_2(\mathbf{r}_0,\mathbf{r}_1) = e^{\mathrm{i}\mathbf{q}\cdot\mathbf{r}_0} v_{\mathbf{q}}^{(2)}(\mathbf{r}_1-\mathbf{r}_0), \quad \delta u_3(\mathbf{r}_0,\mathbf{r}_1) = e^{\mathrm{i}\mathbf{q}\cdot\mathbf{r}_0} u_{\mathbf{q}}^{(3)}(\mathbf{r}_1-\mathbf{r}_0,\mathbf{r}_2-\mathbf{r}_0)\,. \tag{18}$$

Going to momentum space, we make the distinction between background (i.e., host fluid) and impurity quantities by appending the superscripts (I) and (B). The one-body equation (11) becomes

$$\hbar\omega\delta v_1 = \frac{\hbar^2 q^2}{2m_I}\delta v_1 + \frac{\hbar^2}{2m_I}\int\frac{d^3p}{(2\pi)^3\rho}\mathbf{q}\cdot\mathbf{p}\tilde{h}^{(I)}(p)\tilde{v}_{\mathbf{q}}^{(2)}(\mathbf{p}) \equiv \left[\frac{\hbar^2q^2}{2m_I} + \Sigma^{(I)}(q,\omega)\right]\delta v_1\,. \tag{19}$$

Inserting the formal solution of Eq. (17), we can define the self-energy as

$$\begin{aligned}\Sigma^{(I)}(q,\omega) &= \left[1 + I(q,\omega)\right]\Sigma_u^{(I)}(q,\omega)\,,\\ \Sigma_u^{(I)}(q,\omega) &= \left(\frac{\hbar^2}{2m_I}\right)^2\int\frac{d^3p_1d^3p_2}{(2\pi)^6\rho^2}\mathbf{q}\cdot\mathbf{p}_1\tilde{h}^{(I)}(p_1)\tilde{\mathcal{T}}_{\mathbf{q}}^{-1}(\mathbf{p}_1,\mathbf{p}_2)\mathbf{q}\cdot\mathbf{p}_2\tilde{h}^{(I)}(p_2)\end{aligned} \tag{20}$$

where $\tilde{\mathcal{T}}_{\mathbf{q}}(\mathbf{p}_1,\mathbf{p}_2)$ is the momentum space representation of the $\mathcal{T}(\omega)$ with variables spelled out more explicitly, and $I(q,\omega)$ is to be defined in Eq. (21). The inverse in the integrand of (20) is understood in the sense of the convolution product. This equation defines the "unrenormalized" self energy, which leads to Owen's "unrenormalized" effective mass[7]. On the other hand, we have

$$\Sigma^{(I)}(q,\omega)v_1 = \frac{\hbar^2}{2m_I}\int\frac{d^3p}{(2\pi)^3\rho}\mathbf{q}\cdot\mathbf{p}\tilde{h}(p)\tilde{u}_{\mathbf{q}}^{(2)}(\mathbf{p}) \equiv \frac{\hbar^2q^2}{2m_I}I(q,\omega)v_1 \tag{21}$$

and hence

$$\Sigma^{(I)}(q,\omega) = \frac{\Sigma_u^{(I)}(q,\omega)}{1-\frac{2m_I}{\hbar^2q^2}\Sigma_u^{(I)}(q,\omega)}\,. \tag{22}$$

The final form of the dispersion relation, expressed in terms of $\Sigma_u^{(I)}(q,\omega)$, is then

$$\frac{\frac{\hbar^2 q^2}{2m_I}}{1-\frac{2m_I}{\hbar^2 q^2}\Sigma_u^{(I)}(q,\omega)} = \hbar\omega\,. \tag{23}$$

To make contact with previous theories, if one sets δu_3 to zero and makes the uniform limit approximation in the two-body equation (17), the second-order Brillouin-Wigner form of the self-energy is obtained:

$$\Sigma_u^{(I)}(q,\omega) = \frac{1}{N}\sum_{\mathbf{p},\mathbf{p}'}\delta_{\mathbf{q}+\mathbf{p}+\mathbf{p}',0}\,\frac{\left|V^{(I;I,B)}(\mathbf{q};\mathbf{p},\mathbf{p}')\right|^2}{\hbar\omega - t_I(p) - \varepsilon(p') + \mathrm{i}\eta} \tag{24}$$

where $V^{(I;I,B)}(\mathbf{q};\mathbf{p},\mathbf{p}')$ is the coupling matrix element of a "Feynman" impurity state to a two excitation state comprising a "Feynman" impurity state of wavenumber $\mathbf{p}$ and a Feynman host-excitation (phonon-roton spectrum) of the host ^{4}He fluid; $t_I(q)$ is the impurity free particle dispersion relation and $\varepsilon(p')$ is the Feynman phonon-roton spectrum.

When triplet fluctuations are included at the same level of approximation, only the energy denominator changes. In the denominator of Eq. (24) self-energy corrections appear which are, in turn, simply self-energies:

$$\begin{aligned} t_{\mathrm{I}}(p) &\to t_{\mathrm{I}}(p) + \Sigma^{(\mathrm{I})}(p, \hbar\omega - \varepsilon(p)) \\ \varepsilon(p') &\to \varepsilon(p') + \Sigma^{(\mathrm{CBF})}(p', \hbar\omega - t_{\mathrm{I}}(p')) \end{aligned} \tag{25}$$

The structure of the denominator of (24) illustrates the problem referred to in the introduction: the presence of the unrenormalized energy spectra would cause any damping to occur at the wrong wavenumbers. A renormalization of the energy denominator by self-energy corrections is generally expected, only the energy and momenta at which those have to be taken is non-trivial.

We emphasize that, with modern methods, one is not limited to the uniform limit approximation nor to Brillouin-Wigner perturbation theory; the equations of motion, including triplet fluctuations, are imminently doable.

The inclusion of pair fluctuations in the pure Bose quantum fluids without the uniform limit approximation appears to be well within reach of modern computational methods. The full set of equations of motion including triplet fluctuations, which requires an accurate approximation for g_6, is also quite doable.

Acknowledgments

We would like to thank David Taqqu for information about the experimental interest in the problem of muonic helium impurities. One of us (CEC) would like to acknowledge partial travel support provided through a grant from the U.S. Army Research Office, Research Triangle Park, to Southern Illinois University.

References

1. B. E. Clements, E. Krotscheck and C. J. Tymczak, *Phys. Rev. B* **53**, 12253 (1996).
2. H. W. Jackson and E. Feenberg, *Ann. Phys. (NY)* **15**, 266 (1961).
3. H. W. Jackson and E. Feenberg, *Rev. Mod. Phys.* **34**, 686 (1962).
4. C. C. Chang and C. E. Campbell, *Phys. Rev. B* **13**, 3779 (1976).
5. E. Krotscheck and C. E. Campbell, Static and dynamic many-body correlations, in *Recent Progress in Many-Body Theories*, ed. J. Boronat (World Scientific Publishing Co. Pte. Ltd., Singapore, 2008).
6. C. E. Campbell, *Phys. Lett. A* **44**, 471 (1973).
7. J. C. Owen, *Phys. Rev. B* **23**, 5815 (1981).
8. E. Krotscheck, J. Paaso, M. Saarela, K. Schörkhuber and R. Zillich, *Phys. Rev. B* **58**, 12282 (1998).
9. D. Taqqu, (2007), (private communication).

AN EFFICIENT DENSITY FUNCTIONAL ALGORITHM IN A STRONG MAGNETIC FIELD

S. JANECEK and E. KROTSCHECK*

Institute for Theoretical Physics, JKU Linz,
A-4040 LINZ, Austria
* *eckhard.krotscheck@jku.at*

Received 31 July 2008

We describe the implementation of a manifestly gauge invariant configuration space method for *ab initio* electronic structure calculations in an arbitrarily strong external magnetic field. To be able to reproduce empirical data for realistic systems, we will also formulate our real-space algorithm in magnetic fields for non-local ionic pseudo-potentials. Numerical applications focus on two issues, namely a careful assessment of the convergence properties of our algorithm and in particular the implications of our gauge invariant formulation, *and* the calculation of NMR shifts for a number of typical molecules.

1. Introduction

We describe here further developments of our real-space method[?,?,?] for solving the Kohn-Sham[?] equations of density functional theory (DFT). We have recently shown[1] how local one-body Schrödinger equations can be solved in real space for strong uniform magnetic fields with practically no computational overhead compared to the field-free case. This paper is concerned with a method that is gauge invariant independent of the discretization. Our extension is motivated by experimental prospects to reach field strengths of up to 100 Tesla. Interesting physical effects are found in the band structure of materials, conduction properties of rare earth metals, and organic magnets, see Refs. 2–5. Our new method also provides improvements over previous work in the regime of linear magnetic response.

An important aspect connected with magnetic fields is the gauge invariance of any approximate scheme. In quantum chemistry the term "gauge origin problem" has been coined for this aspect. The problem has been discussed in the literature, see, for example, the review by Helgaker, Jazuński and Ruud[6]. These authors state explicitly that *within a finite linear variational subspace, gauge invariance can never be obtained exactly, only approximately for small displacements of the gauge origin.*

A second aspect has to do with the fact that local Hamiltonians are in most cases insufficient to reproduce the properties of realistic molecules and clusters. Therefore, non-local "pseudo-potentials" have been introduced [7]; the gauge invariant construction of such pseudo-potentials is an entirely different issue.

2. Real Space Diffusion Algorithm for Solving Schrödinger Equations in a Uniform Magnetic Field

2.1. *Diffusion method*

A fast and conceptually simple method for calculating the lowest n solutions of the one-body Schrödinger equation is the diffusion algorithm. The strategy is to apply the evolution operator in imaginary time,

$$\mathcal{T}(\epsilon) \equiv e^{-\epsilon \hat{H}}, \quad \hat{H} = \hat{T} + \hat{V} \tag{1}$$

repeatedly to a set of states $\{\psi_j(\mathbf{r}),\ 1 \le j \le n\}$, and to orthogonalize the states after every step. This procedure converges towards the lowest n eigensolutions of the Hamiltonian $\hat{H}$.

The evolution operator (1) has to be approximated, e.g. by factorization formulas. A well-known second order factorization of the evolution operator corresponding to the Hamiltonian (1) is the Trotter formula

$$\mathcal{T}(\epsilon) \equiv e^{-\epsilon(\hat{T}+\hat{V})} = e^{-\frac{1}{2}\epsilon\hat{V}} e^{-\epsilon\hat{T}} e^{-\frac{1}{2}\epsilon\hat{V}} + \mathcal{O}(\epsilon^3). \tag{2}$$

Factorizing $\mathcal{T}(\epsilon)$ reduces the problem of calculating the exponential of the Hamiltonian to the problem of dealing with the different factors separately. If $\hat{T}$ is a local operator in momentum space, and $\hat{V}$ is local in coordinate space, then calculating $e^{-\frac{1}{2}\epsilon\hat{V}}\psi$ and $e^{-\epsilon\hat{T}}\psi$ is almost trivial.

In the last decade, fourth order forward factorization schemes have become available for solving imaginary time evolution equations[8,9,10,?,?]. These lead to very rapid convergence for realistic systems at large imaginary time steps and therefore provide a powerful method for solving typical problems in density functional theory. A specific fourth–order factorization is[9,11]

$$\mathcal{T}^{(4)}(\epsilon) \equiv e^{-\frac{1}{6}\epsilon\hat{V}} e^{-\frac{1}{2}\epsilon\hat{T}} e^{-\frac{2}{3}\epsilon\widetilde{V}} e^{-\frac{1}{2}\epsilon\hat{T}} e^{-\frac{1}{6}\epsilon\hat{V}} = \mathcal{T}(\epsilon) + \mathcal{O}(\epsilon^5) \tag{3}$$

where

$$\widetilde{V} = \hat{V} + \frac{1}{48}\epsilon^2 [\hat{V}, [\hat{T}, \hat{V}]]. \tag{4}$$

The method is particularly simple for local potentials, $\hat{V} = V(\mathbf{r})$, because the double commutator is just another local potential.

In the presence of a vector potential $\mathbf{A}(\mathbf{r})$, the momentum operator $\mathbf{p} = -i\hbar\nabla$ is replaced by the canonical momentum operator

$$\mathbf{\Pi} = -i\hbar\nabla + e\mathbf{A}_{\rm ext}(\mathbf{r}), \tag{5}$$

and the kinetic energy operator is

$$\hat{T} = \frac{1}{2m}\left(-i\hbar\nabla + e\mathbf{A}_{\rm ext}(\mathbf{r})\right)^2 = \frac{1}{2m}\left[\Pi_x^2 + \Pi_y^2 + \Pi_z^2\right]. \tag{6}$$

The key to the inclusion of a uniform and arbitrarily strong magnetic field is the evaluation of the exponential of the *canonical* kinetic energy operator (6). To be specific, we assume that the magnetic field is in z-direction. It is computationally most efficient to work in linear gauge, then

$$\mathbf{A}(\mathbf{r}) = -By\mathbf{e}_x\,. \tag{7}$$

In analogy to the harmonic oscillator [12], the density matrix $e^{-\epsilon\hat{T}}$ can be calculated exactly:

$$e^{-\frac{\epsilon}{2m}(\Pi_x^2+\Pi_y^2+\Pi_z^2)} = e^{-\frac{\epsilon}{2m}C_x(\xi)\Pi_x^2}\, e^{-\frac{\epsilon}{2m}C_y(\xi)\Pi_y^2}\, e^{-\frac{\epsilon}{2m}C_x(\xi)\Pi_x^2}\, e^{-\frac{\epsilon}{2m}\Pi_z^2}\,, \tag{8}$$

where $\xi = \epsilon\hbar eB/m$, and

$$C_x(\xi) = \frac{\cosh(\xi)-1}{\xi\sinh(\xi)} \quad \text{and} \quad C_y(\xi) = \frac{\sinh(\xi)}{\xi}\,. \tag{9}$$

In linear gauge, the z-component factorizes trivially due to

$$[\Pi_x, \Pi_z] = [\Pi_y, \Pi_z] = 0\,. \tag{10}$$

A *gauge transformation*, i.e., adding a gradient to the vector potential and attaching a phase factor to the wave function,

$$\boldsymbol{A}(\mathbf{r}) \to \boldsymbol{A}'(\mathbf{r}) = \boldsymbol{A}(\mathbf{r}) + \nabla\chi(\mathbf{r}) \qquad \psi(\mathbf{r}) \to \psi'(\mathbf{r}) = e^{-\mathrm{i}\frac{e}{\hbar}\chi(\mathbf{r})}\psi(\mathbf{r})\,, \tag{11}$$

does not change the physics of the system.[13] This is true, however, only if discretization errors can be neglected. In realistic calculations one wants to work, of course, on a mesh that is as coarse as possible, and therefore discretization errors must be avoided. This is done by introducing the *covariant derivative*

$$\Pi_j\psi(\mathbf{r}) = -\mathrm{i}\hbar\left[\frac{\partial}{\partial x_j} + \frac{\mathrm{i}e}{\hbar}A_j(\mathbf{r})\right]\psi(\mathbf{r}) = e^{-\mathrm{i}\frac{e}{\hbar}f_j(\mathbf{r})}\left[-\mathrm{i}\hbar\frac{\partial}{\partial x_j}\right]e^{+\mathrm{i}\frac{e}{\hbar}f_j(\mathbf{r})}\,\psi(\mathbf{r})\,, \tag{12}$$

where

$$f_j(\mathbf{r}) \equiv \int^{x_j} A_j(\mathbf{r})dx_j\,. \tag{13}$$

A gauge transformation (11) simply transforms the functions $f_j(\mathbf{r})$ as

$$f_j(\mathbf{r}) \to f_j'(\mathbf{r}) = f_j(\mathbf{r}) + \chi(\mathbf{r}), \tag{14}$$

and cancels the phase factor in Eq. (11). Thus, casting the canonical kinetic energy operator in the form

$$\frac{1}{2m}\boldsymbol{\Pi}^2 = -\frac{\hbar^2}{2m}\sum_j e^{-\mathrm{i}\frac{e}{\hbar}f_j(\mathbf{r})}\frac{\partial^2}{\partial x_j^2}e^{\mathrm{i}\frac{e}{\hbar}f_j(\mathbf{r})}\,, \tag{15}$$

the theory is gauge invariant independent of any errors induced by discrete approximations of the derivative operators.

2.2. *Non-local potentials*

Accurate results of electronic structure calculations for realistic systems can normally only be obtained by describing the ion cores by non-local potentials. In that case, the Hamiltonian consists of three non-commuting terms:

$$\hat{V} = \hat{V}_{\rm loc} + \hat{V}_{\rm nl}, \quad \hat{H} = \hat{T} + \hat{V}_{\rm loc} + \hat{V}_{\rm nl} \equiv \hat{H}_{\rm loc} + \hat{V}_{\rm nl}\,, \tag{16}$$

where $\hat{V}_{\rm loc}$ gathers all local contributions to the potential (Hartree, exchange-correlation and local part of the pseudo-potentials in case of density functional theory), while $\hat{V}_{\rm nl}$ represents the non-local part of the pseudo-potentials used to describe the interaction of the valence electrons with the ion cores. A popular choice for $V_{\rm nl}$ is the Kleinman-Bylander [7] separable form:

$$V_{\rm nl}(\mathbf{r},\mathbf{r}')=\sum_{i=1}^{N} v_{\rm nl}^{(i)}(\mathbf{r}-\mathbf{R}_i,\mathbf{r}'-\mathbf{R}_i)=\sum_{i=1}^{N}\sum_{\ell m} A_\ell^{(i)}\left\langle \mathbf{r}-\mathbf{R}_i \middle| P_{\ell m}^{(i)}\right\rangle\left\langle P_{\ell m}^{(i)}\middle|\mathbf{r}'-\mathbf{R}_i\right\rangle. \tag{17}$$

Above, the $A_\ell^{(i)}$ are numerical constants characterizing the pseudo-potential, and the projectors the $P_{\ell m}^{(i)}$ are $P_{\ell m}^{(i)} = R_\ell^{(i)}(r)Y_{\ell m}$, where $R_\ell(r)$ is a radial function, and $Y_{\ell m}$ are spherical harmonics. The $\mathbf{R}_i$ are the positions of the ions, and the superscript (i) indicates that the $A_\ell^{(i)}$, $P_{\ell m}^{(i)}$ and $R_\ell^{(i)}$ depend on the species of atom i. The range of the non-local pseudo-potential extends only up to a radius $r_c^{(i)}$ around each atomic position $\mathbf{R}_i$. Normally the core radii $r_c^{(i)}$ are chosen so that there is no overlap between the different core spheres. Therefore, the pseudo-potentials centered on different atoms commute.

The diffusion operator can, among others, be factorized as follows:

$$e^{-\epsilon\hat{H}} = e^{-\frac{1}{6}\epsilon\hat{V}_{\rm nl}}e^{-\frac{1}{2}\epsilon\hat{H}_{\rm loc}}e^{-\frac{2}{3}\epsilon\tilde{V}_{\rm nl}}e^{-\frac{1}{2}\epsilon\hat{H}_{\rm loc}}e^{-\frac{1}{6}\epsilon\hat{V}_{\rm nl}} + \mathcal{O}(\epsilon^5)\,. \tag{18}$$

The operator $e^{-\frac{1}{2}\epsilon\hat{H}_{\rm loc}}$ is factorized using (3), and the non-local double commutator is

$$\tilde{V}_{\rm nl} = \hat{V}_{\rm nl} + \frac{\epsilon^2}{48}\,[\hat{V}_{\rm nl},[\hat{H}_{\rm loc},\hat{V}_{\rm nl}]]\,. \tag{19}$$

For a theory to be gauge invariant, all operators in the Hamiltonian must transform appropriately. This can be accomplished by *postulating* that the projector functions transform as

$$\left\langle \mathbf{r}-\mathbf{R}_i\middle|P_{\ell m}^{\mathbf{A},(i)}\right\rangle \to \left\langle \mathbf{r}-\mathbf{R}_i\middle|P_{\ell m}^{\mathbf{A}',(i)}\right\rangle = e^{-{\rm i}\frac{e}{\hbar}\chi(\mathbf{r})}\left\langle \mathbf{r}-\mathbf{R}_i\middle|P_{\ell m}^{\mathbf{A},(i)}\right\rangle. \tag{20}$$

Then, the non-local potential operator (17) is a gauge invariant, and, hence, the non-local part of the potential transforms under gauge transformation like

$$\begin{aligned}\hat{V}_{\rm nl}(\mathbf{r},\mathbf{r}') \to \hat{V}'_{\rm nl}(\mathbf{r},\mathbf{r}') &= e^{-{\rm i}\frac{e}{\hbar}\chi(\mathbf{r})}\hat{V}_{\rm nl}(\mathbf{r},\mathbf{r}')e^{{\rm i}\frac{e}{\hbar}\chi(\mathbf{r})}\\ &= e^{-{\rm i}\frac{e}{\hbar}\chi(\mathbf{r})}\sum_i v_{\rm nl}^{(i)}(\mathbf{r}-\mathbf{R}_i,\mathbf{r}'-\mathbf{R}_i)e^{{\rm i}\frac{e}{\hbar}\chi(\mathbf{r})}\,.\end{aligned} \tag{21}$$

3. Convergence and Stability Tests

The numerical calculations to be discussed in this section have been performed for a homogeneous magnetic field in the z-direction. We have used Landau gauge, $\mathbf{A}(\mathbf{r}) = B(x - x_0)\mathbf{e}_y$ where x_0 is the position of the gauge origin. The functions $f_j(\mathbf{r})$ defined in (13) thus are $f_y(\mathbf{r}) = B(x - x_0)y$, $f_x(\mathbf{r}) = f_z(\mathbf{r}) = 0$. The "gauge origin" x_0 is moved to an arbitrary location by the gauge function $\chi(\mathbf{r}) = -Bx_0y$.

To show the effect of a gauge invariant discretization of the problem, we have compared the following discretizations of the canonical momentum operator:

(1) gauge invariant discretization:

$$\mathbf{\Pi}^2 = -\hbar^2 \left[\left[\frac{\partial^2}{\partial x^2}\right]_n + e^{-\mathrm{i}\frac{e}{\hbar}B(x-x_0)y} \left[\frac{\partial^2}{\partial y^2}\right]_n e^{+\mathrm{i}\frac{e}{\hbar}B(x-x_0)y} + \left[\frac{\partial^2}{\partial z^2}\right]_n\right] \tag{22}$$

(2) Lagrange discretization:

$$\mathbf{\Pi}^2 = -\hbar^2 \left[\frac{\partial^2}{\partial x^2} + \frac{\partial^2}{\partial y^2} + \frac{\partial^2}{\partial z^2}\right]_n - 2\mathrm{i}e\hbar B(x - x_0) \left[\frac{\partial}{\partial y}\right]_n + e^2B^2(x - x_0)^2 . \tag{23}$$

In both cases, the derivatives have been approximated by the finite-difference representations of order $n = 9$.

3.1. *The Fock-Darwin model*

Discretization errors are best estimated by solving a problem that has an exact solution. We have therefore studied the three-dimensional Fock-Darwin model[14,15], i.e., an electron with an effective mass m^* in a uniform magnetic field and a harmonic potential. The Hamiltonian is

$$\hat{H} = \frac{1}{2m^*}\left(-\mathrm{i}\hbar\nabla + e\mathbf{A}_{\mathrm{ext}}(\mathbf{r})\right)^2 + \frac{m^*}{2}\omega_0^2 r^2 . \tag{24}$$

The eigenvalues of the Schrödinger equation are

$$E_{nm\ell} = \hbar\omega_+\left(n + \frac{1}{2}\right) + \hbar\omega_-\left(m + \frac{1}{2}\right) + \hbar\omega_0\left(\ell + \frac{1}{2}\right), \tag{25}$$

with

$$\omega_\pm = \Omega \pm \frac{\omega_c}{2}, \qquad \Omega = \sqrt{\omega_0^2 + \frac{\omega_c^2}{4}}, \qquad \omega_c = \frac{eB}{m^*} . \tag{26}$$

We have used GaAs material parameters, i.e., an effective mass $m^* = 0.066m_e$, and the harmonic potential has been chosen such that the level spacing at zero field is 1.57 meV. The spatial extent of the resulting system is roughly 200 nm. Figure 1 shows the convergence of the lowest eigenvalue of the Hamiltonian (24) as a function of the imaginary time-step ϵ for the factorizations (2) and (3). The calculations start at a large time step with a simple set of initial wave functions. These are iterated to convergence, then the time step is reduced, and the process is repeated. Figure 1 shows that we have perfect fourth order convergence practically independently of

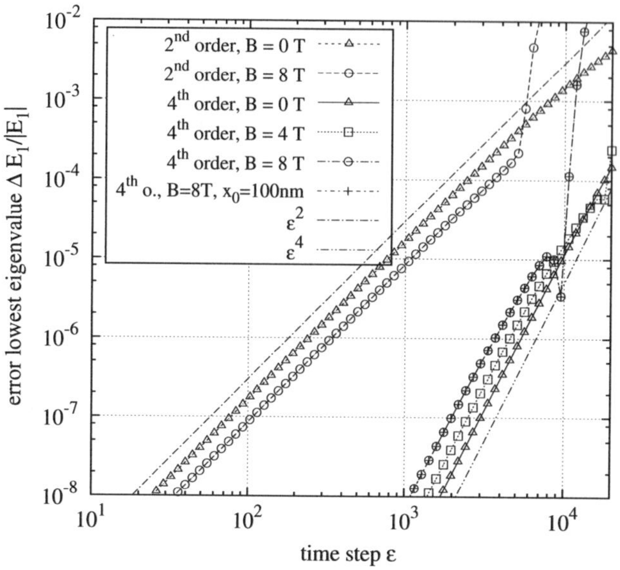

Fig. 1. The relative error of the lowest eigenvalue of non-interacting electrons in a parabolic potential as a function of the imaginary time step ϵ. Results are shown for three different magnetic fields ($B = 0, 4, 8\,T$). For $B = 8\,T$ we also show the results obtained when the gauge origin is shifted about 100 nm in x-direction. The functions ϵ^2 and ϵ^4 are shown as guides to verify the power-law convergence. (From Ref. 16)

the magnetic field strength, and no visible change of the convergence rate as a function of the gauge origin.

Figure 2 compares the spectra obtained using the Lagrange (23) and gauge invariant (22) discretizations with the exact formula (25). The natural choice of the gauge origin would be the minimum of the parabolic potential, $x_0 = 0$, but in more realistic systems there is no such obvious choice. To mimic the problems posed by realistic calculations, while still being able to compare the numerical results to the exact formula, we have performed the calculation of the Fock-Darwin spectra with a gauge origin shifted by about 100 nm away from the center of the dot. While the results obtained by the gauge invariant discretization are indistinguishable from the exact data, Lagrange discretization yields energy levels that deviate drastically from the exact values at fields above 2 T.

For realistic systems, the deviation of the local density around the ion cores, as well as the terms $e^{-i\frac{\epsilon}{\hbar}f_j(\mathbf{r})}$, from spherical symmetry can destroy the exact fourth order convergence of the algorithm. We found,[?] however, at time steps $\epsilon < 10^{-6}$, slight indications for the expected effect that the fourth order power law is violated when the local Kohn-Sham potential is not spherically symmetric in the vicinity of the ion cores. On the other hand, we have practically no effect of the magnetic field even up to unreasonably high field strengths.

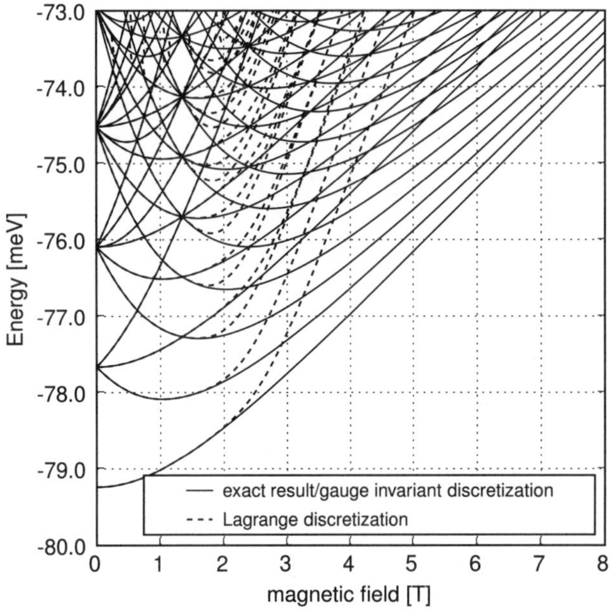

Fig. 2. Energies of an electron in a three-dimensional parabolic quantum dot in an external magnetic field. The solid lines show the exact results, these are indistinguishable from the results of the gauge invariant discretization. The dashed lines denote eigenvalues calculated using the Lagrange discretization. (From Ref. 16)

4. Magnetic Susceptibility and NMR Shifts

The current density corresponding to the Kohn-Sham orbitals $\psi_j(\mathbf{r})$ is

$$\mathbf{j}(\mathbf{r}) = \frac{e}{2m}\sum_j \left[\psi_j^* \boldsymbol{\Pi}\psi_j + \psi_j \left(\boldsymbol{\Pi}\psi_j\right)^*\right] = \frac{e}{m}\sum_j \Re e\left\{\psi_j^* \boldsymbol{\Pi}\psi_j\right\}, \tag{27}$$

where the sum goes over all occupied states.

The magnetic susceptibility tensor χ_{ij} is the weak-field limit of the ratio between the magnetic field $\mathbf{H}$ and the magnetization $\mathbf{M} = \frac{1}{\mu_0}\mathbf{B} - \mathbf{H}$. Since the magnetic susceptibilities are usually very small compared to unity,

$$\frac{\partial B_j}{\partial H_j} = \mu_0\left(1 + \frac{\partial M_j}{\partial B_j}\right) = \mu_0(1+\chi_{jj}) \approx \mu_0, \tag{28}$$

where μ_0 is the vacuum permeability, we can calculate the susceptibility tensor as

$$\chi_{ij} = \mu_0 \frac{\partial M_i}{\partial B_j}. \tag{29}$$

The magnetic dipole moment generated by the current density (27) is

$$\mathbf{m} = \frac{1}{2}\int \mathbf{r}\times\mathbf{j}(\mathbf{r})\, d^3\mathbf{r}. \tag{30}$$

For a density of $n = N/V$ molecules per unit volume, the bulk susceptibility is then

$$\chi_{ij} = \frac{N\mu_0}{V}\frac{\partial}{\partial B_j}m_i. \tag{31}$$

The chemical shift tensor is defined as the ratio between the induced magnetic field and the external field,

$$\sigma_{ij}(\mathbf{R}) = \frac{\partial B_i^{\mathrm{ind}}(\mathbf{R})}{\partial B_j^{\mathrm{ext}}}. \tag{32}$$

The induced field is determined from the current density (27) by the Biot-Savart law,

$$\mathbf{B}^{\mathrm{ind}}(\mathbf{r}) = \frac{\mu_0}{4\pi}\int d^3r'\,\frac{\mathbf{r}'-\mathbf{r}}{|\mathbf{r}'-\mathbf{r}|^3}\times\mathbf{j}(\mathbf{r}'). \tag{33}$$

Equation (33) can be easily evaluated in reciprocal space,[17]

$$\mathbf{B}^{\mathrm{ind}}(\mathbf{G}\neq 0) = -\mu_0 \mathrm{i}\frac{\mathbf{G}}{|\mathbf{G}|^2}\times\mathbf{j}(\mathbf{G}). \tag{34}$$

The $\mathbf{G} = 0$ component of the induced field depends on the bulk magnetic susceptibility, and a prefactor κ depending on the shape of the sample[17]

$$\mathbf{B}^{\mathrm{ind}}(\mathbf{G}=0) = \kappa\chi\mathbf{B}^{\mathrm{ext}}. \tag{35}$$

Experiments often only measure the isotropic susceptibilities and NMR-shifts

$$\bar{\chi} = \frac{1}{3}\mathrm{tr}\,\chi_{ij}, \qquad \bar{\sigma}(\mathbf{R}) = \frac{1}{3}\mathrm{tr}\,\sigma_{ij}(\mathbf{R}). \tag{36}$$

4.1. *Test calculations*

We first assess the importance of a gauge invariant treatment on the currents in a C_6H_6 molecule. The left pane of Fig. 3 shows the current density in the symmetry plane of a C_6H_6 molecule in an external field of 1 T perpendicular to the plane (pointing inward), using the gauge invariant discretization. Different gauge origins between $x_0 = 0.0\,\mathrm{a}_0$ (center of the molecule) and $x_0 = 50\,\mathrm{a}_0$ all yield the same result. One can see the paramagnetic ring current on the inside of the ring, and a diamagnetic current flowing around the outside of the molecule. There are additional current loops alongside the C-C bonds, and around the individual ions. From the direction of these currents, one would conclude a deshielding (*i.e.* a diamagnetic shift) of the Hydrogen atoms, and a positive shift of the Carbon atoms. The right pane shows the same quantity calculated using Lagrange discretization with the gauge origin at the center of the molecule. It is obvious that that the structure of the current density is strongly distorted.

Figure 4 shows the Hydrogen NMR-shifts calculated from the current densities depicted in Fig. 3 as a function of the gauge origin. Gauge invariant discretization yields the same values for all (chemically equivalent) Hydrogen atoms, while in Lagrange discretization the shift strongly depends on the distance of the atom to the gauge origin.

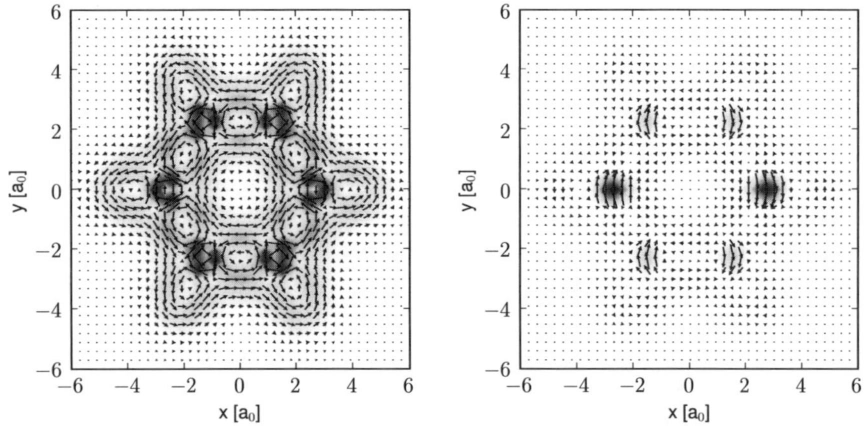

Fig. 3. (Color on line) Current density in a C_6H_6 molecule in the symmetry plane, in an external magnetic field of 1T perpendicular to this plane, pointing into the plane of the figure. Left pane: The gauge invariant discretization scheme (22) has been used to perform several calculations in Landau gauge with different gauge origins between $x_0 = 0.0\,a_0$ (center of the molecule) and $x_0 = 50.0\,a_0$, all showing the same result. Right pane: The same calculation with Lagrange discretization; here the gauge origin is in the center of the molecule. (From Ref. 16)

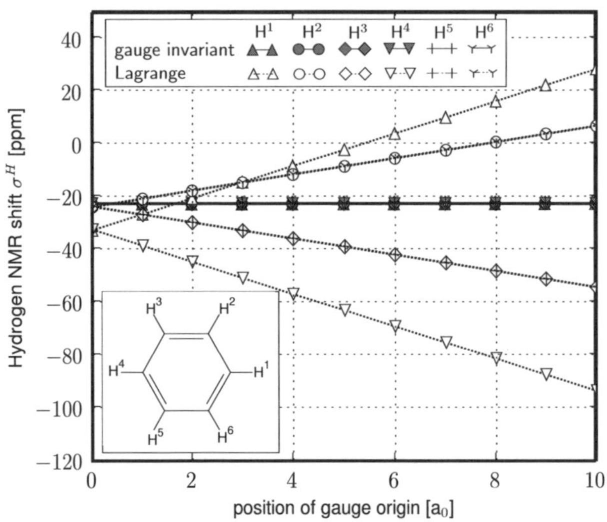

Fig. 4. Hydrogen NMR shifts σ^H for a C_6H_6 molecule, calculated using the Lagrange (23) and gauge invariant (22) discretization schemes. The numbering of the Hydrogen atoms is shown in the inset. The gauge invariant calculation yields the same NMR shifts for all (chemically equivalent) Hydrogen atoms, irrespective of their position relative to the gauge origin. The results obtained using Lagrange discretization strongly depend on the location of the gauge origin. (From Ref. 16)

4.2. *Sample molecules*

We now compare our results to experimental values and to different numerical calculations. We have implemented the gauge invariant discretization of the canonical momentum operator our real-space DFT code `limerec`[18] using the fourth order diffusion method to solve the Kohn-Sham equations. We have used Troullier-Martins[19] pseudo-potentials generated by the program FHI98PP,[20] and the Perdew-Wang LDA density functional.[21] All molecular geometries have been annealed before calculating the magnetic properties.

Sebastiani[17] has used DFT with a BLYP gradient corrected functional, Goedecker pseudo-potentials and the CSGT method. He provides two different sets of results: "full calculation", which is only suitable for isolated systems, and "$\Delta j = 0$", which also works for periodic systems. Keith[22] has performed all-electron calculations using the "Coupled Perturbed Hartree-Fock" (CPHF) method, and compares the CONV,[23,24] IGAIM[25] and CSGT[22] methods to deal with the gauge origin problem. He obtained Carbon shifts using an additional empirical "damping factor", a method termed CSDGT[26] (continuous set of damped gauge transforms). Mauri *et al.*[26] have used DFT, the local density approximation, Troullier-Martins pseudo-potentials and the MPL method.

Figure 5 compares the Hydrogen and Carbon NMR-shifts σ^H for the different methods. The agreement of the Hydrogen shifts with experiment is quite satisfactory, but one can see that the shifts for molecules containing only single bonds (H_2, CH_4 and C_2H_6) are predicted much more accurately than for those with double and

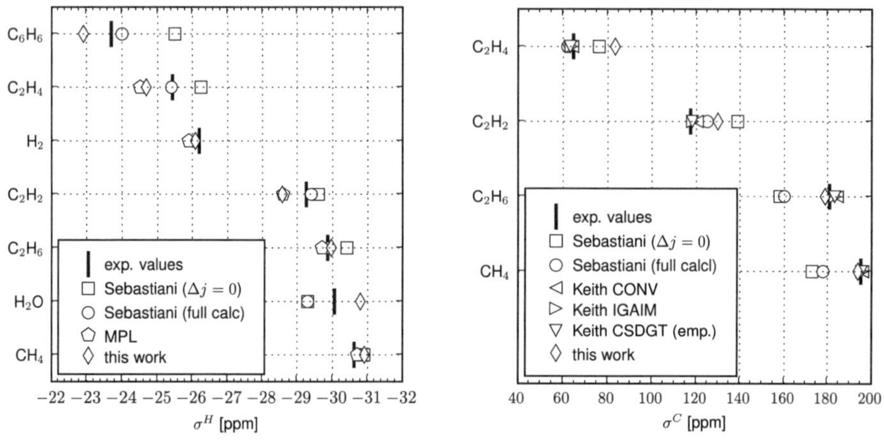

Fig. 5. Calculated and experimental values of Hydrogen NMR shifts (left pane) and Carbon NMR shifts (right pane) for a set of small molecules. Squares and circles denote values calculated by Sebastiani [17] using CSGT, triangles denote results obtained by Keith [22] in an all-electron CPHF calculation using the CONV, IGAIM and CSGT methods, diamonds show the results of this work. (From Ref. 16)

triple bonds (C_2H_4, C_2H_2 and C_6H_6). Since the results of Mauri are quite close to our results, we assume this is due to the failure of LDA in describing double and triple bonds.

The agreement of the calculated Carbon shifts with experiment is less satisfactory for most methods. Here, the limits of the pseudo-potential approximation become apparent: the rigid additive correction for the pseudo-potential was calculated using TMS (tetramethylsilane) as a reference. For similar chemical environments (CH_4, C_2H_6), the carbon shifts are predicted very well. But the rigid correction fails to describe the change in coordination when going to different hybridization states of the carbon atom, an effect that adds up to the error due to LDA described above. Consequently, the all-electron calculations of Keith yield much more accurate predictions for the Carbon shifts than the pseudopotential-based methods.

5. Conclusions

We have presented in this article an implementation of DFT in an arbitrarily strong, homogeneous magnetic field, employing a real space grid to represent Kohn-Sham orbitals, electron density and potential. The eigenvalue/eigenstate pairs for a fixed potential are determined using a fourth order factorization of the evolution operator in imaginary time. A judicious discretization of the kinetic energy operator has permitted a formulation that is gauge invariant in any discretization. The correspondence between the harmonic oscillator and the particle in a homogeneous magnetic field permitted an exact calculation of the exponential of the canonical kinetic energy operator[12] with practically no computational overhead.

Our results are consistent with earlier DFT calculations cited above. Remaining differences with experiments should be attributed to the local density and the pseudopotential approximation but are free from any issues concerned with the choice of the gauge origin. We should again point out the simplicity of the method: the core of the algorithm, namely the propagation of the states, is a matter of a few lines of code,[27] the most complicated part of the method is the propagation with non-local pseudopotentials. In the present problem, the real space formulation has the further significant advantage that the problem can be formulated in a manifestly gauge invariant way. This advantage is immediately lost when the wave functions are represented in a basis other than a real space grid or a plane-wave basis.

Acknowledgments

This work was supported by the Austrian Science Fund FWF under Grant No. P18134 as well as the Austrian Academic Exchange Services ÖAD (11/2006) and the Spanish MEC (HU2005-0033) through a bilateral collaborative project. We also would like to thank S. A. Chin and E. Hérnandez for useful discussions and correspondence.

References

1. M. Aichinger, S. A. Chin and E. Krotscheck, *Computer Physics Communications* **71**, 197 (2005).
2. J. Wosnitza, J. Hagel, N. Kozlova, D. Eckert, K.-H. Müller, C. H. Mielke, G. Goll, T. Yoshino and T. Takabatake, *Physica B* **346-347**, 127 (2004).
3. N. Kozlova, J. Hagel, M. Doerr, J. Wosnitza, D. Eckert, K.-H. Müller, L. Schultz, I. Opahle, S. Elgazzar, M. Richter, G. Goll, H. v. Löhneysen, G. Zwicknagel, T. Yoshino and T. Takabatake, *Phys. Rev. Lett.* **95**, 086403 (2005).
4. J. Wosnitza, G. Goll, A. D. Bianchi, B. Bergk, N. Kozlova, I. Opahle, S. Elgazzar, M. Richter, O. Stockert, H. v. Löhneysen, T. Yoshino and T. Takabatake, *New J. Phys.* **8**, 174 (2006).
5. Z. Jiang, Y. Zhang, Y.-W. Tan, J. A. Jaszczak, H. L. Stormer and P. Kim, *Int. J. Mod. Phys. B* **21**, 1123 (2007).
6. T. Helgaker, M. Jazuński and K. Ruud, *Chem. Rev.* **99**, 293 (1999).
7. L. Kleinman and D. M. Bylander, *Phys. Rev. Lett.* **48**, 1425 (1982).
8. M. Suzuki, New scheme of hybrid exponential product formulas with applications to quantum monte carlo simulations, in *Computer Simulation Studies in Condensed Matter Physics*, eds. D. P. Landau, K. K. Mon and H.-B. Schüttler (Springer, Berlin, 1996) pp. 1–6.
9. S. A. Chin, *Phys. Lett. A* **226**, 344 (1997).
10. S. A. Chin and C. R. Chen, *J. Chem. Phys.* **117**, 1409 (2002).
11. S. Jang, S. Jang and G. A. Voth, *J. Chem. Phys.* **115**, 7832 (2001).
12. R. P. Feynman, *Statistical Mechanics* (Advanced Books Program, Westview Press, Boulder, Oxford, 1998).
13. C. Cohen-Tannoudji, B. Diu and F. Laloe, *Quantum Mechanics* (Wiley, New York, 1977).
14. V. Fock, *Z. Physik* **47**, 446 (1928).
15. C. G. Darwin, *Proc. Camb. Phil. Soc.* **27**, 86 (1930).
16. S. Janecek and E. Krotscheck, (2008), submitted.
17. D. Sebastiani and M. Parrinello, *J. Phys. Chem. A* **2001**, 1951 (2000).
18. `http://www.limerec.net`.
19. N. Troullier and J. L. Martins, *Phys. Rev. B* **43**, 1993 (January 1991).
20. M. Fuchs and M. Scheffler, *Comp. Phys. Comm.* **119**, 67 (1999).
21. J. P. Perdew and Y. Wang, *Phys. Rev. B* **45**, 13244 (1992).
22. T. A. Keith and R. F. W. Bader, *Chem. Phys. Lett.* **210**, 233 (1993).
23. R. M. Stevens, R. M. Pitzer and W. N. Lipscomb, *J. Chem. Phys.* **38**, 550 (1963).
24. W. N. Lipscomb, *Adv. Magn. Res.* **2**, 137 (1966).
25. T. A. Keith and R. F. W. Bader, *Chem. Phys. Lett.* **194**, 1 (1999).
26. F. Mauri, B. G. Pfrommer and S. G. Louie, *Phys. Rev. Lett.* **77**, 5300 (1996).
27. S. Janecek and E. Krotscheck, (2008), in press.

STRONGLY CORRELATED BOSONS AT NONZERO TEMPERATURES

K. A. GERNOTH

Institut für Theoretische Physik, Johannes-Kepler-Universität Linz
A-4040 Linz, Austria
and
School of Physics and Astronomy, The University of Manchester
Manchester, UK

M. SERHAN

Department of Physics, Al al-Bayt University, Jordan

M. L. RISTIG

Institut für Theoretische Physik, Universität zu Köln
D–50937 Köln, Germany

Received 31 July 2008

We investigate the structure of strongly correlated normal Bose liquids and fluids by employing correlated density-matrix (CDM) theory, a generalization of correlated basis functions (CBF) theory to nonzero temperatures. The formalism is applied in a study of structural properties of various correlation functions that characterize the one-body and two-body reduced density matrix elements of a boson system and the associated Fourier transforms such as the static structure function, the exchange structure function, the momentum distribution, and related thermodynamic quantities. We perform a numerical analysis for liquid para-hydrogen, close to the triple point, supercritical ^{4}He gas at temperatures $T \geq 12\,$K, and fluid and liquid ^{4}He in the normal phase below $T = 12\,$K. The results reveal that H_2 and ^{4}He, at $T > 12\,$K, are typical quantum Boltzmann systems following classical statistics. In contrast, liquid ^{4}He exhibits non-classical effects of particle exchange correlations, when the temperature is lowered toward the Bose-Einstein transition regime.

1. Introduction

Correlated Density-Matrix (CDM) theory provides a powerful tool for analyzing the structural properties of quantum liquids and gases that are represented by the N-body density matrix

$$\varrho_N(\mathbf{R}, \mathbf{R}') = \Phi(\mathbf{R}) Q(\mathbf{R}, \mathbf{R}') \Phi(\mathbf{R}'), \tag{1}$$

where $\mathbf{R} = (\mathbf{r}_1, \mathbf{r}_2, \ldots, \mathbf{r}_N)$, with the particle coordinates $\mathbf{r}_i$ confined to volume V. In this description a normal Bose gas of non-interacting bosons is exactly characterized by the specialization $\Phi(\mathbf{R}) = 1$ and by an appropriate N-body permanent, $Q(\mathbf{R}, \mathbf{R}') = \mathrm{Perm}_{ij}\,[\Gamma_{\rm cc}(|\mathbf{r}_i - \mathbf{r}_j|)]$. The statistical factor $\Gamma_{\rm cc}(r)$ and its Fourier transform $\Gamma_{\rm cc}(k)$ are of the exponential form [1]

$$\Gamma_{\rm cc}(r) = \exp\left[-\pi\left(\frac{r}{\lambda}\right)^2\right], \tag{2}$$

$$\Gamma_{\rm cc}(k) = \varrho\lambda^3 \exp\left[-\beta\,\varepsilon_0(k)\right]. \tag{3}$$

Expressions (2) and (3) are determined by the thermal wavelength $\lambda = 2\pi\beta\hbar^2/m$, the particle density $\varrho = N/V$ in the thermodynamic limit ($N, V \longrightarrow \infty$, $\varrho = const.$), the inverse temperature $\beta = (k_{\rm B}T)^{-1}$, and the kinetic energy $\varepsilon_0(k) = \hbar^2k^2/2m$ of a single boson of mass m.

For a homogeneous normal system of interacting bosons at nonzero temperatures the incoherence factor $Q(\mathbf{R}, \mathbf{R}')$ is still assumed to be a permanent. However, the element $\Gamma_{\rm cc}(r)$ may be of a more general form that is determined by a suitable optimization procedure [2] or by choosing an appropriate trial ansatz.

To account for the dynamical effects generated by the interaction potential between a pair of bosons, CDM theory adopts a correlated wave function $\Phi(\mathbf{R})$ of Jastrow form [2]

$$\Phi(\mathbf{R}) = \exp\left[\sum_{i<j}^{N} u\left(|\mathbf{r}_i - \mathbf{r}_j|\right)\right]. \tag{4}$$

The pseudopotential $u(r)$ is then optimally determined in CDM theory by solving an effective Schrödinger equation [3]. If necessary, the ansatz (4) may be systematically improved by adopting a more general Feenberg function [4].

The CDM formalism developed in Refs. 1, 2, 5, and 6 then provides explicit analytic expressions for the reduced density matrices, in particular, for the one-body density-matrix elements $\varrho_1(r) = \varrho n(r)$ and the diagonal two-body density-matrix elements $\varrho_2(r) = \varrho^2 g(r)$ corresponding to the N-body ansatz (1) with input quantities $u(r)$ and $\Gamma_{\rm cc}(r)$. Furthermore, the CDM formalism provides the associated sets of equations that permit to evaluate the radial distribution function $g(r)$ and the static structure function $S(k)$ as well as the quantity $n(r)$ and its Fourier transform $n(k)$, which in turn may be identified with the experimentally measurable single-particle momentum distribution.

We note that the CDM formalism may be also employed to analyze the Bose-Einstein condensed phase of correlated bosons [7], normal Fermi liquids, and gases at temperatures $T \geq 0$ [2]. Further, CDM theory is capable to deal with collective excitations in such quantum systems [2], quantum mixtures, and impurity admixtures at nonzero temperatures [8]. For these systems adequate generalizations [9] of the N-body density-matrix (1) and of the ansätze for the input functions $\Phi(\mathbf{R})$ and $Q(\mathbf{R}, \mathbf{R}')$ are available.

CDM theory permits to study the structure of weakly or strongly correlated quantum fluids composed of charged or neutral particles. In this contribution we confine ourselves to a formal and numerical CDM study of the normal phase of a homogeneous Bose system in thermal equilibrium above the Bose-Einstein transition temperature $T_{\rm BE}$. Specifically, we focus on the properties of liquid para-hydrogen, supercritical ^{4}He vapor, and liquid He I.

Section 2 reports formal and numerical results on the radial distribution function $g(r)$ of liquid para-hydrogen close to the triple point and of supercritical gaseous ^{4}He. We then proceed to study liquid ^{4}He at temperatures between the critical temperature $T_{\rm c} = 5.2\,{\rm K}$ and the Bose-Einstein transition temperature $T_{\rm BE} = 2.17\,{\rm K}$. Our numerical results obtained within CDM theory demonstrate the almost complete suppression of statistical correlations in the distribution function $g(r)$ and in the static structure function $S(k)$ of the normal phase of fluid ^{4}He.

Sections 3 and 4 draw attention to the properties of the one-body reduced density-matrix $n(r)$, the momentum distribution $n(k)$, and the kinetic energy of liquid ^{4}He. For temperatures $T < 12\,{\rm K}$ the results document an increasing influence of statistical correlations, i.e., one observes a transition from Boltzmann statistics to quantum statistics if one lowers the temperature towards $T_{\rm BE}$. In this region the ^{4}He constituents are no longer completely distinguishable. This affects, e.g., the temperature dependence of the entropy of the system and the shape of functions $n(r)$ and $n(k)$. We finally present numerical results and a discussion of the total kinetic energy.

Section 5 concludes the present work with a brief summary of the most characteristic results on the structure of strongly interacting bosons.

2. Screening of Particle Exchange

Starting from the ansatz (1) for the correlated N-body density-matrix ϱ_N, CDM analysis yields a product representation of the radial distribution function [3,10] of the form

$$g(r) = [1 + G_{\rm dd}(r)]\, F(r) \tag{5}$$

with the dynamic (dd) component $G_{\rm dd}(r)$ and the exchange factor $F(r)$. The latter function is given in terms of cyclic (cc) exchange components, direct-exchange (de) portions, and exchange-exchange (ee) parts, which are all defined within the hypernetted-chain (HNC) classification scheme. Explicit expressions for these terms are given in Refs. 3, 2, and 10. We note that the factorization (5) holds not only for a normal Bose system but is valid also for Bose-Einstein condensed fluids and for normal Fermi systems with appropriately generalized expressions for the exchange factor $F(r)$.

Our numerical CDM calculations employ the HNC equations reported in Ref. 2 taking account of all elementary components in so-called HNC/4 approximation conserving certain sum rules [10]. The input pseudopotential $u(r)$ is optimally determined by solving the associated Schrödinger equation for function $g(r)$ (cf. Eq. (5)

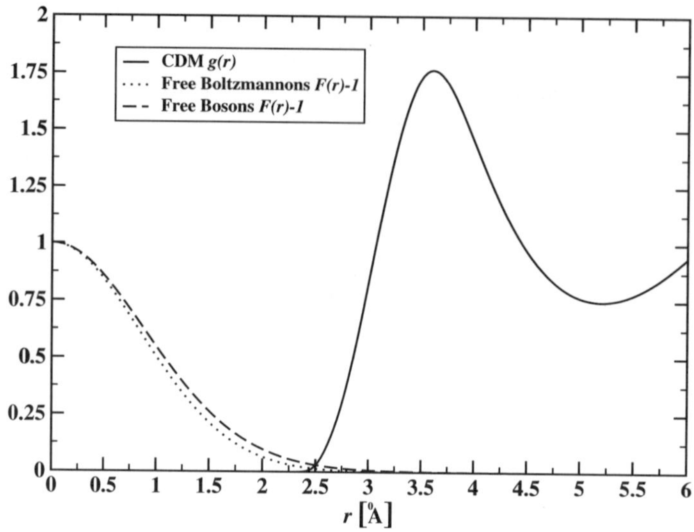

Fig. 1. Displayed are numerical results on the radial distribution function $g(r)$ and on the exchange factor $F(r)$ for liquid para-hydrogen at $T = 16.5\,\mathrm{K}$ and pressure 1 bar. The function $F(r)-1$ does not overlap with $g(r)$, thus confirming the complete suppression of particle exchange effects.

of Ref. 3). The statistical factor $\Gamma_{\rm cc}(k)$ is of the form

$$\Gamma_{\rm cc}(k) = \Gamma_{\rm cc}(0)\exp\left[-\beta\,\varepsilon(k)\right] \tag{6}$$

with $\epsilon(k) = \hbar^2k^2/2m^*(T)$. Ansatz (6) is an appropriate generalization of the exchange factor (2) and (3). For free bosons Eq. (6) specializes to $\Gamma_{\rm cc}(0) = \varrho\lambda^3$ and the effective mass $m^*(T)$ is simply the bare boson mass m. In this case no dynamical correlations appear, since the bosons do not interact with each other. All properties are solely described by statistical correlations generated via the factor (2) or (3). We therefore find

$$g(r) = F(r)\,. \tag{7}$$

The properties of liquid para-hydrogen are in complete contrast to those of the free Bose gas. Studying such a strongly correlated boson system in CDM theory, we find at various temperatures and pressures the result

$$g(r) = 1 + G_{\rm dd}(r)\,, \tag{8}$$

but still $m^*(T) = m$, the bare mass of hydrogen in this case. We point out that these results for the radial distribution function $g(r)$ are in excellent agreement with data computed by means of the Fourier Path Integral Monte Carlo (FPIMC) methods outlined in Ref. 11. This fact demonstrates the high accuracy achieved by CDM (for details see Ref. 10). The complete screening of exchange effects in liquid H_2 is shown in Fig. 1. Plotted are the numerical results on the radial distribution function $g(r)$ together with the data on the statistical portion $F(r)-1$ for liquid para-H_2 at

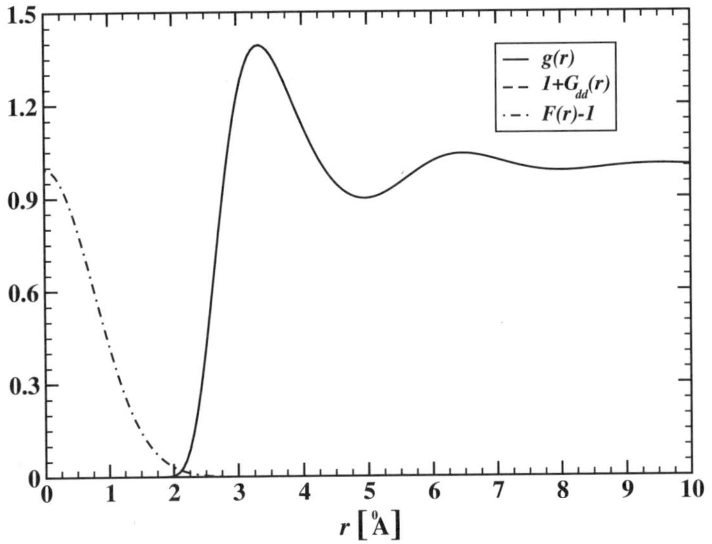

Fig. 2. Radial distribution function $g(r)$, $1 + G_{\rm dd}(r)$, and $F(r) - 1$ for supercritical ^{4}He vapor at $T = 12\,$K and particle density $\varrho = 0.02185\,\AA^{-3}$. The results demonstrate perfect screening of exchange correlations.

a temperature $T = 16.5\,$K and a pressure 1 bar [3]. These functions do not overlap, thus confirming result (8). Consequently, we may say that this quantum liquid is characterized by the absence of statistical (exchange) correlations in $g(r)$ and $S(k)$ in stark contrast to non-interacting bosons.

Numerical CDM calculations for the radial distribution function of supercritical ^{4}He gas and liquid ^{4}He reveal essentially the same screening of particle exchange in the entire normal phase. We have carried out CDM calculations along an isochore at a bulk density $\varrho = 0.02185\,\AA^{-3}$ for a variety of temperatures between $T = 12\,$K and $T = 2.17\,$K. Some results are displayed in Figs. 2 and 3. Functions $1 + G_{\rm dd}(r)$ and $F(r) - 1$ at $T = 12\,$K (Fig. 2) exhibit the complete suppression of particle exchange, since these quantities do not overlap. Even at $T = 2.17\,$K (Fig. 3) the particle screening is still almost complete. The small overlap between the two functions generates only an insignificant decrease in the height of the radial distribution function $g(r)$ compared to the result for the component $1 + G_{\rm dd}(r)$. We should add that excellent agreement is found between the CDM results and FPIMC simulation data on quantity $g(r)$, as is shown in Fig. 4.

It is useful to report here further details of the CDM formalism as applied to the normal ^{4}He system. The exchange generator (6) in the supercritical region $T > 12\,$K is characterized by $m^*(T) \approx m$, just as in the case of free bosons. However, at lower temperatures in the range $2.17\,{\rm K} < T < 12\,$K the effective mass $m^*(T)$ deviates from the ^{4}He mass m. We find that an increase of $m^*(T)$ with decreasing temperature T is necessary to ensure that the helium fluid remains in the normal phase before the

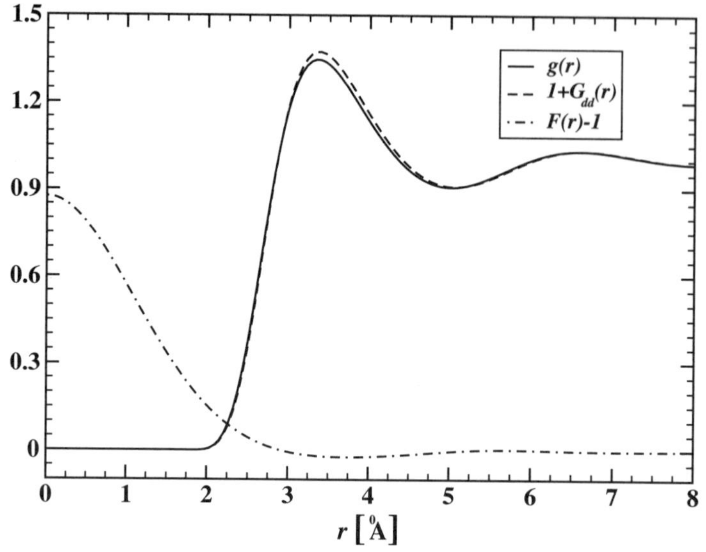

Fig. 3. Numerical results for $g(r)$, the direct-direct factor $1+G_{\rm dd}(r)$, and $F(r)-1$ for liquid ^{4}He I at the Bose-Einstein transition temperature $T_{\rm BE} = 2.17\,$K. At this temperature a rather small overlap between $F(r)-1$ and $1+G_{\rm dd}(r)$ exists, indicating a rather weak influence of particle exchange correlations on the radial distribution function $g(r)$.

experimental Bose-Einstein transition temperature $T_{\rm BE} = 2.17\,$K is reached. This behavior is matched by the following ansatz for the effective mass:

$$m^*(T)/m = 1 + \exp\left(1 - T/T_{\rm BE}\right) . \tag{9}$$

The mass effect (9) is reflected in the non-monotonous temperature dependence of the entropy of the ^{4}He liquid close to the transition temperature $T_{\rm BE}$ (see Sec. 3).

3. From Boltzmann to Quantum Statistics

In Sec. 2 we have shown that particle exchange correlations do not significantly affect the diagonal elements of the reduced density-matrices describing the spatial structure of liquid para-H_2, supercritical ^{4}He vapor, and even liquid ^{4}He. However, statistical exchange correlations become apparent in the off-diagonal matrix elements of the one-body reduced density-matrix of normal ^{4}He. These correlations are characterized by the exchange structure function $S_{\rm cc}(k)$ and its Fourier inverse $G_{\rm cc}(r)$ introduced in Ref. 2. In analogy to the factorized expression (5) CDM theory yields the product form [3]

$$G_{\rm cc}(r) = \left[1 + G_{\rm dd}(r)\right] F_{\rm cc}(r) . \tag{10}$$

The cyclic exchange factor $F_{\rm cc}(r)$ is given explicitly in Ref. 3. The factor $F_{\rm cc}(r)$ in Eq. (10) vanishes in the high-temperature limit $T \longrightarrow \infty$. For non-interacting bosons its Fourier transform $F_{\rm cc}(k)$ is given by the familiar Bose function $\Gamma_0(k) = \Gamma_{\rm cc}(k)\left(1 - \Gamma_{\rm cc}(k)\right)^{-1}$.

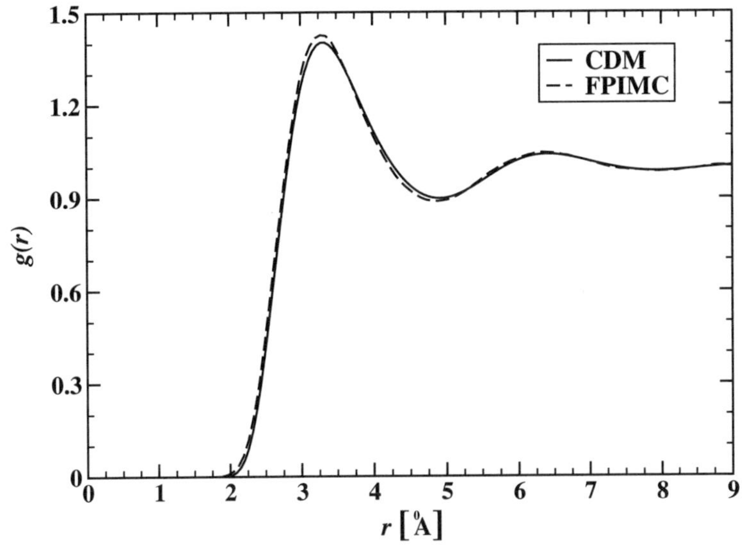

Fig. 4. CDM and FPIMC results for the radial distribution function $g(r)$ at $T = 12\,\mathrm{K}$ of supercritical ^{4}He vapor. The two curves agree almost with each other, indicating the high accuracy of CDM calculations.

For strongly interacting bosons function $G_{cc}(r)$ may differ distinctly from $\Gamma_0(r)$, the Fourier inverse of $\Gamma_0(k)$. The difference originates in the usually existing strong repulsion between the constituents of a many-boson system at short interparticle distances. For example, this applies to para-hydrogen and helium interacting via the Silvera-Goldman potential [12] and Aziz potentials [13], respectively.

Some of our results on the exchange correlation function $G_{cc}(r)$ are plotted in Fig. 5. Shown are numerical data for liquid ^{4}He at various temperatures in the interval $2.17\,\mathrm{K} \leq T \leq 4.5\,\mathrm{K}$ along the isochoric line $\varrho = 0.02185\,\AA^{-3}$. The exchange correlations are completely screened in the strongly repulsive region $0 \leq r \leq 2\,\AA$ and have a peak around the attractive minimum of the ^{4}He–^{4}He potential. The correlations decrease with increasing temperature and vanish completely at temperatures $T > 12\,\mathrm{K}$.

At higher temperatures, $T > 12\,\mathrm{K}$, supercritical ^{4}He gas does not develop exchange correlations, i.e., $G_{cc}(r) \approx 0$. The same behavior is seen in liquid para-hydrogen [3,14].

The entropy $s(T)$ per particle of a normal Bose system is defined by the expression [2]

$$s(T) = \frac{1}{N} \sum_{\mathbf{k}} \left\{ [1 + n_{qp}(k)] \ln [1 + n_{qp}(k)] - n_{qp}(k) \ln [n_{qp}(k)] \right\} \tag{11}$$

with the quasiparticle momentum distribution

$$n_{qp}(k) = \Gamma_{cc}(k) \, [1 + S_{cc}(k)] \ . \tag{12}$$

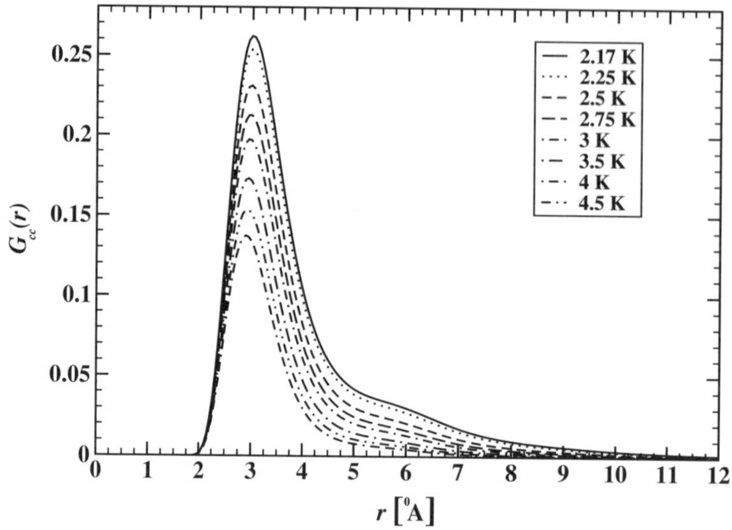

Fig. 5. The exchange correlation function $G_{\rm cc}(r)$ of liquid ^{4}He at various temperatures along the isochore $\varrho = 0.02185\,\AA^{-3}$. No exchange correlations exist at small relative distances r. The exchange correlations are largest at the transition temperature $T_{\rm BE} = 2.17\,{\rm K}$, decrease in height with increasing temperature, and disappear rapidly in supercritical vapor.

At temperatures $T > 12\,{\rm K}$, at which $S_{\rm cc}(k) \approx 0$, we get for the helium system $n_{\rm qp}(k) \approx \Gamma_{\rm cc}(k) = \varrho\lambda^3 \exp\left[-\beta\hbar^2k^2/2m\right]$. Consequently, the quasiparticles are free particles with mass m and their momentum distribution obeys classical Boltzmann statistics in the supercritical region of phase space ($T > 12\,{\rm K}$). The associated entropy, however, still follows the familiar entropy formula (11) of free bosons. At low temperatures the distribution (12) deviates, of course, from the Gaussian form, since the exchange structure function $S_{\rm cc}(k)$ is nonzero within a finite momentum interval $0 \leq k < k_0$. It differs, however, from function $\Gamma_0(k)$.

We may appropriately measure the transition from the classical Boltzmann region ($T > 12\,{\rm K}$) to the quantum-statistical region [10,14] by the kinetic energy distribution of the quasiparticles minus the corresponding Maxwell-Gauß distribution,

$$\Delta_{\rm cc}(k) = \Delta_{\rm qp}(k) = \varepsilon_0(k)\left[n_{\rm qp}(k) - n_{\rm qp}^{\rm M}(k)\right]\,. \tag{13}$$

Function (13) is displayed in Fig. 6. The deviations are large close to the Bose-Einstein transition temperature $T_{\rm BE} = 2.17\,{\rm K}$ and disappear around $T = 12\,{\rm K}$ in the supercritical ^{4}He vapor region.

The evolution of the quantum statistics with decreasing temperature becomes apparent also in the dependence of the specific heat $c_V(T)$ per particle at constant density of liquid ^{4}He on temperature (Fig. 7),

$$c_V(T) = T\frac{\partial s(T)}{\partial T}\,. \tag{14}$$

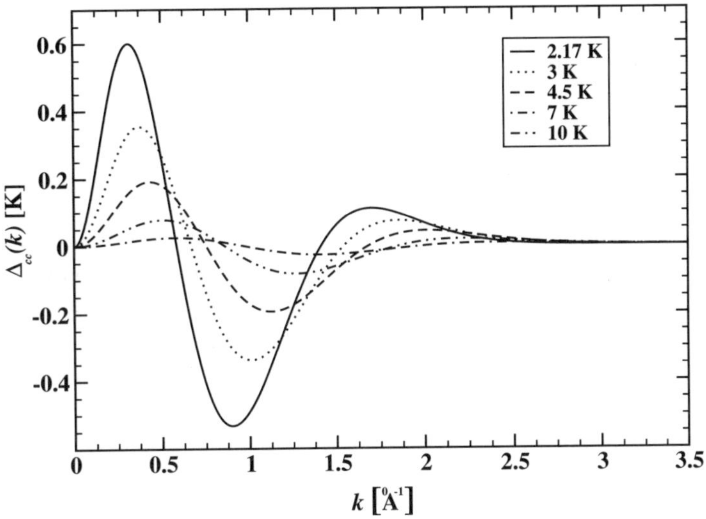

Fig. 6. CDM results on function $\Delta_{\rm qp}(k) = \Delta_{\rm cc}(k)$ that measures the transition from classical Boltzmann to quantum statistics. The measure $\Delta_{\rm qp}(k)$ vanishes identically at temperatures $T \geq 12\,{\rm K}$, where $n_{\rm qp}(k) \approx n_{\rm qp}^{\rm M}(k)$.

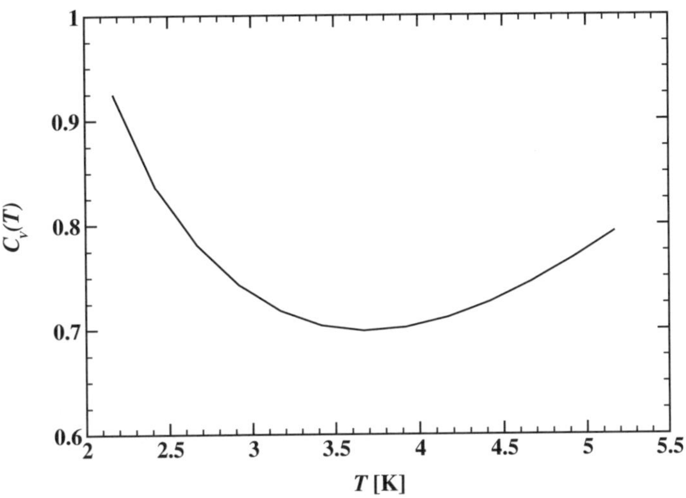

Fig. 7. CDM results on the specific heat $c_V(T)$ per particle at constant density, Eq. (14), of liquid ^{4}He. The increase of $c_V(T)$ is seen also in experimental data.

In contrast to a free Bose gas, the specific heat develops a minimum at $T \approx 3.5\,{\rm K}$ and increases by further decreasing the temperature. This feature is also observed in the experimental data on quantity (14), see Ref. 15.

4. Off-Diagonal Elements and Kinetic Energy

Let us next turn to a detailed study of the one-body reduced density-matrix. This quantity is determined by particle exchange and phase-phase correlations. CDM theory yields the generic factorization [5]

$$n(r) = N_0(r) n_c \exp\left[-Q(r)\right] . \tag{15}$$

The factor $N_0(r)$ embodies the particle exchange effects. The phase-phase correlations of a quantum fluid are incorporated in the dynamical function $Q(r)$. Actually, the product representation (15) holds for normal and condensed systems of bosons and for normal Fermi fluids. For normal systems of bosons considered here the exchange factor $N_0(r)$ is of short range and is closely related to the Fourier inverse $n_{qp}(r)$ of the quasiparticle momentum distribution (12).

The phase-phase correlation function $Q(r)$ is of short range, $Q(r \to \infty) = 0$; the associated long-range portion is absorbed in the strength factor $n_c = \exp\left[Q(0)\right]$. For boson systems with $Q(0) \neq 0$ we may normalize function $Q(r)$ by introducing $P(r) = Q(r)/Q(0)$. The strength factor n_c is then expressed by means of $n_c = \exp\left[Q(0)\right] = \exp[-p_0]$ and we arrive at the identity [3,10] $n_c \exp\left[-Q(r)\right] = \exp\left\{p_0 \left[P(r) - 1\right]\right\}$. Explicit expressions for $Q(r)$ or $P(r)$ and p_0 are reported in Refs. 2, 3, and 5.

Functions $N_0(r)$ and $Q(r)$ determine the total kinetic energy of the boson system by means of the sum [3,10]

$$E_{\text{kin}}(T) = E_0 + p_0 E_{\text{p}} \tag{16}$$

with the exchange energy component

$$E_0(T) = \sum_{\mathbf{k}} \varepsilon_0(k) N_0(k) \tag{17}$$

and the dynamic component

$$E_{\text{p}} = \sum_{\mathbf{k}} \varepsilon_0(k) P(k) , \tag{18}$$

where $N_0(k)$ and $P(k)$ are the Fourier transforms of $N_0(r)$ and $P(r)$, respectively.

Numerical results within CDM theory on quantities (16)–(18) for liquid parahydrogen are reported in Ref. 3. Therein we stated that this system at temperatures $T \geq 16.5\,\text{K}$ can be described as a quantum Boltzmann liquid, since $S_{\text{cc}}(k) \approx 0$ and function $N_0(k)$ specializes to $N_0(k) = n_{\text{cc}}(k) = \Gamma_{\text{cc}}(k) = \Gamma_0(k)$. Thus the distribution $N_0(k)$ is of the classical Gaussian form for distinguishable particles of mass m. Consequently, the kinetic exchange energy per particle is just given by the classical equipartition law $E_0/N = \frac{3}{2}T$.

Present numerical calculations of supercritical ^{4}He vapor at $T > 12\,\text{K}$ yield similar results on functions $N_0(r)$, $Q(r)$, $n(r)$, $n(k)$, etc. We may therefore classify this system as another perfect quantum Boltzmann liquid.

Finally we list in Table 1 some data on various components of the total kinetic energy and on the potential energy of fluid ^{4}He at $T = 12\,\text{K}$ and $\varrho = 0.02185\,\text{Å}^{-3}$.

The potential energy corresponding to the HFDB2 Aziz potential [13] is calculated by means of CDM theory and compared with the FPIMC data. The agreement is excellent. The total kinetic energy is calculated by means of FPIMC simulations and, exploiting the kinetic sum rule (16), used to extract the strength factor n_c and the associated strength p_0. The CDM results on the exchange kinetic energy E_0 and on the quasiparticle energy E_{qp} are almost identical. The agreement provides a further demonstration that supercritical ^{4}He fluid is a quantum Boltzmann system. A more detailed report on results for the kinetic energy, its components, and its distribution in momentum space at temperatures $T < 12\,K$, and, in particular on the normal ^{4}He liquid will be presented elsewhere.

Table 1. Potential and kinetic energies per particle of ^{4}He vapor at $T = 12\,K$ and $\varrho = 0.02185\,Å^{-3}$. Energies are in units of Kelvin. Shown are the results on the exchange portion E_0 and the phase-phase component E_p of the kinetic energy E_{kin} with the strength factor p_0 or, equivalently, n_c. The agreement between the exchange energy E_0 and the quasiparticle energy E_{qp} is a consequence of the complete screening of statistical exchange correlations at $T \geq 12\,K$.

	CDM	FPIMC
E_{pot}/N	-21.47	-21.46
E_{qp}/N	18.16	
E_0/N	18.19	
E_{kin}/N		28.01
E_p/N	5.58	
p_0	1.76	
n_c	0.17	

5. Summary

CDM theory has been successfully applied to analyze the correlation structure of liquid para-hydrogen and normal fluid ^{4}He in thermal equilibrium. The numerical results characterize liquid para-hydrogen at temperatures $T \geq 16.5\,K$ and supercritical ^{4}He vapor at temperatures $T > 12\,K$ as perfect quantum Boltzmann systems. Particular emphasis has been placed on normal helium at low temperatures to study the transition from Boltzmann to quantum statistics in the temperature range $T_{BE} = 2.17\,K < T < 12\,K$ by decreasing the temperature along the isochore

$\varrho = 0.02185\,\text{Å}^{-3}$. Even at temperatures close to the Bose-Einstein temperature $T_{\rm BE}$ the strongly repulsive ^{4}He–^{4}He interactions at short relative particle distances are able to suppress effects of quantum-statistical correlations in the structure of the liquid as described by the quantities $g(r)$ and $S(k)$.

This screening property contrasts with the properties of the exchange structure function $S_{\rm cc}(k)$, the quasiparticle momentum distribution $n_{\rm qp}(k)$, the particle momentum distribution $n(k)$ through the exchange function $N_0(k)$ and related physical quantities. These quantities display the presence of strong exchange correlations in liquid ^{4}He close to the transition to the Bose-Einstein condensed phase, which in turn affects macroscopic quantities such as the specific heat, the total kinetic energy, and its distribution in coordinate and momentum space.

References

1. G. Senger, M. L. Ristig, K. E. Kürten and C. E. Campbell, *Phys. Rev. B* **33**, 7562 (1986).
2. G. Senger, M. L. Ristig, C. E. Campbell and J. W. Clark, *Ann. Phys. (N.Y.)* **218**, 160 (1992).
3. K. A. Gernoth, T. Lindenau and M. L. Ristig, *Phys. Rev. B* **75**, 174204 (2007).
4. C. E. Campbell, *Phys. Lett.* **44A**, 471 (1973).
5. R. Pantförder, T. Lindenau and M. L. Ristig, *J. Low Temp. Phys.* **108**, 245 (1997).
6. M. L. Ristig, T. Lindenau, M. Serhan and J. W. Clark, *J. Low Temp. Phys.* **114**, 317 (1999).
7. M. L. Ristig, G. Senger, M. Serhan and J. W. Clark, *Ann. Phys. (N.Y.)* **243**, 247 (1995).
8. K. E. Kürten and M. L. Ristig, *Phys. Rev. B* **37**, 3359 (1988).
9. K. E. Kürten and M. L. Ristig, *Phys. Rev. B* **31**, 1346 (1985).
10. K. A. Gernoth, M. L. Ristig and T. Lindenau, *Int. J. Mod. Phys.* **21**, 257 (2007); in *Condensed Matter Theories*, Vol. 22, edited by H. Reinholz, G. Röpke and M. de Llano (World Scientific, Singapore, 2007), pp. 117–128.
11. K. A. Gernoth, *Ann. Phys. (N.Y.)* **285**, 61 (2000); *Ann. Phys. (N.Y.)* **291**, 202 (2001); *Z. Kristallogr.* **218**, 651 (2003).
12. I. Silvera and V. Goldman, *J. Chem. Phys.* **69**, 4209 (1978).
13. R. A. Aziz, M. J. Slaman, A. Koide, A. R. Allnat and W. J. Meath, *Mol. Phys.* **77**, 321 (1992).
14. J. Dawidowski, F. J. Bermejo, M. L. Ristig, C. Cabrillo and S. M. Bennington, *Phys. Rev. B* **73**, 144203 (2006).
15. E. W. Lemmon, M. O. McLinden and D. G. Friend, "Thermophysical properties of fluid systems," in *NIST Chemistry Webbook, NIST Standard Reference Database Number 69*, eds. P. J. Linstrom and W. G. Mallard, June 2005, National Institute of Standards and Technology, Gaithersburg MD, 20899 (http://webbook.nist.gov).

DYNAMIC STRUCTURE FUNCTION OF QUANTUM BOSE SYSTEMS; CONDENSATE FRACTION AND MOMENTUM DISTRIBUTION

M. SAARELA

Department of Physical Sciences, University of Oulu,
P.O.Box 3000, FIN-90014 University of Oulu, Finland

Mikko.Saarela@oulu.fi

F. MAZZANTI

Departament de Fısica i Enginyeria Nuclear, Universitat Politècnica de Catalunya
Comte Urgell 187, E-08036 Barcelona, Spain

Ferran.Mazzanti@upc.edu

V. APAJA

Department of Physics, University of Jyväskylä
Jyväskylä, Finland

Vesa.Apaja@gmail.com

We present results on the behavior of the dynamic structure function in the short wave length limit using the equation of motion method. Within this framework we study the linear response of a quantum system to an infinitesimal external perturbation by direct minimization of the action integral. As a result we get a set of coupled continuity equations which define the self-energy. We evaluate the self-energy and the dynamic structure function in the short wavelength limit and show that sum rules up to the third moment are fulfilled. This implies, for instance, that the self-energy at short wavelengths and zero frequency is proportional to the kinetic energy per particle. An essential feature in this derivation is that the short range behavior of the two-particle distribution and the long wavelength phonon induced scattering are exactly satisfied. We calculate the condensate fraction and show that our results agree very well with the Monte Carlo simulations.

Keywords: Dynamic structure function; condensate fraction; equation of motion; momentum distribution; hard core bosons.

1. Introduction

The dynamic structure function $S(k, \omega)$ reveals information on the excitations of the system under study. Experimentally it can be determined by analyzing inelastic scattering data.[1,2] The best studied bosonic quantum system is the liquid ^{4}He. With neutron scattering a wealth of data has been accumulated. At low energy and low momentum transfers the elementary excitation modes, phonon, roton and maxon dispersions are accurately determined. Their structure is also well understood theoretically using many-body techniques like Monte Carlo[3] or variational methods.[4,5,6,7]

The behavior of the dynamic structure function in the high energy and high momentum transfers has been a long term challenge. In this limit the impulse approximation[8,9] holds and $S(k,\omega)$ can be written in terms of the *Compton Profile* $J(y)$

$$S(k,\omega) \to \frac{1}{\hbar v} J(y) \quad \text{when} \quad k \to \infty\,, \tag{1}$$

which is a function of a single variable y known as the West scaling variable,[10]

$$y = \frac{1}{v}\left(\omega - \frac{\hbar k^2}{2m}\right), \tag{2}$$

and $v = \hbar k/m$ is the velocity of the helium atom with momentum $\hbar k$. The leading contribution to the impulse approximation is determined by the one-body density matrix $\rho_1(z)$,

$$J(y) = \frac{1}{2\pi}\int_{-\infty}^{\infty} dz e^{-iyz}\rho_1(z)\,. \tag{3}$$

The momentum distribution, which is the Fourier transform of the one-body density matrix, is not a directly measurable quantity in strongly correlated systems like liquid helium, although it can be accessed indirectly by inspecting the Compton profile.[1,11,12]

In Bose systems the special interest is in determining the fraction of particles in the condensate state, $n_c = N_0/N$, where N is the total number of particles and N_0 the number of particles in the zero momentum state. In scattering experiments at high momentum transfers the condensate fraction should show up as a delta-like peak on top of the non-condensate distribution of the Compton profile broadened by final state effects and the instrumental resolution. Theoretical methods to calculate the one-body density matrix[13,14,15] and from that $J(y)$ are based on summation of infinite series of diagrams starting from the Jastrow-Feenberg wave function or the spectral function analysis where integration over all excited state energies is required.[16] Also the diffusion Monte Carlo method has been successfully implemented for the one-body density matrix calculation.[17,18]

In this work we want to approach the problem from a different point of view and calculate directly the high energy and high momentum limit of $S(k,\omega)$ using equations of motions and the linear response theory.[19,20,21,22] The key quantity in this method is the self-energy, $\Sigma(k,\omega)$, which is a complex quantity that scales in a similar way as the dynamic structure function

$$\Sigma(k,\omega) \to \hbar v(R(y) + iI(y)) \quad \text{when} \quad k \to \infty\,. \tag{4}$$

As an important check we use the second moment sum rule, which tells that at $k \to \infty$ and $\omega = 0$ the self-energy should be proportional to the kinetic energy of the system.

2. Linear Response and the Self-Energy

Let us assume that a quantum fluid at zero temperature is driven out of the ground state by an infinitesimal external interaction $\tilde{U}_{\rm ext}(k,\omega)$, with a given frequency ω and wave number k. It induces a change in the density, $\delta\tilde{\rho}_1(k,\omega)$, of an originally homogeneous system. The response of the system is assumed to be linear and the information on its dynamic properties, excitations and decay modes, is contained in the density–density linear–response function defined as

$$\chi(k,\omega) = \frac{\delta\tilde{\rho}_1(k,\omega)}{\rho_0\tilde{U}_{\rm ext}(k,\omega)} \,. \tag{5}$$

This is a complex quantity and its imaginary part is connected to the dynamic structure function by the fluctuation–dissipation theorem

$$S(k,\omega) = -\frac{1}{\pi}\Im m\left[\chi(k,\omega)\right]. \tag{6}$$

Hence, $S(k,\omega)$ is known once the relation between the one-body density fluctuations and the perturbation has been established, and one can then compare the theoretical results with experiments.

2.1. *Elementary excitation modes*

At low momenta the dynamic structure function typically consists of a low–frequency sharp peak and a high–frequency broad contribution.[23,2] It is therefore customary to write

$$S(k,\omega) = Z(k)\delta(\hbar\omega - \hbar\omega_0(k)) + S_{\rm mp}(k,\omega)\,. \tag{7}$$

This immediately suggests that the density–density response function can be written in a form familiar from the random phase approximation,

$$\chi(k,\omega) = S(k)\left[\frac{1}{\hbar\omega - \varepsilon(k) - \Sigma(k,\omega)} - \frac{1}{\hbar\omega + \varepsilon(k) + \Sigma^*(k,-\omega)}\right], \tag{8}$$

where $\varepsilon(k)$ is the energy of a single collective mode and $\Sigma(k,\omega)$ is the complex self-energy,

$$\Sigma(k,\omega) = A(k,\omega) - iB(k,\omega)\,. \tag{9}$$

The elementary excitation modes are determined by the positive frequency poles $\omega = \omega_0(k)$ of Eq. (8), which lie on the real axis

$$\hbar\omega_0(k) = \varepsilon(k) + A(k,\omega_0(k)) \tag{10}$$

and $B(k,\omega_0(k)) = 0$. The strength $Z(k)$ of that pole can then be evaluated from the derivative of the self-energy,

$$Z(k) = S(k)\left[1 - \left.\frac{dA(k,\omega)}{d(\hbar\omega)}\right|_{\omega=\omega_0}\right]^{-1}. \tag{11}$$

The second term in Eq. (7), $S_{\rm mp}(k,\omega)$, is the multi-phonon background, i.e., the contribution in which a projectile like a neutron probing the system exchanges energy with two or more excitation modes. This contribution is determined by the imaginary part of the self-energy

$$S_{\rm mp}(k,\omega) = \frac{1}{\pi}S(k)\frac{B(k,\omega)}{[\hbar\omega - \varepsilon(k) - A(k,\omega)]^2 + [B(k,\omega)]^2}\,. \tag{12}$$

The dynamic structure function is a positive function for positive values of ω and that is why $B(k,\omega)$ must also be a non-negative function. The condition $B(k,\omega) > 0$ implies that the energy $\hbar\omega$ transferred into the system has created excitation modes that can decay into other lower lying excitations within finite lifetime.

In addition, the relative weight, $Z(k)/S(k)$, gives the efficiency of scattering processes from a single collective excitation, as seen from the (zeroth–moment) sum rule

$$\int_0^\infty S(k,\omega)d(\hbar\omega) = S(k)\,. \tag{13}$$

In other words, it gives the fraction of available scattering processes going through a single collective mode at a given wave vector. If the only excitation in the system were the collective mode $\varepsilon(k)$, like in the Feynman approximation, the ratio $Z(k)/S(k)$ would be one.

2.2. *Impulse approximation*

In the high momentum and energy transfers, $k \to \infty$, $\hbar\omega \to \frac{\hbar^2k^2}{2m}$, the dynamic structure function, $S(k,\omega)$, is completely determined by the one-body density matrix as in Eq. (3) or equivalently by its Fourier transform, the momentum distribution,[8]

$$S^{IA}(k,\omega) = \int \frac{d^3p}{(2\pi)^3\rho_0} n(\mathbf{p})\delta(t(|\mathbf{k}+\mathbf{p}|) - t(k) - \hbar\omega) \tag{14}$$

with $t(k) = \hbar^2k^2/2m$.

The presence of the long range order in the one-body density matrix is due to the macroscopic occupation of the Bose condensate and that reveals the condensate fraction n_c

$$\lim_{r\to\infty} \rho_1(r) = \rho_0 n_c = \frac{N_0}{V}\,. \tag{15}$$

In the momentum distribution that shows up as a δ-function at $p = 0$, which is separated from the non-condensate part of the distribution

$$n(\mathbf{p}) = (2\pi)^3\rho_0 n_c\delta(\mathbf{p}) + n_{nc}(p)\,, \tag{16}$$

we have chosen here the following normalizations

$$\int \frac{d^3p}{(2\pi)^3\rho_0} n(\mathbf{p}) = 1\,, \tag{17}$$

$$\int \frac{d^3p}{(2\pi)^3\rho_0} n_{nc}(p) = 1 - n_c\,, \tag{18}$$

With these definitions the impulse approximation $S^{IA}(k,\omega)$ of Eq. (14) splits into two parts

$$S^{IA}(k,\omega) = n_c\delta(\hbar\omega - t(k)) + S^{NC}(k,\omega) \ , \tag{19}$$

where the first term comes from the fraction of particles in the condensate and the second term gives the contribution of the rest.

In the short wavelength limit $S^{NC}(k,\omega)$ scales as a function of the West variable (2) and the Compton profile can be written in the following form

$$J(y) \equiv \lim_{k\to\infty} \frac{\hbar^2 k}{m} S(k,\omega) = n_c\delta(y) + J^{NC}(y) \,, \tag{20}$$

where the second term comes from the non-condensate part of the momentum distribution

$$J^{NC}(y) = \frac{1}{(2\pi)^2\rho_0} \int_{|y|}^{\infty} pdp \; n_{nc}(p) \ . \tag{21}$$

The complex self-energy scales also in a similar way (4) and we split it into real and imaginary parts in the full expression of $J(y)$

$$J(y) = -\frac{1}{\pi}\Im m \frac{1}{y - R(y) + iI(y)} \ . \tag{22}$$

Equation (20) assumes that $J(y)$ has a pole at $y = 0$. The strength of the pole gives the condensate fraction

$$\frac{1}{n_c} = 1 - \frac{d}{dy}R(y)\bigg|_{y=0} \ , \tag{23}$$

and the non-condensate contribution becomes

$$J^{NC}(y) = \frac{1}{\pi}\frac{I(y)}{(y - R(y))^2 + I^2(y)} \ . \tag{24}$$

Equation (21) shows that $J^{NC}(y)$ is a symmetric function of y and differentiating it with respect to y for $y > 0$ we get the non-condensate part of the momentum distribution

$$n_{nc}(y) = -(2\pi)^2\rho_0 \frac{1}{y}\frac{dJ^{NC}(y)}{dy} \quad \text{when} \quad y > 0 \,, \tag{25}$$

which shows that the complete momentum distribution of the system can be calculated once the real and imaginary parts of the scaled self-energy are known. From the result of Eq. (25) we can also easily integrate the sum rules using the normalization (18) and the definition of the kinetic energy per particle

$$\int_{-\infty}^{\infty} dy J^{NC}(y) = \int \frac{d^3}{(2\pi)^3\rho_0} n_{nc}(p) = 1 - n_c \tag{26}$$

$$\int_{-\infty}^{\infty} dy y J^{NC}(y) = = 0 \tag{27}$$

$$\int_{-\infty}^{\infty} dy\; y^2 J^{NC}(y) = \frac{1}{3}\int \frac{d^3p}{(2\pi)^3\rho_0} p^2 n_{nc}(p) = \frac{2m}{3\hbar^2}\frac{T_{\rm kin}}{N} \tag{28}$$

$$\int_{-\infty}^{\infty} dy y^3 J^{NC}(y) = 0\,. \tag{29}$$

All the odd moments are zero because of the antisymmetry of the integrands.

Using the original formulation of the dynamic response (8) and the analytic properties of the self-energy (9) following the derivation in Ref. [24] one can write the second moment sum rule in the short wavelength limit, but at zero frequency

$$\lim_{k\to\infty} \Sigma(k,\omega=0) = -\frac{4}{3}\frac{T_{\rm kin}}{N}\,, \tag{30}$$

which we shall use for the consistency check.

It is important to notice that in the impulse approximation the pole in the dynamic structure function appears in the middle of the continuum. This means that the imaginary part $I(y)$ must vanish at the origin like y^2 or faster and the real part $R(y)$ must be linear at the origin. The location of this pole is different from the location of the poles generated by elementary excitations at low k which appear below the continuum, where the imaginary part $B(k,\omega)$ is zero.

3. Self-Energy from Equations of Motion

In this chapter we derive the self-energy from the equations of motion and assume that the time evolution of the wave function is governed by the least–action principle[25,26]

$$\delta\mathcal{S} = \delta\int_{t_0}^{t} dt\mathcal{L}(t) \equiv \delta\int_{t_0}^{t} dt\left\langle \Psi(t)\middle| H(t) - i\hbar\frac{\partial}{\partial t}\middle|\Psi(t)\right\rangle = 0 \tag{31}$$

where the Hamiltonian

$$H(t) = H_0 + \hat{U}_{\rm ext}(t) \tag{32}$$

contains the ground-state Hamiltonian H_0 and a time-dependent one-body external potential $\hat{U}_{\rm ext}(t) = \sum_{i=1}^{N} U_{\rm ext}(\mathbf{r}_i;t)$ that creates an infinitesimal disturbance into the system. This is exactly the term that provides the relation between the one-body density fluctuations and the perturbation, needed to calculate the dynamic linear response.

The presence of a time-dependent perturbation modifies the ground state wave function. The time dependence of the new wave function can be separated into three parts: a phase that comes from the time evolution of the unperturbed ground state $\Psi_0(\mathbf{r}_1,\mathbf{r}_2,\ldots,\mathbf{r}_N)$, a normalization factor that is modified due to the perturbation and a set of correlation functions that become time dependent. Putting all this together we get

$$\Psi(\mathbf{r}_1,\ldots,\mathbf{r}_N;t) = e^{-iE_0t/\hbar}\Phi(t) = e^{-iE_0t/\hbar}\frac{1}{\sqrt{\mathcal{N}(t)}}\, e^{\frac{1}{2}\delta U(\mathbf{r}_1,\ldots,\mathbf{r}_N;t)}\Psi_0(\mathbf{r}_1,\ldots,\mathbf{r}_N)\,. \tag{33}$$

One can think about $\delta U(\mathbf{r}_1, ..., \mathbf{r}_N; t)$ as a complex-valued excitation operator

$$\delta U(\mathbf{r}_1, ..., \mathbf{r}_N; t) = \sum_i \delta u_1(\mathbf{r}_i; t) + \sum_{i<j} \delta u_2(\mathbf{r}_i, \mathbf{r}_j; t) \tag{34}$$

expanded in terms of one- and two-body correlation functions. The time-dependent one-body function $\delta u_1(\mathbf{r}_i; t)$ must be included in the description because dynamics will normally break the translational invariance of the system, but restricting the time dependence to the one-body component only would lead directly to the Feynman theory of one-phonon excitations. The time-dependent two-body component is significant in situations where the external field excites fluctuations of wavelengths comparable to the inter–particle distance.

Using the ground-state Schrödinger equation we can write the least action principle in the form

$$\int d^3r_{n+1} \dots d^3r_N \Big[-\frac{\hbar^2}{4m} \sum_{j=1}^{n} \nabla_j \cdot \left(|\Psi|^2 \nabla_j \delta U(t) \right) \tag{35}$$
$$+ |\Psi|^2 \left(-\frac{i\hbar}{2} \delta \dot{U} + \hat{U}_{ext} - \left\langle \Psi \middle| -\frac{i\hbar}{2} \delta \dot{U} + \hat{U}_{ext} \middle| \Psi \right\rangle \right) \Big] = 0\,.$$

which defines n-particle equations of motion

$$\sum_{i=1}^{n} \nabla_i \cdot \mathbf{j}_n^{(i)}(\mathbf{r}_1, \dots \mathbf{r}_n; t) + \delta \dot{\rho}_n(\mathbf{r}_1, \dots \mathbf{r}_n; t) = D_n(\mathbf{r}_1, \dots \mathbf{r}_n; t)\,. \tag{36}$$

The terms which depend on the gradient are collected into the current. The terms with the time derivative give the density fluctuations and the terms which depend on the external potential are included in $D_n(\mathbf{r}_1, \dots \mathbf{r}_n; t)$.

The continuity equations for a homogeneous system are more easily solved in momentum space. Then the one-particle continuity equation can then be written in a more familiar form

$$\hbar\omega - \varepsilon(k) - \Sigma(k, \omega) = \frac{2S(k)\tilde{U}_{\rm ext}(k, \omega)}{\tilde{\xi}(k, \omega)} \tag{37}$$

with the collective reference spectrum $\varepsilon(k)$, the static structure function $S(k)$ and the one-particle density fluctuation $\xi(k, \omega) = \delta\tilde{\rho}(k, \omega)/\rho_0$.

In the short wavelength limit the self-energy becomes

$$\Sigma(\mathbf{k}, \omega) = -\frac{\hbar^2}{2m} \int \frac{d^3q}{(2\pi)^3 \rho_0}\, \mathbf{k} \cdot \mathbf{q} \tilde{G}(q) \frac{\beta(\mathbf{k}, \mathbf{q}; \omega)}{\hbar\omega - t(\mathbf{k} - \mathbf{q}) - t(q))}\,. \tag{38}$$

Here $G(q)$ is the shorthand notation for the Fourier transform

$$\mathbf{q}G(q) = F[\sqrt{g_2(r)} \nabla u_2(r)]\,, \tag{39}$$

and $\beta(\mathbf{k}, \mathbf{q}; \omega)$ is the solution of the two-particle continuity equation,

$$\beta(\mathbf{k}, \mathbf{q}; \omega) = -\frac{\hbar^2}{m} \mathbf{k} \cdot \mathbf{q} \tilde{L}(q) \tag{40}$$
$$+ \frac{\hbar^2}{2m} \frac{S(q) - 1}{(2\pi)^3 \rho_0} \int d^3q' \mathbf{k} \cdot (\mathbf{q} - \mathbf{q}') \tilde{G}(|\mathbf{q} - \mathbf{q}'|) \frac{\beta(\mathbf{k}, \mathbf{q}'; \omega)}{\hbar\omega - t(\mathbf{k} - \mathbf{q}') - t(q')}$$

where $L(q)$ is the Fourier transform of $F[\sqrt{g_2(r)} - 1]$.

4. Results for the Kinetic Energy and Condensate Fraction

Here we show our result for two interesting limits within the high momentum transfer, $k \to \infty$. In the $\omega \to 0$ limit the self-energy must be proportional to the kinetic energy per particle $T_{\rm kin}/N$,

$$\lim_{k\to\infty} \Sigma(k, \omega = 0) = -\frac{\hbar^2}{6m} \int \frac{d^3q}{(2\pi)^3\rho} q^2 G(q)\tilde{L}(q) = -\frac{4}{3}\langle T_{\rm kin}\rangle \,. \tag{41}$$

This result is exact if we assume that the ground state wave function contains only the Jastrow two-particle correlation function $u_2(r)$. In that case

$$\frac{T_{\rm kin}}{N} = \frac{\hbar^2}{8m}\rho_0 \int d^3r \nabla g_2(r) \cdot \nabla u_2(r) = \frac{\hbar^2}{4m}\rho_0 \int d^3r \nabla \sqrt{g_2(r)} \cdot \sqrt{g_2(r)} \nabla u_2(r) \tag{42}$$

which has the Fourier transform of Eq. (41).

The second limit is the impulse approximation where we replace the energy $\hbar\omega$ by $(\hbar^2 k/m)y - \hbar^2 k^2/2m$ and let k go to infinity. We need to keep now both the quadratic and linear terms in k and solve the full integral equation (40). The angle dependence there is most conveniently taken into account by expansion in Legendre polynomials

$$\frac{m}{\hbar^2 k}\beta(\mathbf{k}, \mathbf{q}; \omega) = \sum_{n=0} \beta_n(q, y) P_n(t) \tag{43}$$

where $t = \frac{\mathbf{k}\cdot\mathbf{q}}{kq}$. It turns out that $\beta_n(q, y)$ will be independent of k when we solve the second continuity equation.

We can now write the expansion for the scaled self-energy

$$\frac{m}{\hbar^2 k}\Sigma(\mathbf{k}, y) = -\frac{1}{8\pi^2\rho_0} \int q^3 dq G(q) \sum_{n=0} \beta_n(q, y) \int_{-1}^{1} \frac{tP_n(t)dt}{y - qt} \,. \tag{44}$$

The angle integration can be expressed in terms of the complex Legendre function of the second kind, $Q_n(z)$, giving

$$\frac{m}{\hbar^2 k}\Sigma(\mathbf{k}, y) = \frac{1}{4\pi^2\rho_0} \int q^2 dq G(q) \left[\beta_0(q, y) - \frac{y}{q} \sum_{n=0} \beta_n(q, y) Q_n(y/q)\right] \,. \tag{45}$$

From the real part of the self-energy we calculate the strength of the pole, which is proportional to the condensate fraction

$$\frac{1}{n_c} = 1 - \frac{1}{4\pi^2\rho_0} \int q dq G(q) \left[q \frac{d\beta_0^R(q, y)}{dy}\bigg|_{y=0} - \sum_{n=1,3,\ldots} \beta_n^R(q, 0) Q_n(0)\right] \,. \tag{46}$$

The summation runs over odd values of n, because with even values of n the Legendre functions of the second kind are zero at $y/q = 0$.

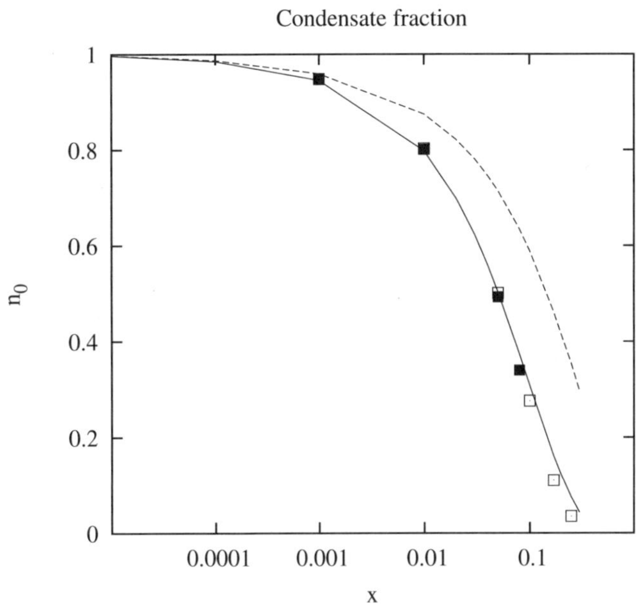

Fig. 1. The condensate fraction of the hard core potential as a function of the gas parameter x. The solid curve is the present result. The comparison is made with Monte Carlo simulations[27] (open squares), HNC-calculations of the one-body density matrix[28] (solid squares) and the CBF-result (dashed curve).

Putting together the real and imaginary parts we get the non-condensate part of the Compton profile

$$J^{NC}(y) = -\frac{1}{n_c \mathcal{D}(y)} \frac{1}{8\pi^2 \rho_0} \int_{|y|}^{\infty} qdq\ G(q) \sum_{n=1,3\ldots} \beta_n^R(q,y) P_n(y/q) \tag{47}$$

with

$$\mathcal{D}(y) = \left[1 - \frac{1}{y} R(y)\right]^2 + \left[\frac{1}{y} I(y)\right]^2 . \tag{48}$$

Differentiating $J^{NC}(y)$ with respect to y and using Eq. (25) one then gets the non-condensate part of the momentum distribution while taking the one-dimensional Fourier transform gives the one-body density matrix in Eq. (3).

As an application of the above approach we have calculated the condensate fraction of the hard core Bose gas as a function of the gas parameter $x = \rho_0 a^3$ where a is the hard core radius. Our results are shown in Fig. 1. They agree well with the Monte Carlo results[27] for the whole range of available x-values. In the figure we also show the HNC results based on the calculation of the one-particle density matrix[28] as well as the CBF-approximation, which reproduces well the elementary excitations.

5. Summary

We have shown in this work how to connect two different methods to calculate the dynamic structure function. The variational method based on equations of motion built from the correlated wave function has been very successful in describing the low momentum excitations. In the high momentum limit the impulse approximation is strictly valid and assumes knowledge of the one-body density matrix. Its systematic diagrammatic expansion has been carefully studied[13,14] and the one-dimensional Fourier transform gives the scaled dynamic structure function, $J(y)$. We have shown that $J(y)$ can also be evaluated if the $k \to \infty$ limit of the self-energy as a function of y is known. The accuracy of the method has been tested at two frequencies. At $\omega = 0$ we get the kinetic energy, which is equal to the one calculated directly from the Jastrow-type correlated wave function. At $\omega = \hbar k^2/(2m)$ we get the condensate fraction, which is in good agreement with the Monte Carlo simulations and the HNC-results using the one-body density matrix.

Acknowledgments

We thank Eckhard Krotscheck and Chuck Campbell for many lively discussions we have had during this work.

References

1. P. C. Hohenberg and P. M. Platzman, *Phys. Rev.* **152**, 198 (1966).
2. H. Glyde, *Excitations in Liquid and Solid Helium* (Oxford University Press, Oxford, 1994).
3. J. Boronat and J. Casulleras, *Europhys. Lett.* **38**, 291 (1997).
4. C. C. Chang and C. E. Campbell, *Phys. Rev. B* **13**, 3779 (1976).
5. V. Apaja and M. Saarela, *Phys. Rev. B* **57**, 5358 (1998).
6. W. Wu, S. A. Vitiello, L. Reatto and M. H. Kalos, *Phys. Rev. Lett.* **67**, 1446 (1991).
7. F. Pistolesi, *Phys. Rev. Lett.* **81**, 397 (1998).
8. H. A. Gersch, L. J. Rodriguez and P. N. Smith, *Phys. Rev. A* **5**, 1547 (1972).
9. S. W. Lovessey, *Theory of Neutron Scattering from Condensed Matter* (Clarendon Press, Oxford, 1984).
10. G. B. West, *Phys. Rep. C* **18**, 263 (1975).
11. C. Carraro and S. E. Koonin, *Phys. Rev. Lett.* **65**, 2792 (1990).
12. F. Mazzanti, J. Boronat and A. Polls, *Phys. Rev. B* **53**, 5661 (1996).
13. M. L. Ristig and J. W. Clark, *Phys. Rev. B* **14**, 2875 (1976).
14. S. Fantoni, *Nuovo Cimento* **44A**, 191 (1978).
15. E. Krotscheck, *Phys. Rev. B* **32**, 5713 (1985).
16. O. Benhar, A. Fabrocini and S. Fantoni, *Nucl.Phys. A* **505**, 267 (1989).
17. P. M. Whitlock and R. M. Panoff, *Can. J. Phys.* **65**, 1409 (1987).
18. E. L. Pollock and D. M. Ceperley, *Phys. Rev. B* **36**, 8343 (1987).
19. M. Saarela and J. Suominen, Dynamic structure function of the quantum boson fluid in the Jastrow theory, in *Condensed Matter Theories*, eds. J. S. Arponen, R. F. Bishop and M. Manninen (Plenum, New York, 1988).
20. B. E. Clements, H. Forbert, E. Krotscheck, H. J. Lauter, M. Saarela and C. J. Tymczak, *Phys. Rev. B* **50**, 6958 (1994).

21. B. E. Clements, E. Krotscheck and C. J. Tymczak, *Phys. Rev. B* **53**, 12253 (1996).
22. M. Saarela, Elementary excitations and dynamic structure of quantum fluids, in *Introduction to Modern Methods of Quantum Many-Body Theory and Their Applications*, eds. A. Fabrocini, S. Fantoni and E. Krotscheck (World-Scientific, London, 2002), pp. 205–264.
23. R. A. Cowley and A. D. B. Woods, *Can. J. Phys.* **49**, 177 (1971).
24. H. W. Jackson, *Phys. Rev. A* **9**, 964 (1974).
25. A. K. Kerman and S. E. Koonin, *Ann. Phys. (NY)* **100**, 332 (1976).
26. P. Kramer and M. Saraceno, *Geometry of the time-dependent variational principle in quantum mechanics*, Lecture Notes in Physics, Vol. 140 (Springer, Berlin, Heidelberg, and New York, 1981).
27. S. Giorgini, J. Boronat and J. Casulleras, *Phys. Rev. A* **60**, 5129 (1999).
28. F. Mazzanti, A. Polls and A. Fabrocini, *Phys. Rev. A* **67**, 063615 (2003).

HELIUM IN PORES AND IRREGULAR SURFACES

E. S. HERNÁNDEZ

Departmento de Física, Facultad de Ciencias Exactas y Naturales, Universidad de Buenos Aires and Comsejo Nacional de Investigaciones Científicas y Técnicas, Ciudad Universitaria, Pabellón I, 1428 Buenos Aires, Argentina

shernand@df.uba.ar

A. HERNANDO*, R. MAYOL† and M. PI‡

Departament d'Estructura i Components de la Matèria, Facultat de Física, Universidad de Barcelona and IN², E-808028 Barcelona, Spain

**alberto@ecm.ub.es*

†ricardo@ecm.ub.es

‡marti@ecm.ub.es

Received 31 July 2008

We report studies of adsorption of helium in translationally invariant polygonal pores at zero temperature, with emphasis on the route to capillary condensation and the appearance of metastable states. We analyze hysteresis and hysterectic-like phenomena associated to the existence of multiple equilibrium states in a rhombic pore and examine the effects of the angular geometry, as opposed to the smooth curvature of cylindrical tubes.

Keywords: Helium; confinement; hysteresis.

1. Introduction

The current research focuses on adsorption of helium inside nanopores and on real surfaces which are not flat within the atomic scale. It is well known that imperfections on a physical adsorber such as steps, stripes, cavities and grooves give rise to corrugation and roughness, which modify the thermodynamics and dynamics of wetting with respect to a planar surface. In particular, recent experiments[1,2] of wetting of superfluid ^{4}He on rough Cs show memory effects and hysterectic dynamics of the contact line, attributed to the persistence of filled micropuddles after dewetting. Although planar Cs and cesiated surfaces are known to be heliophobic at temperatures below 2 K, the presence of cavities limits Cs heliophobicity, as is the case, i.e., in the interior of hollow spheres and cylinders[3] and at the inner angle of planar wedges.[4] It is a general characteristic of confinement potentials that a concave curvature or angle increases the adatom-adsorber attraction and may change heliophobicity into heliophilicity. A similar enhancement of the adhesive forces could be expected in the case of an irregular Cs surface, for which a simplified model consists of a collection of heliophilic sites on a heliophobic surface, as realized in a recent experiment[5] of adsorption of hydrocarbon liquids on silicon surfaces patterned with an array of parabolic nanocavities.

Our present project and emerging studies include, as a first step, adsorption of helium in translationally invariant polygonal pores, that is expected to give rise to capillary condensation (CC) with adsorption–desorption branches arranged into a hysteresis loop, as for classical fluids. The interest of polygonal pores lies in the presence of angles, which permit the appearance of density distributions with either high or low symmetry. The latter implies the simultaneous appearance of various meniscus shapes, which can be associated to drops and bubbles on the walls, in contrast with the much simpler configurations that show up in the popular cylindrical tubes. Furthermore, we intend to investigate adsorption of He on planar Cs at finite temperatures, involving the construction of adsorption isotherms with subsequent appearance of metastability and instability regions, and adsorption of He in finite pores on a flat surface at nonvanishing temperatures, that would originate changes in the wetting-prewetting phase diagram with respect to the case of a perfect planar adsorber.

In this report we provide a brief account of the current state-of-the-art regarding adsorption of classical fluids in nanopores (Section 2) and in Section 3 we describe our calculations and results, which are summarized in Section 4.

2. Classical Fluids in Nanopores

Although nanoporous materials have been used since antiquity as adsorbents and filters, the industrial production of activated carbon began in 1900 and of zeolites around 1950; presently, various processes permit to manufacture porous materials with multiple chemical, geological and geophysical, and industry applications. Very useful reviews on production, classification, uses, and topics of general interest regarding nanopores can be found in Refs. 6, 7. In general, one refers to these environments as micropores if their diameter D is below 2 nm, mesopores if D is between 2 and 50 nm, and macropores for diameters larger than 50 nm. The adsorption sequence of a fluid in a pore permits one to identify two clearly different regimes or "phases", namely a low-density, vapor-like configuration where a thin layer of fluid adheres to the inner walls, and a high-density, liquid-like one with the fluid homogeneously condensed in the tube, corresponding to CC.

A well-studied phenomenon occurring along the CC regime of the adsorption isotherm is hysteresis. In addition to abundant experimental evidence, a very clear picture of hysteresis can be attained by means of Monte Carlo simulations as reported, for example, in Ref. 8. The classical explanation of this phenomenon goes back as far as 1911[9] and received stronger support in 1938,[10] with simple thermodynamic arguments that permit to relate the adsorption and desorption pressures to the saturation properties of the unconfined fluid. The general understanding is that the vapor pressure at the adsorption branch is close to the spinodal pressure of vapor in the pore, while the lower desorption pressure is that of liquid–gas coexistence under confinement. However, this simple scenario can be strongly modified by the geometry of the pore and/or the pore array.

Studies of CC of classical fluids have been performed employing various techniques such as grand canonical and canonical Monte Carlo simulations, nonlocal density functionals, lattice gas models, molecular dynamics and other methods.[6,11] In particular, it has been proven that nonlocal density functional theory (NLDFT), which has been applied to pore physics for more than 20 years, with appropriate choice of the force parameters provides results in excellent agreement with both simulations and experiments for CC in nanopores,[11] and bridges the gap between simulations carried on pores of around 2 nm diameter and the thermodynamic limit. It is important to note that a well-known feature of both theory and experiments of classical fluids in pores, that has been already discussed at length in the literature,[6] is the possible existence of multiple metastable states with grand potential above the equilibrium value at a given fluid density according to the choice of the initial condition. This is the basis for the appearance of hysterectic loops, although such features may also show up below the CC regime, under proper choice of the initial condition for the construction of the mass distribution.

In this work, we concentrate on adsorption of superfluid helium in polygonal pores, as predicted by a specific NLDFT described, i.e., in Ref. 4 and references cited therein. Previous studies by our group[12,13] have undertaken hexagonal pores, for which calculations based on rigorous many-body techniques exist in the literature. However, these approaches make use of a rather crude approximation for the confinement potential, such as the summation of the planar Lenard–Jones potentials of six planes meeting at angles of sixty degrees. Here we consider a rhombic pore, an environment that can be regarded as a double-wedge and has been investigated from various viewpoints. On the one hand, a symmetry-breaking localization–delocalization transition of the liquid–vapor interface in a double wedge has been identified.[14] Moreover, we have shown that helium in wedges can adopt various meniscus shapes[4] and that helium samples adsorbed near a corner grow with accumulation of quasi-onedimensional stripes.[15] Finally, rhombic pores, which appear naturally in mineral crystals and are manufactured in metallic and organic materials since the 60's,[16] have been realized experimentally.[17] The rhombic pore is then a simple and feasible candidate to investigate a variety of aspects of the morphology and the thermodynamics of adsorption in a tube, such as the coexistence of different meniscus shapes at the smallest and the largest angle, the competition of quasi-onedimensional stripes and quasi-twodimensional layers in the process of filling, and the sensitivity of the equilibrium density profile to the initial condition, among others.

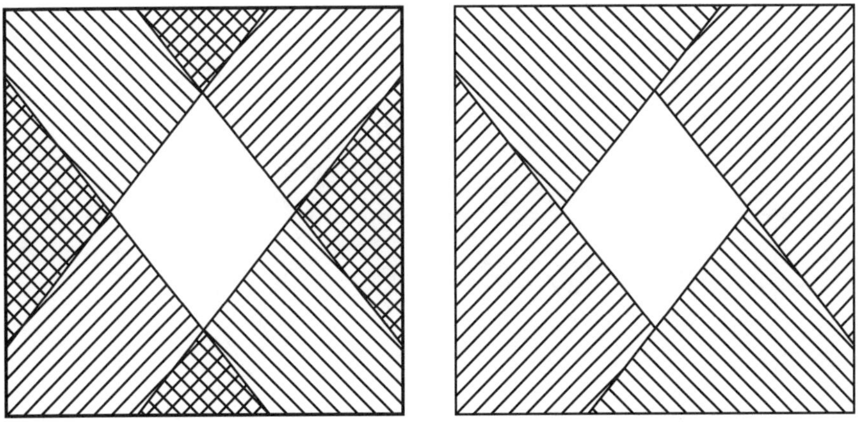

Fig. 1. Left panel: Four half-solids meeting at corners to form a rhombic cavity, overlapping at the vertices. Right panel: Same environment created by four nonoverlapping cusps.

3. Calculations and Results

Our calculations employ grand canonical NLDFT for superfluid helium at zero temperature, which amounts to solving a nonlinear Schrödinger-like equation for the two-dimensional density profile $\rho(x,z)$ (we assume the pore infinity along the y-axis) of the form

$$\left[-\frac{\hbar^2}{2m}\nabla^2 + U(\rho) + U(x,z)\right]\sqrt{\rho(x,z)} = \mu\sqrt{\rho(x,z)}\,. \tag{1}$$

Here $U(\rho)$ is the mean field potential described by the density functional, $U(x,z)$ the adsorbing potential of the pore walls and μ the chemical potential of the helium atoms. The density is normalized to a fixed number of particles per unit length $n = N/L$.

In earlier descriptions of polygonal pores and walls meeting at a corner, the adsorbing potential has been approximated by summing the contributions from infinite planar surfaces. In this context, four half-solids make a rhombic pore with strong overlap at the vertices — thus overestimating the strength of the confinement — as indicated with the doubly hatched regions in the left panel of Fig. 1. In a recent paper[18] we have shown that this approach can be substantially improved by solving first an inverse problem: in fact, given a planar potential $U_{ref}(z)$, it is possible to find a pair potential or source field $V(\mathbf{r}-\mathbf{r}')$ such that the reference field is the convolution

$$U_{ref}(z) = \rho_0 \int_{-\infty}^{0} dz' \int\int dx'dy'\, V(\mathbf{r}-\mathbf{r}')\,. \tag{2}$$

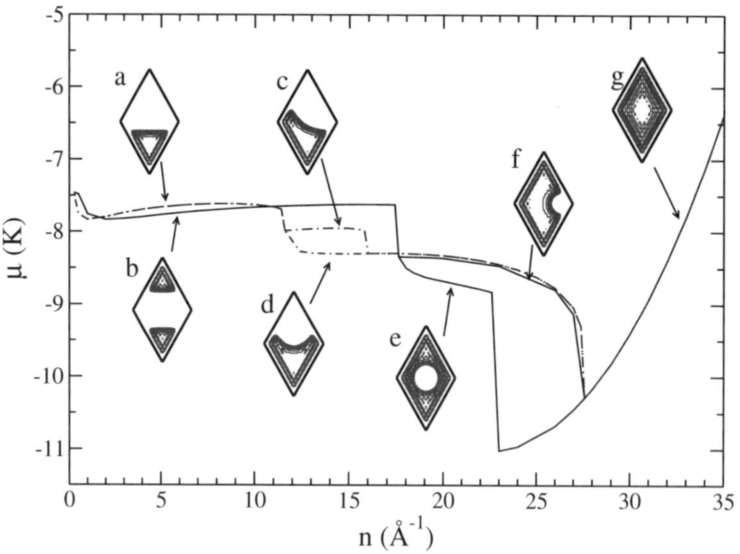

Fig. 2. Grand potential per particle Ω/N along the proposed paths (see text for details) in full and dashed-dotted lines, respectively, in a rhombic Cs pore with small angle of 60° and 50 Å side. The insets (a) to (g) have been respectively drawn for linear densities $n = 9$, 9, 12, 16, 20, 26 and 29 Å^{-1}. The arrows point to the path (not to the value of n).

For a half-solid with bulk density ρ_0, one finds the exact solution

$$V(z) = \frac{U''_{ref}(z)}{2\pi\rho_0 z} \tag{3}$$

so that for other matter distributions $n(\mathbf{r}')$ one can compute

$$U(\mathbf{r}) = \int d\mathbf{r}'\, n(\mathbf{r}')\, \frac{U''_{ref}(|\mathbf{r}-\mathbf{r}'|)}{2\pi\rho_0|\mathbf{r}-\mathbf{r}'|}\,. \tag{4}$$

As shown in Ref. 18, a convenient, infinite unit cell is the cusp, — i.e., the complement of a wedge — for which the adsorbing field $U(\mathbf{r})$ can be computed out of the potential for a plane using Eq. (4). A rhombic pore can thus be viewed as surrounded by four nonoverlapping cusps, as illustrated in the right panel of Fig. 2.

We have computed the equation of state (EOS) for helium atoms in a rhombic Cs mesopore with sides of 50 Å and small angle of 60°, solving Eq. (1) by means of an imaginary-time evolution technique, whose details can be found in Ref. 4 and references cited therein. The adsorption field $U(\mathbf{r})$ has been obtained from the planar *ab–initio* potential U_{ref} for Cs developed in Ref. 19. The outcome of these calculations is usually presented as a zero-temperature isotherm $\Omega(n)$ where $\Omega = E/L - \mu n$ is the grand potential per unit length, or as an isotherm of the form $\mu(n)$. At finite temperatures, the latter can be related to the experimental

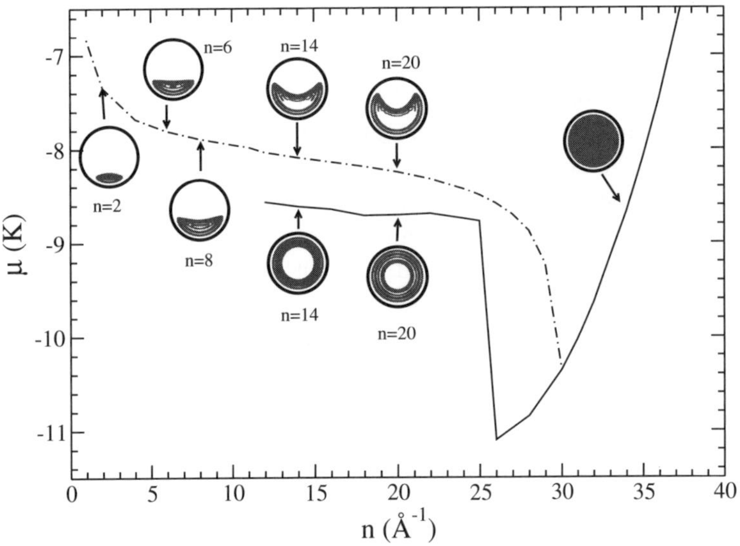

Fig. 3. Same as Fig. 2 for a cylinder with the same transverse area as the rhombic pore.

adsorption isotherms, generally expressing the vapor pressure as a function of the amount of adsorbed material.

The appearance of hysteresis and hysterectic-like loops can be due to two different situations. One can construct adsorption and desorption branches by a step-by-step procedure that computes smooth changes in Ω, μ and the mass distribution due to small variations in the linear density n, giving rise to "true" hysteresis. As an alternative, one can evolve different initial density distributions for a fixed linear density. In this case it is possible to find multiple equilibrium configurations; if this happens, examination of the grand potential permits to identify the thermodynamic equilibrium states and the metastable states.

In Fig. 2 we illustrate the curves $\mu(n)$ obtained by imaginary-time evolution of two different initial conditions, with the insets depicting the equilibrium density profiles in the tube. Configurations (a), (d), (c) and (f) correspond to an initial mass distribution with helium filling uniformly the lower half-pore, while patterns (b) and (e) are obtained from the fluid filling the whole pore. These two trajectories merge together at a linear density near 28 Å^{-1}, showing that for larger helium samples the fluid is capillary condensed in the pore according to profile (g).

The stability analysis by observation of the grand potential Ω shows that stable thermodynamic equilibrium does not always correspond to the configuration with the highest symmetry.[13] Typical symmetry–breaking patterns such as (c) and (d) have been generated by adding noise to the initial steps of the imaginary-time evolution and are thermodynamically unstable. For the lowest densities the mass distribution, either stable or metastable, represents fluid located on the pore walls.

The crossing at the two trajectories at 17.5 Å^{-1} for a chemical potential of -8.5 K is the vapor-like endpoint of a Maxwell construction with rightmost end at 32.5 Å^{-1}. Within this density range both the high and low symmetry patterns are thermodynamically unstable, while for higher densities the fluid is capillary condensed in a stable fashion. We note that although configurations (e) and (g) are similar in appearance, only the latter corresponds to CC with a central density close to bulk saturation, while the former contains a central void.

The behavior above described can be contrasted with the equivalent zero-temperature isotherms for a cylindrical tube with the same transverse area, plotted in Fig. 3. In this case the amount of symmetry-breaking is reduced, due to the absence of corners which precludes patterns like (c) and (f) in Fig. 2; moreover, a high symmetry pattern like (b) in the rhombic pore does not occur in the cylindrical tube. One can recognize as well that the density distributions in the cylinder are of type (a) and (d) at the lowest linear densities, while the configurations for $n =$ 14 and 20 Å^{-1} present a bubble at the center, similar to the rhombic distribution (e). The lowest branch, that resembles the adsorption curve in a true adsorption isotherm, contains the cylindrically symmetric configurations, while those with broken symmetry can be associated to the desorption branch.

4. Summary and Perspectives

In this paper we have reported studies of adsorption of helium in (translationally invariant) polygonal pores at zero temperature, with emphasis on CC and appearance of metastable states. We believe that at the moment we possess the appropriate instruments to study and understand the road to CC, with accompanying hysteresis and hysterectic-like phenomena associated to the existence of multiple equilibrium states in a pore.

We plan to carry finite temperature calculations of adsorption isotherms on a heliophobic Cs surface, employing a temperature-dependent density functional, previous to undertaking a more stringent investigation of adsorption on a rough surface. Since these isotherms present instability regions (i.e., $d\mu/dn < 0$) an important phenomenon that deserves further investigation is nucleation and cavitation of helium in the presence of a wall. We hope to present these results in the near future, as well as those of adsorption of He in finite pores on a flat surface, pointing at examining the changes in the wetting-prewetting phase diagram due to the presence of interfacial irregularities.

Acknowledgments

This work has been performed under grants FIS2005-01414 from DGI, Spain (FEDER), 2005SGR00343 from Generalitat de Catalunya and in Argentina, PIP 5138/05 from CONICET, PICT 31980/05 from ANPCT and X298 from University of Buenos Aires.

References

1. J. Klier, P. Leiderer, D. Reinelt and A. F. G. Wyatt, *Phys. Rev. B* **72**, 245410 (2005).
2. D. Reinelt, V. Iov, P. Leiderer and J. Klier, *J. Phys.: Condens. Matter* **17**, S403 (2005).
3. L. Szybisz and S. M. Gatica, *Phys. Rev. B* **64**, 224523 (2001).
4. E. S. Hernández, F. Ancilotto, M. Barranco, R. Mayol and M. Pi, *Phys. Rev. B* **73**, 2454061 (2006).
5. O. Gang, K. J. Alvine, M. Fukuto, P. S. Pershan, C. T. Black and B. M. Ocko, *Phys. Rev. Lett.* **95**, 217801 (2005).
6. L. D. Gelb, K. E. Gubbins, R. Radhakrishnan and M. Sliwinska-Bartkowiak, *Rep. Prog. Phys.* **62**, 1573–1659 (1999).
7. E. Robens, P. Staszczuk, A. Dabrowsky and M. Barczak, *J. Thermal Analysis and Calorimetry* **79**, 499 (2005).
8. L. D. Gelb, *Molec. Phys.* **100**, 2049 (2002); R. Paul and H. Rieger, *J. Chem. Phys.* **123**, 024708 (2005).
9. R. Zigsmondy, *Z. Anorg. Chem.* **71**, 356 (1911).
10. L. H. Cohan, *J. Am. Chem. Soc.* **60**, 433 (1938).
11. A. V. Neimark, P. I. Ravikovitch and A. Vishnyakov, *Phys. Rev. E* **62**, R1493 (2000).
12. E. S. Hernández, A. Hernando, R. Mayol and M. Pi, *Proceedings of the 14th International Conference on Recent Progress in Many-Body Theory*, Barcelona, Spain, July 2007, eds. G. Astrakharchik, J. Boronat and F. Mazzanti, to be published.
13. A. Hernando, E. S. Hernández, R. Mayol and M. Pi, submitted for publication.
14. A. Milchev, M. Müller, K. Binder and D. P. Landau, *Phys. Rev. Lett.* **90**, 136101 (2003).
15. R. Mayol, F. Ancilotto, M. Barranco, E. S. Hernández and M. Pi, *J. Low Temp. Phys.* **148**, 851 (2007).
16. M. A. M. Beerlage, J. M. M. Peeters, J. A. M. Nolten, M. H. V. Mulder and H. Strathmann, *J. Appl. Pol. Sci.* **75**, 1180 (1999); R. G. Harrison, N. K.Dally and A. Y. Nazarenko, *Chem. Commun.* 1387 (2000); S. Matthias, F. Müller and U. Gösele, *J. Apl. Phys.* **98**, 023524 (2005).
17. P. F. McKenzie, R. M. Weber and J. L. Anderson, *Langmuir* **10**, 1539 (1994).
18. A. Hernando, E. S. Hernández, R. Mayol and M. Pi, *Phys. Rev. B* **76**, 1154291 (2007).
19. A. Chizmeshya, M. W. Cole and E. Zaremba, *J. Low Temp. Phys.* 110, 677 (1998).

Condensed Matter

COLLECTIVE MODES OF TRAPPED INTERACTING BOSONS

G. GNANAPRAGASAM and M. P. DAS*

Department of Theoretical Physics,
Research School of Physical Sciences and Engineering,
The Australian National University,
Canberra, ACT 0200, Australia
**mpd105@rsphysse.anu.edu.au*

Received 31 July 2008

The derivation for collective modes of an interacting Bose gas trapped by an isotropic harmonic oscillator potential is presented using field-theoretic method. The presence of the two-body scattering term beyond the mean-field is seen to appear inevitably in the calculations, even in the simplest approximation. As a result we see the occurrence of a small number of non-condensate atoms in the ground state density fluctuations.

Keywords: Collective modes; harmonically confined; interacting Bose gas.

1. Introduction

After the experimental observation of Bose condensates in dilute gases of trapped alkali atoms, the study of collective excitations has also become a popular area of research in ultracold gases. From the experimental side the measurement of the collective frequencies have been carried out with high precision, aimed at pointing out the role of the interactions and of quantum correlations.[1,2,3] From the theoretical side new challenging features evolve because of the non-uniform nature of trapped systems. Such theoretical studies of trapped condensate in the dilute regime at zero temperature have been carried out by several authors using variational methods,[4,5,6] numerical methods,[7,8,9,10] and also by other analytic methods[11,12,13] within the framework of the mean-field approximation.

The properties of elementary excitations have been investigated by several authors, using the hydrodynamic formulation, by considering small deviations of the state of the gas from equilibrium, for example, fluctuations in density and current. In most of the earlier calculations at zero temperature, the interaction effects arising from the atoms out of the condensate were neglected. Recently, we studied interacting bosons confined by an external potential and at finite temperature, using a microscopic theory.[14] For this system, in order to account for the correlations between the condensate and the non-condensate atoms, we have considered interaction terms beyond the usual mean-field theories. The aim of the current work is to obtain the collective modes of such a system in the presence of harmonic confinement. This is done using field theoretic method and at low temperatures where quantum depletion is very small. When the number of atoms in the condensate is very large, the kinetic energy associated with the confinement of atoms within the cloud, which gives rise to the quantum pressure can be neglected — this is the Thomas-Fermi (TF) approximation. From the calculations we see that, even at low temperature, two-body interactions beyond the mean-field are inevitable. Owing to the complexity reflected in the interaction Hamiltonian,[14] we have discussed a simple case, where we take into account, only one interstate scattering, which gives rise to a small quantum depletion.

Consider a 3D system of interacting bosons described by the Hamiltonian

$$H=\int d\mathbf{x}[\hat{\psi}^\dagger(\mathbf{x})\,(T(\mathbf{x})+V_{trap}(\mathbf{x}))\,\hat{\psi}(\mathbf{x})]+\frac{1}{2}\int\!\!\int d\mathbf{x}\,d\mathbf{x}'\,\hat{\psi}^\dagger(\mathbf{x})\hat{\psi}^\dagger(\mathbf{x}')V^I(\mathbf{x},\mathbf{x}')\hat{\psi}(\mathbf{x}')\hat{\psi}(\mathbf{x}) \quad (1)$$

where $T(\mathbf{x}) = -\frac{\hbar^2}{2m}\nabla^2$ is the kinetic energy, $V_{trap} = \frac{1}{2}m\omega_0^2\mathbf{x}^2$ is the trapping potential of frequency ω_0, V^I is the two-body interaction potential. Here $\hat{\psi}(\mathbf{r})$ and $\hat{\psi}^\dagger(\mathbf{r})$ are the Bose field operators in a Hilbert space which destroy and create, respectively, bosons at position $\mathbf{r}$. They obey the commutation relations

$$[\hat{\psi}(\mathbf{r}), \hat{\psi}^\dagger(\mathbf{r}')] = \delta^{(3)}(\mathbf{r}-\mathbf{r}'),$$
$$[\hat{\psi}(\mathbf{r}), \hat{\psi}(\mathbf{r}')] = [\hat{\psi}^\dagger(\mathbf{r}), \hat{\psi}^\dagger(\mathbf{r}')] = 0. \quad (2)$$

2. Equation of Motion for Density Fluctuation

To discuss collective modes in a trapped Bose gas in the limit $T \to 0$, we first begin with the continuity equation, which is

$$\frac{\partial \hat{n}}{\partial t} + \nabla\cdot\hat{\mathbf{j}} = 0, \quad (3)$$

where the number density operator $\hat{n} = \hat{\psi}^\dagger(\mathbf{x},t)\hat{\psi}(\mathbf{x},t)$ and the current density operator $\hat{\mathbf{j}}$ is

$$\hat{\mathbf{j}}(\mathbf{x},t) = \frac{\hbar}{2mi}\left[\hat{\psi}^\dagger(\mathbf{x},t)\nabla\hat{\psi}(\mathbf{x},t) - (\nabla\hat{\psi}^\dagger(\mathbf{x},t))\hat{\psi}(\mathbf{x},t)\right]. \quad (4)$$

Differentiating Eq. (3) partially with respect to time we get:

$$\frac{\partial^2 \hat{n}}{\partial t^2} + \nabla\cdot\frac{\partial \hat{\mathbf{j}}}{\partial t} = 0. \quad (5)$$

The equation of motion for $\hat{\mathbf{j}}$ in the Heisenberg picture is

$$\frac{\partial \hat{\mathbf{j}}(\mathbf{x},t)}{\partial t} = \frac{1}{i\hbar}[\hat{\mathbf{j}}(\mathbf{x},t), \hat{H}], \quad (6)$$

or

$$\frac{\partial \hat{\mathbf{j}}(\mathbf{x},t)}{\partial t} = \frac{1}{i\hbar}e^{iEt/\hbar}[\hat{\mathbf{j}}(\mathbf{x}), \hat{H}]e^{-iEt/\hbar}. \quad (7)$$

It can be shown that[15]

$$[\hat{\mathbf{j}}(\mathbf{x}), \hat{H}] = \frac{-i\hbar}{m}\Bigg(\hat{\psi}^\dagger(\mathbf{x})\boldsymbol{\nabla}\left(T(\mathbf{x}) + V_{trap}(\mathbf{x})\right)\hat{\psi}(\mathbf{x})$$

$$+ \int d\mathbf{x}'\ \hat{\psi}^\dagger(\mathbf{x})\hat{\psi}^\dagger(\mathbf{x}')\boldsymbol{\nabla}(V^I(\mathbf{x},\mathbf{x}'))\hat{\psi}(\mathbf{x}')\hat{\psi}(\mathbf{x})\Bigg). \tag{8}$$

Thus, Eq. (5) becomes,

$$\frac{\partial^2 \hat{n}(\mathbf{x},t)}{\partial t^2} = \frac{1}{m}\boldsymbol{\nabla}\cdot\Bigg(\hat{\psi}^\dagger(\mathbf{x},t)\boldsymbol{\nabla}\left(T(\mathbf{x}) + V_{trap}(\mathbf{x})\right)\hat{\psi}(\mathbf{x},t)$$

$$+ \int d\mathbf{x}'\ \hat{\psi}^\dagger(\mathbf{x},t)\hat{\psi}^\dagger(\mathbf{x}',t)\boldsymbol{\nabla}(V^I(\mathbf{x},\mathbf{x}'))\hat{\psi}(\mathbf{x}',t)\hat{\psi}(\mathbf{x},t)\Bigg). \tag{9}$$

The Bose field operators are expanded in terms of the eigenstates $\{\phi\}$ of the unperturbed harmonic oscillator, with the creation and annihilation operators as the expansion coefficients.

The expectation value of Eq. (9) in the basis of the number operator is to be found. For this, we consider the case where there are only two levels, the ground state and the first excited state. Thus,

$$|n\rangle = |n_0, n_1, 0, 0, 0...\rangle.$$

The one-body term of Eq. (9) in the TF approximation and in the above basis yields four terms. Of these four terms we neglect the term involving n_1 which is infinitesimally small and the other two off-diagonal terms. Thus we get in a straightforward manner, the matrix element of the one-body term, for the condensate to be

$$n_0(\mathbf{x})\boldsymbol{\nabla}V_{trap}(\mathbf{x}), \tag{10}$$

where $n_0(\mathbf{x}) = n_0\phi_0^*(\mathbf{x})\phi_0(\mathbf{x})$, is the density as it depends on the spatial coordinate $\mathbf{x}$.

Now we consider the interaction term. Let

$$\hat{I} = \int d\mathbf{x}'\ \hat{\psi}^\dagger(\mathbf{x},t)\hat{\psi}^\dagger(\mathbf{x}',t)\boldsymbol{\nabla}(V^I(\mathbf{x},\mathbf{x}'))\hat{\psi}(\mathbf{x}',t)\hat{\psi}(\mathbf{x},t) \tag{11}$$

$$= \int d\mathbf{x}' \sum_{ijkl} \phi_i^*(\mathbf{x})\phi_j^*(\mathbf{x}')\boldsymbol{\nabla}(V^I(\mathbf{x},\mathbf{x}'))\phi_k(\mathbf{x}')\phi_l(\mathbf{x})a_i^\dagger(t)a_j^\dagger(t)a_k(t)a_l(t). \tag{12}$$

Here we take into account only two scattering processes, one when all the atoms are in the condensate which is the intrastate scattering and the second, an interstate scattering between the condensate and the non-condensate atoms in the first excited state. Thus,

$$\hat{I} = \int d\mathbf{x}'\boldsymbol{\nabla}(V^I(\mathbf{x},\mathbf{x}'))\phi_0^*(\mathbf{x})a_0^\dagger(t)\phi_0^*(\mathbf{x}')a_0^\dagger(t)\phi_0(\mathbf{x}')a_0(t)\phi_0(\mathbf{x})a_0(t)$$

$$+ \int d\mathbf{x}'\boldsymbol{\nabla}(V^I(\mathbf{x},\mathbf{x}'))\phi_0^*(\mathbf{x})a_0^\dagger(t)\phi_1^*(\mathbf{x}')a_1^\dagger(t)\phi_1(\mathbf{x}')a_1(t)\phi_0(\mathbf{x})a_0(t). \tag{13}$$

When the above acts on $|n\rangle$, the first term gives

$$n_0(\mathbf{x})n_0 \int d\mathbf{x}' \; \phi_0^*(\mathbf{x}')\boldsymbol{\nabla}(V^I(\mathbf{x},\mathbf{x}'))\phi_0(\mathbf{x}'),$$

where we have used the approximation $n_0 - 1 \approx n_0$. Similarly the second term gives

$$n_0(\mathbf{x})n_1 \int d\mathbf{x}' \; \phi_1^*(\mathbf{x}')\boldsymbol{\nabla}(V^I(\mathbf{x},\mathbf{x}'))\phi_1(\mathbf{x}').$$

Let $\int d\mathbf{x}' \; \phi_0^*(\mathbf{x}')V^I(\mathbf{x},\mathbf{x}')\phi_0(\mathbf{x}') = V_0(\mathbf{x})$ and $\int d\mathbf{x}' \; \phi_1^*(\mathbf{x}')V^I(\mathbf{x},\mathbf{x}')\phi_1(\mathbf{x}') = V_{0110}(\mathbf{x})$. Thus,

$$\langle n|\hat{I}|n\rangle = n_0(\mathbf{x})n_0\boldsymbol{\nabla}V_0(\mathbf{x}) + n_0(\mathbf{x})n_1\boldsymbol{\nabla}V_{0110}(\mathbf{x}). \tag{14}$$

Also,

$$\frac{\partial^2}{\partial t^2}\langle n|\hat{n}(\mathbf{x},t)|n\rangle = \frac{\partial^2 n_0(\mathbf{x})}{\partial t^2}, \tag{15}$$

for the condensate. From Eqs. (10), (14) and (15) the expectation value of Eq. (9) is

$$\frac{\partial^2 n_0(\mathbf{x})}{\partial t^2} - \frac{1}{m}\boldsymbol{\nabla}\cdot\left[n_0(\mathbf{x})\left(\boldsymbol{\nabla}V_{trap}(\mathbf{x}) + n_0\boldsymbol{\nabla}V_0(\mathbf{x}) + n_1\boldsymbol{\nabla}V_{0110}(\mathbf{x})\right)\right] = 0, \tag{16}$$

which could be written as

$$\frac{\partial^2 n_0(\mathbf{x})}{\partial t^2} - \frac{1}{m}\boldsymbol{\nabla}\cdot\left[n_0(\mathbf{x})\boldsymbol{\nabla}(V_{trap}(\mathbf{x}) + n_0V_0(\mathbf{x}) + n_1V_{0110}(\mathbf{x}))\right] = 0. \tag{17}$$

With $n_0(\mathbf{x}) = n_0^{eq}(\mathbf{x}) + \delta n_0(\mathbf{x},t)$, where $n_0^{eq}(\mathbf{x})$ is the equilibrium density, (henceforth referred as $n_0(\mathbf{x})$) and $\delta n_0(\mathbf{x},t)$ denotes the density variations, the above equation for a small variation in energy, becomes

$$\begin{aligned} m\frac{\partial^2 \delta n_0(\mathbf{x},t)}{\partial t^2} &= \boldsymbol{\nabla}n_0(\mathbf{x})\cdot\boldsymbol{\nabla}\delta(V_{trap}(\mathbf{x}) + n_0V_0(\mathbf{x}) + n_1V_{0110}(\mathbf{x})) \\ &\quad + n_0(\mathbf{x})\boldsymbol{\nabla}\cdot\boldsymbol{\nabla}\delta(V_{trap}(\mathbf{x}) + n_0V_0(\mathbf{x}) + n_1V_{0110}(\mathbf{x})). \end{aligned} \tag{18}$$

In the right hand side of the above equation we have neglected $\boldsymbol{\nabla}(\delta n_0(\mathbf{x},t))$ and $\delta n_0(\mathbf{x},t)$ in the first and the second term respectively, as the fluctuations are very small.

The total energy $\epsilon = V_{trap}(\mathbf{x}) + n_0V_0(\mathbf{x}) + n_1V_{0110}(\mathbf{x})$, in the TF approximation. Therefore,

$$n_0 = \frac{\epsilon - (V_{trap}(\mathbf{x}) + n_1V_{0110}(\mathbf{x}))}{V_0(\mathbf{x})}.$$

Thus, we see that n_0 can be treated as an implicit function of $\mathbf{x}$ and thus, we have

$$\delta(V_{trap}(\mathbf{x}) + n_0V_0(\mathbf{x}) + n_1V_{0110}(\mathbf{x})) = V_0(\mathbf{x})\delta n_0(\mathbf{x},t).$$

We consider oscillations with time dependence : $\delta n_0(t) \propto e^{-i\omega_c t}$. Thus, Eq. (18) becomes

$$\begin{aligned} -m\omega_c^2\delta n_0(\mathbf{x}) &= \boldsymbol{\nabla}\left[\frac{\epsilon - (V_{trap}(\mathbf{x}) + n_1V_{0110}(\mathbf{x}))}{V_0(\mathbf{x})}\right]\cdot\boldsymbol{\nabla}(V_0(\mathbf{x})\delta n_0(\mathbf{x})) \\ &\quad + \left[\frac{\epsilon - (V_{trap}(\mathbf{x}) + n_1V_{0110}(\mathbf{x}))}{V_0(\mathbf{x})}\right]\boldsymbol{\nabla}\cdot\boldsymbol{\nabla}(V_0(\mathbf{x})\delta n_0(\mathbf{x})) \end{aligned} \tag{19}$$

$$\approx \nabla\left[\epsilon - (V_{trap}(\mathbf{x}) + n_1 V_{0110}(\mathbf{x}))\right]\cdot\nabla\delta n_0(\mathbf{x})$$
$$+[\epsilon - (V_{trap}(\mathbf{x}) + n_1 V_{0110}(\mathbf{x}))]\nabla^2\delta n_0(\mathbf{x}), \tag{20}$$

where we have dropped the gradients of the interaction potential as an approximation. The total energy $\epsilon \approx \frac{1}{2}m\omega_0^2R^2$, where R is the range of the trapping potential and thus, $\nabla(\frac{1}{2}m\omega_0^2R^2) = 0$. Again as an approximation we drop the term $\nabla V_{0110}(\mathbf{x})$. Thus, the equation for density oscillations is

$$m\omega_c^2\delta n_0(\mathbf{x}) = \nabla V_{trap}(\mathbf{x})\cdot\nabla\delta n_0(\mathbf{x})$$
$$-[\epsilon - (V_{trap}(\mathbf{x}) + n_1 V_{0110}(\mathbf{x}))]\nabla^2\delta n_0(\mathbf{x}). \tag{21}$$

We consider two-body interaction of the form $C_6/|\mathbf{x}-\mathbf{x}'|^6$ and the integrated value of V_{0110} is found out to be $-\frac{1}{2}(kR_0C_6)m\omega_0^2$, where $k = \frac{24\alpha^3}{\sqrt{\pi}\hbar^2}$, R_0 is the range of the two-body interaction potential and $\alpha = \sqrt{\frac{m\omega_0}{\hbar}}$ is the harmonic oscillator length inverse.[15] When the harmonic trapping is isotropic then Eq. (21) in spherical polar coordinates, becomes

$$\omega_c^2\delta n_0(r) = \omega_0^2 r\frac{\partial}{\partial r}\delta n_0(r) - \frac{1}{2}\omega_0^2\left[R^2 - r^2 + kR_0C_6n_1\right]\nabla^2\delta n_0(r), \tag{22}$$

or

$$\eta\,\delta n_0(r) = r\frac{\partial}{\partial r}\delta n_0(r) - \frac{1}{2}\left[R^2 - r^2 + kR_0C_6n_1\right]\nabla^2\delta n_0(r), \tag{23}$$

where $\eta = \omega_c^2/\omega_0^2$.

3. Frequency Spectrum

We take ground state density variations of the form

$$\delta n_0(r) = r^l G(r) Y_{lm}(\theta,\phi). \tag{24}$$

Considering the radial part alone and using

$$\nabla^2 = \frac{\partial^2}{\partial r^2} + \frac{2}{r}\frac{\partial}{\partial r} - \frac{l(l+1)}{r^2},$$

where the third term is the centrifugal barrier term, Eq. (23) becomes

$$\frac{1}{2}\left[R^2 - r^2 + kR_0C_6n_1\right](G'' + \frac{2(l+1)}{r}G') - rG'(r) + (\eta - l)G(r) = 0. \tag{25}$$

Defining a variable $u \equiv \frac{r^2}{R^2}$ the above equation is written as

$$u(1-u)G'' + \left(\left(\frac{2l+3}{2}\right)(1+A) - \left(\frac{2l+5}{2}\right)u\right)G' + \left(\frac{\eta - l}{2}\right)G = 0, \tag{26}$$

where $A = \frac{kR_0C_6n_1}{R^2}$, also appears as the coefficient of G'' and has been dropped as an approximation. Equation (26) is in the standard form for the hypergeometric function $F(\alpha,\beta,\gamma,u)$. Thus, we get

$$u(1-u)F''(u) + [\gamma - (\alpha+\beta+1)u]F'(u) - \alpha\beta F(u) = 0, \tag{27}$$

where $\alpha = -n$, $\beta = l + n + \frac{3}{2}$, $\gamma = (l+\frac{3}{2})(1+A)$ and $\eta - l = -2\alpha\beta$. With this we get the frequency spectrum

$$\omega_c^2 = \omega_0^2(l + 3n + 2nl + 2n^2). \tag{28}$$

We note that the frequency of oscillation is the same as in the case of a contact potential that describes the interaction between the atoms in the condensate alone. However, we mention that Eq. (25) contains the combined effect of the strength of the interaction potential C_6 and the presence of the non-condensate atoms n_1. It may be possible that if Eq. (25) is solved either analytically or numerically we might obtain a new expression for the frequency. A complete solution to this equation as such is not within the scope of the current work. Therefore, in order to discuss the simplest possible case we have to neglect the term $kR_0C_6n_1$ in the coefficient of G'', so that the equation under consideration reduces to the standard hypergeometric equation. The interstate interaction in this case is reflected in the term A present in γ.

The low-lying excitation frequencies in the spectrum given by Eq. (28) are shown in Fig. 1 and are compared with that of an ideal gas. For an ideal gas in a harmonic trap, mode frequencies correspond to those of a free particle if the mean free path is large when compared to the size of the cloud.

In the absence of interaction, the low-lying excitation frequencies for a harmonic oscillator are

$$\omega_{HO} = \omega_0(2n + l). \tag{29}$$

We see that for $l = 1$ which is the dipole case, both the hydrodynamic and harmonic oscillator predictions coincide with the oscillator frequency ω_0. This follows from the fact that in an external harmonic potential the lowest dipole mode corresponds to the center of mass oscillations, and consequently, is unaffected by the interatomic forces. When the number of atoms in the condensate is fixed, the prediction of Eq. (28) becomes diminishing for large values of n and l. For the case of the high energy states the kinetic energy of the confined atoms can no longer be neglected as density varies rapidly in space.

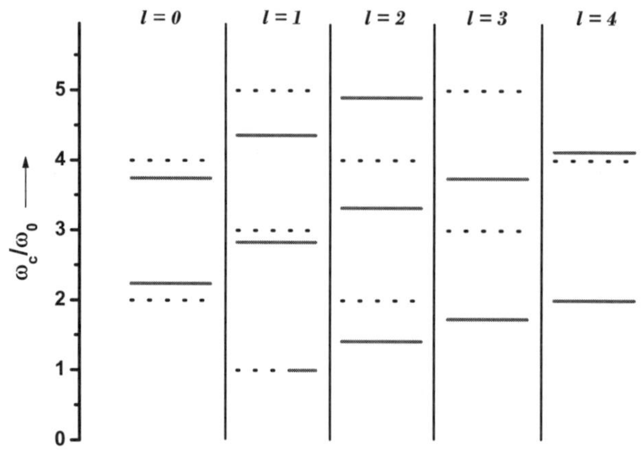

Fig. 1. Excitation frequencies of an interacting condensate in an isotropic harmonic trap(full lines). The dotted lines represent the same, in the absence of interactions.

The normal modes of the cloud are given by

$$\delta n_0(r,t) = Cr^l F(-n, l+n+\frac{3}{2}, (l+\frac{3}{2})(1+A), \frac{r^2}{R^2}) Y_{lm}(\theta,\phi) e^{-i\omega_c t}. \tag{30}$$

The standard series expansion for F is

$$F(\alpha,\beta,\gamma,u) = 1 + \frac{\alpha\beta}{\gamma}\frac{u}{1!} + \frac{\alpha(\alpha+1)\beta(\beta+1)}{\gamma(\gamma+1)}\frac{u^2}{2!} + \ldots \quad . \tag{31}$$

We note that $\gamma = (l+\frac{3}{2})(1+A)$ is present in F. The quantity $A \propto n_1$ is a very small quantity which corresponds to quantum depletion. In the simplest approximation, it has no effect on the frequency spectrum of the collective modes, however it does affect the ground state fluctuations as can be seen from Eq. (30).

4. Solution

One simple solution of Eq. (22) is

$$\delta n_0(r) = Cr^l Y_{lm}(\theta,\phi), \tag{32}$$

where C is an arbitrary constant. For $n = 0$, the modes have no radial nodes and with increasing l these modes become more localized near the surface of the cloud and hence they are called surface modes. Since the function (32) satisfies the Laplace equation, the last term in Eq. (22) vanishes and as a result $\omega_c^2 = l\omega_0^2$. We see that the $l = 0$ mode is trivial because it represents a change in density that is constant everywhere and as a result the frequency of the mode is zero. The mode with $l = 0$ and $n = 1$ is spherically symmetric and the radial velocity has the same sign everywhere. It is therefore referred to as the breathing mode. The three $l = 1$ modes correspond to translational motion of the entire cloud with no change in the internal structure. For the $l = 1$, $m = 0$ mode, the density variation from Eq. (32) is proportional to $rY_{10} \propto z$. The equilibrium density profile is $n_0(r) \propto (1 - r^2/R^2)$ and therefore if the center of the cloud is moved through a distance κ along the z direction, then the change in the density is given by $\delta n_0 = -\kappa\partial n_0/\partial z \propto z$.

The physics of the $l = 1$ modes, for an external harmonic potential, is that the oscillations describe the center of mass movements along the main axes of the trap. The motion of this center of mass, $\mathbf{r}_{cm}$, is essentially that of a free particle of mass μ moving in a fixed potential $\mu\omega_0^2 r_{cm}^2/2$. These modes have the same frequency as that for the motion of a single particle and are referred to as the Kohn modes. It was first noted by Kohn[16] in 1961, that the electron–electron interaction does not change the cyclotron resonance frequency in a bulk three-dimensional electron gas. After the discovery of Bose condensation these modes were also found in trapped dilute Bose gases for many different model approximations.[17,18,19,20] Stringari and others[11,21] have noted Kohn modes in their calculations in which they have used only the contact potential and they have obtained the same frequency spectrum as given in Eq. (28). In the present case, we have included the interaction between condensate and non-condensate atoms and we have found that our results are in conformity with their observation. In Bose condensates, the single-particle and the density oscillation spectra are coupled at low temperatures. The Kohn mode which arises out of density oscillations, must therefore have a single-particle counterpart which disappears from the spectrum when the temperature is raised. Modes with higher values of l have larger number of nodes and higher frequencies.

Collective modes in a trapped Bose gas beyond the mean-field approximation has been studied earlier by Pitaevskii and Stringari[22]; and Braaten and Pearson.[23] They have found that the monopole frequency of breathing oscillations (in a spherical trap), has a shift proportional to the gas parameter $a^3 n(0)$, where a is the s-wave scattering length and $n(0)$ is the density at the center of the trap. This correction showed the same dependence on the gas parameter as the quantum depletion of the condensate, although the proportionality constant slightly differs in the two cases. This shift is of the order of 1%. Also they have discussed finite size effects for isotropic trapping, which brings about a downward shift of $\sim 0.1\%$ which is much smaller than the non-mean-field correction.

5. Conclusion

In this work we have derived using field theoretic method, the collective excitations of an interacting Bose gas trapped by an isotropic harmonic oscillator. This is done in the low temperature limit using the simplest approximation. As a result of the interstate scattering, (described beyond the mean-field Gross-Pitaevskii theory), we get a completely new differential equation, solution to which as such is not attempted in the current work. Nevertheless, in order to obtain standard solution we have made simple approximations. With this we show that though the frequency spectrum remains the same as in the case of a contact potential, a very small contribution from the non-condensate atoms to the ground state density fluctuations is present even at low temperatures. With the present calculations we have discussed the surface modes, breathing modes and Kohn modes of the trapped atoms.

Acknowledgments

GG thanks the Research School of Physical Sciences and Engineering of the Australian National University, Canberra for providing a post-graduate scholarship. MPD thanks the Institute of Physics, Bhubaneshwar, India and the Institute of Mathematical Sciences of the National University of Singapore where part of this work was done.

References

1. D. S. Jin, J. R. Ensher, M. R. Matthews, C. E. Wieman and E. A. Cornell, *Phys. Rev. Lett.* **77**, 420 (1996).
2. M.-O. Mewes, M. R. Andrews, N. J. van Druten, D. M. Kurn, D. S. Durfee, C. G. Townsend and W. Ketterle, *Phys. Rev. Lett.* **77**, 988 (1996).
3. M. R. Andrews, D. M. Kurn, H.-J. Miesner, D. S. Durfee, C. G. Townsend, S. Inouye and W. Ketterle, *Phys. Rev. Lett.* **79**, 553 (1997).
4. A. L. Fetter, *Phys. Rev. A* **53**, 4245 (1996).
5. K. G. Singh and D. S. Rokhsar, *Phys. Rev. Lett.* **77**, 1667 (1996).
6. Y. Kagan, E. L. Surkov and G. V. Shlyapnikov, *Phys. Rev. A* **55**, R18 (1997).
7. P. A. Ruprecht, M. Edwards, K. Burnett and C. W. Clark, *Phys. Rev. A* **54**, 4178 (1996).
8. M. Edwards, P. A. Ruprecht, K. Burnett, R. J. Dodd and C. W. Clark, *Phys. Rev. Lett.* **77**, 1671 (1996).
9. B. D. Esry, *Phys. Rev. A* **55**, 1147 (1997).
10. D. A. W. Hutchinson and E. Zaremba, *Phys. Rev. A* **57**, 1280 (1998).
11. S. Stringari, *Phys. Rev. Lett.* **77**, 2360 (1996).
12. P. Öhberg, E. L. Surkov, I. Tittonen, S. Stenholm, M. Wilkens and G. V. Shlyapnikov, *Phys. Rev. A* **56**, R3346 (1997).
13. M. Fliesser, A. Csordás, P. Szépfalusy and R. Graham, *Phys. Rev. A* **56**, R2533 (1997).
14. G. Gnanapragasam and M. P. Das, *Int. J. Mod. Phys. A* **22**, 4923 (2007).

15. G. Gnanapragasam, Ph.D. Thesis, The Australian National Univ., Canberra, 2008.
16. W. Kohn, *Phys. Rev.* **123**, 1242 (1961).
17. A. L. Fetter and D. Rokhsar, *Phys. Rev. A* **57**, 1191 (1998).
18. E. Zaremba, T. Nikuni and A. Griffin, *J. Low Temp. Phys.* **116**, 277 (1999).
19. M. J. Bijlsma and H. T. C. Stoof, *Phys. Rev. A* **60**, 3973 (1999).
20. A. Minguzzi and M. P. Tosi, *J. Phys.: Condens. Matter* **9**, 10211 (1997).
21. C. J. Pethick and H. Smith, *Bose-Einstein Condensation in Dilute Gases* (Cambridge University Press, Cambridge, 2002).
22. L. Pitaevskii and S. Stringari, *Phys. Rev. Lett.* **81**, 4541 (1998).
23. E. Braaten and J. Pearson, *Phys. Rev. Lett.* **82**, 255 (1999).

p-WAVE PAIRING IN FERMI SYSTEMS WITH UNEQUAL POPULATION NEAR FESHBACH RESONANCE

KHANDKER F. QUADER, RENYUAN LIAO and FLORENTIN POPESCU
Department of Physics, Kent State University,
Kent, OH 44242, USA

Received 31 July 2008

We explore p-wave pairing in a single-channel two-component Fermi system with unequal population near Feshbach resonance. Our analytical and numerical study reveal a rich superfluid (SF) ground state structure as a function of imbalance. In addition to the state $\Delta_{\pm 1} \propto Y_{1\pm 1}$, a multitude of "mixed" SF states formed of linear combinations of Y_{1m}'s give global energy minimum under a phase stability condition; these states exhibit variation in energy with the relative phase between the constituent gap amplitudes. States with local energy minimum are also obtained. We provide a geometric representation of the states. A $T = 0$ polarization vs. p-wave coupling phase diagram is constructed across the BEC-BCS regimes. With increased polarization, the global minimum SF state may undergo a quantum phase transition to the local minimum SF state.

Keywords: p-wave pairing; unequal population; Feshbach resonance; quantum phase transition.

1. Introduction

Discovery of s-wave paired fermion condensates in cold atoms[1] subjected to s-wave Feshbach resonance (FR) has led to explorations in several fascinating directions. One is the study of superfluidity in asymmetrical Fermi systems, a subject of intense experimental and theoretical research.[2,3,4] Another is the prospect for attaining fermion condensates with pairs having non-zero relative orbital angular momentum. Recent observations of p-wave FR[5,6,7,8] in ^6Li and ^{40}K have raised the possibility of observing p-wave superfluidity in cold atoms, leading to renewed theoretical interest. p-wave superfluidity in the BEC-BCS crossover region has been studied using fermion-boson models applied to single-component Fermi gas (where FR occurs in the same hyperfine state),[9,10] and using fermion-only models at $T = 0$ both in two-component Fermi gas (where FR occurs between different hyperfine states)[11] and in single-component case.[12] Ohashi[13] considered both cases at finite-T. While there have been fervent theoretical efforts[3,4] on pairing in asymmetrical Fermi systems subject to s-wave FR, work on the corresponding p-wave case is lacking.

In this paper, within a fermion-only model, we study p-wave superfluidity near a p-wave FR in a two-component Fermi system with *unequal* populations in two hyperfine states. It is conceivable that in future, superfluidity would be observable in population imbalanced systems subjected to p-wave FR, similar to that in the s-wave FR case.

It has been pointed out[7,11] that unlike liquid ^3He, pairing interaction in cold atoms is highly anisotropic in "spin"-space. ("spin" referring to hyperfine states). For example, in ^6Li, when the hyperfine pair $|m_s, m_s' >= |1/2, -1/2>$ is at resonance, the pairs $|1/2, 1/2>$ and $|-1/2, -1/2>$ would not be. So, in studying pairing in this two-component system, intra-spin interactions need not be considered. Pairs in p-wave superfluids with unlike "spin" components can however have different orbital angular momentum content, so the gap parameters Δ_{lm} are related to spherical harmonics, Y_{1m} $(m = \pm 1, 0)$.

Our analytic and numerical considerations of the population imbalanced system lead to several interesting results and predictions: (i) We find a rich superfluid ground state (GS) structure: The p-wave state Δ_1(hence Δ_{-1} by symmetry) give a GS *global* minimum energy. ($\Delta_{1m} \equiv \Delta_m$ hereforth). Interestingly, under a specific condition involving the relative phases of the three pairing amplitudes Δ_m's, a multitude of "mixed" states of the form $a\Delta_0 + b\Delta_1 + c\Delta_{-1}$ are degenerate with $\Delta_{\pm 1}$. In addition, we find states with *local* energy minimum. (ii) We provide a geometric representation of the p-wave superfluid states (Fig. 1): The states exhibiting global minimum lie on a "semicircle" formed by the intersection of the surface of the sphere formed by $\Delta_1, \Delta_{-1}, \Delta_0$ with a plane defined by $|\Delta_1| + |\Delta_{-1}| =$ const. (iii) We obtain polarization (P) vs. p-wave coupling phase diagram at $T = 0$. (Fig. 2). The superfluid phase SF1 comprises of states with global minimum, while SF2 of states with the local minimum. We also find a region of phase separation PS between SF2 and the normal phase N. In this two-component system, PS persists onto full polarization. (iv) These raise the intriguing possibility for a $T = 0$ quantum phase transition from SF1 to SF2 at finite polarization (Fig. 2). Additionally, transitions at finite-T would also occur. (v) In the limit $P \to 0$, we find that the ground state structure of $P \neq 0$, is preserved, hence richer than that obtained earlier.[11]

2. Model

We consider pairing in a two-component Fermi system with unequal "spin" $((\uparrow, \downarrow) \equiv (1/2, -1/2))$ population; the unlike-spin fermions are taken to have equal masses. Pairing interactions in all three $\ell = 1$ orbital channels $(m = 0, \pm 1)$ are taken to be equal. Following the rationale above, interactions between like-spin ($\uparrow\uparrow$ or $\downarrow\downarrow$) are set to zero. Since we adjust self-consistently the chemical potential with the strength and sign of the coupling, in our fermion-only model, molecules would naturally appear as 2-fermion bound states. The $S = 1, m_s = 0$ triplet pairing Hamiltonian is then given by:

$$\mathcal{H} = \sum_{\mathbf{k}\sigma} \xi_{\mathbf{k}\sigma} c^\dagger_{\mathbf{k}\sigma} c_{\mathbf{k}\sigma} + \sum_{\mathbf{k}\mathbf{k}'\mathbf{q}} V_{\mathbf{k}\mathbf{k}'} c^\dagger_{\mathbf{k}+\mathbf{q}/2\uparrow} c^\dagger_{-\mathbf{k}+\mathbf{q}/2\downarrow} c_{-\mathbf{k}'+\mathbf{q}/2\downarrow} c_{\mathbf{k}'+\mathbf{q}/2\uparrow} \tag{1}$$

where $c_{\mathbf{k}\sigma}$ ($c^\dagger_{\mathbf{k}\sigma}$) is the annihilation (creation) operator for a fermion with momentum $\mathbf{k}$, kinetic energy $\xi_{\mathbf{k}\sigma}=\epsilon_{\mathbf{k}}-\mu_\sigma$, and spin σ; μ_σ is the chemical potential of each component, and $\epsilon_{\mathbf{k}}=\hbar^2k^2/2m$. $V_{\mathbf{k}\mathbf{k}'}$ is the pairing interaction.

We consider condensate pairs with zero center-of-mass momentum, ($\mathbf{q}$=0). $\mathcal{H}$ is mean-field (MF) decoupled via the $S = 1, m_s = 0$ "spin"-triplet pairing gap function $\Delta_{\downarrow\uparrow}(\mathbf{k}) \equiv \Delta(\mathbf{k})=-\sum_{\mathbf{k}'} V_{\mathbf{k}\mathbf{k}'}\langle c_{-\mathbf{k}'\downarrow}c_{\mathbf{k}'\uparrow}\rangle$ giving:

$$\begin{aligned}\mathcal{H}^{\mathrm{MF}} &= \sum_{\mathbf{k}\sigma} \xi_{\mathbf{k}\sigma} c^\dagger_{\mathbf{k}\sigma} c_{\mathbf{k}\sigma} \\ &-\sum_{\mathbf{k}} \left[\Delta(\mathbf{k}) c^\dagger_{\mathbf{k}\uparrow} c^\dagger_{-\mathbf{k}\downarrow} + \mathrm{H.c.}\right] \\ &-\sum_{\mathbf{k},\mathbf{k}'} V_{\mathbf{k}\mathbf{k}'} < c^\dagger_{\mathbf{k}\uparrow} c^\dagger_{-\mathbf{k}\downarrow} >< c_{-\mathbf{k}'\downarrow} c_{\mathbf{k}'\uparrow} > \end{aligned} \tag{2}$$

To obtain the ground state energy of the MF Hamiltonian (2) we first use the equation-of-motion method[4] for the imaginary-time ($\tau = it$) normal ($G_{\sigma\sigma'}$) and anomalous ($F_{\sigma\sigma'}$) Green's functions:

$$\begin{aligned}\partial_\tau G_{\sigma\sigma'}(\mathbf{k},\tau) &= -\delta(\tau)\delta_{\sigma\sigma'} - \xi_{\mathbf{k}\sigma} G_{\sigma\sigma'}(\mathbf{k},\tau) \\ &\quad + \Delta_{-\sigma\sigma}(\mathbf{k}) F_{-\sigma\sigma'}(\mathbf{k},\tau), \end{aligned} \tag{3}$$

$$\begin{aligned}\partial_\tau F_{\sigma\sigma'}(\mathbf{k},\tau) &= \xi_{-\mathbf{k}\sigma} F_{\sigma\sigma'}(\mathbf{k},\tau) \\ &\quad + \Delta^*_{\sigma-\sigma}(\mathbf{k}) G_{-\sigma\sigma'}(\mathbf{k},\tau). \end{aligned} \tag{4}$$

Equations (3) and (4) are Fourier transformed with $\tau\rightarrow i\omega_n \equiv \nu$ and $\partial_\tau \rightarrow -\nu$, where $i\omega_n=(2n+1)\pi T$ are Matsubara frequencies. This gives $G_{\sigma\sigma}(\mathbf{k},\nu) = -(\xi_{-\mathbf{k}-\sigma}+\nu)/[(\xi_{\mathbf{k}\sigma}-\nu)(\xi_{-\mathbf{k}-\sigma}+\nu)+|\Delta_{\sigma,-\sigma}(\mathbf{k})|^2]$ and $F_{\sigma-\sigma}(\mathbf{k},\nu)=\Delta^*_{\sigma-\sigma}(\mathbf{k})/[(\xi_{\mathbf{k}-\sigma}-\nu)(\xi_{-\mathbf{k}\sigma}+\nu)+|\Delta_{\sigma-\sigma}(\mathbf{k})|^2] = -F_{-\sigma\sigma}(\mathbf{k},\nu)$.

Next, we separate the radial and angular parts of the interaction: $V_{\mathbf{k}\mathbf{k}'} = (4\pi/3)\sum_m V_{kk'} Y_{1,m}(\hat{\mathbf{k}}) Y^*_{1,m}(\hat{\mathbf{k}'})$ ($\hat{\mathbf{k}}=(\theta,\phi)$). We take for $V_{kk'}$ a separable form convenient for calculations: $V_{kk'} = \lambda w(k) w(k')$. Using $\langle c_{-\mathbf{k}'\downarrow}c_{\mathbf{k}'\uparrow}\rangle \equiv F^*_{\downarrow\uparrow}(\mathbf{k}',\tau=0^-)$ and Fourier transforming, we obtain, $\Delta(\mathbf{k})=-(1/k_BT)\sum_{\nu\mathbf{k}'} V_{\mathbf{k}\mathbf{k}'} F^*_{\downarrow\uparrow}(\mathbf{k}',\nu)e^{\nu 0^+}$. With the separable form of $V_{kk'}$, the gap function can be written as: $\Delta(\mathbf{k}) = \sum_m \Delta_m w(k) Y_{1m}(\hat{\mathbf{k}}) \equiv w(k)\Delta(\hat{k})$, where $\Delta_m = -(1/k_BT)\sum_{\nu\mathbf{k}'}\lambda w(k') \cdot Y^*_{1m}(\hat{\mathbf{k}'}) F^*_{\downarrow\uparrow}(k',\nu) e^{\nu 0^+}$. Obtaining $F_{\downarrow\uparrow}(k,\nu)$ and $G_{\sigma\sigma}$ from Eqs. (3)–(4), we deduce equations for the $\ell = 1; m = 0, \pm 1$ gap amplitudes,

$$\Delta_m = -\lambda \sum_{\mathbf{k}} w(k) Y^*_{1m}(\hat{\mathbf{k}}) \Delta(\mathbf{k}) [n_F(E^\downarrow_k) - n_F(E^\uparrow_k)]/(2E_k) \tag{5}$$

and for the particle number densities,

$$n_\sigma = \sum_k \langle c^+_{k\sigma} c_{k\sigma} \rangle = \sum_k G_{\sigma\sigma}(k, \tau = 0^-). \tag{6}$$

Above, $h = (\xi_{k\downarrow} - \xi_{k\uparrow})/2 = (\mu_\uparrow - \mu_\downarrow)/2$, $E_k = [\xi_k^2 + |\Delta^2(\mathbf{k})|]^{1/2}$, with $\xi_k=(\xi_{k\uparrow}+\xi_{k\downarrow})/2=\epsilon_k - \mu$ and $\mu=(\mu_\uparrow+\mu_\downarrow)/2$; $n_F(E^\pm_k)$ are Fermi functions, with

$E_k^{\pm} = -h \pm E_k$ At $T = 0$ and for $h>0$, the ground state energy E_G is given:

$$E_G \equiv \langle \mathcal{H} - \mu_\uparrow N_\uparrow - \mu_\downarrow N_\downarrow \rangle = \sum_k \left\{ \xi_k - E_k + \frac{|\Delta(\mathbf{k})|^2}{2\epsilon_k} \right\}$$
$$+ \sum_k \{(E_k - h)\theta(-E_k + h)\} - \frac{1}{g} \sum_m |\Delta_m|^2. \tag{7}$$

Following standard practice, the coupling λ has been expressed in terms of the "regularized" interaction g: $1/g = 1/\lambda + (1/(2\pi\hbar)^3) \int w^2(k) d^3\mathbf{k}/2\epsilon_k$.[14] In all our calculations, we adopt the N-SR[15] form: $w(k) = k_0 k/(k_0^2 + k^2)$.

3. Ground State Analysis

First, we carry out an *analytic study* of the p-wave superfluid ground state structure. We assume $0 \le |\Delta(\mathbf{k})| \ll E_k$, and expand E_k in powers of $|\Delta(\mathbf{k})|$ keeping terms to 4th order: $E_k \sim |\xi_k|[1 + |\Delta(\mathbf{k})|^2/2|\xi_k|^2 - |\Delta(\mathbf{k})|^4/8|\xi_k|^4]$. Substituting this into (7) and converting momentum sums to integrals we obtain a simple expression: $E_G = \alpha \int |\Delta(\hat{k})|^2 d\Omega + \beta \int |\Delta(\hat{k})|^4 d\Omega + \gamma$, where $\alpha = (1/(2\pi\hbar)^3) \int_{E_k<h} w^2(k) k^2 dk/2\epsilon_k + (1/(2\pi\hbar)^3) \int_{E_k>h} w^2(k) k^2 dk (1/2\epsilon_k - 1/2|\xi_k|) - 1/g$, $\beta = (1/(2\pi\hbar)^3) \int_{E_k>h} w^4(k) k^2 dk/8|\xi_k|^3$ and $\gamma = (4\pi/(2\pi\hbar)^3) \int_{E_k<h} (\xi_k - h) k^2 dk$. Existence of a superfluid phase requires $\alpha<0$, otherwise minimization of E_G forces the gap to vanishes. The polarization ($P = (n_\uparrow - n_\downarrow)/(n_\uparrow + n_\downarrow)$) dependence of E_G is contained in α, β, γ, which depend on h, μ and $w(k)$; α alone depends explicitly on the coupling g. Accordingly, the Δ_m's are sensitive to changes in P.

For fixed μ and h, and taking E_k (to $\mathcal{O}(\Delta(\mathbf{k})^4)$), E_G can be expressed in terms of the pairing amplitudes Δ_m[16]:

$$E_G = -\frac{\alpha^2}{8\beta} + \gamma + 2\beta \left(|\Delta_0|^2 + |\Delta_1|^2 + |\Delta_{-1}|^2 + \frac{\alpha}{4\beta} \right)^2$$
$$+ \beta(|\Delta_0|^2 - 2|\Delta_1||\Delta_{-1}|)^2 + 4\beta(1-t)|\Delta_0|^2|\Delta_1||\Delta_{-1}| \tag{8}$$

where $t = \cos\theta$, with $\theta = \sphericalangle(\Delta_0\Delta_1^*, \Delta_0^*\Delta_{-1}) = 2\phi_0 - \phi_1 - \phi_{-1}$; ϕ_m's are the phases associated with the gap amplitudes Δ_m. From (8), upon minimization with respect to Δ_m's, we find that for a stable p-wave superfluid phase to exist, the following conditions have to be satisfied simultaneously

$$\begin{aligned} &(a)\ |\Delta_0|^2 + |\Delta_1|^2 + |\Delta_{-1}|^2 = -\alpha/4\beta \\ &(b)\ \ |\Delta_0|^2 - 2|\Delta_1||\Delta_{-1}| \ = 0 \\ &(c)\ (1-t)|\Delta_0|^2|\Delta_1||\Delta_{-1}| = 0\,. \end{aligned} \tag{9}$$

This gives the ground state (GS) *global minimum* energy,

$$E_G = -\alpha^2/8\beta + \gamma. \tag{10}$$

Conditions (a) and (b) together represent a semicircle formed by the intersection of the surface of a sphere of radius R, $R^2 = -\alpha/4\beta$, with a plane defined by $|\Delta_1| +$

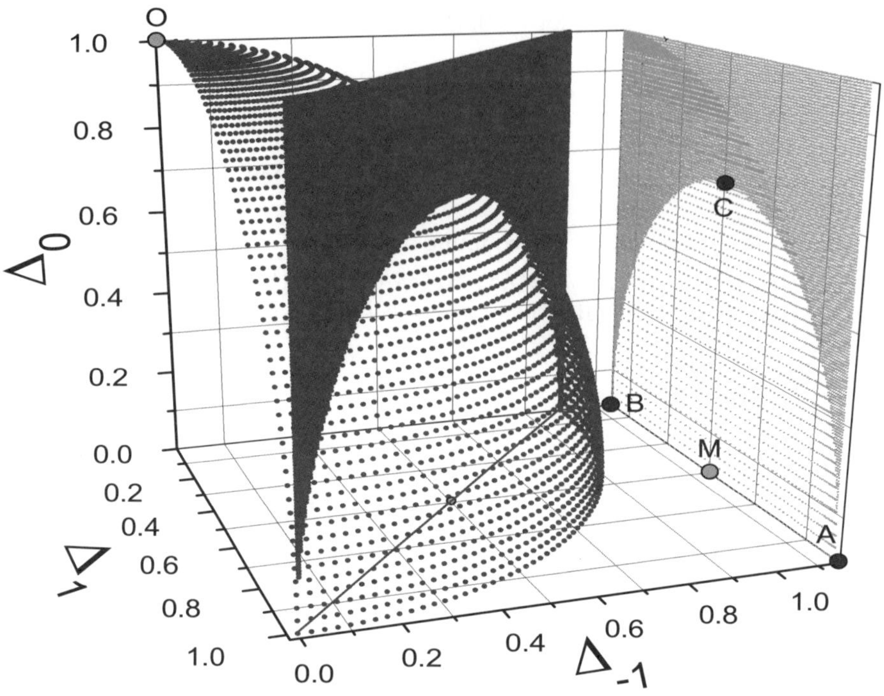

Fig. 1. Geometric representation of p-wave states in two-component population imbalanced system. States (e.g. A, B, C) exhibiting global energy minimum lie on the semicircle formed by the intersection of sphere surface with plane (see text). For clarity, plane containing semicircle is also shown separately (green). O (north pole) and M (on $\Delta_1 - \Delta_{-1}$ plane) are states with local energy minimum. $\Delta_m (m = 0, \pm 1)$'s have been normalized to the sphere radius $\sqrt{-\alpha/4\beta}$; $\alpha < 0$.

$|\Delta_{-1}|=R$. All points on the semicircle must necessarily satisfy the phase constraint (c). They represent a multitude of "mixed" superfluid states of the form $a\Delta_0 + b\Delta_1 + c\Delta_{-1}$. A 3D geometric representation of the states is shown in Fig. 1. The GS global energy minimum condition is satisfied only for $t = 1$ in Eq. 9c, i.e., when the relative phase $\theta = (\phi_0 - \phi_1) - (\phi_{-1} - \phi_0) = 2n\pi$. For the states $\Delta_{\pm 1}$ (points A and B on the semicircle in Fig. 1), this is always satisfied. For the other states on this curve, containing varying admixtures of Δ_m's, E_G fluctuates with θ, and attains the global minimum only for $\theta = 2n\pi$ (Fig. 3a); the fluctuation amplitude depends on the specific state chosen; e.g. C at the top of the projected figure has the maximum amplitude variation. $t = 1$ corresponds to a parallel orientation of vectors $\Delta_0\Delta_1^*$ and $\Delta_0^*\Delta_{-1}$; this gives the same global minimum $E_G = -\alpha^2/8\beta + \gamma$ for three particular cases: (A) $|\Delta_0|=|\Delta_{-1}|=0$ and $|\Delta_1|^2 = -\alpha/4\beta$; (B) $|\Delta_0|=|\Delta_1|=0$ and $|\Delta_{-1}|^2 = -\alpha/4\beta$; (C) $|\Delta_0|^2 = -\alpha/8\beta$ and $|\Delta_1|^2 = |\Delta_{-1}|^2 = -\alpha/16\beta$. These are shown as points A, B, C respectively in Fig. 1. In addition to these, we also find the existence of *local* minimum with $E_G^L = -\alpha^2/12\beta + \gamma$ for: (i) $|\Delta_0|^2 = -\alpha/4\beta$ and $|\Delta_1|=|\Delta_{-1}|=0$; (ii) $|\Delta_0|=0$ and $|\Delta_1|^2 = |\Delta_{-1}|^2 = -\alpha/12\beta$; these are shown as O and

M respectively in Fig. 1. Values for the GS energy are completely determined by α, β, and γ given earlier.

In the limit of zero polarization, $P \to 0$, the GS energy coefficient $\gamma \to 0$, while $\alpha < 0, \beta > 0$ remain finite. Thus the GS structure of the polarized case is preserved for $P \to 0$, with the same expressions for the energy global and local minima (different numerical values). The states A and B (equivalent to A by symmetry) above and in Fig. 1 correspond to the finding of Ref. 11. Our work reveals an additional set of "mixed" states on the intersecting semicircle (Fig. 1) that also have the same global minimum for specific values of the relative phase discussed above.

4. Numerical Calculations

Complementary to, but independent of, the analytic study above, we next carry out detailed numerical study of the p-wave superfluid states. We solve self-consistently the three gap equations (Eq. (5)) and two number equations (Eq. (6)), for fixed population imbalance $P = (n_\uparrow - n_\downarrow)/(n_\uparrow + n_\downarrow)$, and p-wave coupling parameter g. (g is related to the triplet scattering parameter: $g = 25k_F^3 a_t/8\pi$ for a cutoff $k_0 = 10k_F$). This gives us the gap amplitudes Δ_m, and the chemical potentials μ_σ for different P and $1/k_F^3 a_t$. Using these we obtain the ground state energy E_G from Eq. (7), as well as the coefficients α, β, γ that determine E_G given by Eq. (8). The agreement between our numerical and analytical E_G is quite good.

To check for *stability* of the p-wave states obtained from numerical solutions, we enforce that the stability matrix $(\partial^2 E_G/\partial\Delta_{m_i}\partial\Delta_{m_j})$ is positive definite; and that $\delta p/\delta h > 0$. Based on this, we construct a polarization (P)-coupling $(1/k_F^3 a_t)$ *phase diagram* in BEC-BCS crossover regime (Fig. 2). In Fig. 2, SF1 denotes a stable superfluid phase, corresponding to the states on the "semicircle" in Fig. 1 that produce GS *global* minimum. At $T = 0$, with increased polarization, SF1 becomes unstable, and gives way to the superfluid phase SF2, corresponding to states with the *local* minimum discussed above. At even larger polarizations, SF2 becomes unstable to phase separation (PS). SF1, SF2, and PS occupy a relatively much narrower part of the phase diagram on the BCS side compared to the BEC side. In our two-component system with inter-species interaction, PS persists into full polarization, P=1. This is reasonable because at P=1, the system is essentially a one-component system in which the absence of minority species atoms makes inter-species interaction inoperative. Such a system can exhibit superfluidity only under the effect of intra-species interactions. Our results suggest several interesting possibilities: A $T = 0$ quantum phase transition from SF1 to SF2 driven by polarization may be expected. Since the energies of SF2 and the continuum of "mixed" states for $\theta \neq 2n\pi$ lie higher than the GS global minimum, they may be accessed at finite-T. Also, each p-wave state on the semicircle (Fig. 1) (generally formed of a linear combination of the Δ_m's except for the "pure" $\Delta_{\pm 1}$ states A,B), is characterized by a relative phase $2n\pi \leq \theta \leq (2n + 2)\pi$. Hence, for each there is a multitude of

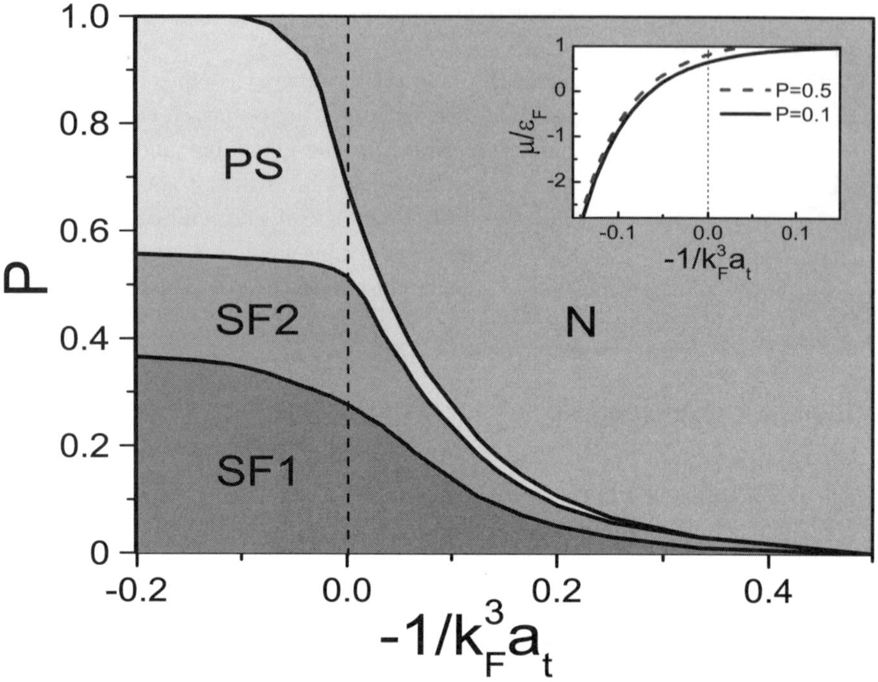

Fig. 2. Polarization P vs. p-wave coupling $-1/(k_F^3 a_t)$ phase diagram of two-component Fermi system with p-wave pairing. Shown are normal, N, and p-wave superfluid phases, SF1 and SF2, and phase separation, PS. Unitarity limit is shown by the dashed vertical line. Inset: Calculated chemical potential vs. coupling across BCS and BEC regimes for $P = 0.1$ (solid line); 0.5 (dashed line).

θ-dependent states (Fig. 3a) with varying E_G that may be interesting to probe with experiments sensitive to such relative phases.

The inset in Fig. 2 shows the calculated behavior of chemical potential μ across the BEC-BCS regime. It deviates significantly from the Fermi energy in a wider region around FR even on the BCS side, and drops much more rapidly to negative values on the BEC side compared to the s-wave FR case. For sufficiently weak coupling in the BCS regime, μ approaches Fermi energy ϵ_F.

Figures 3b and 3c show the variation of E_G with coupling g at fixed polarizations P, and with P for fixed g. E_G is normalized to the Fermi energy $\epsilon_F = \hbar^2 k_F^2/2m$, with $2k_F^3 = k_{F\uparrow}^3 + k_{F\downarrow}^3$. For a given P, E_G becomes less negative as g approaches unitarity; the trend is more noticeable for smaller P's. For a given g, E_G lie higher for smaller P, presumably due to the lower majority-species band becoming progressively more occupied with increasing P, thereby lowering E_G with increasing polarization. For small P, E_G becomes less negative with increasing g. This trend is reversed for $P \geq 0.3$.

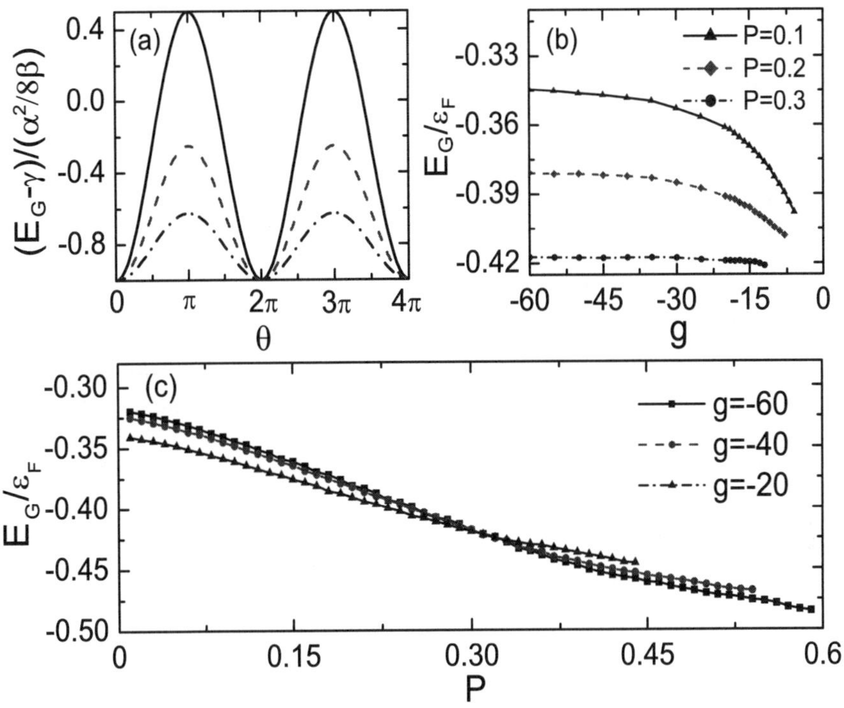

Fig. 3. (a) Calculated ground state energy (scaled to global minimum) vs. relative phase angle θ (see text). Solid curve with maximal fluctuations of E_G corresponds to state C in Fig. 1, ($|\Delta_0|^2{=}-\alpha/8\beta$, $|\Delta_1|^2{=}|\Delta_{-1}|^2{=}-\alpha/16\beta$). The dash and dot-dash curves correspond to two other states on the arc in Fig. 1: $|\Delta_0|^2{=}-\alpha/11.3\beta$, $|\Delta_1|^2{=}-\alpha/76.3\beta$, $|\Delta_{-1}|^2{=}-\alpha/6.73\beta$, and $|\Delta_0|^2{=}-\alpha/16\beta$, $|\Delta_1|^2{=}-\alpha/186\beta$, $|\Delta_{-1}|^2{=}-\alpha/5.49\beta$ respectively. (b) Calculated ground state energy vs. coupling g for different polarizations, P. (c) Calculated ground state energy vs. polarization P for different coupling strengths g.

5. Conclusion

We have presented new results and made several predictions for two-component population imbalanced Fermi systems across the entire BCS-BEC regimes. Future experiments near p-wave FR may be able to provide tests of some of these predictions. Insight into the nature of the orbital part of our superfluid states may be gained from measuring the angular dependence of momentum distributions; from molecular spectroscopy using light radiation; or possibly measurements of zero sound attenuation. The work could also be of interest to other two-component Fermi systems where p-wave intra-species couplings are small or negligible.

Acknowledgments

We acknowledge helpful discussions with Jason Ellis, Randy Hulet, Harry Kojima, and Adriana Moreo. We also acknowledge support from the U.S. Department of Army and ICAM.

References

1. C.A. Regal et al., *Phys. Rev. Lett.* **92**, 040403 (2004); M. W. Zwierlein et al., *Phys. Rev. Lett.* **92**, 120403 (2004); C. Chin et al., *Science* **305**, 1128 (2004); T. Bourdel et al., *Phys. Rev. Lett.* **93**, 050402 (2004).
2. M.W. Zwierlein et al., *Science* **311**, 492 (2006); G.B. Patridge et al., *Science* **311**, 503 (2006); Y. Shin et al., *Phys. Rev. Lett.* **97**, 030401 (2006).
3. W. V. Liu and F. Wilczek, *Phys. Rev. Lett.* **94**, 017001 (2005); J. Carson and S. Reddy, *Phys. Rev. Lett.* **95**, 060401 (2005); C.-H. Pao et al. *Phys. Rev. B* **73**, 132506 (2006); D. E. Sheehy and L. Radzihovsky, *Phys. Rev. Lett.* **96** , 060401 (2006). M.M. Parish, et al., *Phys. Rev. Lett.* **98**, 160402 (2007).
4. R. Liao and K. Quader, *Phys. Rev. B* **76**, 212502 (2007).
5. J. Zhang, et al., *Phys. Rev. A.* **70**, 030702(R) (2004).
6. C.A. Regal, C. Ticknor, J.L. Bohn and D.S. Jin, *Phys. Rev. Lett.* **90**, 053201 (2003).
7. C. Ticknor, C.A. Regal, D.S. Jin and J.L. Bohn, *Phys. Rev. A* **69**, 042712 (2004).
8. C.H. Schunck, et al., *Phys. Rev. A* **71**, 045601 (2005).
9. V. Gurarie, L. Radzihovsky and A.V. Andreev, *Phys. Rev. Lett.* **94**, 230403 (2005).
10. C.-H. Cheng and S.-K. Yip, *Phys. Rev. Lett.* **95**, 070404 (2005).
11. Tin-Lun Ho and Roberto B. Diener, *Phys. Rev. Lett.* **94**, 090402 (2005).
12. S. S. Botelho and C.A.R. Sá de Melo, *J. Low Temp. Phys.* **140**, 409 (2005); M. Iskin and C.A.R. Sá de Melo, *Phys. Rev. Lett.* **96**, 040402 (2006).
13. Y. Ohashi, *Phys. Rev. Lett.* **94**, 050403 (2005).
14. This allows g to be expressed in terms of scattering parameters, viz. triplet scattering volume a_t and an effective range or equivalently a cut-off momentum k_0; see, for example, M. Iskin and C.A.R. Sá de Melo, *Phys. Rev. A* **74**, 013608 (2006).
15. P. Nozières and S. Schmitt-Rink, *J. of Low Temp. Phys.* **59**, 195 (1985).
16. Details will be presented elsewhere.

INTRIGUING ROLE OF HOLE-COOPER-PAIRS IN SUPERCONDUCTORS AND SUPERFLUIDS

M. GRETHER

Facultad de Ciencias, Universidad Nacional Autónoma de México
Apdo. Postal 70-360, 04510 México, DF, Mexico
mdgg@hp.fciencias.unam.mx

M. de LLANO

Instituto de Investigaciones en Materiales, Universidad Nacional Autónoma de México
Apdo. Postal 70-360, 04510 México, DF, Mexico
dellano@servidor.unam.mx

S. RAMÍREZ

Departamento de Física, Universidad Autónoma Metropolitana-Iztapalapa
Apdo. Postal 09-340, 09340 México, DF, Mexico
sdellano@correo.unam.mx

O. ROJO

PESTIC, Secretaría Académica & CINVESTAV, IPN
04430 México DF, Mexico
orojo@cinvestav.mx

Received 31 July 2008

The role in superconductors of *hole*-Cooper-pairs (CPs) are examined and contrasted with the more familiar *electron*-CPs, with special emphasis on their "background" effect in enhancing superconducting transition temperatures T_c — even when *electron*-CPs drive the transition. Both kinds of CPs are, of course, present at *all* temperatures. An analogy is drawn between the hole CPs in any many-fermion system with the antibosons in a relativistic ideal Bose gas that appear in substantial numbers *only* at higher and higher temperatures. Their indispensable role in yielding a lower Helmholtz free energy equilibrium state is established. For superconductors, the problem is viewed in terms of a generalized Bose-Einstein condensation (GBEC) theory that is an extension of the Friedberg-T.D. Lee 1989 boson-fermion BEC theory of high-T_c superconductors in that the GBEC theory includes hole CPs as well as electron-CPs — thereby containing as well as further extending BCS theory to higher temperatures with the same weak-coupling electron-phonon interaction parameters. We show that the Helmholtz free energy of both 2e- and 2h-CP pure condensates has a positive second derivative, and are thus stable equilibrium states. Finally, it is conjectured that the role of hole pairs in ultra-cold fermionic atom gases will likely be negligible because the very low densities involved imply a "shallow" Fermi sea.

Keywords: Superconductivity; fermionic atom superfluidity; hole-Cooper-pairs; Bose-Einstein condensation; electron-phonon coupling.

1. Introduction

Quantum many-body matter comprises a wide variety of both mass ρ and number densities n. For superconductors, superfluids and Bose-Einstein condensates (BECs), a wide range of critical temperatures T_c is also present. For many-fermion systems such as superconductors one has $n = 10^{19} - 10^{23}\ cm^{-3}$ which implies Fermi temperatures $T_F = 70 - 10^5 K$. For neutral-fermion superfluids, such as i) liquid-^3He one has $n \simeq 10^{22}\ cm^{-3}$ (or $\rho \simeq 0.1\ g/cm^3$, i.e., about a 1/10th that of water) and $T_F = 3 - 5K$; ii) neutron-star matter $\rho = 10^{15}\ g/cm^3$ and $T_F = 10^{12}K$; iii) dilute mixtures of ^3He in ^4He; and iv) ultra-cold fermionic BECs with enormous average spacings (when compared with inter-atomic potential ranges) $\rho \simeq 10^{-8}\ g/cm^3$ and $T_F < 10^{-6}K$.

Here, we survey the role of hole-Cooper-pairs in many-fermion gases, starting with their analogy with antiparticles such as antineutrinos and antibosons in general.

2. Role of Antiparticles

2.1. *In high-energy physics*

It is well-known that starting from the Einstein equation $E = m_{rel}c^2$ relating the energy E of a particle with relativistic mass $m_{rel} \equiv m/\sqrt{1 - v^2/c^2}$ where m is the particle rest mass, v its velocity and c the speed of light, as well as the relativistic expression for linear momentum $p = m_{rel}v$, one can readily deduce the relativistic energy-momentum relation $E^2 = p^2c^2 + m^2c^4$ for the particle. Solving for the energy E leads to the *two* solutions $E = \pm(\hbar^2K^2c^2 + m^2c^4)^{1/2} \equiv \pm \mid E_K \mid$. In 1928, Dirac decided *not to exclude* the minus-sign solution and discovered antiparticles, the first example of which was the antielectron or positron that was observed by Anderson in 1932. The simplest instance of the occurrence of antimatter in high-energy physics is perhaps beta decay whereby an isolated neutron n decays into a proton p, an electron e and an *anti*neutrino $\overline{\nu}_e$, and occurs with a half-life of about 15 minutes, namely

$$n \to p + e + \overline{\nu}_e. \tag{1}$$

Here, conservation of lepton number *requires* that the neutrino on the right-hand side be an antiparticle. This contrasts with the inverse process

$$p + e \to n + \nu_e \tag{2}$$

known as electron capture as occurs, e.g., under the influence of the enormous gravitational attractions in the evolution of an ideal white dwarf into a neutron star.

2.2. *In relativistic BEC*

For *any* temperature T the correct energy-momentum relation of identical bosons of mass m in a many-boson gas is, of course, not strictly $E_K = \hbar^2K^2/2m$ but rather

$$|E_K| \equiv \sqrt{c^2\hbar^2K^2 + m^2c^4} = mc^2 + \hbar^2K^2/2m + O(K^4)\ if\ c\hbar K \ll mc^2 \qquad \mathbf{NR}$$

$$= c\hbar K[1 + \frac{1}{2}(mc/\hbar K)^2 + O(K^{-4})]\ if\ c\hbar K \gg mc^2 \quad \mathbf{UR}$$

where NR and UR stand for the nonrelativistic and ultrarelativistic limits. In the UR case, the BEC singularity at $T_c^{N const}$ that follows by imposing constancy of the total number of bosons N given by

$$N \equiv N_0(T) + \sum_{\mathbf{K}\neq 0}[\exp\beta(|\,E_K\,| - \mu) - 1]^{-1} \tag{3}$$

is well-known to be $T_c^{N const} = (\hbar c/k_B)[\pi^2 n/\zeta(3)]^{1/3}$ where $n \equiv N/L^3$ with L the system size and $\zeta(3) \simeq 1.20206$ is the Riemann Zeta function of order 3. Not surprisingly, there is no mass m dependence. On the other hand, in the NR case one gets the familiar formula $T_c^{N const} \simeq 3.31\hbar^2 n^{2/3}/m$ which *does* have a mass dependence. Indeed, both these expressions for $T_c^{N const}$ are special cases of a general dispersion relation $E_K = C_s K^s$, with C_s a constant, in any dimensionality d and implying the general T_c-formula[1]

$$k_B T_c^{N const} = C_s\left[\frac{s\Gamma(d/2)(2\pi)^d n}{2\pi^{d/2}\Gamma(d/s)g_{d/s}(1)}\right]^{s/d} \tag{4}$$

where $\Gamma(x)$ is the gamma function and $g_\sigma(1) \equiv \zeta(\sigma)$ if $\sigma > 1$ but ∞ if $\sigma \leq 1$. This latter property leads to the well-known conclusion that $T_c \equiv 0\ \forall\ d \leq s$ but is otherwise nonzero.

These results, however, ignore the appearance at higher and higher temperatures of substantial numbers of antibosons. At high T's BEC *must account for, say,* $\overline{N}'$ *antibosons* along with N' bosons. Here, $\overline{N}'$ is N' as in (3) but with $+\mu$ instead of $-\mu$. Thus, in general it is not N' that is constant but rather $N' - \overline{N}'$. The latter is given by

$$(N' - \overline{N}') \equiv N \equiv nL^d \equiv \sum_{\mathbf{K}}(n_{\mathbf{K}} - \overline{n_{\mathbf{K}}}) = \sum_{\mathbf{K}}\{[\exp\{\beta(|\,E_K\,| - \mu)\} - 1]^{-1} \tag{5}$$
$$-\mathit{same\ with}\ \mu \to -\mu\}.$$

In the UR limit of this *exact* equation one can extract $T_c^{N'-\overline{N}' const} = (\hbar^{3/2}c^{1/2}/k_B)(3n/m)^{1/2}$ which, surprisingly, now depends on m. The *exact* value of the BEC T_c associated with (5) was found[2] numerically to be *higher* than that associated with (3) for *all* values of the dimensionless number density $\hbar^3 n/m^3c^3$. Their associated Helmholtz free energies were similarly found to be *lower* when antibosons are not neglected as in (5) than when only bosons are included as in (3). This means that the complete system with bosons as well as antibosons is the stable thermodynamic state whereas the system without antibosons is a metastable state.

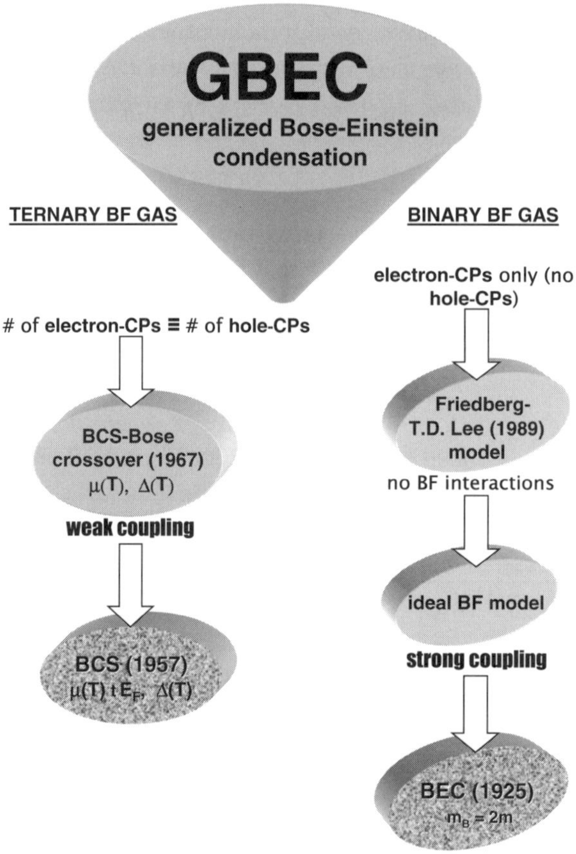

Fig. 1. Flowchart showing how the GBEC theory reduces to: i) the BCS-Bose crossover theory [defined by two coupled transcendental equations for $\Delta(T)$ and $\mu(T)$] whenever the number of 2e-CPs equals the number of 2h-CPs in *both* condensate $n_0(T) = m_0(T)$ *and* noncondensate $n_{B+}(T) = m_{B+}(T)$; ii) the latter reduces to the ordinary BCS theory [defined by the gap equation for $\Delta(T)$] whenever interelectronic coupling λ is sufficiently small so that $\mu(T) \simeq E_F$, the Fermi energy; iii) the three theories shown on the rhs leg correspond to entirely ignoring 2h-CPs and includes the ordinary BEC theory when *all* electrons are paired into bosons of mass $2m$ and number density $n/2$ (i.e., strong coupling).

3. Hole-Cooper-Pairs in a Generalized BEC Theory

A generalized BEC theory (GBEC) of a *ternary* gas mixture — composed of two-electron (2e)[3] Cooper pairs (CPs), two-hole (2h) CPs and unpaired electrons — was described earlier.[4–6] It leads to three coupled transcendental equations for three unknown functions of absolute temperature T which are $n_0(T)$ (the condensate number density of 2e-CPs), $m_0(T)$ (the same for 2h-CPs) and the electron chemical potential $\mu(T)$.

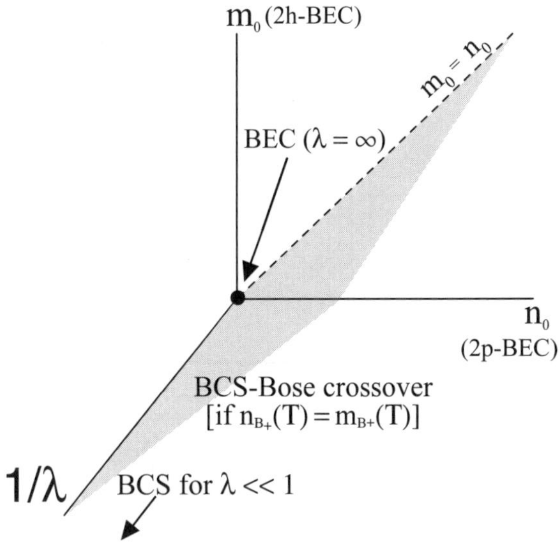

Fig. 2. Parameter octant defined by the two condensate densities $n_0(T) \geqslant 0$ and $m_0(T) \geqslant 0$ as well as the (also nonnegative) *inverse* $1/\lambda \geqslant 0$ of the interelectronic coupling λ, and valid in principle at all temperatures T. The GBEC describes a ternary gas and holds in the *entire* octant. The BCS-Bose crossover theory occurs *only* on the shaded plane defined by $n_0(T) \equiv m_0(T)$ provided the additional restriction $n_{B+}(T) = m_{B+}(T)$ is imposed whereby the total number of 2p noncondensate CPs equals that of 2h CPs. Here 2p refers to two-particle, i.e., two electrons in superconductivity and two fermionic atoms (as opposed to two fermionic holes) in fermionic superfluidity. BCS theory is valid along the forefront where $\lambda \ll 1$ of the shaded BCS-Bose crossover plane. For quadratically-dispersive bosons the usual BEC theory ensues at the origin of the octant where $m_0(T) = 0$ for all T and $n_0(T_c) = 0$, giving there the *implicit* expression $T_c \simeq 3.31\hbar^2 n_B(T_c)^{2/3}/2mk_B$. This result has the same *form* as the standard *explicit* BEC T_c-formula for mass $2m$ bosons and where the boson number density n_B is, of course, independent of T_c.

The original BCS-Bose crossover picture for the electronic gap $\Delta(T)$ and chemical potential $\mu(T)$ is now supplemented by the central relation

$$\Delta(T) = f\sqrt{n_0(T)} = f\sqrt{m_0(T)} \tag{6}$$

where f is a boson-fermion vertex interaction coupling constant inherent to the GBEC theory. All three functions $\Delta(T)$, $n_0(T)$ and $m_0(T)$ have the familiar "half-bell-shaped" forms. Namely, they are zero above a certain critical temperature T_c, and rise monotonically upon cooling (lowering T) to maximum values $\Delta(0)$, $n_0(0)$ and $m_0(0)$ at $T = 0$. The energy gap $\Delta(T)$ is the order parameter describing the superconducting (SC) [or superfluid (SF)] condensed state, while $n_0(T)$ and $m_0(T)$ are the BEC order parameters depicting the macroscopic occupation that occurs below T_c in a BE condensate. This $\Delta(T)$ is precisely the BCS energy gap if the GBEC theory coupling f is made to correspond to $\sqrt{2V\hbar\omega_D}$ where V and $\hbar\omega_D$

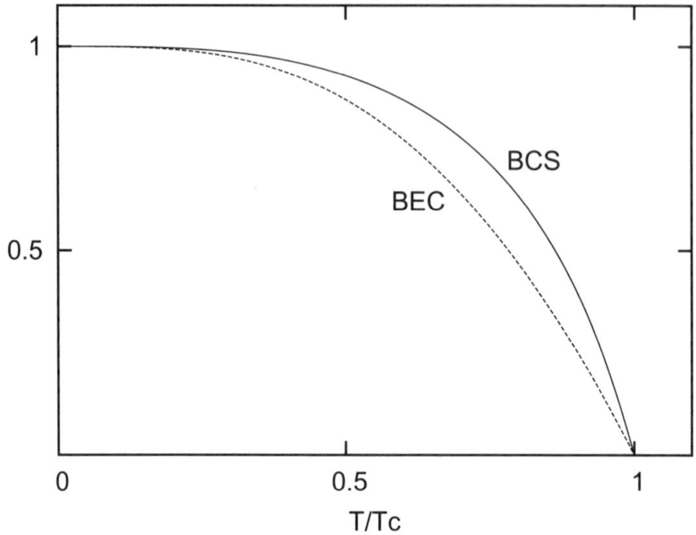

Fig. 3. Order parameters (normalized to unity on both axes) of BCS theory and of the BEC theory as related together by the GBEC theory in accordance with (7).

are the two parameters of the BCS model interelectronic interaction. Evidently, the BCS and BEC T_cs are the same. Writing (6) for $T = 0$ and dividing this into (6) gives the much simpler *f-independent* relation involving order parameters, as well as temperatures T, *normalized* to unity in the interval $[0, 1]$, namely

$$\Delta(T)/\Delta(0) = \sqrt{n_0(T)/n_0(0)} = \sqrt{m_0(T)/m_0(0)} \underset{T\to 0}{\longrightarrow} 1 \textit{ as well as} \underset{T\geq T_c}{\longrightarrow} 0. \quad (7)$$

The first equality, apparently first obtained in Ref. 8, connects in a simple way the two heretofore unrelated "half-bell-shaped" order parameters of the BCS and the BEC theories. The second equality[4−6] implies that a BCS condensate is precisely a BE condensate of equal numbers of 2e- and 2h-CPs. Figure 3 shows the relationship between the normalized order parameters of BCS and BEC. Since (7) is *independent* of the particular two-fermion dynamics of the problem, it can be expected to hold for either SCs and SFs.

The insensitivity of T_c/T_F exhibited in Fig. 3 as to whether BF vertex interactions are present ($f \neq 0$) or not ($f = 0$) is a striking result that suggests how "good" the zeroeth-order Hamiltonian H_0 chosen in the GBEC Ref. 6, Eq. (1), and how "unimportant" the interaction Hamiltonian H_{int} (which vanishes as $f \to 0$) Ref. 6, Eq. (2), turn out to be—at least in determining critical temperatures.

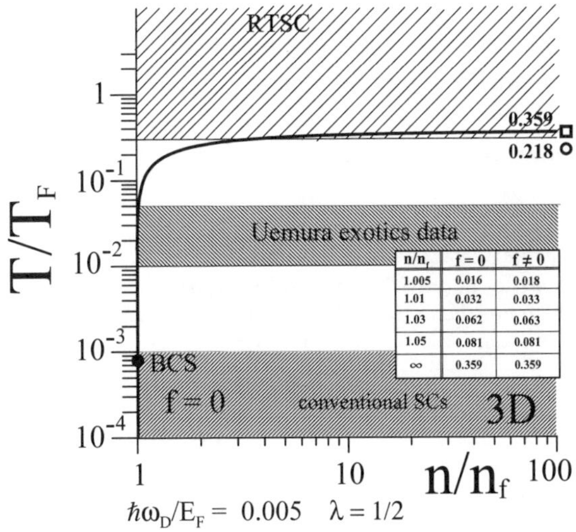

Fig. 4. Phase boundary curve showing net two- to three-order-of-magnitude enhancement, over the BCS prediction of T_c (in units of T_F) for the BCS model interaction with $\lambda = 1/2$ and $\Theta_D/T_F = 0.005$, obtained with the GBEC theory with $f = 0$ for the 2e-CP GBE condensate. Square symbol on the extreme right give $n/n_f \to \infty$ limit of 2e-CP GBE condensate T_c/T_F value (third column of Table 1). The open circle marks the familiar value 0.218 (last column of Table 1) associated with an ideal Fermi gas in which all fermions are pair into bosons of mass $2m$ and number density $n/2$; it is given for reference. RTSC stands for room-temperature superconductivity in a material with $T_F = 10^3 K$. Inset highlights small differences in T_c/T_F when calculations are repeated with $f \neq 0$.

Table 1. GBEC 2e-CP condensate critical temperatures T_c/T_F when $n/n_f \to \infty$. The BCS value (dot in Fig. 4) is given by $T_c/T_F = 1.134(\Theta_D/T_F)\exp(-1/\lambda) \simeq 0.0008$ for $\lambda = 1/2$ and $\Theta_D/T_F = 0.005$ and occurs as a special case of the GBEC phase diagram at $n/n_f = 1$. Note enhancements of T_c/T_F above the familiar value of 0.218 (in bold) corresponding to a many-electron system with all electrons paired of asterisked ($*$) values associated with a hole-CP background present, i.e., $m_{B+}(T) \neq 0$, in the 2e-CP condensate specified within the GBEC by the condition that the 2h-CP number density in $K = 0$ state $m_0(T) \equiv 0$ for all T. Here n is the electron number density, $n_f(T)$ the *unpaired*-electron number density, $\varepsilon_K = C_s K^s$ the bosonic CP dispersion ($s = 1$ or 2 signifying linear or quadratic), $m_{B+}(T)$ the 2h-CP number density in *all* $K > 0$ states, Δ the zero-temperature fermionic gap and 2Δ the zero-temperature bosonic CP gap in the linearly-dispersive $\varepsilon_K = 2\Delta + \lambda\hbar v_F K/4$ Ref. 9 bosons moving in the Fermi sea, while $s = 2$ refers to quadratically-dispersive $\varepsilon_K = \hbar^2 K^2/4m$ bosons (as, e.g., in Refs. 10 and 11) and is obviously independent of coupling λ. Uemura exotics data on both 2D and 3D SCs is taken from Ref. 12 while the lowest shaded area refers to conventional 3D SCs.

3D ($\lambda = 1/2$ and $\Theta_D/T_F = 0.005$)	$s = 1,\ 2\Delta = 0$	$s = 1,\ 2\Delta \neq 0$	$s = 2$
with $n_f(T) = 0$ (i.e., all e's paired)	0.129	0.130	**0.218**
with $m_B(T) \equiv m_0(T) + m_{B+}(T) = 0\ \forall\ T$	0.127	0.127	0.204
with $m_0(T) = 0$ but $m_{B+}(T) \neq 0\ \forall\ T$	0.359	0.361*	0.507*

4. Cold-atom BECs, Bosonic and Fermionic

Based on Ref. 13, p. 21, with additions and modifications, Table 2 reveals ranges over 12 orders of magnitude in number density (in particles cm^3) for several physical systems. Also given, where appropriate, are condensation temperatures T_c. The very-low-density conjectured "Efimov liquid" is based on Ref. 14.

Table 2. Number densities in particles/cm^3 for a diverse variety of many-body systems. Also given where appropriate are the critical temperatures T_c (in K) which refer to superconductor or superfluidity or cold-atom BEC transition temperatures.

Many-body system	**statistics**	**number density** (cm^{-3})	T_c (K)
electron gas in metals	FD	$10^{22} - 10^{23}$	$0 - 23$
liquid ^{4}He	BE	$\sim 10^{22}$	2.2
liquid ^{3}He	FD	$\sim 10^{22}$	2×10^3
exotic SCs (including cuprates)	FD	$10^{21} - 10^{22}$	$1 - 164$
Air (STP) [78% N_2 + 21% O_2 $+\cdots$]	-	$\sim 10^{19}$	-
ultracold Bose gases	BE	$10^{12} - 10^{15}$	$10^{-8} - 10^{-5}$
ultracold Fermi gases	FD	$10^{12} - 10^{13}$	$10^{-7} - 10^{-6}$
conjectured "Efimov liquid"	BE or FD	$\sim 10^{10}$	-

Below are compiled empirical parameters associated with both bosonic and fermionic ultra-cold gases where BEC has been observed. Along with the usual four states of matter—gas, liquid, solid and plasma—they constitute what have sometimes been termed the Vth and VIth states.

Table 3. Ultra-cold bosonic-atom BEC (sometimes dubbed the "*Vth state of matter*") experimental parameters associated with trapped bosonic gases in which BEC has been observed to date, N and N_0 being the number of atoms in the initial cloud and in the condensate, respectively; T_c the BEC transition temperature; n_0 the reported boson (or peak atom) number density at T_c of the condensate in cm^{-3}; $n_0^{-1/3}$is average interbosonic spacing in $\mathring{A}$.

BOSONS	$^{87}_{37}$Rb	$^{7}_{3}$Li	$^{23}_{11}$Na	$^{1}_{1}$H	$^{85}_{37}$Rb
Year/Ref.	1995[15]	1995[16]	1995[17]	1998[18]	2000[19]
N	4×10^4	2×10^5	5×10^5	-	3×10^8
N_0	2×10^3	-	-	10^9	10^4
T_c (μK)	0.17	0.4	2	50	0.015
n_0 (cm^{-3})	2.5×10^{12}	2×10^{12}	1.5×10^{14}	1.8×10^{14}	1×10^{12}
$n_0^{-1/3}$ ($\mathring{A}$)	7,368	7,937	1,882	1,771	10,000

BOSONS	^4_2He	$^{41}_{19}K$	$^{133}_{55}Cs$	$^{174}_{70}Yb^*$	$^{52}_{24}Cr$
Year/Ref.	2001[20]	2001[21]	2003[22]	2003[23]	2005[24]
N	8×10^6	-	2×10^7	10^7	1.3×10^8
N_0	5×10^6	10^4	6.5×10^4	5×10^3	5×10^4
T_c (μK)	4.7	0.16	0.046	0.73	0.7
n_0 (cm^{-3})	3.8×10^{13}	6×10^{11}	1.3×10^{13}	7×10^{14}	-
$n_0^{-1/3}$ $(\AA)$	$2,974$	$11,856$ **max**	$4,253$	$1,126$ **min**	-

*of five stable isotopes.

Table 4. Ultra-cold fermionic-atom BEC (sometimes dubbed the "*VIth state of matter*") experimental parameters associated with trapped fermionic gases in which BEC has been observed to Dec. 2007, N and N_0 being the number of atoms in the initial cloud and in the condensate, respectively; T_c the BEC transition temperature, n_0 the reported boson (or peak atom) number density of the condensate at T_c in cm^{-3} and $n_0^{-1/3}$ is average interbosonic spacing in $\AA$. The lowest recorded temperature $T \simeq 45 \times 10^{-5} \mu K$ Ref. 28 is ~ 0.03 *lower* than lowest BEC critical temperature from Table 2 which is $T_c^{BEC}(^{85}Rb) = 0.015 \mu K$.

FERMIONS	^6_3Li	$^{40}_{19}K$	$^{173}_{70}Yb$ (of 2 stable isotopes)
Year/Ref.	2003 [25]	2003 [26]	2007 [27]
N	3.5×10^7	1.4×10^6	-
N_0	9×10^5	-	-
T_c (μK)	0.6	0.07	cooled to $T/T_F = 0.37$
n_0 (cm^{-3})	7×10^{13}	7×10^{12}	-
$n_0^{-1/3}$ $(\AA)$	$2,426$	$5,228$	-

5. Conclusion

We conclude that hole-Cooper-pairs play a significant role in determining the value of the critical generalized BEC temperature at *all* temperatures, at least in superconductors, just as antibosons do in the relativistic ideal Bose gas problem at higher temperatures where antibosons appear in substantial numbers. However, given that in cold-atom fermion systems densities are so low, we conjecture that their role will be significantly diminished.

Acknowledgments

MdeLl thanks UNAM-DGAPA-PAPIIT (Mexico) IN106401 for partial support, and is grateful for travel support through a grant to Southern Illinois University at

Carbondale from the U.S. Army Research Office. S.R. thanks CONACyT (Mexico) for partial support.

Appendix A. Stability of both 2e- and 2h-CP pure GBEC phases

The Helmholtz free energy per unit volume is by definition

$$F(T, L^d, \mu, n_0, m_0)/L^d \equiv \mu n - P(T, L^d, \mu, n_0, m_0) \tag{A.1}$$

where P is the system pressure. A necessary condition for a stable equilibrium thermodynamic state is a minimum in F with respect to n_0 and m_0, at fixed T, fixed number-density of electrons $n = N/L^d$ and fixed electron chemical potential $\mu = \mu(T, n, n_0)$. Here we show that both $[\partial^2(F/L^d)/\partial n_0^2]_n$ and $[\partial^2(F/L^d)/\partial m_0^2]_n$ are positive definite in the GBEC theory. The thermodynamic potential per unit d-dimensional volume is[6]

$$\begin{aligned}\Omega(T, L^d, \mu,\, n_0,\, m_0)/L^d &\equiv -P(T, L^d, \mu,\, n_0, m_0) = \\ \int_0^\infty d\epsilon N(\epsilon)\,[\epsilon - \mu - E(\epsilon)] &- 2\,k_B T \int_0^\infty d\epsilon\, N(\epsilon)\, \ln\{1 + \exp[-\beta\, E(\epsilon)]\} + \\ +[E_+(0) - 2\,\mu]\, n_0 &+ k_B T \int_{0+}^\infty d\varepsilon\, M(\varepsilon)\, \ln\{1 - \exp[-\beta\{E_+(0) + \varepsilon - 2\mu\}]\} + \\ +[2\,\mu - E_-(0)]\, m_0 &+ k_B T \int_{0+}^\infty d\varepsilon\, M(\varepsilon)\, \ln\{1 - \exp[-\beta\{2\mu - E_-(0) + \varepsilon\}]\}\end{aligned}$$

where $N(\epsilon)$ and $M(\varepsilon)$ are the fermionic and bosonic, respectively, density of states, while $E_+(0)$ and $E_-(0)$ are the phenomenological zero-center-of-mass-momentum 2e- and 2h-CPs, respectively. Here, the fermion spectrum $E(\epsilon)$ and fermion energy gap $\Delta(\epsilon)$ are related according to

$$E(\epsilon) = \sqrt{(\epsilon - \mu)^2 + \Delta^2(\epsilon)} \quad \textit{and} \quad \Delta(\epsilon) \equiv \sqrt{n_0} f_+(\epsilon) + \sqrt{m_0} f_-(\epsilon). \tag{A.2}$$

We first calculate $(\partial P/\partial n_0)_\mu$ and $(\partial P/\partial m_0)_\mu$, which are

$$(\partial P/\partial n_0)_\mu = -[E_+(0) - 2\,\mu] + \frac{f^2}{2} \int_{E_f}^{E_f + \hbar\omega_D} d\epsilon \frac{N(\epsilon)}{E(\epsilon)} \tanh \frac{1}{2}\beta E(\epsilon) \tag{A.3}$$

$$(\partial P/\partial m_0)_\mu = -[2\,\mu - E_-(0)] + \frac{f^2}{2} \int_{E_f - \hbar\omega_D}^{E_f} d\epsilon \frac{N(\epsilon)}{E(\epsilon)} \tanh \frac{1}{2}\beta E(\epsilon) \tag{A.4}$$

where from (A.2) $\partial E(\epsilon)/\partial n_0 = f^2/2E(\epsilon)$ was employed. Next, letting $x \equiv \beta E(\epsilon)/2$ one has

$$(\partial P^2/\partial n_0^2)_\mu = \frac{f^4 \beta^2}{16} \int_{E_f}^{E_f + \hbar\omega_D} d\epsilon \frac{N(\epsilon)}{E(\epsilon)} \frac{\partial}{\partial x} \left[\frac{\tanh x}{x}\right] < 0 \tag{A.5}$$

since $\partial(\tanh x/x)/\partial x$ can be shown[29] to be negative for all $x > 0$. Similarly, one finds that

$$(\partial P^2/\partial m_0^2)_\mu = \frac{f^4 \beta^2}{16} \int_{E_f - \hbar\omega_D}^{E_f} d\epsilon N(\epsilon) \frac{1}{E(\epsilon)} \frac{\partial}{\partial x} \left[\frac{\tanh x}{x}\right] < 0. \tag{A.6}$$

We then calculate

$$\left[\frac{\partial(F/L^d)}{\partial n_0}\right]_n = n\left[\frac{\partial\mu}{\partial n_0}\right]_n - \left[\frac{\partial P}{\partial n_0}\right]_n$$

$$= n\left[\frac{\partial\mu}{\partial n_0}\right]_n - \left[\frac{\partial P}{\partial n_0}\right]_\mu - \left[\frac{\partial P}{\partial \mu}\right]_{n_0}\left[\frac{\partial\mu}{\partial n_0}\right]_n = -\left[\frac{\partial P}{\partial n_0}\right]_\mu \tag{A.7}$$

since $n = [\partial P/\partial\mu]_{n_0}$. Next, for the pure 2e-CP GBEC phase one determines that

$$\left[\frac{\partial^2(F/L^d)}{\partial n_0^2}\right]_n = -\left[\frac{\partial}{\partial n_0}\right]_n\left[\frac{\partial P}{\partial n_0}\right]_\mu = -\left[\frac{\partial^2 P}{\partial n_0^2}\right]_\mu - \left[\frac{\partial\mu}{\partial n_0}\right]_n\left[\frac{\partial}{\partial\mu}\right]_{n_0}\left[\frac{\partial P}{\partial n_0}\right]_\mu$$

$$= -\left[\frac{\partial^2 P}{\partial n_0^2}\right]_\mu + \left[\frac{\partial\mu}{\partial n}\right]_{n_0}\left[\frac{\partial n}{\partial n_0}\right]_\mu\left[\frac{\partial^2 P}{\partial\mu\partial n_0}\right] = -\left[\frac{\partial^2 P}{\partial n_0^2}\right]_\mu + \left[\frac{\partial n}{\partial\mu}\right]_{n_0}^{-1}\left[\frac{\partial n}{\partial n_0}\right]_\mu^2 > 0$$

where (A.5) was employed. The derivation to here closely follows that of Ref. 30.

Finally, we turn to hole-CPs where one can similarly show for the pure 2h-CP GBEC phase that

$$\left[\frac{\partial^2(F/L^d)}{\partial m_0^2}\right]_n = -\left[\frac{\partial^2 P}{\partial m_0^2}\right]_\mu + \left[\frac{\partial n}{\partial\mu}\right]_{m_0}^{-1}\left[\frac{\partial n}{\partial m_0}\right]_\mu^2 > 0$$

where (A.6) was used. **QED**. Thus, both pure phases are stable, equilibrium thermodynamic states. However, the 2h phase exhibits an unacceptable divergent T_c as $n/n_f \to 0$ which will be investigated elsewhere.

References

1. M. Casas, A. Rigo, M. de Llano, O. Rojo and M.A. Solís, *Phys. Lett. A* **245**, 55 (1998).
2. M. Grether, M. de Llano and G.A. Baker, Jr., *Phys. Rev. Lett.* **99**, 200406 (2007).
3. L.N. Cooper, *Phys. Rev.* **104**, 1189 (1956).
4. V.V. Tolmachev, *Phys. Lett. A* **266**, 400 (2000).
5. M. de Llano and V.V. Tolmachev, *Physica A* **317**, 546 (2003).
6. S.K. Adhikari, M. de Llano, F.J. Sevilla, M.A. Solís and J.J. Valencia, *Physica C* **453**, 37 (2007).
7. Q. Chen, J. Stajic, Sh. Tan and K. Levin, *Phys. Repts.* **412**, 1 (2005).
8. J. Ranninger, R. Micnas and S. Robaszkiewicz, *Ann. Phys. Fr.* **13**, 455 (1988).
9. M. Fortes, M.A. Solís, M. de Llano and V.V. Tolmachev, *Physica C* **364,** 95 (2001).
10. R. Friedberg, T.D. Lee and H.-C. Ren, *Phys. Rev. B* **45**, 10732 (1992) and refs. therein.
11. M. de Llano and V.V. Tolmachev, *Physica A* **317**, 546 (2003).
12. Y.J. Uemura, *J. Phys.: Condens. Matter* **16**, S4515 (2004).
13. A.J. Leggett, *Quantum Liquids: Bose Condensation and Cooper Pairing in Condensed-Matter Systems* (Oxford University Press, UK, 2006).
14. A. Bulgac, *Phys. Rev. Lett.* **89**, 050402 (2002).
15. M.H. Anderson, J.R. Ensher, M.R. Wieman and E.A. Cornell, *Science* **269**, 198 (1995).
16. C.C. Bradley, C.A. Sackett, J.J. Tollett and R.G. Hulet, *Phys. Rev. Lett.* **75**, 1687 (1995).

17. K.B. Davis, M.O. Mewes, M.R. Andrews, N.J. van Drutten, D.S. Durfee, D.M. Kurn and W. Ketterle, *Phys. Rev. Lett.* **75**, 3969 (1995).
18. D.G. Fried, T.C. Killian, L. Willmann, D. Landhuis, S.C. Moss, D. Kleppner and T.J. Greytak, *Phys. Rev. Lett.* **81**, 3811 (1998).
19. S.L. Cornish, N.R. Claussen, J.L. Roberts, E.A. Cornell and C.E. Wieman, *Phys. Rev. Lett.* **85**, 1795 (2000).
20. F. Pereira Dos Santos, J. Léonard, J. Wang, C.J. Barrelet, F. Perales, E. Rasel, C.S. Unnikrishnan, M. Leduc and C. Cohen-Tannoudji, *Phys. Rev. Lett.* **86**, 3459 (2001).
21. G. Mondugno, G. Ferrari, G. Roati, R.J. Brecha, A. Simoni and M. Inguscio, *Science* **294**, 1320 (2001).
22. T. Weber, J. Herbig, M. Mark, H.C. Nagel and R. Grimm, *Science* **299**, 232 (2003).
23. Y. Takasu, K. Maki, K. Komori, T. Takano, K. Honda, M. Kumakura, T. Yabuzaki and Y. Takahashi, *Phys. Rev. Lett.* **91**, 040404 (2003).
24. A. Griesmaier, J. Werner, S. Hensler, J. Stuhler and T. Pfau, *Phys. Rev. Lett.* **94**, 160401 and also 183201 (2005).
25. M.W. Zwierlein, C.A. Stan, C.H. Schunck, S.M.F. Raupach, S. Gupta, Z. Hadzibabic and W. Ketterle, *Phys. Rev. Lett.* **91**, 250401 (2003).
26. M. Greiner, C.A. Regal and D.S. Jin, *Nature* **426**, 537 (2003).
27. T. Fukuhara, Y. Takasu, M. Kumakura and Y. Takahashi, *Phys. Rev. Lett.* **98**, 030401 (2007).
28. A.E. Leanhardt, T.A. Pasquini, M. Saba, A. Schirotzek, Y. Shin, D. Kielpinski, D.E. Pritchard and W. Ketterle, *Science* **301**, 1513 (2003).
29. S. Wolfram, *The MATHEMATICA Book*, 3rd. Ed. (Wolfram Media, IL, 1996).
30. R. Friedberg and T. D. Lee, *Phys. Rev. B* **40**, 6745 (1989).

SUPERFLUID TO BOSE METAL TRANSITION IN SYSTEMS WITH RESONANT PAIRING

JULIUS RANNINGER

Institut Néel, CNRS and Université Joseph Fourier,
BP 166, 38042 Grenoble cedex 9, France
julius.ranninger@grenoble.cnrs.fr

Received 31 July 2008

Experiments in thin films whose thickness can be modified and by this way induce a superconductor to insulator transition, seem to suggest that in the quantum critical regime of this phase transition there might be a Bose metal, i.e., uncondensed bosonic carriers with a finite dissipation. This poses a fundamental problem as to our understanding of how such a state could be justified. On the basis of a simple Boson-Fermion model, where bosonic and fermionic degrees of freedom are strongly inter-related via a Boson-Fermion pair exchange coupling g, we illustrate how such a bosonic metal phase could possibly come about. We show that, as we approach the quantum critical point at some critical g_c from the superfluid side, the superfluid phase locking is sustained only for longer and longer spatial scales. On a finite spatial scale, the boson have a quasi-free itinerant behavior with metallic features. At the quantum critical point the systems exhibits a phase separation which shows a ressemblance to that of a He^3–He^4 mixture. This could be the clue to the apparent dilemma of a Bose metal at zero temperature.

Keywords: Superconductor-insulator transition; Bose metal.

1. Introduction

Frustrating the superconducting phase in thin films (Ga deposited on Al substrates) either by changing the film thickness[1] or by creating resistive vortices upon applying a magnetic field in MoGe thin films,[2] induces a superconductor to insulator quantum phase transition. What is exceptional in those experiments is that at this quantum phase transition the systems behaves as metallic down to the lowest temperatures measured: as the temperature is lowered the resistivity saturates. In the first case,[1] the superconductor to insulator transition is driven by changing the degree of boundary scattering, which changes the degree of disorder and which in turn controls the localization versus delocalization in such low dimensional systems. In the second case,[2] the magnetic field creates vortices which, depending on the strength of this field, are localized in the form of a vortex glass for low fields and delocalized above a certain critical value. As a consequence the Cooper pairs are either delocalized or localized.

According to conventional wisdom, a Bose metal should not exist. The ground state should either be an eigenstate of a fixed constant density of carriers or of a fixed value of the phase, which are the two conjugate variables in the system. This then can only result in either an insulator or a phase-locked superfluid phase. The systems mentioned above have fermionic charge carriers. In bulk materials such fermionic charge carriers scatter only weakly on impurities, phonons, etc, and lead to a resistivity ρ =const as $T \to 0$. But in thin films, even non-interacting electrons get localized by an arbitrarily weak disorder via the Anderson localization. If those electrons interact attractively with each other they pair up inside regions which are controlled by the localization length, forming phase uncorrelated bosonic entities. A superconducting states can nevertheless result from such a situation via a phase coherent superposition of states with different occupation numbers in those localization bubbles.

The superconducting ground state is characterized by a macroscopic wave function with the following properties:

$$\langle\Psi(x)\rangle \equiv |\langle\Psi(x)\rangle|e^{i\phi(x)} = \langle\text{BCS}|\psi^\dagger_\uparrow(x)\psi^\dagger_\downarrow(x)|\text{BCS}\rangle$$

$$\langle\Psi(x,t)\Psi(0,t)\rangle \propto x^{-\eta} \text{ as } x \to \infty (0 \leq \eta \leq 1) \text{ for 2D} \tag{1}$$

$$\langle\Psi(x,t)\Psi(0,t)\rangle \propto \text{const as } x \to \infty \text{ for 3D.} \tag{2}$$

The superconducting state can be broken up either by:

(i) interrupting the long range phase locking, resulting in a system with localized bosons: $\langle e^{i\phi}\rangle \to 0$;
(ii) breaking up the bosonic pairs, resulting in localized electrons: $|\langle\Psi(x)\rangle| \to 0$.

A metallic phase in an alloy film would thus require that such a state remains unaffected by the standard localization arguments for fermions as well as for bosons. A scenario,[3] much advanced in this quest for understanding a Bose metal phase, is the following: The charge carriers in the presence of disorder lose their global phase coherence but can sustain on a finite time and length scale such phase locking. This leads to a so-called phase glass picture, characterized by:

$$\langle e^{i\phi(r_i)}\rangle \neq 0 \text{ but } \sum_i \langle e^{i\phi(r_i)}\rangle = 0 \tag{3}$$

$$G(r_i,t) = \langle e^{i\phi(r_i,t)} e^{i\phi(r_i,0)}\rangle = t^{-2}, \tag{4}$$

with a local phase $\phi(r_i,t)$ diffusing very slowly. This would result in a spectral weight of the corresponding excitations in such a phase glass, which has a DOS scaling with the frequency ω, and which would thus outweigh the spectral weight of the superconducting fluctuations ($\propto \omega^2$). It is this which could inhibit the bosons from condensing as well as from localizing and provide them with a "self-generated dissipation", not frozen out as $T \to 0$.[3] All those arguments have so far been only developed on the basis of a purely phenomenological picture, while a microscopic derivation of a Bose metal is still lacking.

2. The Boson-Fermion Model

Motivated by these experiments we shall ask here the question if and how in principles such a Bose metal could be realized. We shall do this on a tracktable model, defined on the basis of a simple Hamiltonian which contains the necessary ingredients for such a potential Bose metal phase. We consider here a bulk system without any extrinsic disorder. Different from the experimental situations, here such a superconductor to insulator transition is driven by intrinsic dynamic disorder rather than the effect of low dimensionality. The intrinsic disorder comes from local amplitude fluctuations of bosonic entities, representing bound fermionic pairs which can momentarily break up into two unbound itinerant fermions and subsequently recombine into such bound pairs. A simple microscopic model on which such physics can be studied is the Boson-Fermion scenario, described by the Hamiltonian:

$$H = -\mu \sum_{i\sigma} c^{\dagger}_{i\sigma} c_{i\sigma} + (\Delta_B - 2\mu) \sum_{i} \rho_i^+ \rho_i^- \\ - \sum_{i \neq j,\, \sigma} t_{ij} (c^{\dagger}_{i\sigma} c_{j\sigma} + H.c.) + g \sum_{i} \left(\rho_i^+ \tau_i^- + \rho_i^- \tau_i^+ \right), \quad (5)$$

$\vec{\rho}_i = \{\rho_{\rm i}^+, \rho_{\rm i}^-, \rho_{\rm i}^z\}$ and $\vec{\tau}_i = \{\tau_{\rm i}^+, \tau_{\rm i}^-, \tau_{\rm i}^z\}$ denote respectively localized tightly bound electron-pairs and pairs of itinerant electrons in real space which are the equivalents of the Cooper-pairs

$$\rho_{\rm i}^+ = d_{\rm i\uparrow}^+ d_{\rm i\downarrow}^+, \quad \rho_{\rm i}^- = d_{\rm i\downarrow} d_{\rm i\uparrow}, \; \rho_{\rm i}^z = \frac{1}{2} - \rho_{\rm i}^+ \rho_{\rm i}^- \quad (6)$$

$$\tau_{\rm i}^+ = c_{\rm i\uparrow}^+ c_{\rm i\downarrow}^+, \quad \tau_{\rm i}^- = c_{\rm i\downarrow} c_{\rm i\uparrow}, \quad \tau_{\rm i}^z = \frac{1}{2} - \tau_{\rm i}^+ \tau_{\rm i}^- . \quad (7)$$

The structure of the Boson-Fermion model permits to track separately the local pairing amplitude and long range phase locking.[4] For weak exchange coupling g the Boson-Fermion model describes a BCS type superconducting state. For large values of g, the local exchange interaction results in locally tightly bound fermion pairs in form of bonding states $b_i^{\dagger}|v\rangle \equiv \frac{1}{\sqrt{2}}[u_i c^{\dagger}_{i\uparrow} c^{\dagger}_{i\downarrow} + v_i \rho_i^+]|0\rangle$ with $u_i = |u_i| e^{i\phi_i/2}$, $v_i = |v_i| e^{-i\phi_i/2}$ and $|0\rangle$ denoting the physical vacuum of the Hamiltonian, Eq. (5). Such bonding states are represented by bonding operators: $b_i^{\dagger}$ with $|v\rangle$ denoting the vacuum state with respective to the operators, creating a bonding pair on a given site i as well as of creating an empty site: $d_i^{\dagger}|v\rangle \equiv |0\rangle$, respectively a completely filled site $f_i^{\dagger}|v\rangle = \rho_i^+ c^{\dagger}_{i\uparrow} c^{\dagger}_{i\downarrow}|0\rangle$. The corresponding global state is a phase uncorrelated liquid of such bonding pairs, given by $\prod_i [u_v + u_b b_i^{\dagger}]|v\rangle$. Establishing a superconducting phase composed of such bonding pairs requires that the phases of the individual local bonding states get spatially correlated with the phases of the unoccupied and the fully occupied local states. This results in a state given by: $\prod_i [u_v + u_d d_i^{\dagger} + u_f f_i^{\dagger} + u_b b_i^{\dagger}|v\rangle$. It however necessitates a locally fluctuating density of such bonding states and this implies breaking them up locally. The competition between such a superconducting and insulating phase has been studied on the basis of such a bond operator representation,[4] within a corresponding mean field analysis. The superconductor to insulator transition is then characterized by the vanishing of the superfluid phase locking. i.e., $\langle d_i^\dagger \rangle \to 0, \langle f_i^\dagger \rangle \to 0$ as g approaches some critical value g_c. Concerning the fermionic properties one finds a semiconducting-like gap for $g \geq g_c$ and which goes over into a superconducting gap in a continuous fashion as g drops below g_c. The questions then arises, if beyond a mean field description of such physics the Bosons can have a gapless spectrum and thus potentially can participate in a metallic conductivity.[4] In order to address this question, one has to examine the highly involved inter-related dynamics of the bosons and the fermions and we sketch here our main results recently obtained.[5] Our aim was to study the excitations of the bosonic entities which will tell us how the Goldstone mode of the superfluid phase is modified as we approach the quantum critical point at g_c, separating the superconducting from the insulating phase. This has been done in terms of a flow equation renormalization procedure for which we outline below its main structure and results.

The idea of the flow equation renormalization[6] is to eliminate in a continuous fashion the exchange coupling term and bring the Hamiltonian into block-diagonal form. For this it is judicious to rewrite our Hamiltonian, Eq. (5) in a k-vector representation

$$H = \sum_{k\sigma} (\varepsilon_k - \mu) c_{k\sigma}^\dagger c_{k\sigma} + \sum_q (E_q - 2\mu) b_q^\dagger b_q + \frac{1}{\sqrt{N}} \sum_{k,p} (g_{k,p} b_{k+p}^\dagger c_{k,\downarrow} c_{p,\uparrow} + g_{k,p}^* b_{k+p} c_{k,\uparrow}^\dagger c_{p,\downarrow}^\dagger) \tag{8}$$

where ε_k and E_q denote the dispersion for the electrons and the bound electron pairs whose creation (annihilation) operators we denote by $b_k^{(\dagger)}$ and which, for the present discussion, we take as ordinary rather than hardcore bosons. The flow of the Hamiltonian is controlled by the flow of the various parameters: the electron dispersion $\varepsilon_k(l)$, the dispersion of the bosons $E_q(l)$ and the exchange coupling $g_{k,p}(l)$, where l indicates the flow parameter. For $l = 0$ we take a simple tight binding spectrum for the electrons, $\varepsilon_k(l = 0) = -2t \cos k$, a q independent dispersion for the bosons, $E_q(l = 0) = \Delta_B$ signaling their intrinsic bare localized behavior and a constant exchange coupling $g_{k,p}(l = 0) = g$. The flow of the Hamiltonian is then controlled in the standard way by

$$\partial_l H(l) = [\eta(l), H(l)]. \tag{9}$$

With the choice of the canonical generator[6] $\eta(l) = [H_0(l), H(l)]$, this implies that at the fixed point of this renormalization flow ($l = \infty$), the Hamiltonian reduces to the structure of its free part, H_0 given by the first two terms in Eq. (8), but with renormalized interdependent Fermion and Boson dispersions. These quantities are controlled by the flow equations

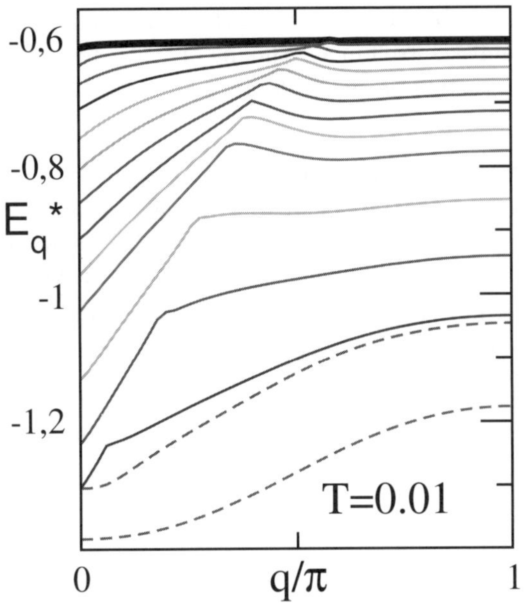

Fig. 1. Evolution of the Boson dispersion as we cross the superconductor to insulator transition at some low temperature $T = 0.01$ in units of the bare electron bandwidth $D = 4t$. The solid black line corresponds to a very weak coupling $g = 0.05$. The other continuous lines showing Goldstone modes correspond to $g = 0.1, 0.15, 0.2, 0.25, 0.3, 0.35, 0.4, 0.45, 0.5, 0.6, 0.7, 0.79$. For $g = 0.8$ and 0.9 (dashed lines), the bosons have quadratic spectrum.

$$\partial_l g_{k,p} = -\alpha_{k,p}^2 g_{k,p}, \qquad \alpha_{k,p} = \varepsilon_k^{\downarrow} + \varepsilon_p^{\uparrow} - E_{k+p} \tag{10}$$

$$\partial_l \varepsilon_k^{\sigma} = \frac{2}{N}\sum_p \left(\alpha_{p,k}|g_{p,k}|^2\delta_{\sigma,\uparrow} + \alpha_{k,p}|g_{k,p}|^2\delta_{\sigma,\downarrow}\right) n_{k+p}^{(BE)} \tag{11}$$

$$\partial_l E_q = \frac{2}{N}\sum_p \{\alpha_{q-p,p}|g_{q-p,p}|^2[-1 + n_{\downarrow,q-p}^{(FD)}] \tag{12}$$

$$+\alpha_{p,q-p}|g_{p,q-p}|^2 n_{\uparrow,q-p}^{(FD)}\}, \tag{13}$$

initially derived for this model by Domanski and the author.[7]

3. Discussion

We illustrate in Fig. 1 the evolution of the Boson dispersion as we cross the superconductor to insulator phase transition at low temperature by gradually increasing the Boson-Fermion exchange coupling g. We notice that as g increases, the initially localized Bosons acquire a linear in q Goldstone mode which spreads in its extent of the range in the Brillouin zone, reaching its maximal extent for $g = 0.5$. Upon further increasing g, we begin to approach the superconductor to insulator phase transition, which manifests itself in a shrinking of this extent of the Goldstone mode to lower and lower q vectors in the Brillouin zone. This behavior is unusual in two ways:

(i) Generally, a phase transition, related to a continuous symmetry breaking is manifest in a softening of the Goldstone mode, indicative of the weakening of the phase rigidity. Here this phase rigidity remains essentially uninfluenced.
(ii) Generally, one expects a breaking down of phase locking on a large spatial extent, but resisting on a smaller spatial scale as one approaches the phase transition. Here the opposite happens. The phase rigidity on the finite length scale is broken down before it affects the long range correlations.

These features indicate that the phase transition is not of a classical second order type. Moreover, the variation of the chemical potential with g at low temperatures shows a non-monotonic S-like shape, which is indicative of a first order superconductor to Bose metal phase transition.[5]

Given these findings, we want to advance the following interpretation of this rather unusual behavior. We adhere to the thesis that our system can only be in either a superconducting or in an insulating state, following the arguments which we developed in Section 2. We thus conclude that the system, upon approaching this phase transition, tries to comply as much as possible with those constraints: it sustains a Goldstone mode at least on a very long spatial scale, i.e., for small q vectors. This shows a certain analogy to He^3–He^4 mixtures which, beyond a certain critical concentration of He^3, can only maintain its superfluidity by undergoing a phase separation into a superfluid and a normal liquid. In our picture, the normal part of the liquid is represented by the Bosons which have finite q vectors. They do not participate in the condensate and can behave as metallic. Work is in progress, examining the spectral properties of these modes in order to determine their coherent versus incoherent components.

Acknowledgments

The present short note on the Bose metal is based on a recent work with T. Stauber, published in Ref. 5 and whom I thank for permitting me to reproduce part of our published results in Fig. 1. I would also like to thank my collaborators in the past, M. Cuoco and T. Domanski for having helped me over the years to gradually understand better the underlying rich physics of the Bose Fermi model.

References

1. C. Christiansen *et al.*, *Phys. Rev. Lett.* **88**, 37004 (2002).
2. A. Yazdani and A. Kapitulnik, *Phys. Rev. Lett.* **74**, 3037 (1995).
3. Ph. Phillips and D. Dalidovich, *Science* **3002**, 243 (2003).
4. M. Cuoco and J. Ranninger, *Phys. Rev. B* **74**, 94511 (2006).

5. T. Stauber and J. Ranninger, *Phys. Rev. Lett.* **99**, 45301 (2007).
6. F. Wegner, *Ann. Physik* **3**, 77 (1994).
7. T. Domanski and J. Ranninger, *Phys. Rev. B* **63**, 134505 (2001).

DC RESISTIVITY OF CHARGED COOPER PAIRS IN A SIMPLE BOSON-FERMION MODEL OF SUPERCONDUCTORS

T. A. MAMEDOV

*Faculty of Engineering, Baskent University, 06530 Ankara, Turkey and Institute of Physics, Academy of Sciences of Azerbaijan, Baku, 370143, Azerbaijan**

tmamedov@baskent.edu.tr

M. DE LLANO

Instituto de Investigaciones en Materiales, Universidad Nacional Autónoma de México Apdo. Postal 70-360, 04510 México, DF, Mexico

dellano@servidor.unam.mx

Received 31 July 2008

An analytic expression for the contribution $\sigma_B(\lambda, T)$ to the conductivity from charged bosonic Cooper pairs (CPs) is derived via two-time Green function techniques as a function of the BCS interelectron interaction model parameter λ and temperature T. Within the framework of a binary boson-fermion gas mixture model, it is shown that a self-consistent description of the resistivity data observed in high-temperature superconductors is possible *only* by assuming the presence of a finite gap between the energy spectra of free fermions and bosonic CPs.

Keywords: High-T_c superconductivity; boson-fermion models; charged boson conductivity; gapped Cooper-pair dispersion.

1. Introduction

Boson-fermion (BF) statistical models of superconductivity (SC) as a Bose-Einstein condensation (BEC) began to be studied in the mid-1950s, predating even the BCS-Bogoliubov theory of SC. However, the successes of the BCS theory in describing properties of traditional low-temperature SCs left BF-models neglected for many years. But the discovery of copper-oxide SCs, and discovering that it is impossible to describe the peculiarities of cuprates within the framework of BCS model, led to revisiting many traditonal SC scenarios. Because of the short coherence length of Cooper pairs (CPs), as well as explaining very naturally the pseudogap phenomenon observed in high-temperature superconductors (HTSCs) in terms of preformed CPs, BF models became attractive candidates to examine the physics of HTSCs. BCS theory merely contemplates the presence of Cooper "*correlations*" of single-fermion

*Permanent address

states. By contrast, BF models[1,2,3] posit the existence of actual bosonic CPs. For example, one may assume that the subsystem of electrons lying within the spherical shell $E_F - \hbar\omega_D \leq \epsilon \leq E_F + \hbar\omega_D$ about the Fermi energy E_F of the ideal Fermi gas in single-electron ϵ energy space, with $\hbar\omega_D$ the Debye energy parameter of the BCS model interaction, consists of two coexisting and dynamically interacting subsystems: Fermi particles (or pairable but *unpaired* fermions), and individual pair bosonic CP entities made up of two mutually confined fermions. The simplest Hamiltonian describing a binary mixture of interacting fermions with bosons as suggested in Refs. 1, 2, 3 has been applied in an effort to understand the properties of HTSCs. This Hamiltonian is

$$H = H_e^o + H_B^o + H_{int} \tag{1}$$

$$H_e^o \equiv \sum_{\mathbf{k},\sigma} \xi_{\mathbf{k}} a_{\mathbf{k}\sigma}^{+} a_{\mathbf{k}\sigma} \qquad \text{and} \qquad H_B^o \equiv \sum_{\mathbf{K}} \mathcal{E}_{\mathbf{K}} b_{\mathbf{K}}^{+} b_{\mathbf{K}} \tag{2}$$

where H_e^o and H_B^o are zeroeth-order Hamiltonians of free (pairable but unpaired) fermions and composite-boson CPs. Here $a_{\mathbf{k}\sigma}^{+}$ and $a_{\mathbf{k}\sigma}$ are the usual fermion creation and annihilation operators for individual electrons of momenta $\mathbf{k}$ and spin $\sigma =\uparrow$ or $\downarrow$ while $b_{\mathbf{K}}^{+}$ and $b_{\mathbf{K}}$ are postulated[4,5] to be bosonic operators associated with CPs of definite total, or center-of-mass momentum (CMM), wavevector $\mathbf{K}$. Fermion $\xi_{\mathbf{k}} \equiv \epsilon_{\mathbf{k}} - \mu$ and boson energies $\mathcal{E}_{\mathbf{K}}$ are measured from μ and 2μ, respectively, where μ is the fermionic chemical potential defined in Ref. 7. Processes of formation/disintegration of bosons are then represented by an interaction Hamiltonian reminiscent of an analogous Fröhlich electron-phonon expression

$$H_{int} \equiv \frac{f}{L^{d/2}} \sum_{\mathbf{q},\mathbf{K}} \left(b_{\mathbf{K}}^{+} a_{\mathbf{q}+\mathbf{K}/2\uparrow} a_{-\mathbf{q}+\mathbf{K}/2\downarrow} + b_{\mathbf{K}} a_{-\mathbf{q}+\mathbf{K}/2\downarrow}^{+} a_{\mathbf{q}+\mathbf{K}/2\uparrow}^{+} \right) \tag{3}$$

where f is a phenomenological BF vertex interaction form factor coupling parameter, related with the attractive interelectron (*four*-fermion) interaction strength V of the s-wave BCS model interaction through[4,5] $f = \sqrt{2\hbar\omega_D V}$. Here L is the system size in d dimensions.

The new ingredient in the BF model (1) comes from Ref. 8 where it was shown that introducing an *attractive* interaction between electrons in the gas of electrons leads to the formation a new type of lower-energy mixture state with bosonic excitations *above the Fermi sea* of unpaired elecrons. Competition between electrons to occupy energy levels below E_F so as to minimize the volume of the Fermi sea leads to pushing away from that sea attractively-interacting charge carrier levels and raising them above E_F. These "raised" carriers appear outside the Fermi sea confined into positive-energy resonant CPs. Processes of pair formation and their subsequent disintegration into two unpaired electrons given by (3) were crucial[8] in getting a BF mixture state with positive-energy bosonic-excitations. Owing to these continual formation/disintegration processes, the total energy E of a mixture becomes *lower* compared with the energy of a single-component Fermi system of interacting fermions *without* explicit bosons. Separation of the initial attractively-interacting-fermion system into bosons and fermions with spectra shifted with respect to each other by the coupling-dependent, positive-energy gap were anticipated in Ref. 9. Linearly-dispersive (in CMM wavevector K) and gapped by $2\Delta(\lambda)$, with $\Delta(\lambda)$ the BCS single-fermion gap, describe our composite-boson CPs as found in Ref. 9 via a Bethe-Salpeter integral equation in the ladder approximation for coupled two-particle and two-hole wave functions. In Ref. 8 it was proved there that such a state with resonant bosons, i.e., bosons rising *above* the Fermi sea, may be energetically favorable if and only if H_{int} is included in (1). It was shown[7] that the separation by $\Delta(\lambda)$ between spectra of bosons and fermions provides, in contrast to BCS theory, a continuous decrease upon cooling of the chemical potential below the chemical potential E_F associated with interactionless fermions at zero temperature T, i.e., the Fermi energy. In other words, the BF mixture state develops gradually as T is lowered. It emerges from the system of unpaired electrons by forming *incoherent* CPs (with a λ- and T-dependent number density) *one by one*, and therefore differing from the BCS state where a phase transition to a *coherent* state occurs abruptly at T_c. Physically, the BF mixture includes *nonzero*-CMM pairs whereas BCS theory does not. Owing to the gapped boson spectrum it was convenient in Ref. 7 to define two characteristic temperatures. Firstly, a depairing temperature T^* below which the electronic chemical potential $\mu(\lambda, T)$ first becomes less than E_F of the interactionless electrons; below this T^* the first CPs begin to appear in the system. The condition $E_F - \mu(\lambda, T^*) = 0$ yields the T^* below which a transition occurs from normal state with no composite bosons to one with such bosons, i.e., the subsystem becomes an incoherent binary BF mixture. Secondly, the BEC temperature T_c at which a singularity in the total number density of bosons $n_B(\lambda, T_c) \equiv (1/L^d) \sum_{\mathbf{K}} (e^{\Omega_{\mathbf{K}}/k_B T_c} - 1)^{-1}$ first occurs, where $\Omega_{\mathbf{K}}$ is the boson energy $\mathcal{E}_{\mathbf{K}}$ renormalized by (3). At temperatures below T^* ($T^* > T_c$), $2e$-charged bosons contribute to a charge current in the presence of an applied electric field. In this sense, the bosons of the present model resemble CP fluctuations above T_c in the Aslamazov-Larkin theory.[10] The perturbation H_{int} (3) in (1) is necessary in yielding a lower-energy ground-state corresponding to the BF mixture; a finite boson lifetime will produce a resistivity in the boson gas. There may be boson scatterings in the subsystem other than those associated with H_{int} in (1), such as scattering from ionic-lattice irregularities. However, these events are important in accounting for the *total* resistivity and are of no direct concern for the mechanism due to the separation of the many-electron system into bosons and fermions; they can be considered as part of an *improper* part of the resistivity. In this paper we investigate the *self*-resistivity $\rho_B(\lambda, T)$ caused solely by the formation/disintegration of bosons without which a BF mixture state (with positive-energy resonant bosons) cannot be realized.

2. Basic Equations

According to (C.3) the Green function $\langle\langle b^+_{\mathbf{p}} b_{\mathbf{p}+\mathbf{k}} \mid \rho_B(-\mathbf{k})\rangle\rangle_\omega$ determines the response of the subsystem of composite CPs to an external electric field, i.e., describes the transport due to $2e$-charged CPs. The limit when $\omega = 0$ and $k \to 0$ defines the bosonic part of the conductivity associated with a static and longwave electric field. The equation to determine this function $\langle\langle b^+_{\mathbf{p}} b_{\mathbf{p}+\mathbf{k}} \mid \rho_B(-\mathbf{k})\rangle\rangle_\omega$ may be obtained by choosing $A \equiv b^+_{\mathbf{p}} b_{\mathbf{p}+\mathbf{k}}$ and $B \equiv \rho_B(-\mathbf{k})$ in

$$\omega\,\langle\langle A\,|B\rangle\rangle_\omega = \langle[A,B]\rangle_H + \langle\langle[A,H]\,|B\rangle\rangle_\omega \tag{4}$$

which is the first of the well-known infinite chain of equations for the Green functions.[12] The relevant expression becomes

$$\omega\,\langle\langle b^+_{\mathbf{p}} b_{\mathbf{p}+\mathbf{k}} \mid \rho_B(-\mathbf{k})\rangle\rangle_\omega = \langle[b^+_{\mathbf{p}} b_{\mathbf{p}+\mathbf{k}}, \rho_B(-\mathbf{k})]\rangle + \langle\langle[b^+_{\mathbf{p}} b_{\mathbf{p}+\mathbf{k}}, H] \mid \rho_B(-\mathbf{k})\rangle\rangle_\omega \tag{5}$$

where ω is henceforth in energy units. Putting (A.3) in the first term on rhs of (5) and inserting (1) in the second one, and using relations **a)** $[b^+_{\mathbf{p}} b_{\mathbf{p}+\mathbf{k}}, \rho_B(-\mathbf{k})] = b^+_{\mathbf{p}} b_{\mathbf{p}} - b^+_{\mathbf{p}+\mathbf{k}} b_{\mathbf{p}+\mathbf{k}}$; **b)** $[b^+_{\mathbf{p}} b_{\mathbf{p}+\mathbf{k}}, H^o_{el}] = 0$; and **c)** $[b^+_{\mathbf{p}} b_{\mathbf{p}+\mathbf{k}}, H^o_B] = (\mathcal{E}_{\mathbf{p}+\mathbf{k}} - \mathcal{E}_{\mathbf{p}}) b^+_{\mathbf{p}} b_{\mathbf{p}+\mathbf{k}}$, transforms (5) into

$$\begin{aligned}(\omega - \mathcal{E}_{\mathbf{pk}})\,\langle\langle b^+_{\mathbf{p}} b_{\mathbf{p}+\mathbf{k}} \mid \rho_B(-\mathbf{k})\rangle\rangle_\omega &= n_B(\mathbf{p}) - n_B(\mathbf{p}+\mathbf{k}) \\ &\quad + \langle\langle[b^+_{\mathbf{p}} b_{\mathbf{p}+\mathbf{k}}, H_{int}] \mid \rho_B(-\mathbf{k})\rangle\rangle_\omega\,.\end{aligned} \tag{6}$$

Here $\mathcal{E}_{\mathbf{pk}} \equiv \mathcal{E}_{\mathbf{p}+\mathbf{k}} - \mathcal{E}_{\mathbf{p}}$ are energy differences associated with transitions between bosonic states with center-of-mass momentum (CMM) wavevectors $\mathbf{p}$ and $\mathbf{p}+\mathbf{k}$ which are caused by the applied electric field of wavevector $\mathbf{k}$; $n_B(\mathbf{p}) \equiv \langle b^+_{\mathbf{p}} b_{\mathbf{p}}\rangle$ and $n_B(\mathbf{p}+\mathbf{k}) \equiv \left\langle b^+_{\mathbf{p}+\mathbf{k}} b_{\mathbf{p}+\mathbf{k}}\right\rangle$ are Bose occupation numbers $[\exp(\mathcal{E}/k_BT) - 1]^{-1}$ of corresponding energy states $\mathcal{E}_{\mathbf{p}}$ and $\mathcal{E}_{\mathbf{p}+\mathbf{k}}$.[7] The formal solution for (6) can be written through a mass operator $M_{\mathbf{p}}(\mathbf{k},\omega)$ defined as

$$\langle\langle[b^+_{\mathbf{p}} b_{\mathbf{p}+\mathbf{k}}, H_{int}] \mid \rho_B(-\mathbf{k})\rangle\rangle_\omega \equiv M_{\mathbf{p}}(\mathbf{k},\omega)\,\langle\langle b^+_{\mathbf{p}} b_{\mathbf{p}+\mathbf{k}} \mid \rho_B(-\mathbf{k})\rangle\rangle_\omega\,. \tag{7}$$

It relates the higher-order Green functions on the rhs of (6) in terms of $\langle\langle b^+_{\mathbf{p}} b_{\mathbf{p}+\mathbf{k}} \mid \rho_B(-\mathbf{k})\rangle\rangle_\omega$. From (6) and (7) we have

$$G_{\mathbf{p}}(\mathbf{k},\omega) \equiv \langle\langle b^+_{\mathbf{p}} b_{\mathbf{p}+\mathbf{k}} \mid \rho_B(-\mathbf{k})\rangle\rangle_\omega = \frac{n_B(\mathbf{p}) - n_B(\mathbf{p}+\mathbf{k})}{\omega - \mathcal{E}_{\mathbf{pk}} - M_{\mathbf{p}}(\mathbf{k},\omega)}. \tag{8}$$

The problem now is to develop some physically reasonable approximation to determine $M_{\mathbf{p}}(\mathbf{k},\omega)$ in (8). To do this, we must write the next equation in the chain of equations (4) for $\langle\langle[b^+_{\mathbf{p}} b_{\mathbf{p}+\mathbf{k}}, H_{int}] \mid \rho_B(-\mathbf{k})\rangle\rangle$ which is on rhs of (6). Calculating $[b^+_{\mathbf{p}} b_{\mathbf{p}+\mathbf{k}}, H_{int}]$ and inserting it into (6) gives

$$\begin{aligned}&(\omega - \mathcal{E}_{\mathbf{pk}})G_{\mathbf{p}}(\mathbf{k},\omega) = n_B(\mathbf{p}) - n_B(\mathbf{p}+\mathbf{k}) \\ &+ fL^{-d/2}\sum_{\mathbf{q}}\Big\{\langle\langle b^+_{\mathbf{p}} a_{\mathbf{q}+(\mathbf{p}+\mathbf{k})/2\uparrow} a_{-\mathbf{q}+(\mathbf{p}+\mathbf{k})/2\downarrow} \mid \rho_B(-\mathbf{k})\rangle\rangle_\omega \\ &- \left\langle\left\langle b_{\mathbf{p}+\mathbf{k}} a^+_{-\mathbf{q}+\mathbf{p}/2\downarrow} a^+_{\mathbf{q}+\mathbf{p}/2\uparrow} \mid \rho_B(-\mathbf{k})\right\rangle\right\rangle_\omega\Big\}.\end{aligned} \tag{9}$$

Equations for the higher-order Green functions appearing on the rhs of (9) may be written by choosing in (4) $B \equiv \rho_B(-\mathbf{k})$, *first* $A \equiv b^+_{\mathbf{p}} a_{\mathbf{q}+(\mathbf{p}+\mathbf{k})/2\uparrow} a_{-\mathbf{q}+(\mathbf{p}+\mathbf{k})/2\downarrow}$ and *then* $A \equiv b_{\mathbf{p}+\mathbf{k}} a^+_{-\mathbf{q}+\mathbf{p}/2\downarrow} a^+_{\mathbf{q}+\mathbf{p}/2\uparrow}$. This leads to

$$\begin{aligned}&\omega \left\langle\left\langle b^+_{\mathbf{p}} a_{\mathbf{q}+(\mathbf{p}+\mathbf{k})/2\uparrow} a_{-\mathbf{q}+(\mathbf{p}+\mathbf{k})/2\downarrow} \mid \rho_B(-\mathbf{k})\right\rangle\right\rangle_\omega = \\ &\left\langle [b^+_{\mathbf{p}}, \rho_B(-\mathbf{k})] a_{\mathbf{q}+(\mathbf{p}+\mathbf{k})/2\uparrow} a_{-\mathbf{q}+(\mathbf{p}+\mathbf{k})/2\downarrow}\right\rangle_H \\ &+ \left\langle\left\langle [b^+_{\mathbf{p}} a_{\mathbf{q}+(\mathbf{p}+\mathbf{k})/2\uparrow} a_{-\mathbf{q}+(\mathbf{p}+\mathbf{k})/2\downarrow}, H] \mid \rho_B(-\mathbf{k})\right\rangle\right\rangle_\omega\end{aligned} \tag{10}$$

$$\begin{aligned}&\omega \left\langle\left\langle b_{\mathbf{p}+\mathbf{k}} a^+_{-\mathbf{q}+\mathbf{p}/2\downarrow} a^+_{\mathbf{q}+\mathbf{p}/2\uparrow} \mid \rho_B(-\mathbf{k})\right\rangle\right\rangle_E = \\ &\left\langle [b_{\mathbf{p}+\mathbf{k}}, \rho_B(-\mathbf{k})] a^+_{-\mathbf{q}+\mathbf{p}/2\downarrow} a^+_{\mathbf{q}+\mathbf{p}/2\uparrow}\right\rangle_H \\ &+ \left\langle\left\langle [b_{\mathbf{p}+\mathbf{k}} a^+_{-\mathbf{q}+\mathbf{p}/2\downarrow} a^+_{\mathbf{q}+\mathbf{p}/2\uparrow}, H] \mid \rho_B(-\mathbf{k})\right\rangle\right\rangle_\omega .\end{aligned} \tag{11}$$

Here, the averages $\langle ...\rangle_H$ on the rhs (10) and (11) are simplified by applying $[b^+_{\mathbf{p}}, \rho_B(-\mathbf{k})] \equiv -b^+_{\mathbf{p}+\mathbf{k}}$ and $[b_{\mathbf{p}+\mathbf{k}}, \rho_B(-\mathbf{k})] \equiv b_{\mathbf{p}}$. The calculation of $[\cdots, H]$ in (10) and (11) are straightforward. In the rhs of (10) and (11) consider the Green functions originating from the terms $[\cdots, H^o_e]$ and $[\cdots, H^o_B]$ separately, and ignore the Green functions stemming from the commutators $[\cdots, H_{int}]$ which produce expressions such as a^+a^+aa and b^+ba^+a and lead to the higher-order Green functions which turned out *not* to be reducible to lower-order functions.[13] The mathematical justification of ignoring Green functions coming from $[\cdots, H_{int}]$ is not trivial. Formally, omitted terms on the rhs of (10) and (11) are $O(f^3)$ and so contain an extra power f with respect to the lhs ones. Long, tedious manupulations finally give

$$\left\langle\left\langle b^+_{\mathbf{p}} a_{\mathbf{q}+(\mathbf{p}+\mathbf{k})/2\uparrow} a_{-\mathbf{q}+(\mathbf{p}+\mathbf{k})/2\downarrow} \mid \rho_B(-\mathbf{k})\right\rangle\right\rangle = \frac{-\left\langle b^+_{\mathbf{p}+\mathbf{k}} a_{\mathbf{q}+(\mathbf{p}+\mathbf{k})/2\uparrow} a_{-\mathbf{q}+(\mathbf{p}+\mathbf{k})/2\downarrow}\right\rangle}{\omega - [\xi_{\mathbf{q}+(\mathbf{p}+\mathbf{k})/2} + \xi_{-\mathbf{q}+(\mathbf{p}+\mathbf{k})/2} - \mathcal{E}_{\mathbf{p}}]} \tag{12}$$

$$\left\langle\left\langle b_{\mathbf{p}+\mathbf{k}} a^+_{-\mathbf{q}+\mathbf{p}/2\downarrow} a^+_{\mathbf{q}+\mathbf{p}/2\uparrow} \mid \rho_B(-\mathbf{k})\right\rangle\right\rangle = \frac{\left\langle b_{\mathbf{p}} a^+_{-\mathbf{q}+\mathbf{p}/2\downarrow} a^+_{\mathbf{q}+\mathbf{p}/2\uparrow}\right\rangle}{\omega + [\xi_{\mathbf{q}+\mathbf{p}/2} + \xi_{-\mathbf{q}+\mathbf{p}/2} - \mathcal{E}_{\mathbf{p}+\mathbf{k}}]} . \tag{13}$$

Putting (12) and (13) into (9) and using (7) leads to

$$\begin{aligned}M_{\mathbf{p}}(\mathbf{k},\omega) G_{\mathbf{p}}(\mathbf{k},\omega) = -f L^{-d/2} \sum_{\mathbf{q}} &\left[\frac{\left\langle b^+_{\mathbf{p}+\mathbf{k}} a_{\mathbf{q}+(\mathbf{p}+\mathbf{k})/2\uparrow} a_{-\mathbf{q}+(\mathbf{p}+\mathbf{k})/2\downarrow}\right\rangle}{\omega - [\xi_{\mathbf{q}+(\mathbf{p}+\mathbf{k})/2} + \xi_{-\mathbf{q}+(\mathbf{p}+\mathbf{k})/2} - \mathcal{E}_{\mathbf{p}}]} \right. \\ &\left. + \frac{\left\langle b_{\mathbf{p}} a^+_{-\mathbf{q}+\mathbf{p}/2\downarrow} a^+_{\mathbf{q}+\mathbf{p}/2\uparrow}\right\rangle}{\omega + [\xi_{\mathbf{q}+\mathbf{p}/2} + \xi_{-\mathbf{q}+\mathbf{p}/2} - \mathcal{E}_{\mathbf{p}+\mathbf{k}}]}\right]\end{aligned} \tag{14}$$

which must be solved together with (8). Some manipulation gives

$$M_{\mathbf{p}}(\mathbf{k},\omega) = \frac{(\omega - \mathcal{E}_{\mathbf{pk}}) L_{\mathbf{p}}(\mathbf{k},\omega)}{n_B(\mathbf{p}) - n_B(\mathbf{p}+\mathbf{k}) + (\omega - \mathcal{E}_{\mathbf{pk}}) L_{\mathbf{p}}(\mathbf{k},\omega)} \tag{15}$$

where the rhs of (14) is designated as $L_{\mathbf{p}}(\mathbf{k},\omega)$. Note that the denominators in $L_{\mathbf{p}}(\mathbf{k},\omega)$ are associated with the transitions between the two single-electron energies $\xi_{\mathbf{q}+\mathbf{Q}/2}$ and $\xi_{-\mathbf{q}+\mathbf{Q}/2}$ as well as boson states of energy $\mathcal{E}_{\mathbf{Q}}$ where $\mathbf{Q}$ takes on values $\mathbf{p}+\mathbf{k}$ and $\mathbf{p}$. In the longwavelength and static limits (14) becomes

$$M_{\mathbf{p}}(\omega) = f[\partial n_B(\mathcal{E})/\partial\mathcal{E}]^{-1}L^{-d/2}\sum_{\mathbf{q}}\left[\frac{\langle b^+_{\mathbf{p}}a_{\mathbf{q}+\mathbf{p}/2\uparrow}a_{-\mathbf{q}+\mathbf{p}/2\downarrow}\rangle}{\omega-[\xi_{\mathbf{q}+\mathbf{p}/2}+\xi_{-\mathbf{q}+\mathbf{p}/2}-\mathcal{E}_{\mathbf{p}}]}\right.$$
$$\left.+\frac{\left\langle b_{\mathbf{p}}a^+_{-\mathbf{q}+\mathbf{p}/2\downarrow}a^+_{\mathbf{q}+\mathbf{p}/2\uparrow}\right\rangle}{\omega+[\xi_{\mathbf{q}+\mathbf{p}/2}+\xi_{-\mathbf{q}+\mathbf{p}/2}-\mathcal{E}_{\mathbf{p}}]}\right] \tag{16}$$

where $M_{\mathbf{p}}(\omega)\equiv M_{\mathbf{p}}(\mathbf{0},\omega)$ since

$$n_B(\mathbf{p})-n_B(\mathbf{p}+\mathbf{k})\simeq-\frac{\partial n_B(\mathcal{E})}{\partial\mathcal{E}}\left(\frac{\partial\mathcal{E}_{\mathbf{p}}}{\partial\mathbf{p}}\cdot\mathbf{k}\right)\qquad\text{and}\qquad\mathcal{E}_{\mathbf{pk}}\equiv\mathcal{E}_{\mathbf{p}+\mathbf{k}}-\mathcal{E}_{\mathbf{p}}\simeq\frac{\partial\mathcal{E}_{\mathbf{p}}}{\partial\mathbf{p}}\cdot\mathbf{k}.$$

To determine $M_{\mathbf{p}}(\omega)$ one must first find the averages $\langle b^+_{\mathbf{p}}a_{\mathbf{q}+\mathbf{p}/2\uparrow}a_{-\mathbf{q}+\mathbf{p}/2\downarrow}\rangle$ and $\langle b_{\mathbf{p}}a^+_{-\mathbf{q}+\mathbf{p}/2\downarrow}a^+_{\mathbf{q}+\mathbf{p}/2\uparrow}\rangle$ in the rhs of (16). These may be expressed in terms of the Green function $\langle\langle a_{\mathbf{q}+\mathbf{Q}/2}a_{-\mathbf{q}+Q/2\downarrow}|b^+_{\mathbf{Q}}\rangle\rangle$ given by Eq. (B9) of Ref. 7. However, to obtain a quick result we approximate $\langle b^+_{\mathbf{p}}a_{\mathbf{q}+\mathbf{p}/2\uparrow}a_{-\mathbf{q}+\mathbf{p}/2\downarrow}\rangle\approx\langle b^+_{\mathbf{p}}\rangle\langle a_{\mathbf{q}+\mathbf{p}/2\uparrow}a_{-\mathbf{q}+\mathbf{p}/2\downarrow}\rangle$ and $\langle b_{\mathbf{p}}a^+_{-\mathbf{q}+\mathbf{p}/2\downarrow}a^+_{\mathbf{q}+\mathbf{p}/2\uparrow}\rangle\approx\langle b_{\mathbf{p}}\rangle\langle a^+_{-\mathbf{q}+\mathbf{p}/2\downarrow}a^+_{\mathbf{q}+\mathbf{p}/2\uparrow}\rangle$ which would otherwise be exact for the non-interacting statistical BF mixture. At temperatures T within the interval $T_c<T<T^*$ where the number density of condensed bosons n_{B0} is negligible, using Eqs. (A14) and (A15) as established in Ref. 7, we obtain

$$\left\langle a^+_{\mathbf{q}+\mathbf{p}/2\uparrow}a^+_{-\mathbf{q}+\mathbf{p}/2\downarrow}\right\rangle=\frac{f}{2L^{d/2}}\frac{\langle b^+_{\mathbf{p}}\rangle}{\xi_{\mathbf{q}+\mathbf{p}/2}}[\tanh(\xi_{\mathbf{q}+\mathbf{p}/2}/2k_BT)$$
$$-\frac{\xi_{-\mathbf{q}+\mathbf{p}/2}\tanh(\xi_{\mathbf{q}+\mathbf{p}/2}/2k_BT)-\xi_{\mathbf{q}+\mathbf{p}/2}\tanh(\xi_{-\mathbf{q}+\mathbf{p}/2}/2k_BT)}{\xi_{\mathbf{q}+\mathbf{p}/2}+\xi_{-\mathbf{q}+\mathbf{p}/2}}]. \tag{17}$$

It is easy to see that contributions to (16) associated with terms such as the second in square brackets in (17) may be neglected. Indeed, these terms with small $\mathbf{q}$ cancel out in (17). As to terms with large $\mathbf{q}$ in (16) from the second term in square brackets in (17), they vanish since $p\ll q$ with p and q being respectively boson CMM and electron wavenumbers. Assuming $p\ll q$ in the main part of the integration interval in (16) we may substitute $\langle b^+_{\mathbf{p}}\rangle/\xi_{\mathbf{q}}$ for the prefactor $\langle b^+_{\mathbf{p}}\rangle/\xi_{\mathbf{q}+\mathbf{p}/2}$ before the square brackets in (17). Then the second term in square brackets in (17) changes its sign when $\mathbf{q}\to-\mathbf{q}$ and therefore does not contribute to (16); the only contribution to it comes from the first term in (17) marked below as $\langle\ldots\rangle^{(1)}_{\mathcal{H}}$, i.e., from

$$\langle a^+_{-\mathbf{q}+\mathbf{p}/2\downarrow}a^+_{\mathbf{q}+\mathbf{p}/2\uparrow}\rangle^{(1)}_{\mathcal{H}}=-\frac{f}{2L^{d/2}}\frac{\langle b^+_{\mathbf{p}}\rangle}{\xi_{\mathbf{q}+\mathbf{p}/2}}\tanh(\xi_{\mathbf{q}+\mathbf{p}/2}/2k_BT) \tag{18}$$

where the sign change is due to Fermi commutation relations and

$$\langle a_{\mathbf{q}+\mathbf{p}/2\uparrow}a_{-\mathbf{q}+\mathbf{p}/2\downarrow}\rangle^{(1)}_{\mathcal{H}}=-\frac{f}{2L^{d/2}}\frac{\langle b_{\mathbf{p}}\rangle}{\xi_{\mathbf{q}+\mathbf{p}/2}}\tanh(\xi_{\mathbf{q}+\mathbf{p}/2}/2k_BT). \tag{19}$$

which can be obtained from (18) by complex conjugation. Substituting (18) and (19) into (16), where we put $n_B(\mathbf{p}) \approx \langle b_{\mathbf{p}}^{+}\rangle \langle b_{\mathbf{p}}\rangle$, and using the fact that $p \ll q$ (p and q being respectively boson CMM and electron wavenumbers) yields

$$M_{\mathbf{p}}(\omega) = \frac{f^2 n_B(\mathbf{p})}{2L^d[\partial n_B(\mathcal{E}_{\mathbf{p}})/\partial \mathcal{E}_{\mathbf{p}}]} \sum_{\mathbf{q}} \xi_{\mathbf{q}}^{-1} \tanh(\xi_{\mathbf{q}}/2k_BT) \left(\frac{1}{\omega - \omega_{\mathbf{qp}}^{BF}} - \frac{1}{\omega + \omega_{\mathbf{qp}}^{BF}} \right) \tag{20}$$

where we define $\omega_{\mathbf{qp}}^{BF} \equiv 2\xi_{\mathbf{q}} - \mathcal{E}_{\mathbf{p}}$. Replacing the summation in (20) over fermion momenta $\mathbf{q}$ by an energy ξ integration and then using the identity $(x \pm i0)^{-1} = P(1/x) \mp \pi i \delta(x)$ one easily finds the *imaginary* part $\Gamma_{\mathbf{p}}(\omega)$ of (20) which is responsible for the resistivity caused by boson depairing, namely

$$\Gamma_{\mathbf{p}}(\omega) = \frac{\pi N(E_F) f^2 n_B(\mathbf{p})}{2[-\partial n_B(\mathcal{E}_{\mathbf{p}})/\partial \mathcal{E}_{\mathbf{p}}]} \left(\frac{\tanh[(\omega + \mathcal{E}_{\mathbf{p}})/4k_BT]}{\omega + \mathcal{E}_{\mathbf{p}}} + \frac{\tanh[(\omega - \mathcal{E}_{\mathbf{p}})/4k_BT]}{\omega - \mathcal{E}_{\mathbf{p}}} \right). \tag{21}$$

Here $N(E_F)$ is the electronic density-of-states (for each spin *and* per unit volume) at the Fermi surface. In the limit of a static external field (21) becomes

$$\Gamma_{\mathbf{p}}(0) = 2\pi\hbar\omega_D\lambda \frac{n_B(\mathbf{p})}{[-\partial n_B(\mathcal{E}_{\mathbf{p}})/\partial \mathcal{E}_{\mathbf{p}}]} \frac{\tanh(\mathcal{E}_{\mathbf{p}}/4k_BT)}{\mathcal{E}_{\mathbf{p}}} \tag{22}$$

where we have put $f = \sqrt{2\hbar\omega_D V}$[4,5] and defined $N(E_F)V \equiv \lambda$. The function $\Gamma_{\mathbf{p}}(0)$ depends weakly on $\mathbf{p}$. This is because the boson-fermion vertex coupling f in (1) which is responsible for the smearing out of bosonic linewidths, is assumed *not* to be CMM $\mathbf{p}$-dependent and because the of structure of $\mathcal{E}_{\mathbf{p}}$ itself. In general, bosons energies may be written as[7,8]

$$\mathcal{E}_{\mathbf{p}} = 2[E_F - \mu(\lambda, T)] + 2\Delta + E(\mathbf{p}). \tag{23}$$

Here $E_F - \mu(\lambda, T)$ is the shift due to bosonization of the chemical potential $\mu(\lambda, T)$ with respect to the E_F of noninteracting $T = 0$ fermions, namely

$$E_F - \mu(\lambda, T) = -\Delta(\lambda) + \frac{\lambda\hbar\omega_D}{2} \int_{-\hbar\omega_D}^{\hbar\omega_D} \frac{dx}{\sqrt{x^2 + f^2 n_{B0}(T)}} \tanh \frac{\sqrt{x^2 + f^2 n_{B0}(T)}}{2k_BT} \tag{24}$$

as established in Ref. 7, Eq. (28), $E(\mathbf{p})$ is the CP dispersion relation, and the generally coupling-dependent 2Δ is a quantity describing the bosonic gap. In the absence of *no* $\mathbf{p}$-dependent leading terms in (23), the sensitivity of $\Gamma_{\mathbf{p}}(0)$ to changes of $\mathbf{p}$ appears significantly reduced. We take $\mathbf{p} = 0$ in (22), i.e., write $\Gamma_{\mathbf{p}}(0) \approx \overline{\Gamma}$ to get

$$\tau_B^{tr} \equiv \hbar/\overline{\Gamma} = \frac{1}{2\pi\lambda\omega_D(k_BT)} \frac{\mathcal{E}_{\mathbf{0}}}{\tanh(\mathcal{E}_{\mathbf{0}}/4k_BT)} \frac{\exp(\mathcal{E}_{\mathbf{0}}/k_BT)}{\exp(\mathcal{E}_{\mathbf{0}}/k_BT) - 1}. \tag{25}$$

This is the mean value for the boson transport relaxation time which in turn determines the boson-induced Drude-like[14] electrical conductivity

$$\sigma_B = \frac{(2e)^2 n_B(\lambda, T)}{M_B} \tau_B^{tr}. \tag{26}$$

Note that $\mathcal{E}_0$ in (25) is independent of the assumed gap Δ in the boson spectrum (23) and, according to (24), is given by

$$\mathcal{E}_0(\lambda, \omega_D, T) = \lambda \hbar\omega_D \int_{-\hbar\omega_D}^{\hbar\omega_D} \frac{dx}{x} \tanh(x/2k_B T). \tag{27}$$

3. Discussion

If $\Delta \leq 0$ in (24) one immediately has $E_F - \mu(\lambda, T) > 0$ which implies that at any arbitrarily large T the Hamiltonian (1) describes a BF mixture. However, only a $\Delta > 0$ in (23) gives a *finite* T^* which as approached from above implies a transition from a normal metal to a BF mixture. Only positive Δ leads to finite bosonic resistivity $\rho_B \equiv 1/\sigma_B$ and which diverges at T^*—therefore, providing the continuous crossing between the resistivity in normal and pseudogapped phases as observed in HTSCs.[15] Analysis, details of which will be given elsewhere, of τ_B^{tr} as a function of the coupling parameter λ for different ratios of Debye-to-Fermi temperatures Θ_D/T_F shows a very strong λ-dependence. A specific feature found from (25) and (26) is a rapid decrease in the mean effective time between succesive collisions as λ increases. In particular, for $\Theta_D/T_F = 0.1$, not atypical of quasi 2D HTSCs, τ_B^{tr} is $\simeq 10^{-12}s$ for small λ and becomes $\lesssim 10^{-14}s$ for $\lambda \simeq 1$. Scattering effects due to bosons are distinct from those due to fermions: τ_B^{tr} rises linearly with normal-state charge density n^o thus reducing the boson scattering rate $1/\tau_B^{tr}$ in the presence of densely-populated fermion states, as compared with the dilute Fermi sea of electrons. In Fig. 1, the temperature behavior of the resistivity ρ_B caused by boson scattering is displayed for zero and nonzero values of Δ for fixed $\lambda = 0.35$, $\Theta_D/T_F = 0.1$ and $n^o = 10^{21} cm^{-3}$. Considered a free parameter $\Delta/k_B T_F$ is chosen to be 0 (full curve), 0.01 (dashed curve) and 0.011 (dotted curve). For $\Delta \neq 0$ (dotted and dashed curves) the number density of CPs $n_B(\lambda, T)$ vanishes at some characteristic T^* marked in Fig. 1 by vertical arrows. For temperatures immediately above T_c (which appears less sensitive to exact Δ than T^*) and for any Δ, as in the Aslamazov-Larkin theory,[10] ρ_B is nearly linear in T . However, depending on the assumed value of Δ, the later deviation of T from T_c leads to qualitatively different ρ_B. For $\Delta = 0$ (full curve in Fig. 1) ρ_B reaches a maximum and then decreases slowly over a broad range of T without any sign of a BF mixture-to-normal metal transition. For $\Delta < 0$ the situation is qualitatively the same as for $\Delta = 0$, but ρ_B is smaller. However, for $\Delta > 0$ (dashed- and dotted-curves in Fig. 1) there is a finite T^* on approaching of which the boson resistivity diverges so that for temperatures above T^* the total resistivity is determined only by the contribution of unpaired electrons, thus reflecting the main peculiarity found in all experiments (see, for

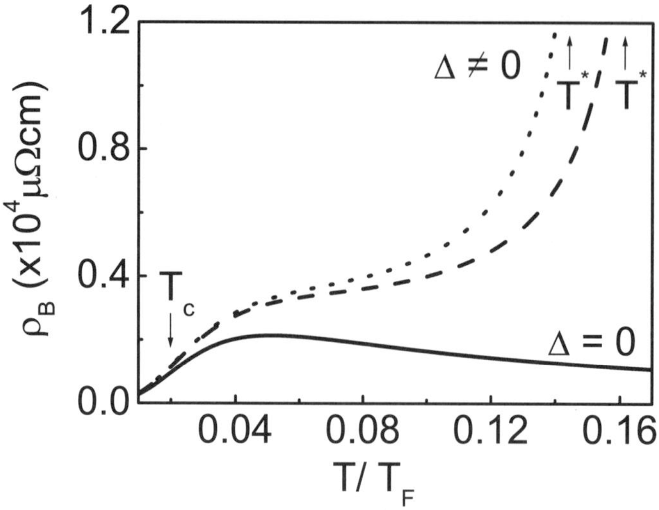

Fig. 1. Temperature behavior (in units of the Fermi temperature T_F) of the resisitivity ρ_B caused by bosonic CPs scatterings for the BCS model interaction with $\lambda = 0.35$, $\Theta_D/T_F = 0.1$ and $n^o = 10^{23} cm^{-3}$. Full, dashed and dotted curves correspond to $\Delta/k_B T_F = 0, 0.01$ and 0.011, respectively.

example, Ref. 15) Also from Fig. 1 a decrease in Δ (though still positive) leads to a shift of T^* to the higher Ts, while for $\Delta \geq 0$, T^* is finite and ρ_B increases with T becoming infinitely large at T^*. The increase of ρ_B with T may be understood in terms of temperature depairing effects.

Furthermore, the number density of bosons $n_B(\lambda, T)$ radically decreases with T, leading to enormous values for ρ_B around the value T^*. However, if there were $\Delta \leq 0$ in (26) then according to (24) the characteristic T^* becomes infinitely large, i.e., not depending how high T is, the system appears in a BF mixture state. In the case of $\Delta \leq 0$ the number density of bosons $n_B(\lambda, T)$ becomes less sensitive to temperature changes than for $\Delta > 0$. For example, if $\Delta \leq 0$ in (23) then at $T = 0$ depairing does not occur at all because of the absence of free fermionic states to be occupied as a result of boson breakups. Owing to the exclusion principle CP breakups occur more rarely for $\Delta \leq 0$. With increasing T the mobility of CPs increases which explains the monotonic decrease in ρ_B (full curve) in Fig. 1.

4. Conclusion

We have shown that within the framework of a binary BF mixture gas model it is possible to get a self-consistent description of the resistivity data observed in HTSCs only by assuming the presence of a finite positive gap between the energy spectra of free fermions and of bosons. Otherwise, calculations contradict with the experimental findings in HTSCs of a finite T^* below which the resistivity differs from that of a normal metal.

Acknowledgments

TAM thanks the Institute of Physics of Azerbaijan for granting him academic leave at the Baskent University of Turkey. MdeLl thanks UNAM-DGAPA-PAPIIT (Mexico) IN106401 for partial support; he is also grateful for travel support through a grant to Southern Illinois University at Carbondale from the U.S. Army Research Office.

Appendix A. Current Density Operator for 2e-Charged Bosons

The second-quantized boson density operator is defined as

$$\rho_B(\mathbf{r}) \equiv \phi^+(\mathbf{r})\phi(\mathbf{r}) \tag{A.1}$$

with $\phi^+(\mathbf{r})$ and $\phi(\mathbf{r})$ being phenomenological local boson field operators. These can be decomposed into Fourier components $b_\mathbf{k}$ in a volume L^d with periodic boundary conditions

$$\phi(\mathbf{r}) = L^{-d/2}\sum_\mathbf{k} b_\mathbf{k} \exp i\mathbf{k}\cdot\mathbf{r}. \tag{A.2}$$

In terms of the boson creation $b^+_\mathbf{k}$ and annihilation $b_\mathbf{k}$ operators, (A.1) becomes

$$\rho_B(\mathbf{r}) = L^{-d}\sum_\mathbf{q} \rho_B(\mathbf{q}) \exp i\mathbf{q}\cdot\mathbf{r} \quad \text{with} \quad \rho_B(\mathbf{q}) \equiv \sum_\mathbf{k} b^+_\mathbf{k} b_{\mathbf{k}+\mathbf{q}} \tag{A.3}$$

being a Fourier component of $\rho_B(\mathbf{r})$. The classical wave functions $\Psi(r)$ become operators $\phi(\mathbf{r})$ in the second quantized representation. In particular, to get a boson current density, $\mathbf{j}_B(\mathbf{r})$, in the second quantized representation, the functions $\Psi(r)$ in the quantum mechanical current density

$$\mathbf{j}(\mathbf{r}) = \alpha\mathbf{Re}(\Psi^*(\mathbf{r})\frac{-i\nabla}{M}\Psi(\mathbf{r})) = \frac{i\alpha}{2M}[\{\nabla\Psi^*(\mathbf{r})\}\Psi(\mathbf{r}) - \Psi^*(\mathbf{r})\nabla\Psi(\mathbf{r})] \tag{A.4}$$

must be replaced by the operators $\phi(\mathbf{r})$. Here α and M are the charge and mass of carriers, respectively. For 2e-charged bosons we have $\alpha \equiv 2e$ with $M = 2m$ the boson mass. Using (A.2) in (A.4), $\mathbf{j}_B(\mathbf{r})$ takes a form

$$\mathbf{j}_B(\mathbf{r}) = L^{-d}\sum_\mathbf{q} \mathbf{j}_B(\mathbf{q}) \exp i\mathbf{q}\cdot\mathbf{r} \quad \text{with} \quad \mathbf{j}_B(\mathbf{q}) = \frac{\alpha}{M}\sum_\mathbf{k}(\mathbf{k}+\frac{\mathbf{q}}{2}) b^+_\mathbf{k} b_{\mathbf{k}+\mathbf{q}} \tag{A.5}$$

where symbols in bold stand for vector quantities.

Appendix B. Subsystem Response to External Field and Green Functions

A longitudinal external electrical field $\mathbf{E}(\mathbf{r},t)$ may then be written in terms of a coordinate- and time-dependent scalar potential $\varphi(\mathbf{r},t)$ as $\mathbf{E}(\mathbf{r},t) = -\nabla\varphi(\mathbf{r},t)$. The Fourier transforms of $\varphi(\mathbf{r},t)$ and $E(\mathbf{r},t)$ are, say,

$$X(\mathbf{r},t) = \frac{e^{\varepsilon t}}{2\pi}\int_{-\infty}^{+\infty} d\omega \exp(-i\omega t) L^{-d}\sum_\mathbf{k} X(\mathbf{k},\omega) \exp i\mathbf{k}\cdot\mathbf{r} \tag{B.1}$$

where $X(\mathbf{r},t)$ can be chosen to be $\varphi(\mathbf{r},t)$ or $E(\mathbf{r},t)$. There is then the simple relation $\mathbf{E}(\mathbf{k},\omega) = -i\mathbf{k}\varphi(\mathbf{k},\omega)$. We assume that $\mathbf{E}(\mathbf{r},t)$ is adiabatically "switched on" at time $t=-\infty$. To get the correct asymptotic behavior for $\mathbf{j}_B(\mathbf{r})$, vanishing at $t=-\infty$, one introduces a prefactor $e^{\varepsilon t}$ with infinitesimal $\varepsilon > 0$ in (B.1).[11] Up to linear-order terms in $\varphi(\mathbf{r},t)$, the Hamiltonian $H^1(t)$ which is associated with the interaction between bosons and the external electric field can be expressed as

$$H^1(t) = \alpha \int d\mathbf{r}\rho_B(\mathbf{r})\varphi(\mathbf{r},t). \tag{B.2}$$

Inserting the expressions for $\rho_B(\mathbf{r})$ and $\varphi(\mathbf{r},t)$ into $H^1(t)$ and changing the order of integrations leaves

$$H^1(t) = \alpha e^{\varepsilon t} \int_{-\infty}^{+\infty} \frac{d\omega}{2\pi} \left\{ L^{-d} \sum_{\mathbf{k}} \varphi(\mathbf{k},\omega)\rho_B(-\mathbf{k}) \right\} \exp(-i\omega t), \quad \varepsilon \to 0. \tag{B.3}$$

We rewrite $H^1(t)$ in terms of its Fourier transform as

$$\begin{aligned} H^1(t) &= e^{\varepsilon t} \int_{-\infty}^{+\infty} \frac{d\omega}{2\pi} H^1(\omega)\exp(-i\omega t), \\ H^1(\omega) &= \alpha L^{-d} \sum_{\mathbf{q}} \varphi(\mathbf{q},\omega)\rho_B(-\mathbf{q}). \end{aligned} \tag{B.4}$$

Consider now the average value $\overline{A(t)} \equiv Tr\{\rho(t)A\}$ of a dynamical operator A defined through the statistical operator $\rho(t)$ which satisfies the Liouville equation $i\partial\rho(t)/\partial t = [H + H^1(t), \rho(t)]$ with the Hamiltonian $H + H^1(t)$, and an equilibrium thermal average $\langle A\rangle_H$ carried out with the time-independent operator $\rho^o \equiv e^{-H/k_BT}/Tr\{e^{-H/k_BT}\}$ which in turn is a solution of $[H,\rho^o] \equiv H\rho^o - \rho^o H = 0$. It can be shown (see, e.g., Ref. 11) that the change $\overline{\Delta A(t)} \equiv \overline{A(t)} - \langle A\rangle_H$ in $\langle A\rangle_H$ due to switching on an interaction $H^1(t)$ is directly related with the Fourier component of the retarded Green function $\langle\langle A|H^1(\omega)\rangle\rangle^{ret}$ by

$$\overline{\Delta A(t)} = e^{\varepsilon t} \int_{-\infty}^{+\infty} \frac{d\omega}{2\pi} \langle\langle A|H^1(\omega)\rangle\rangle^{ret}_{\omega+i0} \exp(-i\omega t). \tag{B.5}$$

That is, the Fourier-component $\overline{\Delta A(\omega)}$ of $\overline{\Delta A(t)}$ is given by

$$\overline{\Delta A(\omega)} \equiv \langle\langle A|H^1(\omega)\rangle\rangle^{ret}_{\omega}. \tag{B.6}$$

Appendix C. Charged-Boson Currents and Conductivity

Let us define the conductivity $\sigma_B(\mathbf{q},\omega)$ associated with the current of CPs in analogy with the conductivity in ordinary electron gas, i.e., as

$$\overline{\mathbf{j}_B(\mathbf{q},\omega)} = \sigma_B(\mathbf{q},\omega)\mathbf{E}(\mathbf{q},\omega). \tag{C.1}$$

Setting $A \equiv \mathbf{j}_B(\mathbf{q})$ in (B.6) we get the Fourier component of the average current density $\overline{\mathbf{j}_B(\mathbf{q},\omega)} \equiv \langle\langle \mathbf{j}_B(\mathbf{q}) | H^1(\omega) \rangle\rangle_\omega^{ret}$ caused by the external electric field. Because of (B.4) this becomes

$$\overline{\mathbf{j}_B(\mathbf{q},\omega)} \equiv \alpha L^{-d} \langle\langle \mathbf{j}_B(\mathbf{q}) | \rho_B(-\mathbf{q}) \rangle\rangle_\omega^{ret} \varphi(\mathbf{q},\omega). \tag{C.2}$$

Here $\mathbf{j}_B(\mathbf{q})$ is defined by (A.5). Comparing (C.1) and (C.2) yields an expression

$$\sigma_B(\mathbf{q},\omega) = \frac{i\alpha^2}{Mq^2} L^{-d} \sum_{\mathbf{p}} \mathbf{q} \cdot (\mathbf{p} + \frac{\mathbf{q}}{2}) \langle\langle b_{\mathbf{p}}^{+} b_{\mathbf{p}+\mathbf{q}} | \rho_B(-\mathbf{q}) \rangle\rangle_\omega^{ret} \tag{C.3}$$

to calculate the conductivity provided by CPs, and where the dot in (C.3) is the usual dot product of two vectors. The applied external electric field $\mathbf{E}(\mathbf{r},t)$ is assumed to change so slowly that its variation over the spatial extent of a CP is negligible.

References

1. J. Ranninger, R. Micnas and S. Robaszkiewicz, *Ann. Phys. Fr.* **13**, 455 (1988).
2. R. Friedberg and T.D. Lee, *Phys. Rev. B* **40**, 6745 (1989).
3. R. Friedberg, T.D. Lee and H.-C. Ren, *Phys. Lett. A* **152**, 417 and 423 (1991).
4. V.V. Tolmachev, *Phys. Lett. A* **266**, 400 (2000).
5. S.K. Adhikari, M. de Llano, F.J. Sevilla, M.A. Solís and J.J. Valencia, *Physica C* **453**, 37 (2007).
6. M. de Llano and V.V. Tolmachev, *Physica A* **317**, 546 (2003).
7. T.A. Mamedov and M. de Llano, *Phys. Rev. B* **75**, 104506 (2007).
8. T.A. Mamedov and M. de Llano, *Int. J. Mod. Phys. B* **21** (2007) 2335.
9. M. Fortes, M.A. Solís, M. de Llano and V.V. Tolmachev, *Physica C* **364**, 95 (2001).
10. L.G. Aslamazov and A.I. Larkin, *Phys. Lett.* **26A**, 238 (1968).
11. D.N. Zubarev, *Sov. Phys. Uspekhi* **3**, 320 (1960) [English trans. of Usp. Fiz. Nauk **71**, 71 (1960)].
12. V.L. Bonch-Bruevich and S.V. Tyablikov, *The Green Function Method in Statistical Mechanics* (North-Holland, Amsterdam, 1962).
13. N.M. Plakida, *High-Temperature Superconductivity* (Springer, Berlin, 1995).
14. N.W. Ashcroft and N.D. Mermin, *Solid State Physics* (Holt, Rinehart & Winston, NY, 1976) p. 16.
15. T. Kondo, T. Takeuchi, S. Tsuda and S. Shin, *Phys. Rev. B* **74**, 224511 (2006).
16. L.N. Cooper, *Phys. Rev.* **104**, 1189 (1956).

GROUND STATE ENERGY OF BOSE-EINSTEIN CONDENSATION IN A DISORDERED SYSTEM

VIRULH SA-YAKANIT and WATTANA LIM

Center of Excellence in Forum for Theoretical Science,
Department of Physics, Faculty of Science,
Chulalongkorn University, Bangkok 10330 Thailand

Received 31 July 2008

A modeled Bose system consisting of N particles with two-body interaction confined within volume V under inhomogeneity of the system is investigated using the Feynman path integral approach. The two-body interaction energy is assumed to be dependent on the two-parameter interacting strength a and the correlation length l. The inhomogeneity of the system or the porosity can be represented as density $\overline{n}$ with interacting strength b and correlation length L. The mean field approximation on the two-body interaction in the Feynman path integrals representation is performed to obtain the one-body interaction. This approximation is equivalent to the Hartree approximation in the many-body electron gas problem. This approximation has shown that the calculation can be reduced to the effective one-body propagator. Performing the variational calculations, we obtain analytical results of the ground state energy which is in agreement with that from Bugoliubov's approach.

Keywords: Bose-Einstein condensation; disordered system; path integral.

1. Introduction

Huang and Meng[1] proposed a model for the three-dimensional dilute Bose gas in a random potential. They formulated the random potential for porous material with the delta-function impurity potential and then analyzed their model using the Bogoliubov transformation and taking the ensemble average. They found that both BEC and superfluidity are depressed by the random potential. They also found that the superfluidity disappears below the critical density even at 0 K and predicted that the superfluid phase enters the normal phase with decreasing temperature. However, the random potential of their model does not include the pore size and thus it is difficult to compare quantitatively with experiment. Recently, Kobayashi and Tsubota[2] had proposed a model to improve Huang and Meng's model by adding the pore size dependence of the random potential. Their model works well, and leads to specific heat calculation.

In this paper, we consider a system of N Bosons confined within the volume V. The two-body interaction is assumed to be a Gaussian function with interacting strength a and the correlation length l. The porosity of the system or the inhomogeneity is represented by density of porosity $\overline{n}$ with interacting strength b and correlation length L. The main idea of this approach is to perform the mean field approximation on the two-body interaction in the Feynman path integral representation. This approximation is equivalent to replacing the two-body interaction into a one-body interaction with the effective random potential. As in the electron gas problem[3] we assume that the static variable are distributed randomly throughout the volume V as the inhomogeneity of the system. We will show that this effective one-body propagator allows us to determine many physical properties such as the ground state energy, the effective mass and the condensate density. In this paper we consider only the ground state energy.

As a consequence of the above assumptions, we will show that physical quantities can be explained using some dimensionless parameters, namely $a^3\overline{N}$ the gas parameter, $\chi \equiv n/N$ the ratio of concentration of the porosity and interacting particle, and $R \equiv \chi\,(b/a)^2$ the strength of disorder. The outline of this paper is the following. In Section 2, we present the model Lagrangian of the system and introduce the non-local harmonic trial action. In Section 3, we present the effective one-body propagator which leads to calculating the ground state energy. The final section is devoted to the discussion and conclusion.

2. The Model Lagrangian

We consider a system with N Bosons interacting via the pair potentials $u\left(\vec{r}_i - \vec{r}_j\right)$ under the influence of n external impurity potentials $v\left(\vec{r}_i - \vec{R}_k\right)$ distributed randomly. The Lagrangian of the system is

$$L=\sum_{i=1}^{N}\frac{1}{2}m\dot{\vec{r}}_i^{\,2} - \sum_{i<j}^{N} u\left(\vec{r}_i - \vec{r}_j\right) - \sum_{i=1}^{N}\sum_{k=1}^{n} v\left(\vec{r}_i - \vec{R}_k\right), \tag{1}$$

where $\vec{R}_k$ are the impurity positions assumed to be completely random, and m is the mass of the Bosons. In terms of Feynman's path integral representation, we can write the propagator associated with this Lagrangian as

$$P(\vec{r}_N(t), \vec{r}_N(0), t; \{\vec{R}_n\}) = \int_{\vec{r}_N(0)}^{\vec{r}_N(t)} D^N\left(\vec{r}_N(\tau)\right) \exp\left[\sum_{i=1}^{N}\frac{i}{\hbar}\int_0^t d\tau\left(\frac{1}{2}m\dot{\vec{r}}_i^{\,2}(\tau)\right)\right]$$

$$\times \exp\left[-\frac{i}{\hbar}\int_0^t d\tau \sum_{i<j}^{N} u\left(\vec{r}_i - \vec{r}_j\right) - \frac{i}{\hbar}\int_0^t d\tau \sum_{i=1}^{N}\sum_{k=1}^{n} v\left(\vec{r}_i - \vec{R}_k\right)\right]. \tag{2}$$

Edwards and Gulyaev[4] have pointed out that the average over all configurations of Eq.(2) can be performed exactly, and the result is

$$P\left(\vec{r}_N(t), \vec{r}_N(0), t\right) = \int_{\vec{r}_N(0)}^{\vec{r}_N(t)} D^N\left(\vec{r}_N(\tau)\right) \exp\left[\sum_{i=1}^{N} \frac{i}{\hbar} \int_0^t d\tau \left(\frac{1}{2} m \dot{\vec{r}}_i^{\,2}(\tau)\right)\right]$$

$$\times \exp\left[\begin{array}{c} -\sum_{i<j}^{N} \frac{i}{\hbar} \int_0^t d\tau u\left(\vec{r}_i(\tau) - \vec{r}_j(\tau)\right) \\ +\sum_{i=1}^{N} \overline{n} \int d\vec{R}_k \left(e^{-\frac{i}{\hbar}\int_0^t d\tau v\left(\vec{r}_i(\tau) - \vec{R}_k\right)} - 1\right) \end{array}\right], \tag{3}$$

where $\overline{n} \equiv n/V$ is the number density of the impurity. To perform the Gaussian approximation of the propagator, we expand the exponential of the impurity potential v in Eq.(3) and keep terms only to the second order of v. The first-order term then gives the average of scattering potential, while the second-order term leads to the fluctuations of the scattering potential. Alternatively, we may think of the scattering potential as a random potential, and the result of Edwards and Gulyaev enables us to explicitly calculate its average and fluctuations. In detail, expanding the exponential leads to

$$\sum_{i=1}^{N} \overline{n} \int d\vec{R}_k \left(e^{-\frac{i}{\hbar}\int_0^t d\tau v\left(\vec{r}_i(\tau) - \vec{R}_k\right)} - 1\right)$$

$$\approx -\sum_{i=1}^{N} \frac{i}{\hbar} \overline{n} \int_0^t d\tau V(\tau) + \sum_{i=1}^{N} \left(-\frac{i}{\hbar}\right)^2 \frac{\overline{n}}{2} \int_0^t \int_0^t d\tau d\sigma W_L\left(\vec{r}_i(\tau) - \vec{r}_i(\sigma)\right). \tag{4}$$

For the impurity potential in the form of a Gaussian function,

$$\overline{n} v\left(\vec{r}(\tau) - \vec{R}_k\right) = \frac{2\overline{n}\pi\hbar^2 b}{m} \left(\pi L^2\right)^{-3/2} \exp\left[-\left(\frac{\vec{r}(\tau) - \vec{R}_k}{L}\right)^2\right], \tag{5}$$

the mean potential is

$$\overline{n} V(\tau) = \overline{n} \int d\vec{R}_k v\left(\vec{r}_i(\tau) - \vec{R}_k\right) = \frac{\xi_b}{2} \tag{6}$$

and the correlation takes the form

$$\begin{aligned} \frac{\overline{n}}{2} W_L\left(\vec{r}_i(\tau) - \vec{r}_i(\sigma)\right) &= \frac{\overline{n}}{2} \int d\vec{R}_k v\left(\vec{r}_i(\tau) - \vec{R}_k\right) v\left(\vec{r}_i(\sigma) - \vec{R}_k\right) \\ &= \xi_L \exp\left[-\frac{\left(\vec{r}_i(\tau) - \vec{r}_i(\sigma)\right)^2}{2L^2}\right], \end{aligned} \tag{7}$$

where b and L, respectively, represent the interacting strength and the correlation length, $\xi_b = 4\overline{n}\pi\hbar^2 b/m$ and $\xi_L = (\overline{n}/2)\left(2\pi\hbar^2 b/m\right)^2 \left(\pi L^2\right)^{-3/2}$ with the dimension of energy squared. It will be seen later that the correlation length is comparable with the pore size of the porous material as discussed by Kobayashi and Tsubota.[2]

3. The Effective One-Body Propagator

In order to compare our approach with other theories, we must first transform the two-body interaction into a one-body interaction. This can be achieved by replacing one of the dynamical variable $\vec{r}_j(\tau)$ of the two-body interaction into the one-body with the static parameter $\vec{R}_j$. We assume that the $\vec{R}_j$ distribute completely randomly. Physically, this approximation resembles the Hartree approximation in electron gas problem. In this case, one assumes that the wave function of other particles are classical. Therefore, one can replace other wave functions as electrostatic charges distributed throughout the solid. In the Feynman path integral approach, we work on the configuration representation instead of the wave function. Performing the mean field random average, we can write the propagator as

$$P\left(\vec{r}_N(t), \vec{r}_N(0), t\right) = \int_{\vec{r}_N(0)}^{\vec{r}_N(t)} D^N\left(\vec{r}_N(\tau)\right) \exp\left[\sum_{i=1}^{N} \frac{i}{\hbar} \int_0^t d\tau \left(\frac{1}{2} m \dot{\vec{r}}_i^{\,2}(\tau)\right)\right]$$

$$\times \exp\left[\begin{array}{c} \sum\limits_{i=1}^{N} \frac{\overline{N}}{2} \int d\vec{R}_j \left(e^{-\frac{i}{\hbar}\int_0^t d\tau u\left(\vec{r}_i(\tau)-\vec{R}_j\right)} - 1\right) \\ -\frac{i}{\hbar} N \frac{\xi_b}{2} t + \sum\limits_{i=1}^{N} \left(-\frac{i}{\hbar}\right)^2 \frac{\overline{n}}{2} \int_0^t \int_0^t d\tau d\sigma W_L\left(\vec{r}_i(\tau) - \vec{r}_i(\sigma)\right) \end{array}\right]. \quad (8)$$

The two-particle interacting potential is assumed to be a Gaussian function potential of the form

$$\overline{N} u\left(\vec{r}(\tau) - \vec{R}_j\right) = \frac{4\overline{N}\pi\hbar^2 a}{m} \left(\pi l^2\right)^{-3/2} \exp\left[-\left(\frac{\vec{r}(\tau) - \vec{R}_j}{l}\right)^2\right]. \quad (9)$$

The two parameters, a and l, represent the interacting strength and the correlation length of two-body interaction, respectively. Again, expanding the exponential term, we obtain the first mean field contribution. The second term gives rise to the fluctuation of the mean field contribution. The same method used to obtain the porosity is applied to this problem. Inserting the interacting potential, we have finally the one-body propagator. Therefore, the effective one-body propagator is

$$\begin{aligned} &P\left(\vec{r}_N(t), \vec{r}_N(0), t\right) \\ &= \int_{\vec{r}_N(0)}^{\vec{r}_N(t)} D^N\left(\vec{r}_N(\tau)\right) \exp\left[\frac{i}{\hbar} \int_0^t d\tau \sum_{i=1}^{N} \frac{1}{2} m \dot{\vec{r}}_i^{\,2}(\tau)\right] \\ &\quad \times \exp\left[-\frac{i}{\hbar} N \frac{\xi_a}{2} t + \sum_{i=1}^{N} \left(-\frac{i}{\hbar}\right)^2 \frac{\overline{N}}{4} \int_0^t \int_0^t d\tau d\sigma W_l\left(\vec{r}_i(\tau) - \vec{r}_i(\sigma)\right)\right] \\ &\quad \times \exp\left[-\frac{i}{\hbar} N \frac{\xi_b}{2} t + \sum_{i=1}^{N} \left(-\frac{i}{\hbar}\right)^2 \frac{\overline{n}}{2} \int_0^t \int_0^t d\tau d\sigma W_L\left(\vec{r}_i(\tau) - \vec{r}_i(\sigma)\right)\right], \end{aligned} \quad (10)$$

where $\xi_a = 4\overline{N}\pi\hbar^2 a/m$ and $\xi_l = \left(\overline{N}/4\right)\left(4\pi\hbar^2 a/m\right)^2 \left(2\pi l^2\right)^{-3/2}$. In order to perform the calculation, we follow the method we developed to study the electron in random potential. The main idea is to introduce a trial action with non-local harmonic action given as

$$S_0 = \sum_{i=1}^{N} \int_0^t d\tau \frac{1}{2} m \dot{\vec{r}}_i^{\,2}(\tau) - \sum_{i=1}^{N} \frac{m\omega^2}{4t} \int_0^t \int_0^t d\tau d\sigma \left(\vec{r}_i(\tau) - \vec{r}_i(\sigma)\right)^2, \tag{11}$$

here ω is a variational parameter. Fortunately, this model can be solved exactly. The result is

$$P_0\left(\vec{r}_N(t), \vec{r}_N(0), t\right) = \left(\frac{m}{2\pi i\hbar t}\right)^{\frac{3}{2}N} \left(\frac{\omega t}{2\sin\left(\frac{\omega t}{2}\right)}\right)^{3N} \times \exp\left[\frac{i}{\hbar}\sum_{i=1}^{N}\frac{m\omega}{4}\cot\left(\frac{\omega t}{2}\right)\left(\vec{r}_i(t) - \vec{r}_i(0)\right)^2\right]. \tag{12}$$

We can rewrite the effective one-body propagator in terms of the trial propagator as

$$P\left(\vec{r}_N(t), \vec{r}_N(0), t\right) = P_0\left(\vec{r}_N(t), \vec{r}_N(0), t\right)\left\langle \exp\left[\frac{i}{\hbar}(S - S_0)\right]\right\rangle_{S_0}, \tag{13}$$

where the symbol $\langle\ \rangle_{S_0}$ represents the average with respect to the trial action S_0. Expanding the above average in terms of cumulants and keeping only the first cumulant,[3] we have

$$P\left(\vec{r}_N(t), \vec{r}_N(0), t\right) \simeq P_0\left(\vec{r}_N(t), \vec{r}_N(0), t\right)\exp\left[\frac{i}{\hbar}\langle S - S_0\rangle_{S_0}\right]. \tag{14}$$

The propagator P_0 is given in Eq. (12). The next step is to evaluate the action difference $\frac{i}{\hbar}\langle S - S_0\rangle_{S_0}$ which can be written as

$$\begin{aligned}
\frac{i}{\hbar}\langle S - S_0\rangle_{S_0}
&= -\frac{i}{\hbar}N\frac{\xi_a}{2}t + \left[\sum_{i=1}^{N}\left(\frac{i}{\hbar}\right)^2 \int_0^t\int_0^\tau d\tau d\sigma \xi_l \frac{1}{A_l^{3/2}(\tau,\sigma)}\exp\left[-\frac{\vec{R}^2(\tau,\sigma)}{2l^2 A_l^{3/2}(\tau,\sigma)}\right]\right] \\
&\quad -\frac{i}{\hbar}N\frac{\xi_b}{2}t + \left[\sum_{i=1}^{N}\left(\frac{i}{\hbar}\right)^2 \int_0^t\int_0^\tau d\tau d\sigma \xi_L \frac{1}{A_L^{3/2}(\tau,\sigma)}\exp\left[-\frac{\vec{R}^2(\tau,\sigma)}{2L^2 A_L^{3/2}(\tau,\sigma)}\right]\right] \\
&\quad -\frac{i}{\hbar}\langle S_0\rangle_{S_0},
\end{aligned} \tag{15}$$

where

$$\begin{aligned}
\vec{R}^2(\tau,\sigma) &= \langle \vec{r}_i(\tau) - \vec{r}_i(\sigma)\rangle_{S_0}^2 \\
&= \left(\frac{\cos\left(\frac{\omega}{2}(t - |\tau+\sigma|)\right)\sin\left(\frac{\omega}{2}(\tau-\sigma)\right)}{\sin\left(\frac{\omega}{2}t\right)}\right)^2 \left(\vec{r}(t) - \vec{r}(0)\right)^2,
\end{aligned} \tag{16}$$

and

$$A_l^{3/2}(\tau,\sigma) = \left(1 + \frac{2}{l^2}G(\tau,\sigma)\right)^{3/2};\ A_L^{3/2}(\tau,\sigma) = \left(1 + \frac{2}{L^2}G(\tau,\sigma)\right)^{3/2}, \tag{17}$$

where Green function is given by

$$G(\tau,\sigma) = \frac{i\hbar \sin\left(\frac{\omega(\tau-\sigma)}{2}\right)\sin\left(\frac{\omega(t-(\tau-\sigma))}{2}\right)}{m\omega\sin\left(\frac{\omega t}{2}\right)}. \tag{18}$$

The average $\langle S_0\rangle_{S_0}$ in Eq. (15) can be evaluated exactly. The result has been reported by one of us.[3] Collecting all contributions, we obtain the approximated propagator with first cumulant approximation as

$$P(\vec{r}_N(t), \vec{r}_N(0), t) = \left(\frac{m}{2\pi i\hbar t}\right)^{\frac{3}{2}N}\left(\frac{\omega t}{2\sin\left(\frac{\omega t}{2}\right)}\right)^{3N}$$

$$\times \exp\left[\sum_{i=1}^{N}\frac{i}{\hbar}m\left[\frac{1}{4}\omega t\cot\left(\frac{1}{2}\omega t\right) - \left(\frac{1}{2}\omega t\csc\left(\frac{1}{2}\omega t\right)\right)^2\right]\frac{\left(\vec{r_i}(t)-\vec{r_i}(0)\right)^2}{2t}\right]$$

$$\times \exp\left[\frac{3N}{2}\left(\frac{1}{2}\omega t\cot\left(\frac{1}{2}\nu t\right)-1\right)\right]\exp\left[-\frac{i}{\hbar}N\left(\frac{\xi_a}{2}+\frac{\xi_b}{2}\right)t\right]$$

$$\times \exp\left[\sum_{i=1}^{N}\left(\frac{i}{\hbar}\right)^2\int_0^t\int_0^\tau d\tau d\sigma \xi_l \frac{1}{A_l^{3/2}(\tau,\sigma)}\exp\left[-\frac{\vec{R}^2(\tau,\sigma)}{2l^2A_l^{3/2}(\tau,\sigma)}\right]\right]$$

$$\times \exp\left[\sum_{i=1}^{N}\left(\frac{i}{\hbar}\right)^2\int_0^t\int_0^\tau d\tau d\sigma \xi_L \frac{1}{A_L^{3/2}(\tau,\sigma)}\exp\left[-\frac{\vec{R}^2(\tau,\sigma)}{2L^2A_L^{3/2}(\tau,\sigma)}\right]\right] \tag{19}$$

3.1. *Ground state energy*

To obtain the ground state energy, we replace t by $-i\hbar\beta$. This transformation allow us to turn the propagator into the density matrix. For large β the one-body density matrix can be written as

$$\rho(\vec{r}(\beta), \vec{r}(0), \beta) \underset{\beta\to\infty}{=} \left(\frac{m^*}{2\pi\hbar^2\beta}\right)^{3/2}\exp\left[-E_0\beta - \frac{m^*}{2\hbar^2\beta}\left(\vec{r}(\beta)-\vec{r}(0)\right)^2\right]. \tag{20}$$

The leading term gives the ground state energy and the second term in the exponential gives the wave function of the system. To be able to use the Feynman variational principle, we take the trace of the density matrix. Since the system is translational invariant, only the diagonal part contributes to the ground states energy. The trace gives right to the volume of the system

$$\mathrm{Tr}\rho(\vec{r}(\beta), \vec{r}(0), \beta) \underset{\beta\to\infty}{=} V\left(\frac{m^*}{2\pi\hbar^2\beta}\right)^{3/2}\exp\left[-E_0\beta\right], \tag{21}$$

where the ground state energy is

$$E_0 = \frac{3}{4}E_\omega + 4\pi a^3\overline{N}E_a\left(1+\frac{b}{a}\frac{n}{N}\right) + \frac{2\xi_l}{\theta_l^3 E_\omega}\left(1-\theta_l+\ln\left(\frac{1}{2}+\frac{1}{2\theta_l}\right)\right)$$
$$+\frac{2\xi_L}{\theta_L^3 E_\omega}\left(1-\theta_L+\ln\left(\frac{1}{2}+\frac{1}{2\theta_L}\right)\right), \tag{22}$$

where $\theta_l = \sqrt{1+2E_l/E_\omega}, \theta_L = \sqrt{1+2E_L/E_\omega}, E_a = \hbar^2/2ma^2, E_l = \hbar^2/2ml^2, E_L = \hbar^2/2mL^2$ and $E_\omega = \hbar\omega$ is the variational parameter to be determined.

3.1.1. *White noise limit*

Taking the small l and L limit so that $2E_l/E_\omega >> 1$ and $2E_L/E_\omega >> 1$. We can write the ground state energy in the term of the dimensionless parameters as

$$E_0 = 4\pi a^3\overline{N}E_a\left(1+\frac{b}{a}\frac{n}{N}\right) - 4\pi a^3\overline{N}E_a\left(\sqrt{\frac{2}{\pi}}\frac{a}{l}+\frac{b}{a}\frac{n}{N}\frac{1}{\sqrt{2\pi}}\frac{b}{L}\right)$$
$$+\frac{3}{4}E_\omega + 2\left(1-\ln 2\right)a^3\overline{N}\sqrt{E_a}\sqrt{E_\omega}\sqrt{\pi}\left(2+R\right), \tag{23}$$

here $R = \chi\left(b/a\right)^2$ represents the strength of disorder and $\chi = n/N$. Minimizing the ground state energy by solving $dE_0/dE_\omega = 0$, we get

$$E_\omega = \left(\frac{4}{3}\sqrt{\pi}\left(1-\ln 2\right)a^3\overline{N}\left(2+R\right)\right)^2 E_a. \tag{24}$$

In order to avoid the divergency in taking the "white noise" limit, we set $l = a$ and $L = r_p$. The physical meaning of the hard-sphere model is that the correlation length of the interacting particle and the impurity cannot be less than diameter of particle a and diameter of impurity or pore size r_p. Substituting Eq. (24) into Eq. (23), we obtain

$$\frac{E_0}{E_a} = 4\pi a^3\overline{N}\left(1+\frac{b}{a}\frac{n}{N}\right) + 4\pi\left(1-\ln 2\right)^2\left(a^3\overline{N}\left(2+R\right)\right)^2$$
$$-4\pi a^3\overline{N}\left(\sqrt{\frac{2}{\pi}}+\frac{R}{\sqrt{2\pi}}\frac{a}{r_p}\right). \tag{25}$$

We note that for white noise limit, a and b are scattering lengths of the interacting particles and of impurities. This result can be compared with the result of Kobayashi and Tsubota[2] for the case of $l \to 0$ and $L = r_p$ as

$$\frac{E_0}{E_a} = 4\pi a^3\overline{N}\left(1+\frac{b}{a}\frac{n}{N}\right) + \left(a^3\overline{N}\right)^{3/2}\frac{512\sqrt{\pi}}{15}$$
$$+2\left(a^3\overline{N}\right)^{3/2}\pi^{3/2}R\left[\begin{array}{c}-e^{2\alpha}\left(5+4\alpha\right)\left\{1-\mathrm{erf}\sqrt{2\alpha}\right\}\\ +\sqrt{\frac{2}{\pi\alpha}}\left(1+\alpha\right)\end{array}\right], \tag{26}$$

where $\alpha = \left(a^3\overline{N}/\pi\right)\left(r_p^2/a^2\right)$. By examining Eq. (25) and Eq. (26arises from the Bose gases without disorder. To include the disorder, we find that both approaches are slightly different due to different approaches. For the case of Kobayashi and Tsubota, the approach is based on the extension of Huang and Meng theory[1] for finite correlation length. The method of Huang and Meng is the generalization of the Bogoliubov model which is represented in the momentum representation. In our approach, we start with the Feyman path integral approach which is written in the configuration representation. In principle, both approaches should give the same result if there is no approximation. Our result can be compared with the result of Astrakharchik *et al.*[5] for the case of both the two-body interaction and the random interaction are taken the delta potential functions

$$\frac{E_0}{E_a} = 4\pi a^3\overline{N}\left(1+\frac{b}{a}\frac{n}{N}\right)+\left(a^3\overline{N}\right)^{3/2}\left(\frac{512\sqrt{\pi}}{15}+16\pi^{3/2}R\right). \tag{27}$$

4. Conclusion

In this paper, we consider Bose systems in a disordered system with a finite correlation length L and with a pair potential of interacting particles with correlation length l using the Feynman path integral approach. The two-body interacting strength is taken as a and the impurity is taken as b. For the white noise limit both a and b are scattering lengths. The main idea of this approach is to perform the mean field approximation in the Feynman path integrals. We replace one of the dynamic variables of the two-body interaction into a static parameter and assume that the static parameters are completely random distribution throughout the sample. This average over all configurations of the static parameters can be compared with the Hartree approximation of the many electron gas problem. The advantage of using the Feynman path integral is that we can take care the divergence arising from using the delta potential. For white noise limit $l \to a$ and $L = r_p$, we can compare with the result of Kobayashi and Tsubota with $l \to 0$ and $L = r_p$. For the case of $l \to 0$ and $L \to 0$, we can compare with the result of Astrakharchik *et al.* for the mean field contribution. For the random potential contribution, it is different due to different approaches.

Acknowledgments

The authors would like to thank the financial support from the Royal Golden Jubilee Ph.D. Program of the Thailand Research Fund (TRF).

References

1. K. Huang and H.F. Meng, *Phys. Rev. Lett.* **69**, 644 (1992).
2. M. Kobayashi and M. Tsubota, *Phys. Rev. B* **66**, 174516 (2002).
3. V. Sa-yakanit, *J. Phys C.* **7**, 2849 (1974).

4. S. F. Edwards and V. B. Gulyaev, *Proc. Phys. Soc.* **83** 495 (1964).
5. G. E. Astrakharchik, J. Boronat, J. Casulleras and S. Giorgini, *Phys. Rev. A* **66**, 023603-2 (2002).

IMPLICATIONS OF RELATIVISTIC CONFIGURATIONS AND BAND STRUCTURES IN THE PHYSICS OF BIO-MOLECULES AND SOLIDS*

M. FHOKRUL ISLAM

Physics Department, University of Texas,
Arlington, TX 76019, USA

islamm@uta.edu

HENRIK G. BOHR

Quantum Protein Centre, Physics Department, Danish Technical University
(DK-2800) Kgs. Lyngby, Denmark

hbohr@fysik.dtu.dk

F. B. MALIK

Physics Department, Southern Illinois University
Carbondale, IL 62901, USA

fbmalik@physics.siu.edu

Beyond the second row of elements in the Mendeleev periodic table, the consideration of the relativistic effect is important in determining proper configurations of atoms and ions, in many cases. Many important quantities of interest in determining physical and chemical properties of matter, such as the effective charge, root mean square radii, and higher moments of radii used in many calculations, e.g. in the determinations of legend stabilization bond energies depend on whether the treatment is relativistic or not. In general, these quantities for a given l-orbital having two different j-values, e.g. $d_{\frac{3}{2}}$ and $d_{\frac{5}{2}}$, differ from each other, hence, making it necessary to treat them as separate orbitals. This also necessitates characterizing bands with their j-values in many instants and not l-values, particularly for "d" and f-orbitals. For example, in Au, $5d_{\frac{3}{2}}$ and $5d_{\frac{5}{2}}$ are to be dealt with as two distinct bands. The observed enhancement of laser induced field emission in W, which is not understood in terms of non-relativistic band-structures, can be explained in terms of the expected relativistic band structure. Spin-orbit coupling, which is the manifestation of the relativistic effect, is a prime factor in facilitating intersystem crossing in bio-molecules.

Keywords: Relativistic treatment of atomic configuration; effective charges; legend stabilization energies; band structures; hyperfine coupling constants and intersystem crossing.

*Dedicated to celebrate the 80th birthday of His Majesty King Bhumibol Adulyadej whose foresight and wisdom have led to the emergence of Thailand as a modern nation. In particular his leadership to democratize his country has been of critical importance.

1. Introduction

It is now well established that the physical properties of atoms involving inner-shell electrons are best described relativistically, particularly for medium or heavy-elements, e.g. inner-shell ionization cross section by electron impact.[1,2,3] Although it was pointed out as early as 1966[4] that the inclusion of relativistic effects may be important for detailed descriptions of physical properties of atoms starting from the third rows of Mendeleev's periodic table, i.e., atoms heavier than Ne, this has been broadly overlooked in most investigations. In this investigations, we present a number of examples indicating that simple phenomena such as valence configurations of atoms which are the cornerstone in understanding chemical bonding and band structures of solids are distinctly affected in relativistic treatment.

Although this is illustrated in the next section for elements K to Zn, it is valid throughout the periodic table including super-heavy elements. The configurations determined in the relativistic treatment seem to be in accord with the observation in all cases, but not those obtained by the non-relativistic method.

More importantly, the effective charge, used widely in calculating many atomic, chemical and solid-state properties, e.g. in the tight binding approximation and in Hückel theory, are dependent on whether the treatment is relativistic or not. In fact, the effective charges or shielding constants for $j = l \pm \frac{1}{2}$ states may differ from each other significantly, e.g. in relativistic treatment the effective charges for the $d_{\frac{5}{2}}$ and $d_{\frac{3}{2}}$ orbitals differ from each other. This is discussed in Section 3. The consequences of these are important in understanding many properties involving bio-molecules, e.g. ligand field stabilization energies, which play important roles in metalloproteins, an example of which is presented in Section 4.

The long standing puzzle of sharp enhancement in field emission in laser exposed tungsten, W, can well be understood in terms of relativistic configuration of W, but not if this is calculated non-relativistically. This is discussed in Section 5. Lastly, a case is made in Section 6 for computing band structures in solids, with relativistically determined configuration using Au as an example. The important role of the spin-orbit interaction in electronic excitation involving organic molecules is discussed in Section 7.

2. Relativistic and Non-Relativistic Electronic Configurations of First Transition Metals, Doubly Ionized Ions and in Rare Earth

Experimentally, the information of the number of (3d) electrons, N, in the doubly ionized atoms of Ca, Sc, Ti, V, Cr, Mn, Fe, Co, Ni, Cu and Zn are, respectively, 0, 1, 2, 3, 4, 5, 6, 7, 8, 9 and 10.[5] This is in accord with relativistic Dirac Hartree-Slater (DHS)[6] and Dirac Hartree Fock (DHF)[7] calculations of their configurations as shown in Table 1. The valence configuration of Ni calculated using non-relativistic Hartree-Fock-Slater approach[8] has, however, some ambiguities, e.g. the calculated configuration of Ni in the Hartree-Fock-Slater approximation could be either $[Ar]3d^8 4s^2$ or $[Ar]3d^9 4s^8$. Similarly, the configurations of these elements noted in Ref. 9 are misleading. Thus, chemical bond structure and band structure calculations are to be done using relativistically calculated orbitals.

Table 1. Relativistic configurations of first transition metals[6] and the number of d-electrons in (2+) metallic ions, N(th) and the observed ones N(expt)

Configuration of neutral	N(th)	N(expt)
$(4s_{\frac{1}{2}})^2$	0	0
$(3d_{\frac{3}{2}})(4s_{\frac{1}{2}})^2$	1	1
$(3d_{\frac{3}{2}})^2(4s_{\frac{1}{2}})^2$	2	2
$(3d_{\frac{3}{2}})^3(4s_{\frac{1}{2}})^2$	3	3
$(3d_{\frac{3}{2}})^4(3d_{\frac{5}{2}})(4s_{\frac{1}{2}})$	4	4
$(3d_{\frac{3}{2}})^4(3d_{\frac{5}{2}})(4s_{\frac{1}{2}})^2$	5	5
$(3d_{\frac{3}{2}})^4(3d_{\frac{5}{2}})^2(4s_{\frac{1}{2}})^2$	6	6
$(3d_{\frac{3}{2}})^4(3d_{\frac{5}{2}})^3(4s_{\frac{1}{2}})^2$	7	7
$(3d_{\frac{3}{2}})^4(3d_{\frac{5}{2}})^4(4s_{\frac{1}{2}})^2$	8	8
$(3d_{\frac{3}{2}})^4(3d_{\frac{5}{2}})^6(4s_{\frac{1}{2}})$	9	9
$(3d_{\frac{3}{2}})^4(3d_{\frac{5}{2}})^8(4s_{\frac{1}{2}})^2$	10	10

The difference in the ordering of relativistically and non-relativistically calculated configurations persists all throughout the periodic table. Table 2 indicates the difference in the configurations of La, Ce and Yb noted in Ref. 9 and determined from relativistic approximation.[6] The latter ones agree with the experimentally determined ones. Clearly, their properties involving valence electrons, e.g. chemical bond structures and band calculations, will differ from each other significantly in the two cases. The energy splitting between $4f_{\frac{5}{2}}$ and $4f_{\frac{7}{2}}$ configurations in (DHS) is about 1.45 eV in Yb which makes it necessary to consider them as two separate bands or treat them as two different entities in chemical bond calculations.

Table 2. Non-relativistic (NON-RELV)[9] and relativistic valence configurations (RELV)[5] for La, Ce and Yb are shown in columns 2 and 3, respectively. The energy difference between $4f_{\frac{5}{2}}$ and $4f_{\frac{7}{2}}$ levels is 1.45 eV. The core configuration in both cases is that of Xe.

	NON-RELV.	RELV.
La ($Z = 57$):	$6s^2 5d$	$(5d_{\frac{3}{2}})(6s_{\frac{1}{2}})^2$
Ce ($Z = 58$):	$6s^2 4f 5d$	$(4f_{\frac{5}{2}})(5d_{\frac{3}{2}})(6s_{\frac{1}{2}})^2$
Yb ($Z = 70$):	$6s^2 4f^{14}$	$(4f_{\frac{5}{2}})^6(4f_{\frac{7}{2}})^8(6s_{\frac{1}{2}})^2$

This large spin-orbit splitting also raises the possibility that the electrons in these two sub-shells may have to be assigned two different effective charges as discussed in the next section.

3. Effective Charge and Shielding Constant in Relativistic Case

Effective charge, Z(eff), or corresponding shielding constant, σ, used in estimating various properties in atoms and molecules is defined as

$$Z(eff) = Z - \sigma = \langle r_{Hi} \rangle / \langle r_i \rangle. \tag{1}$$

In (1), Z is the atomic number of an atom or ion in consideration, $\langle r_{Hi} \rangle$ is the mean radius of the ith orbital in the hydrogen atom, and $\langle r_i \rangle$ is the mean radius of the ith orbital calculated in some mean-field approximation, e.g. (HF), i.e.,

$$\langle r_i \rangle = \int_0^\infty R_i^*(r) r R_i(r) r^2 dr \tag{2}$$

where $R_i(r)$ is the radial wave function of the ith orbital in mean field, e.g. Hartree-Fock potential.

In the relativistic case, (2) is to be replaced by

$$\langle r_i \rangle = \int_0^\infty (F_i^2(r) + G_i^2(r)) r r^2 dr \tag{3}$$

where $F_i(r)$ and $G_i(r)$ are components of the radial wave function calculated in the (DHS) approximation. The calculated screening constants using (1) and (3) could be un-physical. This is evident from Fig. 1. σ calculated using (1) both for the non-relativistic $\langle r_i \rangle$ defined by (2) and relativistic defined by (3), are shown respectively by dashed and solid lines. They differ from each other significantly already for Z = 20. With increasing Z, the relativistic σ becomes negative for $1s$, $2p_{\frac{1}{2}}$, and $2s$ orbitals, which, being negative, is unphysical! Furthermore, the non-relativistic σ varies very slowly with Z as expected, but the relativistic ones depend strongly on Z.

The latter problem can be solved by defining effective charge for the relativistic case, $Z(eff)_r$, as follows

$$Z(eff)_r = Z - \sigma = Z \langle r_{DHi} \rangle / \langle r_i \rangle, \tag{4}$$

In (4), $\langle r_{DHi} \rangle$ is the mean radius of the ith orbital for the hydrogen atom calculated using the Dirac equation. The shielding constants using the definition (4) are plotted in Fig. 2 for $1s_{\frac{1}{2}}$, $2s_{\frac{1}{2}}$, $2p_{\frac{1}{2}}$ and $2p_{\frac{3}{2}}$ orbitals. Although they still depend slightly on atomic number, the dependence is not strong and they are now always positive. Two features, however, stand out strikingly from Fig. 1: (i) Whereas the values of the shielding parameters for $(1s)$ orbitals are close both in the relativistic and non-relativistic cases, they differ significantly for $(2s)$ and $(1p)$ orbitals, and (ii) for $Z \geq 50$, the values of σ for $1p_{\frac{1}{2}}$ and $1p_{\frac{3}{2}}$ differ significantly. This difference in σ between j orbitals for a particular "l" value is even more striking for "d" and "f" states. This is a consequence of the fact that the average radius for $d_{\frac{3}{2}}$, $d_{\frac{5}{2}}$, $f_{\frac{5}{2}}$ and $f_{\frac{7}{2}}$ are different for the (DHS) case. This difference is reflected also in higher moments of mean radii requiring re-evaluation of observed quantities involving them; a particular case of which, involving bio-molecules, is discussed in the next section.

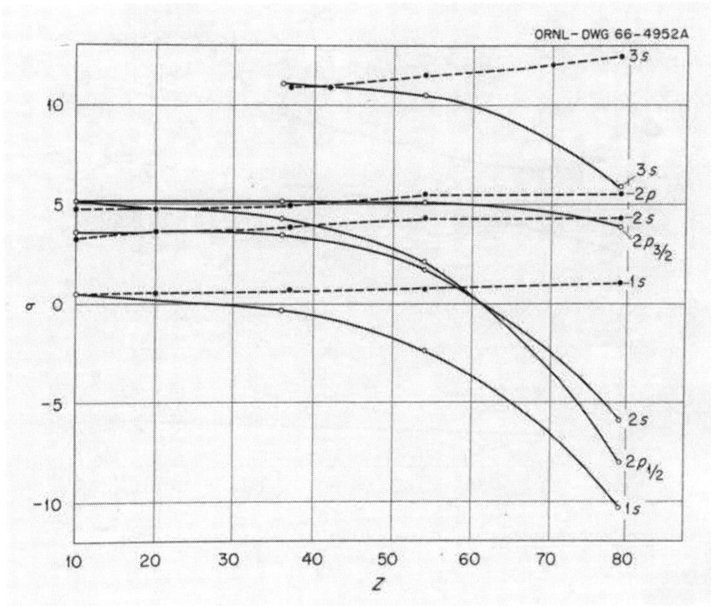

Fig. 1. Relativistic and non-relativistic screening constant, σ, calculated using (1) with wave functions obtained from (DHS) and (HF) approximation, respectively, are shown as solid and dashed lines, respectively, as a function of atomic number, Z [4].

4. Electrostatic Interaction between Ligands

In many bio-molecules, e.g. metalloproteins, a large number of hydrocarbons are quite often attached to an ionized metal at the center. For example, a doubly ionized metal such as Zn can be attached to six legends in an octahedral configuration or eight ligands with Zn^{2+} at the center. The electrostatic potential holding them together is the ionic one generated by polarization. In the non-relativistic case, this potential involves d-electrons of Zn because the valence orbitals of Zn are $(3d)$.[10] Using Mulliken's classification of d-orbitals characterized by e, referring to doubly degenerate $d_{3z^2-2r^2}$ orbitals and t, to triply degenerate d_{xy}, d_{yz} and d_{zx} orbitals, the electrostatic interactions are given by[9–11]:

$$V(t) = \frac{6Q}{R}(1 - (1/9)\langle r^4\rangle/R^4) \tag{5}$$

$$V(t) = \frac{6Q}{R}(1 + (1/6)\langle r^4\rangle/R^4). \tag{6}$$

In (5) and (6), Q is the point charge in each ligand, R, the distance between Zn^{2+} and a ligand and $\langle r^4\rangle$, the mean value of r^4 in the unit of $(a_0/Z_e)^4$, Z_e being the effective charge on the ligand.

The fourth moment of r is different for the relativistic and non-relativistic case. Moreover, $\langle r^4\rangle$ for $d_{\frac{3}{2}}$ and $d_{\frac{5}{2}}$ differ from each other in the relativistic case, as noted

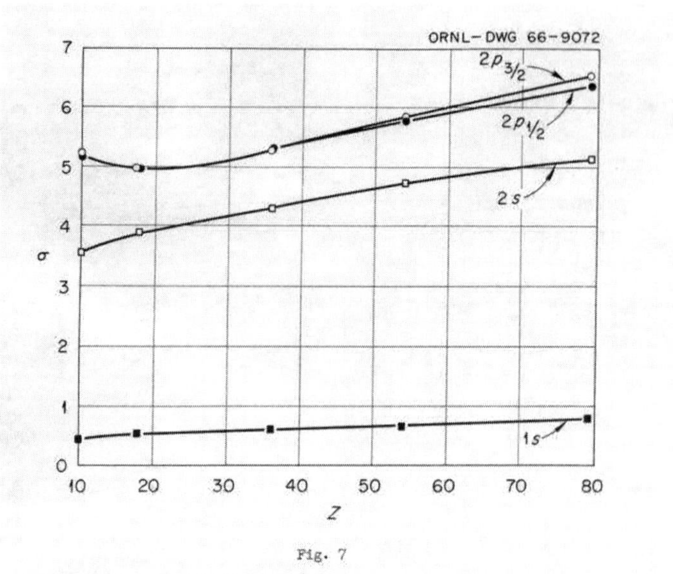

Fig. 2. Screening constants for $(1s)$, $(2s)$, $(2p_{\frac{1}{2}})$ and $(2p_{\frac{3}{2}})$ calculated using (1) and (4) and wave function in the (DHS) approximation are shown as a function of atomic number, Z.[4]

in Table 3. *Thus, the ligand stabilization bond energies are to be re-evaluated, using the relativistic approach.*

Table 3. Valence configurations, ionization potential, IP, first and fourth moments of radius, R in Bohr radius, a_0 and a_0^4, respectively, for Cu and Zn.[6]

		IP(eV)	$\langle R\rangle(a_0)$	$\langle R^4\rangle(a_0^4)$
Cu:[Ar]	$(3d_{\frac{3}{2}})^4$	10.106	0.956	5.491
	$(3d_{\frac{5}{2}})^4$	9.799	0.966	5.824
	$(4s_{\frac{1}{2}})$	7.140	3.006	2.173
Zn:[Ar]	$(3d_{\frac{3}{2}})^4$	16.848	0.846	2.556
	$(3d_{\frac{5}{2}})^4$	16.447	0.852	2.566
	$(4s_{\frac{1}{2}})^2$	8.635	2.633	12.632

Mn is an important metal holding many ligands in photosynthesis. $\langle r^4\rangle$ for $d_{\frac{3}{2}}$ and $d_{\frac{5}{2}}$ orbitals for Mn in the unit of for a_0^4 are, respectively, 7.24 and 7.40, which differ significantly from each other and from the one used in Ref. 9. Its ligand band stabilization energy needs to be re-evaluated.

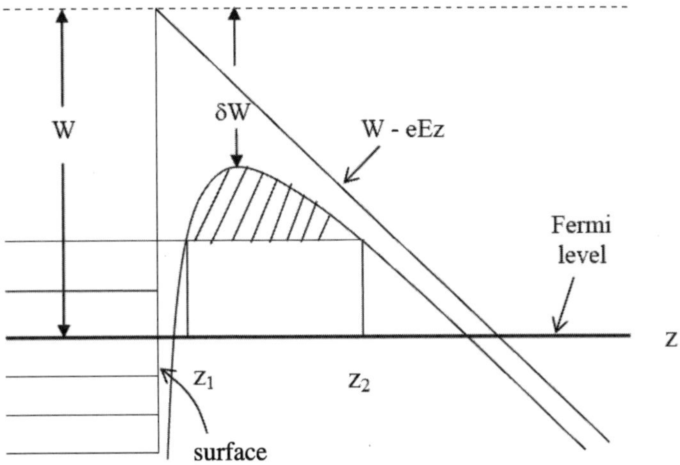

Fig. 3. Schematic diagram depicting field emission of electrons from a metal in the direction of applied electric field, E, which is taken to be z-direction, W and δW are, respectively, the work function and reduction of it due to the external electric field $(-eEz)$. z_1 and z_2 are the two turning points of the barrier.

5. Anomalous Enhancement in Laser Induced Field Emission in Tungsten

Brau and his collaborators[12,13] have reported the enhancement of about two orders of magnitude in field emission when W is exposed to laser beams compared to the non-exposed case. The field, emission occurs due to the lowering of Fermi surface in the presence of an electrostatic field schematically shown in Fig. 3. Investigations[12–14] indicate that in the presence of laser beam, the lowering of the Fermi surface can be, at the utmost, about 3.4×10^{-3} eV, hence it does not affect the field emission noticeably. Neither the needle temperature nor the needle geometry affects the emission rate significantly. One can, however, understand the enhancement in field-emission rate in W induced by laser by noting the difference in configurations of W calculated relativistically and non-relativistically.

The non-relativistic and relativistic configurations of the valence and first two unoccupied orbitals of W, are, respectively, $[Xe]6s^2 4f^{14} 5d^4 6p^0 5f^0$ [9] and $[Xe](4f_{\frac{5}{2}})^6(4f_{\frac{7}{2}})^8(5d_{\frac{3}{2}})^4(6s_{\frac{1}{2}})^2(5d_{\frac{5}{2}})^0(6p_{\frac{1}{2}})^0(6p_{\frac{3}{2}})^0$ [6] as shown, schematically, in Fig. 4. In the non-relativistic case, the valence $(5d)$ electrons are excited by allowed dipole transition, E1, to $6p$ and $5f$ orbitals when exposed to laser which can, immediately, de-excite to the ground state by E1 transition. In the relativistic case, laser light induces E1 transitions from the ground state of $(6s_{\frac{1}{2}})$ to both $(6p_{\frac{1}{2}})$ and $(6p_{\frac{3}{2}})$ vacant states which are above the unoccupied $(5d_{\frac{5}{2}})$ state. These electrons then de-excite to both the ground state $(6s_{\frac{1}{2}})$ and first excited state $(5d_{\frac{3}{2}})$ by electric dipole transition. Electrons in $(5d_{\frac{3}{2}})$ state cannot decay back to $(6s_{\frac{1}{2}})$ instantaneously because of the forbidden nature of this transition. This populated $(5d_{\frac{5}{2}})$ level is meta-stable

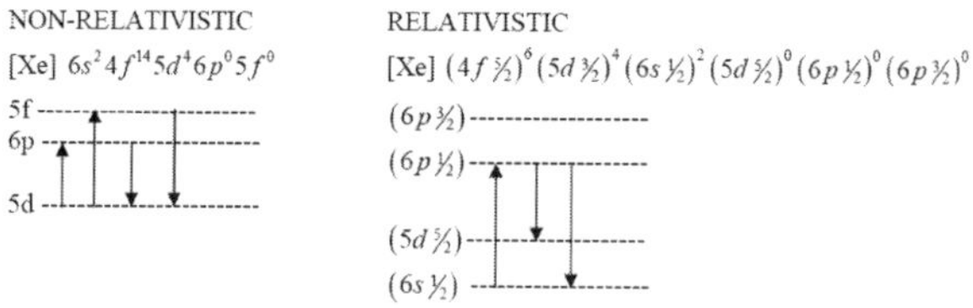

Fig. 4. Schematic diagram of dipole excitation and de-excitation of W exposed to laser in non-relativistic[9] and relativistic case.[6]

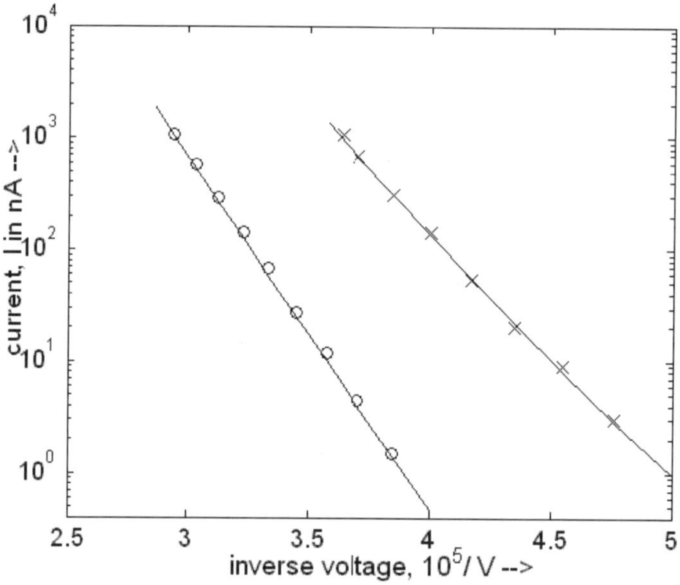

Fig. 5. Observed current in nA when W is not exposed to laser (open circle)[12,13] and when exposed to laser beam (crosses) as a function of inverse voltage of amplified field. Solid lines are calculated emission rates using Fowler-Nordheim theory[20] using work function for the ground and meta-stable first excited state at 0.546 eV.

and has smaller work function and hence, field emission from this meta-stable state would be enhanced. However, this field emission rate competes with the forbidden transition rates to ground states. The latter is typically greater than a millisecond, whereas the former for the experimental set up is of the order of a microsecond. Hence, the electron from this meta-stable state will primarily be emitted by field emission, and the emission is enhanced, since the work function of this meta-stable state is smaller than the one from the ground state.

In Fig. 5, the observed emission current in the off and on position of the laser are shown, respectively, by open circle and crosses. The solid lines are theoretical values of emitted current using the known work function for the ground state and if this excited ($5d_{\frac{3}{2}}$) would have been located at 0.5 eV higher energy above the ground state. The latter is in accord with the theoretically expected location of 0.64 eV estimated from the energy difference of ($5d_{\frac{5}{2}}$) orbital in the neighboring atom and the location of ($6s_{\frac{1}{2}}$) orbital in W.

Thus, the observed enhancement in field emission is primarily due to tunneling from the first excited meta-stable state in W and the role of laser is to populate this state.

6. Hyperfine Coupling Constants and Relativistic Bands

The computation of hyperfine coupling constants (hfc) involves the charge density of electrons, quite often the valence ones, at the nucleus. The study of (hfc) of Au, by Tucker *et al.*[15] indicates that (a) the relativistically determined configurations are important in calculating (hfc), a point subsequently well established by Lindren and Rosen[16] and (b) the determination of the structure of valence orbital could differ for a free atom and an atom embedded in solids. As noted in Table 4, in the (DHS) approximation the ordering of three valence orbitals for a gold atom in free space characterized by letting the wave function go to zero at infinity and an atom confined to a Wigner-Seitz cell of 3.019 Bohr radius characterized by imposing the boundary condition that the derivative of the wave function at the cell surface go to zero are, respectively, $(5d_{\frac{3}{2}})^4(5d_{\frac{5}{2}})^6(6s_{\frac{1}{2}})^{1}$ [6] and $(6s_{\frac{1}{2}})^2(5d_{\frac{3}{2}})^4(5d_{\frac{5}{2}})^5$.[15] Non-relativistically calculated configuration is $(5d)^{10}6s$ for a free atom. (hfc) calculated non-relativistically does not even reproduce the proper signature of the observed (hfc) but calculated (hfc) using relativistic wave function of a gold atom in metals with a Wigner-Seitz cell of about 3.02 Bohr radius leads to proper signature and magnitude close to the observed ones. Whereas, the valence configuration for gold in free space is ($6s_{\frac{1}{2}}$), it is ($5d_{\frac{5}{2}}$) in gold embedded in metal. Thus, the conduction band in metallic gold is ($5d$) and not ($6s$). Furthermore, the splitting between ($5d_{\frac{5}{2}}$) and ($5d_{\frac{3}{2}}$) orbitals is 1.73 and 0.98 eV, respectively, for gold in free space and gold in metal. Hence, Au in metal *must have two distinct conduction bands*, one $5d_{\frac{3}{2}}$ and another $5d_{\frac{5}{2}}$. It would be very interesting to establish that experimentally. The splitting between these two distinct bands is likely to be temperature and pressure dependent because both of these would change the size and shape of Wigner-Seitz cell. This illustrates that (a) the determination of valence structure may differ for an atom in free space from the one in confined environment and (b) one needs to characterize bands in the *j*-*j* coupling, particularly those associated with $d_{\frac{3}{2}}$, $d_{\frac{5}{2}}$, $f_{\frac{5}{2}}$ and $f_{\frac{7}{2}}$ orbitals.

7. Intersystem Crossing (ISC) and Relativistic Effect

Many aromatic, non-aromatic and bio-molecules being exposed to UV radiation, undergoes electronic transition from the singlet ground state to an excited electronic state via electric dipole transition and then makes a crossover transition to a triplet state. This is known as intersystem crossing (ISC).[17–19] The timescale for this (ISC) is typically microseconds to nanoseconds, i.e., very fast, and the primary interaction responsible for it is the spin-orbit one. Hence, the relativistic effect plays the primary role causing this transition, which plays a critical role in determining the physical and chemical properties of UV induced electronic-excited biomolecules.

Table 4. Eigen energies and valence orbitals of gold in free space[6] and confined in a Wigner-Seitz cell.[15] The spin-orbit energy difference between *d*-orbitals are also noted.

FREE GOLD		GOLD IN METAL: CELL: $3.019a_0$	
$(5d_{\frac{3}{2}})^4$	12.190	$(6s_{\frac{1}{2}})^2$	11.799
$(5d_{\frac{5}{2}})^6$	10.459	$(5d_{\frac{3}{2}})^4$	10.815
$(6s_{\frac{1}{2}})^1$	8.357	$(5d_{\frac{5}{2}})^5$	9.459
$E(5d_{\frac{5}{2}}) - E(5d_{\frac{3}{2}}) = 1.731eV$		0.984eV	

8. Conclusion

Although the velocity of valence electrons is much less than that of light in a vacuum, this study indicates that the properties of the valence electrons relevant to many solid state and chemical features are to be investigated within the context of a relativistic theory. In particular, the relativistically and non-relativistically determined atomic configurations, even in some light elements differ from each other. Moreover, mean values of r and its higher power are also different in the two cases. This affects a number of important properties such as band structures, ligand-field stabilization energies, valence bond energies, effective charge, the nature of electronic transition in resonant absorption. For the calculations of hyperfine coupling constant, and conduction bands involving particularly $d_{\frac{3}{2}}$, $d_{\frac{5}{2}}$, $f_{\frac{5}{2}}$ and $f_{\frac{7}{2}}$ orbitals, a relativistic approach is desirable. The density functional approach used in determining ground state configuration and associate properties of many complex molecules including bio-molecules is to be modified to extend the Kohn-Sham orbitals to the relativistic ones.

Acknowledgments

FBM thankfully acknowledges a travel grant from the U.S. Army Research Office and the National Science Foundation that enabled him to present this paper at the workshop.

References

1. J.H. Scofield, *Phys. Rev.* **A18**, 963 (1978).
2. J.T. Ndefru and F.B. Malik, *Jour. Phys.* **B13**, 2117 (1980).
3. A.K.F. Haque, M.A. Uddin, M.A.R. Patoany, A.K. Basak, M.R. Talukder, B.C. Saha, K.R. Karim and F.B. Malik, *Eur. Phys. Jour.* **D42**, 203 (2007).

4. C.W. Nestor, Jr., T.C. Tucker, T.A. Carlson, L.D. Roberts, F.B. Malik and C. Froese, Relativistic and non-relativistic SCF wave functions for atoms and ions from Z = 2 to 80 together with calculations of binding energies, mean radii, screening constants, charge distributions and electron shake-off probabilities. Oak Ridge National Laboratory Report ORNL-4027 (1966).
5. H. Haken and H.C. Wolf, *Atomic and Quantum Physics* (Springer Verlag, 1984).
6. C.C. Lu, T.A. Carlson, F.B. Malik, T.C. Tucker and C.W. Nestor, Jr., *Atomic Data* **3**, 1 (1971).
7. J.P. Desclaux, *Atomic Data* **12**, 311 (1973).
8. F. Herman and S. Skillman, *Atomic Structure Calculations* (Prentice Hall, Englewood Cliffs, N.J., 1963), Ch. 6.
9. M. Karplus and R.N. Porter, *Atoms and Molecules* (The Benjamin/Cummings Publishing Co., 1970).
10. H. Basch, A. Vista and H.B. Gray, *J. Chem. Phys.* **44**, 10 (1966).
11. L.E. Orgel, *An Introduction to Transition-Metal Chemistry* (Wiley, 1966).
12. M. Bousoukaya et al., *Nucl. Instr. and Meth.* **A279**, 405 (1985).
13. C. Hernandez-Garcia and C.A. Brau, *Nucl. Instr. and Meth.* **A429** 257(1–3) (1999).
14. M. Fhokrul Islam, M.S. thesis, Southern Illinois University at Carbondale, (Unpublished, 2001).
15. T.C. Tucker, L.D. Roberts, C.W. Nestor, Jr., T.A. Carlson and F.B. Malik, *Phys. Rev.* **173**, 998 (1969).
16. A. Rosen and I. Lindgren, *Phys. Scr.* **6**, 109 (1972).
17. S.P. McGlynn, T. Azumi and M. Kinoshita, *Molecular Spectroscopy of Triplet State* (Prentice Hall, 1969).
18. M. Schwoerer and H.C. Wolf, *Organic Molecular Solids* (Wiley-VCH Verlag, 2007).
19. H.G. Bohr and F.B. Malik, *Phys. Lett.* **A362**, 460 (2007).
20. R.H. Fowler and L.W. Nordheim, *Proc. Roy. Soc. (Lond.)* **A119**, 173 (1928).

THE SAWTOOTH CHAIN: FROM HEISENBERG SPINS TO HUBBARD ELECTRONS

J. RICHTER

Institut für Theoretische Physik, Otto-von-Guericke Universität Magdeburg, P.O. Box 4120, 39016 Magdeburg, Germany

www.uni-magdeburg.de/itp

O. DERZHKO

Institute for Condensed Matter Physics of the National Academy of Sciences of Ukraine, 1 Svientsitskii Street, L'viv-11, 79011, Ukraine

A. HONECKER

Institut für Theoretische Physik, Georg-August-Universität Göttingen, 37077 Göttingen, Germany

Received 31 July 2008

We report on recent studies of the spin-half Heisenberg and the Hubbard model on the sawtooth chain. For both models we construct a class of exact eigenstates which are localized due to the frustrating geometry of the lattice for a certain relation of the exchange (hopping) integrals. Although these eigenstates differ in details for the two models because of the different statistics, they share some characteristic features. The localized eigenstates are highly degenerate and become ground states in high magnetic fields (Heisenberg model) or at certain electron fillings (Hubbard model), respectively. They may dominate the low-temperature thermodynamics and lead to an extra low-temperature maximum in the specific heat. The ground-state degeneracy can be calculated exactly by a mapping of the manifold of localized ground states onto a classical hard-dimer problem, and explicit expressions for thermodynamic quantities can be derived which are valid at low temperatures near the saturation field for the Heisenberg model or around a certain value of the chemical potential for the Hubbard model, respectively.

Keywords: Frustration; Heisenberg model; Hubbard model; localized eigenstates.

1. Introduction

Frustrated lattices play an important role in the search for exotic quantum states of condensed matter. The term "frustration" was introduced in physics in the 1970s by Toulouse[1] in the context of spin glasses[2] and describes a situation where exchange interactions are in competition with each other. The studies on spin glasses have demonstrated that frustration may have an enormous influence on ground-state and thermodynamic properties of spin systems.[2]

In the 1970s, Anderson and Fazekas[3] first considered the quantum spin-1/2 Heisenberg antiferromagnet on the geometrically frustrated triangular lattice and proposed a liquid-like ground state without magnetic long-range order. Although later on it was found that the spin-1/2 Heisenberg antiferromagnet on the triangular lattice possesses semi-classical three-sublattice Néel order (see, e.g., Refs. 4, 5 for recent reviews), Anderson's suggestion was the starting point to search for exotic quantum ground states in frustrated spin systems.

The recent progress in synthesizing frustrated magnetic materials with strong quantum fluctuations[6] and the rich behavior of such magnetic systems have stimulated an enormous interest in frustrated quantum magnets, see, e.g., Refs. 7–11. There are many compounds which correspond to quantum antiferromagnetic Heisenberg models with frustrated spin interactions. We mention as examples the frustrated spin-1/2 $J_1 - J_2$ chains ($Rb_2Cu_2Mo_3O_{12}$, $LiCuVO_4$, Li_2ZrCuO_4)[12] and the kagomé lattice ($ZnCu_3(OH)_6Cl_2$).[13] There are also compounds which correspond to electronic (Hubbard, $t - J$, periodic Anderson) models on geometrically frustrated lattices. We mention as examples cobaltates,[14] $CeRh_3B_2$,[15] as well as artificial crystals from quantum dots.[16]

In this paper we will focus on a special property of the Heisenberg and the Hubbard model on a particular geometrically frustrated lattice (the sawtooth chain, see Fig. 1), namely the existence of localized eigenstates (on a perfect lattice) and their relevance for the low-temperature physics of those correlated systems. Note, however, that arguments and calculations presented in this paper can in principle be applied to wide class of frustrated lattices, see the discussion below and Refs. 17–28.

In general, for perfect lattices an elementary excitation as a non-interacting quasiparticle is spread over the entire lattice. For example, for a simple hypercubic lattice a magnon or electron wave function is extended over all lattice sites due to a hopping term in the Hamiltonian. However, for some lattice geometries a wave function of an elementary excitation in a quantum system may have amplitudes which are non-zero only in a restricted area owing to destructive quantum interference. We call such excitations localized excitations (for example, localized magnons[17–19] or localized electron states.[25–30]) Due to the local character of these excitations exact many-particle eigenstates of the Hamiltonian can be built by n independent localized excitations (i.e., they have a sufficiently large separation between each other) even in the presence of interactions. The number n of localized excitations cannot exceed a certain maximal value $n_{\max}$ which depends on the specific lattice under consideration, where $n_{\max}$ is proportional to the system size N.[18,28] If the localized excitation is the lowest-energy eigenstate of the Hamiltonian in the one-particle subspace one may expect that a state with n independent (isolated) localized excitations is the lowest-energy eigenstate of the Hamiltonian in the corresponding n-particle subspace[17,18,28,31] provided there is no attractive interaction. The localized eigenstates may become ground states in high magnetic fields (Heisenberg model) or at certain electron fillings (Hubbard model), respectively. Therefore they may substantially contribute to or even completely dominate the low-temperature thermodynamic properties of the system.

In the present paper we discuss the effect of localized elementary excitations on the low-temperature thermodynamics focusing on the quantum Heisenberg antiferromagnet and the Hubbard model on the sawtooth chain. We follow the lines which have been developed in a series of papers on localized eigenstates for the Heisenberg model[5,17–24,31–42] and for electronic models..[25–30]

To be specific we consider the Heisenberg antiferromagnet of N spins with quantum number $s = 1/2$ in a magnetic field h

$$H = \sum_{\langle i,j \rangle} J_{ij} \vec{s}_i \cdot \vec{s}_j - hS^z \tag{1}$$

and the Hubbard model of N lattice sites

$$H = \sum_{\substack{\langle i,j \rangle \\ \sigma=\uparrow,\downarrow}} t_{ij} \left(c^\dagger_{i,\sigma} c_{j,\sigma} + c^\dagger_{j,\sigma} c_{i,\sigma} \right) + U \sum_i n_{i,\uparrow} n_{i,\downarrow} + \mu \sum_{i,\sigma=\uparrow,\downarrow} n_{i,\sigma}. \tag{2}$$

In (1) and (2) the first sum runs over all neighboring sites on the lattice under consideration, $J_{ij} > 0$ is the antiferromagnetic isotropic Heisenberg exchange interaction between the sites i and j, and $S^z = \sum_i s^z_i$ is the z-component of the total spin. In (2) $t_{ij} > 0$ is the hopping matrix element between the nearest-neighbor sites i and j, $U > 0$ is the on-site Coulomb repulsion, μ is the chemical potential, and $n_{i,\sigma} = c^\dagger_{i,\sigma} c_{i,\sigma}$. For electronic models the chemical potential μ plays the role of the magnetic field h. While for the Heisenberg antiferromagnet h controls the magnetization $M = S^z$, μ controls the average number of electrons in the system for the Hubbard model.

In what follows we first consider the frustrated quantum Heisenberg antiferromagnet on the sawtooth chain (Fig. 1) and discuss some generic properties of the model which are caused by the localized magnon states. In particular, we consider the magnetization process, calculate the ground-state degeneracy of the localized eigenstates leading to a finite residual entropy and discuss the low-temperature

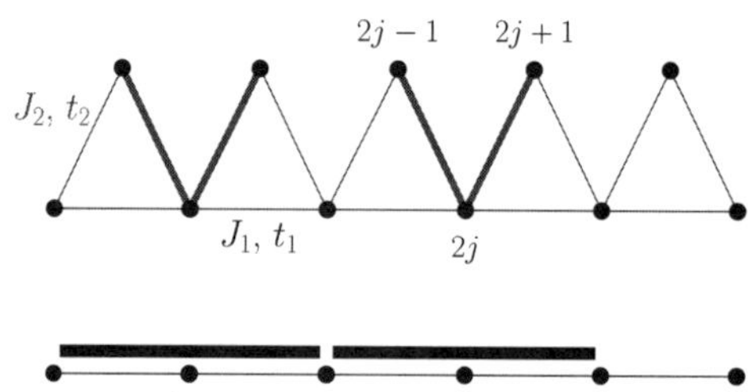

Fig. 1. (Color online) Upper part: the sawtooth chain. Filled circles indicate the lattice sites, lines indicate the exchange/hopping bonds. Two trapping cells occupied by localized magnons/electrons are indicated by bold lines. Note that for the sawtooth chain one has two kinds of bonds of different strength. The lower part of the figure indicates the corresponding hard-dimer model (two hard dimers on a linear chain).

thermodynamics for magnetic fields in the vicinity of the saturation field (Sec. 2). Then we illustrate the application of the concepts elaborated for the spin system to the Hubbard model on the sawtooth chain in Sec. 3. Section 4 presents a short summary of our discussion.

2. Localized Magnon States in the Heisenberg Antiferromagnet on the Sawtooth Chain

2.1. *Flat bands and localized eigenstates*

In this section we illustrate how the localized magnon states emerge for the frustrated quantum Heisenberg antiferromagnet (1). The fact that S^z commutes with the Hamiltonian (1) permits us to consider the eigenstates separately in each subspace with different values of $S^z = N/2, N/2-1, \ldots$. In the subspace with $S^z = N/2$ the only eigenstate is the fully polarized ferromagnetic state, $|\mathrm{FM}\rangle = |\uparrow\uparrow\uparrow\uparrow\uparrow\ldots\rangle$, which plays the role of the vacuum state for the magnon excitations.

In the one-magnon subspace ($S^z = N/2 - 1$) it is simple to calculate the eigenstates given by $|1_\kappa\rangle = \sum_{l=0}^{1} c_l \sum_{j=1}^{N/2} \mathrm{e}^{\mathrm{i}\kappa j} s_{2j+l}^{-}|\mathrm{FM}\rangle$; $H|1_\kappa\rangle = \varepsilon_{\pm}(\kappa)|1_\kappa\rangle$. The two one-magnon branches are given by

$$\varepsilon_{\pm}(\kappa) = h - \frac{J_1 + 2J_2}{2} + \frac{1}{2}\left[J_1 \cos\kappa \pm \sqrt{J_1^2\left(-1+\cos\kappa\right)^2 + 2J_2^2\left(1+\cos\kappa\right)}\right]. \quad (3)$$

For $J_2 = 2J_1$, the lower magnon band becomes completely flat, i.e., $\varepsilon_{-}(\kappa) = \varepsilon_{-} = h - 4J_1$, see the left panel of Fig. 2. Let us now focus on the case $J_2 = 2J_1$. A dispersionless band allows one to construct localized excitations given here by

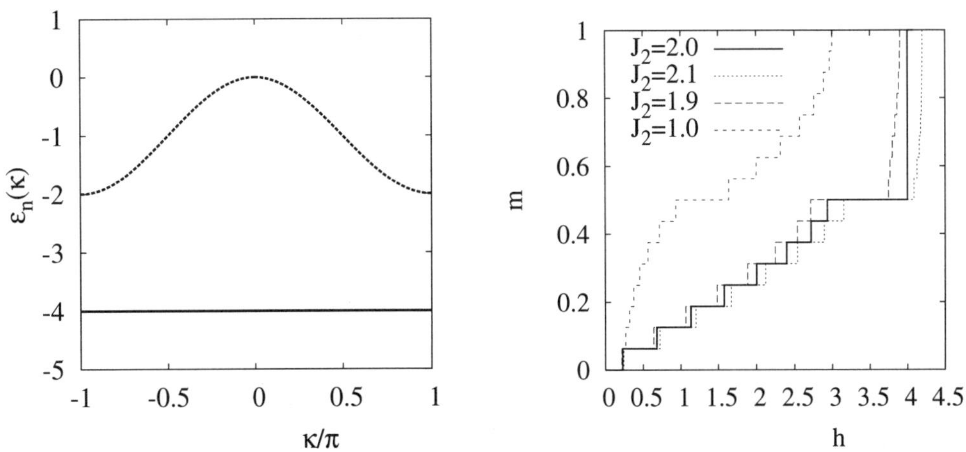

Fig. 2. Left: One-magnon dispersion for the spin-1/2 Heisenberg antiferromagnet on the sawtooth chain with $J_2 = 2$, $J_1 = 1$ and $h = 0$ (cf. Eq. (3)). Right: Ground-state magnetization curves $m(h) = M(h)/M_{\max}$ for the spin-1/2 Heisenberg antiferromagnet on the sawtooth chain for various values of J_2 and $J_1 = 1$.

$|1\mathrm{lm}\rangle = l_{2j}^{\dagger}|\mathrm{FM}\rangle$, where $l_{2j}^{\dagger} = (1/\sqrt{6})\left(s_{2j-1}^{-} - 2s_{2j}^{-} + s_{2j+1}^{-}\right)$ creates a spin excitation (magnon) localized in a valley (trapping cell) indicated by bold lines in the upper part of Fig. 1. Note that a typical geometrical feature of a lattice leading to the possibility to localize eigenstates is a triangular configuration of antiferromagnetic bonds, where the triangle is built by one bond of the trapping cell (here a valley) and two bonds attached to the trapping cell,[18,19] see Fig. 1.

Let us consider the n-magnon subspace with $S^z = N/2 - n$. In this subspace the construction of the eigenstates of the Heisenberg model is, generally, a difficult many-body problem. However, for a lattice which supports localized magnon states, a state $|n\mathrm{lm}\rangle$ consisting of n independent (i.e., isolated) localized magnons is an exact eigenstate of the Hamiltonian (1). Using the $l^{\dagger}$-operators introduced above these states can be written as $|n\mathrm{lm}\rangle = l_{i_1}^{\dagger} l_{i_2}^{\dagger} \ldots l_{i_n}^{\dagger}|\mathrm{FM}\rangle$, where the i_l are sufficiently separated lattice sites. For the sawtooth chain all n localized magnons are independent (isolated) if they do not occupy neighboring valleys (hard core rule). This constraint immediately leads to a maximum number of localized magnons $n_{\mathrm{max}} = N/4$. The energy of the n-particle state $|n\mathrm{lm}\rangle$ is

$$E_{n\mathrm{lm}} = E_{\mathrm{FM}} - \frac{N}{2}h + n(h - 4J_1), \tag{4}$$

i.e., at $h = h_1 = 4J_1$ all localized magnon states are degenerate. It is important to note that the localized magnon states are the lowest eigenstates in all sectors of $S^z = N/2-1, N/2-2, \ldots, N/2-n_{\mathrm{max}}$.[17,31] Hence these states become ground states in an appropriate magnetic field. Furthermore, it can be shown that all localized magnon states are linearly independent for the sawtooth chain[38] and that the localized magnon states present the complete manifold of ground states in all relevant sectors of S^z.[21,22,24]

In the following sections we will discuss how the localized eigenstates influence the physical properties of frustrated lattices.

2.2. *Plateaus and jumps in the magnetization curve*

First we consider the relevance of the localized magnon states for the magnetization process. For the calculation of the magnetization $M = S^z$ at $T = 0$ it is sufficient to find the lowest energy levels $E(M)$ in the subspaces with different $M = N/2, N/2 - 1, \ldots$ for $h = 0$. The energy in the presence of an external magnetic field h is given by $E(M, h) = E(M) - hM$, where the magnetization M should acquire a value which minimizes $E(M, h)$. Hence M can be determined from the equation $\mathrm{d}E(M)/\mathrm{d}M = h$ which finally gives the magnetization curve $m(h)$ where $m = M/M_{\mathrm{max}}$, $M_{\mathrm{max}} = N/2$. For a classical non-frustrated Heisenberg antiferromagnet one typically finds a parabolic relation $E(M) \propto M^2$ resulting in a straight-line behavior $M \propto h$. Often quantum fluctuations lead only to small deviations from a linear $M - h$ relation, see, e.g., Refs. 4, 5, 43. However, in the presence of frustration and quantum fluctuations more exotic magnetization curves, e.g., curves with plateaus, can be observed.[4,5,43] Another spectacular feature observed in magnetization curves of frustrated quantum spin systems consists in discontinuous jumps related to a linear relation $E \propto M$.[5,17–19,24,44] As discussed in the previous section we find such a linear $E - M$ relation for the sawtooth Heisenberg antiferromagnet with $J_2 = 2J_1$ for values of the magnetization for which the lowest eigenstates are localized states, see Eq. (4). This leads to a magnetization jump from $m = 1/2$ directly to saturation $m = 1$ at the saturation field $h_1 = 4J_1$, see the right panel of Fig. 2. In addition there is wide plateau preceding the jump. This plateau state represents a regular pattern of alternately occupied and empty valleys and is two-fold degenerate. Magnetization curves with a jump to saturation for other lattices can be found, e.g., in Refs. 17–19, 33, 37, 40, 43–45. We emphasize that the jump is macroscopic, and that there is no finite-size effect. Furthermore, we mention that a jump to saturation can be found also for the sawtooth Heisenberg antiferromagnet with higher spin quantum number $s > 1/2$. However, the height of the jump decreases with $1/s$, i.e., the jump is a true quantum effect and disappears in the classical limit $s \to \infty$.

Finally, let us discuss deviations from the ideal parameter constellation $J_2 = 2J_1$ for which the localized magnon states are true eigenstates. The right panel of Fig. 2 shows that small deviations (e.g., $J_2 = 1.9J_1$, $J_2 = 2.1J_1$) do not change the magnetization curve drastically, whereas the model with uniform bonds $J_2 = J_1$ exhibits a qualitatively different $m(h)$ behavior.

2.3. *Ground-state residual entropy and low-temperature thermodynamics*

It has been shown above that the energy of the n-magnon state in a magnetic field is $E_{\mathrm{FM}} - Nh/2 + n(h - 4J_1)$, cf. Eq. (4). Obviously, for $n < n_{\max}$ this energy level is highly degenerate, since there are many ways to place n independent localized magnons on a lattice. The degeneracy further increases at the saturation field $h_1 = 4J_1$, since the energies of the states with different numbers of localized magnons $n = 0, 1, \ldots, n_{\max}$ become equal, namely $E_{\mathrm{FM}} - Nh_1/2$. We denote this degeneracy at $h = h_1$ by $\mathcal{W}$. Since all localized magnon states are linearly independent,[38] they span a highly degenerate ground-state manifold at $h = h_1$. The degree of degeneracy can be calculated by taking into account the hard-core rule (simultaneous occupation of neighboring valleys by localized magnons is forbidden). The remaining counting problem can be solved by mapping the localized magnon problem on the sawtooth chain with N sites onto a hard-dimer problem (simultaneous occupation of neighboring sites by dimers is forbidden) on a simple linear chain with $\mathcal{N} = N/2$ sites, see the lower part of Fig. 1 and also Refs. 20–22, 24. Taking the number of hard-dimer distributions from the literature[46] we can use this mapping to find the ground-state degeneracy at the saturation field $\mathcal{W}$. For $\mathcal{N} \to \infty$ one finds $\mathcal{W} = \left((1+\sqrt{5})/2\right)^{\mathcal{N}} \approx \exp(0.4812\mathcal{N})$ leading to a finite residual entropy of $S/k_{\mathrm{B}}N = (1/2)\ln\left((1+\sqrt{5})/2\right) \approx 0.2406$ for the sawtooth chain with $J_2 = 2J_1$ at $h = h_1$.[20–22]

In addition, we can use the correspondence between the localized magnon states and the spatial configurations of hard dimers to calculate the contribution of the localized magnon states to the thermodynamic quantities following the lines given, e.g., in Refs. 46, 47. This contribution may dominate the low-temperature thermodynamics and therefore we may find predictions for the low-temperature behavior of the magnetic quantities in the vicinity of the saturation field h_1. The contribution of the localized states to the partition function of the spin model can be written as

$$\begin{aligned} Z_{\mathrm{lm}}(T,h,N) &= \exp\left(-\frac{E_{\mathrm{FM}} - h\frac{N}{2}}{k_{\mathrm{B}}T}\right) \sum_{n=0}^{n_{\max}} g_N(n) \exp\left(\frac{h_1 - h}{k_{\mathrm{B}}T} n\right) \\ &= \exp\left(-\frac{E_{\mathrm{FM}} - h\frac{N}{2}}{k_{\mathrm{B}}T}\right) \Xi(T,\mu,\mathcal{N})\;; \quad \mathcal{N} = \frac{N}{2}. \end{aligned} \tag{5}$$

Here $g_N(n)$ is the degeneracy of the ground state of the spin model with N sites in the sector with n localized magnons, i.e., with $M = S^z = N/2 - n$. In the hard-dimer description $g_N(n)$ corresponds to the canonical partition function $Z(n,\mathcal{N})$ of the classical hard-dimer model. $h_1 - h = \mu$ is the chemical potential of the hard dimers and $\Xi(T,\mu,\mathcal{N})$ (or $\Xi(z,\mathcal{N})$, $z = \exp(\mu/k_{\mathrm{B}}T)$) is the grand-canonical partition function of the one-dimensional hard-dimer lattice gas given by

$$\Xi(T,\mu,\mathcal{N}) = \lambda_1^{\mathcal{N}} + \lambda_2^{\mathcal{N}}, \quad \lambda_{1,2} = \frac{1}{2} \pm \sqrt{\frac{1}{4} + \exp x}, \quad x = \frac{\mu}{k_{\mathrm{B}}T}. \tag{6}$$

Formula (5) describes the low-temperature thermodynamics of the spin model near the saturation field accurately, i.e., $Z(T,h,N) \approx Z_{\mathrm{lm}}(T,h,N)$, because of the huge degeneracy of the ground state at $h = h_1$ (note that there are no other ground states apart from the considered localized-magnon states in the corresponding sectors of S^z). We mention that similar considerations are possible for other frustrated lattices.[21–24,39,41,42] The contribution of the localized magnon states to the Helmholtz free energy F of the spin model is given by $F_{\mathrm{lm}}(T,h,N)/N = E_{\mathrm{FM}}/N - h/2 - k_{\mathrm{B}}T \ln \Xi(z,\mathcal{N})/N$. The entropy S, the specific heat C, the magnetization M and the susceptibility χ follow from $F_{\mathrm{lm}}(T,h,N)$ according to usual relations $S_{\mathrm{lm}}(T,h,N) = -\partial F_{\mathrm{lm}}(T,h,N)/\partial T$, $C_{\mathrm{lm}}(T,h,N) = T\partial S_{\mathrm{lm}}(T,h,N)/\partial T$, $M_{\mathrm{lm}}(T,h,N) = N/2 - \langle n \rangle = N/2 - k_{\mathrm{B}}T\partial \ln \Xi(T,\mu,\mathcal{N})/\partial\mu$, $\chi_{\mathrm{lm}}(T,h,N) = \partial M_{\mathrm{lm}}(T,h,N)/\partial h$. In the limit $N \to \infty$ this leads to[20,23,24]

$$\frac{S_{\mathrm{lm}}(T,h,N)}{k_{\mathrm{B}}N} = \frac{1}{2}\left[\ln\left(\frac{1}{2} + \sqrt{\frac{1}{4} + \exp x}\right) - x\left(\frac{1}{2} - \frac{1}{4\sqrt{\frac{1}{4} + \exp x}}\right)\right],$$

$$\frac{C_{\mathrm{lm}}(T,h,N)}{k_{\mathrm{B}}N} = \frac{1}{16}\frac{x^2 \exp x}{\left(\frac{1}{4} + \exp x\right)^{\frac{3}{2}}},$$

$$\frac{M_{\mathrm{lm}}(T,h,N)}{\frac{N}{2}} = 1 - \left(\frac{1}{2} - \frac{1}{4\sqrt{\frac{1}{4} + \exp x}}\right),$$

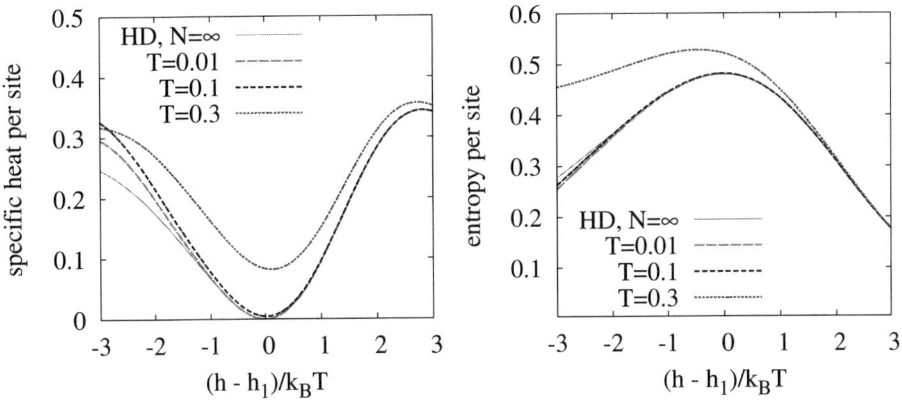

Fig. 3. (Color online) The specific heat (left) and the entropy (right) in dependence on $x = (h_1 - h)/k_BT$ for the spin-1/2 Heisenberg antiferromagnet on the sawtooth chain with $N = 20$, $J_1 = 1$, $J_2 = 2$ and $k_BT = 0.01, 0.1, 0.3$ in comparison with the one-dimensional hard-dimer (HD) gas, Eq. (7).

$$\frac{k_{\rm B}T\chi_{\rm lm}(T,h,N)}{N} = \frac{1}{16}\frac{\exp x}{\left(\frac{1}{4}+\exp x\right)^{\frac{3}{2}}}\,; \quad x = \frac{\mu}{k_{\rm B}T} = \frac{h_1 - h}{k_{\rm B}T}. \tag{7}$$

The thermodynamic quantities depend on T and h via the universal parameter $x = (h_1 - h)/k_BT$ only. Corresponding formulas for finite systems can be found using $\Xi(T, \mu, \mathcal{N})$ from Eq. (6) in combination with the relation between $F_{\rm lm}(T, h, N)$ and $\Xi(T, \mu, \mathcal{N})$ given above.

Figure 3 shows a comparison of the entropy and the specific heat of the spin model in dependence on the universal parameter $x = (h_1 - h)/k_BT$ with the hard-dimer formulas. In addition, we show the specific heat as an important measurable quantity in dependence on the temperature for magnetic fields slightly above and below the saturation field in Fig. 4.

We emphasize here some prominent features: an extra low-temperature peak in the dependence C vs. T for fields slightly below or slightly above h_1 (Figs. 3 and 4) and an enhanced entropy at h_1 at low temperatures (Fig. 3). Note that C in Eq. (7) is zero at $x = 0$ and consequently there is no extra peak in $C(T)$ for $h = h_1$, see also Fig. 3 (left). Furthermore from Figs. 3 and 4, it becomes evident that the hard-dimer description works excellently for temperatures up to 10% of the exchange coupling and reproduces qualitatively the characteristic features of the spin model for higher temperatures up to about $0.3J_1$.

Similar as for the magnetization curve, we consider now the influence of deviations from the ideal parameter constellation $J_2 = 2J_1$ (for which the localized magnon states are true exact eigenstates) on thermodynamic quantities. From Fig. 4 it is obvious that only large deviations suppress the extra low-temperature peak in $C(T)$. This behavior can be explained by inspection of the low-energy spectrum. For small deviations the energy is only slightly changed and the originally highly

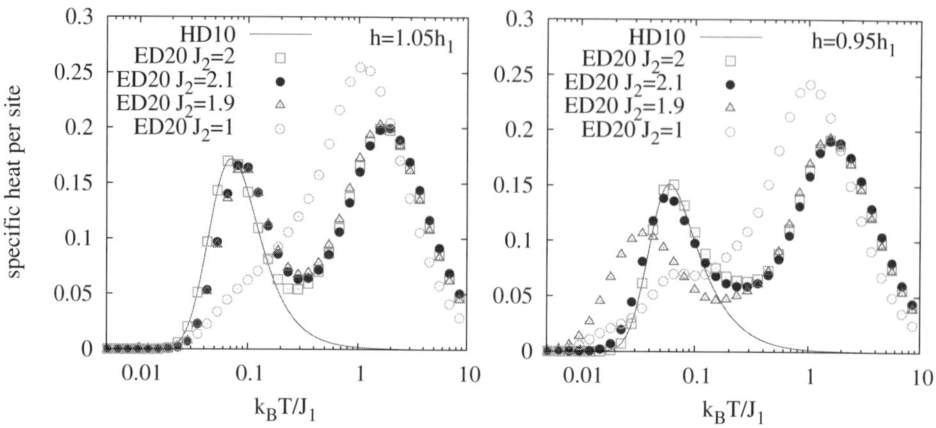

Fig. 4. (Color online) Temperature dependence of the specific heat for the spin-1/2 Heisenberg antiferromagnet on the sawtooth chain with $N = 20$, $J_1 = 1$ and various J_2 for magnetic fields slightly above ($h = 1.05h_1$, left) and below ($h = 0.95h_1$, right) the saturation field ($h_1 = 4$, 4.2, 3.9 and 3 for $J_2 = 2$, 2.1, 1.9 and 1, respectively). For comparison we show the results for the one-dimensional hard-dimer (HD) gas with $\mathcal{N} = 10$ sites.

degenerate ground-state manifold becomes quasi-degenerate. As a result, the δ-peak present in the low-energy density of states for $J_2 = 2J_1$ is broadened but there is still a well-pronounced maximum in the density of states leading to the extra low-T peak in $C(T)$.

Let us very briefly discuss an aspect of the localized magnon scenario which might have some relevance for a possible application of highly frustrated magnets. Due to the huge degeneracy of the localized magnon states and the resulting residual entropy at $h = h_1$ there is a well-pronounced low-temperature peak in the entropy S versus field h curve, see Fig. 3 (right). It has been pointed out first by Zhitomirsky[48] considering the classical kagomé Heisenberg antiferromagnet that such a degeneracy leads to an enhanced magnetocaloric effect. Later on this point has been discussed for quantum spin systems, e.g., in Refs. 20, 23, 24, 49.

3. Hubbard Electrons on the Sawtooth Chain

3.1. *Flat one-electron band and localized electron eigenstates*

We consider now the Hubbard model (2) on a sawtooth chain. The specific Hamiltonian reads

$$H = \sum_{j=0}^{\frac{N}{2}-1} \sum_{\sigma=\uparrow,\downarrow} \Big[t_1 c_{2j,\sigma}^{\dagger} c_{2j+2,\sigma} + t_2 \left(c_{2j,\sigma}^{\dagger} c_{2j+1,\sigma} + c_{2j+1,\sigma}^{\dagger} c_{2j+2,\sigma} \right) + \text{h.c.}$$

$$+\mu \left(n_{2j,\sigma} + n_{2j+1,\sigma} \right) \Big] + U \sum_{j=0}^{\frac{N}{2}-1} \left(n_{2j,\uparrow} n_{2j,\downarrow} + n_{2j+1,\uparrow} n_{2j+1,\downarrow} \right), \qquad (8)$$

where $t_1 > 0$ and $t_2 > 0$ are the hopping integrals along the base line and the zig-zag path, respectively (see the upper part of Fig. 1), and $U > 0$ is the on-site Coulomb repulsion. The sawtooth-chain Hubbard model has attracted much attention since the 1990s.[50] Here we focus on a special aspect, namely the existence of localized ground states and their consequences for the low-temperature physics of the model. On the one-particle level the description of the electron system is the same as of the XY spin system.[26–28] The one-electron dispersion reads

$$\varepsilon_{\pm}(\kappa) = \mu + t_1 \cos\kappa \pm \sqrt{t_1^2 \cos^2\kappa + 2t_2^2 (1 + \cos\kappa)} \ . \tag{9}$$

Thus, if $t_2 = \sqrt{2}t_1$ the lowest single electron energy becomes $\varepsilon_- = \mu - 2t_1$, i.e., it is completely flat. Similar to the Heisenberg model we can construct N localized one-electron ground states, given by $l^{\dagger}_{2j,\sigma}|0\rangle$, $l^{\dagger}_{2j,\sigma} = (1/2)(c^{\dagger}_{2j-1,\sigma} - \sqrt{2}c^{\dagger}_{2j,\sigma} + c^{\dagger}_{2j+1,\sigma})$ (i.e., the electron is localized in any of the $N/2$ valleys labeled by the index $2j$ and having either spin up or spin down) with energy $\varepsilon_- = -2t_1 + \mu$. Note that the indices of the $l^{\dagger}$ and $c^{\dagger}$ operators correspond to the lattice sites as illustrated in Fig. 1.

The Hubbard repulsion becomes relevant in the two-electron subspace. Obviously, a two-particle ground state can be constructed by two independent localized electrons with arbitrary spin trapped on two valleys which do not touch each other. However, in contrast to the Heisenberg model there is no “hard-core rule”, i.e., there are further two-particle ground states with two electrons trapped on two neighboring valleys, e.g., with indices $2j$ and $2j+2$. The energy of the corresponding eigenstates $l^{\dagger}_{2j,\uparrow}l^{\dagger}_{2j+2,\uparrow}|0\rangle$ and $l^{\dagger}_{2j,\downarrow}l^{\dagger}_{2j+2,\downarrow}|0\rangle$ is also independent of U, since both electrons have the same spin and therefore the Pauli principle forbids the simultaneous occupation of the site $2j+1$ belonging to both valleys. In addition, a straightforward direct calculation shows that for two electrons having different spin the linear combination

$$l^{\dagger}_{2j,\uparrow}l^{\dagger}_{2j+2,\downarrow}|0\rangle + l^{\dagger}_{2j,\downarrow}l^{\dagger}_{2j+2,\uparrow}|0\rangle, \tag{10}$$

is also a ground state in the two-electron subspace with an energy independent of U. This can be seen also by using the SU(2) symmetry of the Hubbard Hamiltonian: the state (10) and the states $l^{\dagger}_{2j,\uparrow}l^{\dagger}_{2j+2,\uparrow}|0\rangle$ and $l^{\dagger}_{2j,\downarrow}l^{\dagger}_{2j+2,\downarrow}|0\rangle$ form a triplet, i.e., (10) can be obtained by acting with the total spin lowering operator $S^- = \sum_i c^{\dagger}_{i,\downarrow}c_{i,\uparrow}$ on the state $l^{\dagger}_{2j,\uparrow}l^{\dagger}_{2j+2,\uparrow}|0\rangle$. Of course, all states belonging to one triplet have the same energy $2(\mu - 2t_1)$.

We can generalize this procedure to construct the ground states in the subspaces with $n = 3, \ldots, N/2$ electrons

$$|\varphi^{\uparrow}_n\rangle \propto l^{\dagger}_{2i_n,\uparrow}\cdots l^{\dagger}_{2i_1,\uparrow}|0\rangle \quad ; \quad H|\varphi^{\uparrow}_n\rangle = n(-2t_1 + \mu)|\varphi^{\uparrow}_n\rangle. \tag{11}$$

They are all degenerate for $\mu = \mu_0 = 2t_1$ and do not feel U. Evidently, they are fully polarized

$$S^z|\varphi^{\uparrow}_n\rangle = \frac{n}{2}|\varphi^{\uparrow}_n\rangle \quad ; \quad \vec{S}^2|\varphi^{\uparrow}_n\rangle = \frac{n}{2}\left(\frac{n}{2}+1\right)|\varphi^{\uparrow}_n\rangle. \tag{12}$$

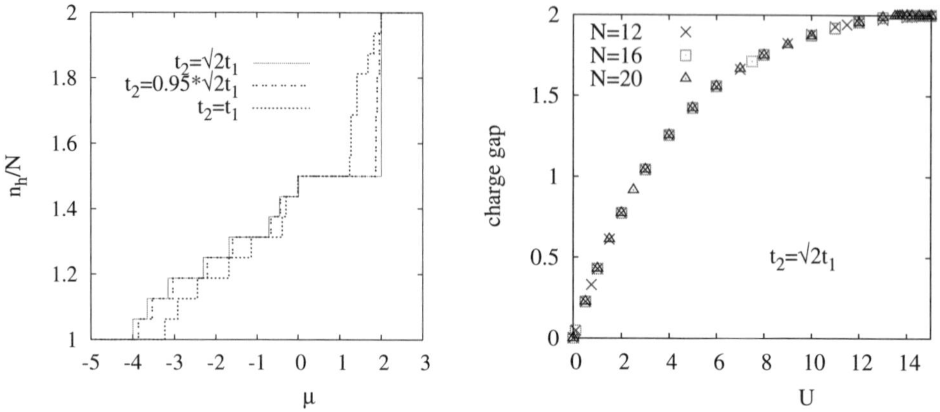

Fig. 5. (Color online) Left: Hole concentration $n_{\rm h}/N = 2 - n/N$ versus chemical potential μ for $t_2 = \sqrt{2}t_1$ (localized-electron regime), $t_2 = 0.95\sqrt{2}t_1$ and $t_2 = t_1$ for a finite sawtooth chain of $N = 16$ sites (periodic boundary conditions) and $U \to \infty$, $t_1 = 1$. Right: Charge gap $\Delta\mu = E(N/2+1) - 2\,E(N/2) + E(N/2-1)$ at quarter filling versus U for $t_2 = \sqrt{2}t_1$, $t_1 = 1$ and $N = 12, 16, 20$.

Again the application of S^- yields new eigenstates with the same energy and the same $\vec{S}^2$, but with $S^z(S^-)^k|\varphi_n^\uparrow\rangle = (n/2-k)(S^-)^k|\varphi_n^\uparrow\rangle$. Note that $l_{2i_n,\uparrow}^\dagger \cdots l_{2i_k,\downarrow}^\dagger \cdots l_{2i_1,\uparrow}^\dagger|0\rangle$, where $i_1,\ldots,i_k,\ldots,i_n$ denote n contiguous valleys, is not an eigenstate. Since there is no hard-core rule the maximum filling with localized electrons is $n_{\rm max} = N/2$, i.e., it is twice as large as for localized magnons.

In the next step we use the fully polarized n-electron states $|\varphi_n^\uparrow\rangle$ to construct the complete set of ground states for $0 \le n \le N/2$. The $|\varphi_n^\uparrow\rangle$ can be grouped into two classes, namely in one-cluster states and in multi-cluster states. While for the one-cluster states the electrons occupy a cluster of contiguous valleys, for a multi-cluster state the electrons occupy two or more clusters, where each cluster is built by contiguous valleys and the different clusters are separated by one or more empty valleys. The key observation is that further ground states can be constructed by application of a certain cluster spin flip operator $S_{\rm clust}^- = \sum_{i\in{\rm clust}} c_{i,\downarrow}^\dagger c_{i,\uparrow}$ on a multi-cluster n-electron ground state $|\varphi_n^\uparrow\rangle$. The resulting new states are not fully polarized and complete the set of ground states in each sector n.[28]

3.2. *Hole concentration in dependence on the chemical potential*

In correspondence to the $m(h)$ curve of the Heisenberg model, we consider now the hole concentration $n_{\rm h}/N = 2 - n/N$ in dependence on the chemical potential μ (Fig. 5 (left)). Like for spin systems, see Sec. 2.2, the main characteristics for the system with localized eigenstates (i.e., for $t_2 = \sqrt{2}t_1$) are a size-independent jump of $n_{\rm h}/N$ from 3/2 to 2 and a plateau at $n_{\rm h}/N = 3/2$. This plateau determines the range of validity of the localized-electron picture at $T = 0$. The right panel of Fig. 5 presents the plateau width, i.e., the size of the charge gap, versus U for $N = 12$,

16, and 20. One observes that there is almost no finite-size dependence. Since the charge gap is zero for $U = 0$ and increases with U we conclude that its appearance is due to the on-site repulsion. Small deviations from the ideal parameter values $t_2 = \sqrt{2}t_1$ do not change the $n_{\rm h}/N$ versus μ curve substantially, as illustrated for the case $t_2 = 0.95\sqrt{2}t_1$ in Fig. 5 (left), whereas for the model with uniform hopping integrals $t_2 = t_1$ the charge gap is significantly smaller and there is no indication of a jump from the plateau at $n_{\rm h}/N = 3/2$ to $n_{\rm h}/N = 2$.

3.3. *Ground-state residual entropy and low-temperature thermodynamics*

The localized-electron states are *linearly independent*, which is connected with the fact (as in the case of spin systems, see Ref. 38) that the middle site is unique to each valley. Therefore all these highly degenerate states contribute to the partition function. Now the question arises whether the ground state degeneracy can be calculated. Due to the different statistics of Hubbard electrons and Heisenberg spins there are some differences in the construction rules of localized eigenstates (e.g., the occupation of neighboring valleys is forbidden for spins but allowed for electrons, see above). Hence it is not surprising that the ground state degeneracy $g_N(n)$ for n electrons on the N-site sawtooth chain does not coincide with the one for the Heisenberg sawtooth chain (which was equal to the canonical partition functions of n hard dimers on a simple chain of $\mathcal{N} = N/2$ sites, see Sec. 2.3). Nevertheless, $g_N(n)$, $n = 0, 1, 2, \ldots, N/2$ for the Hubbard sawtooth chain can also be found by a mapping of the localized-electron degrees of freedom onto the one-dimensional hard-dimer problem. However, this mapping is more intricate and hard dimers have to be considered on a simple chain of N sites (instead of $N/2$ sites as for Heisenberg spins), for details see Ref. 28. One finds $g_N(n) = Z(n, N)$ for $n = 0, 1, \ldots, N/2 - 1$ and $g_N(N/2) = N/2 + 1 = Z(N/2, N) + N/2 - 1$ where $Z(n, N)$ is the canonical partition function of the classical one-dimensional hard-dimer model.[46,47] As for spin systems we can calculate the contribution of localized electron states to the partition function by using this mapping. Again we can present analytical formulas for the low-temperature thermodynamic quantities for a non-trivial quantum many-body problem. The grand-canonical partition function Ξ of the electron system for a chemical potential μ in the vicinity of $\mu_0 = 2t_1$ takes the form

$$\Xi(T, \mu, N) = \lambda_1^N + \lambda_2^N + \lambda_3^N,$$
$$\lambda_{1,2} = \frac{1}{2} \pm \sqrt{\frac{1}{4} + \exp x}, \quad \lambda_3 = \left(\frac{N}{2} - 1\right)^{\frac{1}{N}} \exp\frac{x}{2}, \quad x = \frac{2t - \mu}{k_{\rm B}T}. \quad (13)$$

In the thermodynamic limit $N \to \infty$ only the largest eigenvalue λ_1 of the transfer matrix survives and, using the definitions $S(T, \mu, N) = k_B \partial \left(T \ln \Xi(T, \mu, N)\right) / \partial T$, $C(T, \mu, N) = T\, \partial S(T, \mu, N)/\partial T$, we obtain the following results for the thermody-

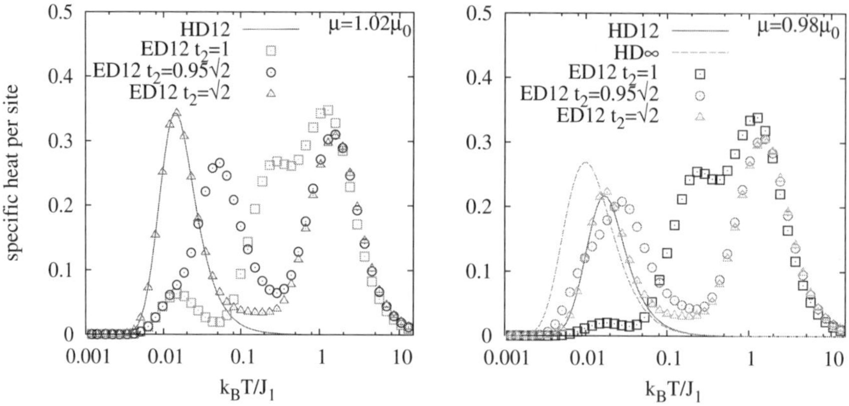

Fig. 6. (Color online) Grand-canonical specific heat per site $C(T,\mu,N)/k_{\mathrm{B}}N$ vs. temperature for the sawtooth Hubbard chain of $N=12$ sites for two values of μ, $U=\infty$ and $t_1=1$, $t_2=\sqrt{2}$, $0.95\sqrt{2}$ and 1 (symbols). Note that $\mu_0=2$ for $t_2=\sqrt{2}$, $0.95\sqrt{2}$ and 1. For comparison we show the hard-dimer data for $N=12$ (solid line) which follows from Eq. (13) and for $N=\infty$ (dashed line, Eq. (14)). Note that for $\mu=1.02\mu_0$ the hard-dimer data for $N=12$ and $N=\infty$ practically coincide.

namics of one-dimensional hard dimers (see also Ref. 28)

$$\begin{aligned}
\frac{S(T,\mu,N)}{k_{\mathrm{B}}N} &= \ln\left(\frac{1}{2}+\sqrt{\frac{1}{4}+\exp x}\right) - x\left(\frac{1}{2}-\frac{1}{4\sqrt{\frac{1}{4}+\exp x}}\right),\\
\frac{C(T,\mu,N)}{k_{\mathrm{B}}N} &= \frac{x^2\exp x}{8\left(\frac{1}{4}+\exp x\right)^{\frac{3}{2}}},\\
\frac{\langle n\rangle}{N} &= \frac{1}{2}-\frac{1}{4\sqrt{\frac{1}{4}+\exp x}},
\end{aligned} \tag{14}$$

which are quite similar to the corresponding expressions for Heisenberg spins, see Eq. (7). Again we have a finite residual entropy $S/k_{\mathrm{B}}N=\ln((1+\sqrt{5})/2)\approx 0.4812$, which is twice as large as for the Heisenberg model.

Results for the low-temperature grand-canonical specific heat are shown in Fig. 6 for two values of the chemical potential slightly above and below μ_0. Similar to the spin system we see (i) that the hard-dimer model, Eqs. (13) and (14), yields a good description of the electronic model at low temperatures and (ii) that there is an extra low-temperature maximum in the grand-canonical specific heat due to the manifold of localized electron ground states. Again this additional low-temperature maximum in $C(T)$ disappears at $\mu=\mu_0$ as can be read off from Eq. (14) (note that $C(x=0)=0$).

At the end of this section we would like to mention a relation to the so-called flat-band ferromagnetism in the Hubbard model found by Mielke and Tasaki in the early 1990s.[25] In particular, the ground states belonging to the plateau at $n = N/2$, see Sec. 3.2, are fully polarized ferromagnetic states. For further details of flat-band ferromagnetism in the sawtooth-chain Hubbard model the interested reader is referred to the original papers of Tasaki[25] but also to Ref. 28.

4. Summary

To summarize, we have illustrated some basic concepts of localized eigenstates in correlated systems on highly frustrated lattices and their effect on the low-temperature thermodynamics. As a rule non-interacting electrons or magnons on a lattice are delocalized, i.e., are described by a wave function distributed over the whole lattice. Electrons or magnons may become localized due to randomness or after switching on interactions. As we have discussed on this paper, a frustrating lattice topology may lead to another mechanism for localization. Localized states may survive in the presence of interactions and under certain conditions they can determine the properties of the system at low temperatures.

Acknowledgments

The authors would like to thank J. Jędrzejewski, T. Krokhmalskii, R. Moessner, H.-J. Schmidt, J. Schnack, J. Schulenburg and M. E. Zhitomirsky for useful discussions and fruitful collaboration in this field. A. H. acknowledges financial support by the Deutsche Forschungsgemeinschaft through a Heisenberg fellowship (grant HO 2325/4-1). We mention that most of the numerical results presented in this article were obtained using J. Schulenburg's *spinpack*.

References

1. G. Toulouse, *Commun. Phys.* **2**, 115 (1977).
2. K. Binder and A. P. Young, *Rev. Mod. Phys.* **58**, 801 (1986).
3. P. W. Anderson, *Mater. Res. Bull.* **8**, 153 (1973); P. W. Anderson and P. Fazekas, *Phil. Mag.* **30**, 423 (1974).
4. C. Lhuillier and G. Misguich, in *High Magnetic Fields*, eds. C. Berthier, L.P. Lévy, G. Martinez (Lecture Notes in Physics, **595**) (Springer, Berlin, 2001), pp. 161–190.
5. J. Richter, J. Schulenburg and A. Honecker, in *Quantum Magnetism*, U. Schollwöck, J. Richter, eds. D. J. J. Farnell, R. F. Bishop (Lecture Notes in Physics, **645**) (Springer, Berlin, 2004), pp. 85–153.
6. P. Lemmens and P. Millet, in *Quantum Magnetism*, eds. U. Schollwöck, J. Richter, D. J. J. Farnell, R. F. Bishop (Lecture Notes in Physics, **645**) (Springer, Berlin, 2004), pp. 433–477.
7. P. Schiffer, *Nature* **413**, 48 (2001).
8. R. Moessner, *Can. J. Phys.* **79**, 1283 (2001).
9. R. Moessner and A. P. Ramirez, *Physics Today*, February 2006, p. 24.
10. H. T. Diep (ed.), *Frustrated Spin Systems* (World Scientific, Singapore, 2004).
11. U. Schollwöck, J. Richter, D. J. J. Farnell, R. F. Bishop (eds.), *Quantum Magnetism* (Lecture Notes in Physics, **645**) (Springer, Berlin, 2004).

12. M. Hase, H. Kuroe, K. Ozawa, O. Suzuki, H. Kitazawa, G. Kido and T. Sekine, *Phys. Rev. B* **70**, 104426 (2004); M. Enderle, C. Mukherjee, B. Fåk, R. K. Kremer, J.-M. Broto, H. Rosner, S.-L. Drechsler, J. Richter, J. Malek, A. Prokofiev, W. Assmus, S. Pujol, J.-L. Raggazzoni, H. Rakoto, M. Rheinstädter and H. M. Rønnow, *Europhys. Lett.* **70**, 237 (2005); M. G. Banks, F. Heidrich-Meisner, A. Honecker, H. Rakoto, J.-M. Broto and R. K. Kremer, *J. Phys.: Condens. Matter* **19**, 145227 (2007); S.-L. Drechsler, O. Volkova, A. N. Vasiliev, N. Tristan, J. Richter, M. Schmitt, H. Rosner, J. Málek, R. Klingeler, A. A. Zvyagin and B. Büchner, *Phys. Rev. Lett.* **98**, 077202 (2007).
13. P. Mendels, F. Bert, M. A. de Vries, A. Olariu, A. Harrison, F. Duc, J. C. Trombe, J. S. Lord, A. Amato and C. Baines, *Phys. Rev. Lett.* **98**, 077204 (2007).
14. K. Takada, H. Sakurai, E. Takayama-Muromachi, F. Izumi, R. A. Dilanian and T. Sasaki, *Nature* **422**, 53 (2003); Y. Wang, N. S. Rogado, R. J. Cava and N. P. Ong, *Nature* **423**, 425 (2003); M. L. Foo, Y. Wang, S. Watauchi, H. W. Zandbergen, T. He, R. J. Cava and N. P. Ong, *Phys. Rev. Lett.* **92**, 247001 (2004).
15. H. N. Kono and Y. Kuramoto, *J. Phys. Soc. Jpn.* **75**, 084706 (2006).
16. H. Tamura, K. Shiraishi, T. Kimura and H. Takayanagi, *Phys. Rev. B* **65**, 085324 (2002); R. Arita, K. Kuroki, H. Aoki, A. Yajima, M. Tsukada, S. Watanabe, M. Ichimura, T. Onogi and T. Hashizume, *Phys. Rev. B* **57**, R6854 (1998).
17. J. Schnack, H.-J. Schmidt, J. Richter and J. Schulenburg, *Eur. Phys. J. B* **24**, 475 (2001).
18. J. Schulenburg, A. Honecker, J. Schnack, J. Richter and H.-J. Schmidt, *Phys. Rev. Lett.* **88**, 167207 (2002).
19. J. Richter, J. Schulenburg, A. Honecker, J. Schnack and H.-J. Schmidt, *J. Phys.: Condens. Matter* **16**, S779 (2004).
20. M. E. Zhitomirsky and A. Honecker, *J. Stat. Mech.: Theor. Exp.*, P07012 (2004).
21. M. E. Zhitomirsky and H. Tsunetsugu, *Phys. Rev. B* **70**, 100403(R) (2004).
22. O. Derzhko and J. Richter, *Phys. Rev. B* **70**, 104415 (2004).
23. M. E. Zhitomirsky and H. Tsunetsugu, *Prog. Theor. Phys. Suppl.* **160**, 361 (2005).
24. O. Derzhko and J. Richter, *Eur. Phys. J. B* **52**, 23 (2006).
25. A. Mielke, *J. Phys. A* **24**, L73 (1991); A. Mielke, *J. Phys. A* **24**, 3311 (1991); A. Mielke, *J. Phys. A* **25**, 4335 (1992); H. Tasaki, *Phys. Rev. Lett.* **69**, 1608 (1992); A. Mielke and H. Tasaki, *Commun. Math. Phys.* **158**, 341 (1993); H. Tasaki, *Prog. Theor. Phys.* **99**, 489 (1998).
26. A. Honecker and J. Richter, *Condensed Matter Physics (L'viv)* **8**, 813 (2005).
27. A. Honecker and J. Richter, *J. Magn. Magn. Mater.* **310**, 1331 (2007).
28. O. Derzhko, A. Honecker and J. Richter, *Phys. Rev. B* **76**, 220402(R) (2007).
29. Z. Gulácsi, A. Kampf and D. Vollhardt, *Phys. Rev. Lett.* **99**, 026404 (2007).
30. C. Wu, D. Bergman, L. Balents, and S. Das Sarma, *Phys. Rev. Lett.* **99**, 070401 (2007); D. L. Bergman and L. Balents, arXiv:0803.3628.
31. H.-J. Schmidt, *J. Phys. A* **35**, 6545 (2002).
32. J. Richter, O. Derzhko and J. Schulenburg, *Phys. Rev. Lett.* **93**, 107206 (2004).
33. J. Richter, J. Schulenburg, A. Honecker and D. Schmalfuß, *Phys. Rev. B* **70**, 174454 (2004).
34. J. Richter, J. Schulenburg, P. Tomczak and D. Schmalfuß, arXiv:cond-mat/0411673.
35. R. Schmidt, J. Richter and J. Schnack, *J. Magn. Magn. Mater.* **295**, 164 (2005).
36. O. Derzhko and J. Richter, *Phys. Rev. B* **72**, 094437 (2005).
37. J. Richter, *Fizika Nizkikh Temperatur (Kharkiv)* **31**, 918 (2005) [*Low Temperature Physics* **31**, 695 (2005)].
38. H.-J. Schmidt, J. Richter and R. Moessner, *J. Phys. A* **39**, 10673 (2006).

39. J. Richter, O. Derzhko and T. Krokhmalskii, *Phys. Rev. B* **74**, 144430 (2006); O. Derzhko, J. Richter and T. Krokhmalskii, *Acta Physica Polonica A* **113**, 433 (2008).
40. J. Schnack, H.-J. Schmidt, A. Honecker, J. Schulenburg and J. Richter, *J. Phys.: Conf. Ser.* **51**, 43 (2006).
41. O. Derzhko, J. Richter, A. Honecker and H.-J. Schmidt, *Fizika Nizkikh Temperatur (Kharkiv)* **33**, 982 (2007) [*Low Temperature Physics* **33**, 745 (2007)].
42. M. E. Zhitomirsky and H. Tsunetsugu, *Phys. Rev. B* **75**, 224416 (2007).
43. A. Honecker, J. Schulenburg and J. Richter, *J. Phys.: Condens. Matter* **16**, S749 (2004).
44. A. Honecker, F. Mila and M. Troyer, *Eur. Phys. J. B* **15**, 227 (2000).
45. F. Mila, *Eur. Phys. J. B* **6**, 201 (1998).
46. M. E. Fisher, *Phys. Rev.* **124,** 1664 (1961).
47. R. J. Baxter, *Exactly Solved Models in Statistical Mechanics* (Academic Press, London, 1982).
48. M. E. Zhitomirsky, *Phys. Rev. B* **67**, 104421 (2003).
49. J. Schnack, R. Schmidt and J. Richter, *Phys. Rev. B* **76**, 054413 (2007).
50. K. Penc, H. Shiba, F. Mila and T. Tsukagoshi, *Phys. Rev. B* **54**, 4056 (1996); H. Sakamoto and K. Kubo, *J. Phys. Soc. Jpn.* **65**, 3732 (1996); Y. Watanabe and S. Miyashita, *J. Phys. Soc. Jpn.* **66**, 2123; Y. Watanabe and S. Miyashita, *J. Phys. Soc. Jpn.* **66**, 3981 (1997).

MAGNETORESISTANCE IN COPPER

SHIGEJI FUJITA, NEBI DEMEZ and JEONG-HYUK KIM

Department of Physics, University at Buffalo, State University of New York
Buffalo, New York 14260, USA
fujita@buffalo.edu

H. C. HO

Physics Division, National Center for Theoretical Sciences
Hsinchu 30013, Taiwan
hcho@phys.cts.nthu.edu.tw

The motion of the guiding center of magnetic circulation generates a charge transport. By applying kinetic theory to the guiding center motion, an expression for the magnetoconductivity σ is obtained: $\sigma = e^2 n_c \tau / M^*$, where M^* is the magnetotransport mass distinct from the cyclotron mass, n_c the density of the conduction electrons, and τ the relaxation time. The density n_c depends on the magnetic field direction relative to copper's fcc lattice, when Cu's Fermi surface is nonspherical with "necks". The anisotropic magnetoresistance is analyzed based on a one-parameter model, and compared with experiments. A good fit is obtained.

Keywords: Magnetoresistance; Fermi surface; copper.

1. Introduction

If the Fermi surface is nonspherical, the magnetoresistance (MR) in general is anisotropic. Copper (Cu) has a so-called "open" orbits in the k-space as shown in Fig. 1(b).[1] This "open" orbit contains positive and negative curvatures along the energy contour, and hence, it cannot be traveled by any physical electron. An "electron" ("hole"), by definition, has an energy higher (lower) than the Fermi energy *and* circulates counterclockwise (clockwise) viewed from the tip of the applied magnetic field vector. The static magnetic field cannot supply energy, and hence, no physical electron can travel "electron"-like in one section of the contour and "hole"-like in another. Klauder and Kunzler[2] observed a striking angle-dependent MR as shown in Fig. 2. The MR is over 400 times the zero-field resistance in some directions. The purpose of the present work is to treat the anisotropic MR, using kinetic theory and the Fermi surface.

Traditionally, the electron transport has been treated, using kinetic theory or the Boltzmann equation method. In the presence of a static magnetic field, the classical electron orbit is curved. Then, the basic kinetic theoretical picture in which the

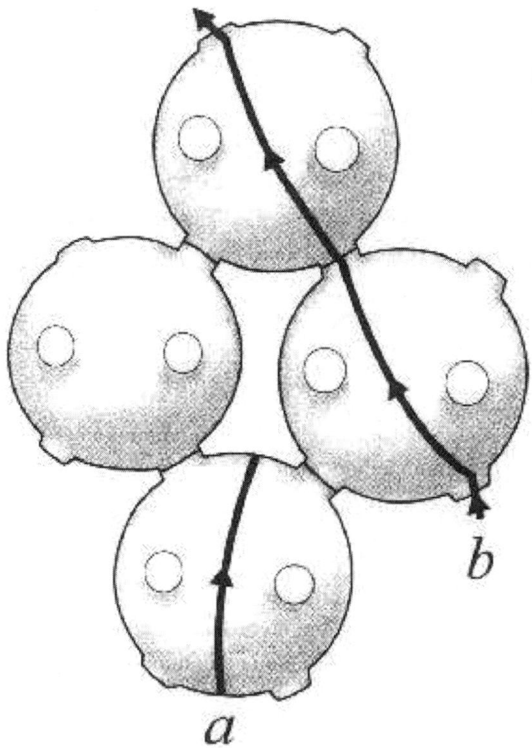

Fig. 1. (a) A closed orbit ***a*** in ***k***-space which can be traced by the electron and (b) an open orbit ***b*** which extends over two Brillouin zones and which cannot be traveled by the electron.

electron moves on a straight line, hits a scatterer (impurity), changes direction, and moves on another straight line, breaks down. Furthermore, the Boltzmann collision term containing the scattering cross section cannot be written down. Fortunately, quantum theory can save the situation. If the magnetic field is applied, the classical electron can continuously change from the straight line motion at zero field to the curved motion at a finite B. When the magnetic field is applied slowly, the energy of the electron does not change but the spiral motion always acts so as to reduce the magnetic field. Hence, the total energy of the electron with its surrounding fields is less than the electron energy plus the unperturbed field energy. The electron "dressed" with the fields is in a bound (negative energy) state, and it is stable against the break-up. The guiding center of the circulation can move in all directions in the absence of the electric field. If a weak electric field is applied in the x-direction, the dressed electron whose position is the guideing center, preferentially jumps in the x-direction, and generate a current. We can then apply kinetic theory to the guiding center motion, and obtain an expression for the electrical conductivity σ[3]:

$$\sigma = (e^2/M^*)n_c\tau, \tag{1}$$

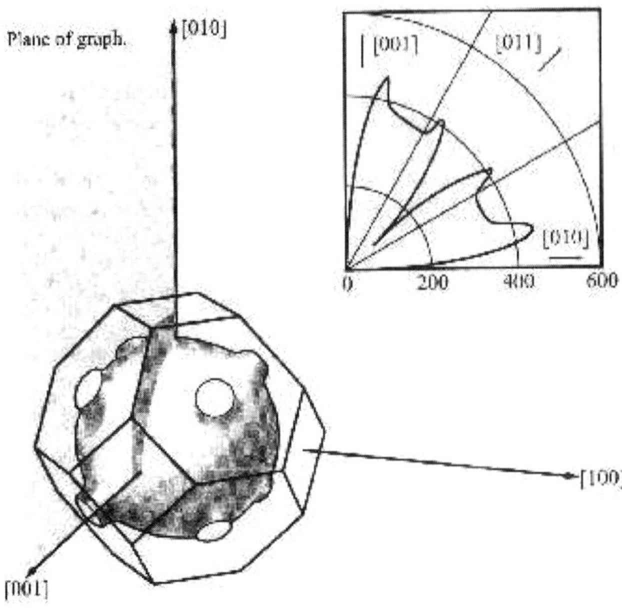

Fig. 2. The striking anisotropy of the MR in Cu, after Klauder and Kunzler.[2] The [001] and [010] directions of the copper crystal are shown, and the current flows in the [100] direction. The magnetic field is in the plane (100). Its magnitude is fixed at 18 kilogauss, and its direction varied continuously from [001] to [010].

where n_c is the density of the dressed electrons, e the charge, M^* the magnetotransport (effective) mass, and τ the relaxation time. The magnetotransport mass M^* is different from the cyclotron mass m^*. Equation (1) can also be obtained by the Boltzmann equation method as was shown earlier by Fujita *et al.*[3] In the Ref. 3, the dressed electron is identified as the composite fermion[4] used in the theory of quantum Hall effect.[5] Briefly, the electron circulates around a finite number of flux quanta (fluxons) intact according to Onsager's flux quantization hypothesis.[6] Applying relativity, we may view that the fluxons circulate around the electron. From this view, the electron is thought to carry a number of fluxons. The dressed electron is, then, a composite of an electron and fluxons. The composite particle moves as a fermion (boson) if it carries an even (odd) number of fluxons. The free-energy minimum consideration favors a population dominance of the c-fermions, each with two fluxons, over the c-bosons, each with one fluxon in the experimental condition at the liquid helium temperatures. The entropy is much higher for the c-fermions than for the c-bosons. The magnetic oscillation, which occurs only with fermionic carriers, is observed in Cu. This experimental feature also supports that carriers in the magnetotransport are c-fermions.

Pippard in his seminal book, *Magnetoresistance in Metal*,[7] argued that the MR for the quasifree electron system vanishes after using the relaxation time approximation in the Boltzmann equation method. But, the MR in actual experimental conduction is always finite. In fact, (1) contains the magnetotransport mass M^* distinct from the cyclotron mass m^*. This fact alone makes the MR nonzero.

The MR is defined by

$$\text{MR} \equiv \Delta\rho/\rho_0 = [\rho(B) - \rho_0]/\rho_0, \tag{2}$$

where $\rho(B)$ is the magnetoresistivity at the field magnitude B and $\rho_0 \equiv \rho(0)$, the resistivity at zero field.

We first consider the narrow limits of the "necks", called *singular points.* There are eight (8) singular points in total, as seen from Fig. 2. If the field **B** is in [001], then there is one plane perpendicular to [001] containing four singular points and another containing four points. The same condition also holds when the field **B** is in [010]. This condition corresponds to the major minimal MR seen in Fig. 2. Consider now the case where the field **B** is in [011]. There are three (3) planes perpendicular to [011] which contain (2,4,2) singular points. This case corresponds to the second minimum in $\Delta\rho$. The broad minima in the data correspond to the case where the field **B** is such that there are four (4) planes perpendicular to **B**, each containing two singular points. There is a range of angles in which this condition holds. Hence, these minima are broad. This singular-points model can explain the three minima. We propose a more realistic model in the following section.

2. Theory

We shall introduce the following theoretical model.

(i) We assume that the magnetoconductivity σ can be calculated based on (1). The effective mass M^* and the relaxation time τ are unlikely to depend on the direction of the field **B**. Only the conduction electron density n_c depends on the **B**-direction relative to the lattice.

(ii) We assume that each "neck" (bad points) is represented by a sphere of radius a centered at the eight points on the ideal Fermi surface. When the magnetic field **B** is applied, the electron, then, circulates perpendicular to **B** in the k-space. If it hits the bad points, then it cannot complete the orbit, and cannot contribute to the conduction.

In Fig. 2, we observe the following five principal features as the magnetic field **B** is rotated in the (100) plane from [001] to [011]. The MR has (a) a deep steep minimum, (b) a greatest maximum, (c) a broad minimum, (d) a second greatest maximum, and (e) a second steep minimum. These features are repeated in the reversed order as the field direction changes from [011] to [001]. We note that the three minima, (a), (c), and (e) were qualitatively explained based on the singular-points model. There are eight (8) bad spheres located at the (spherical) Fermi surface in the directions $\langle 111 \rangle$ from the center 0.

Let us first consider the case (c). In Fig. 3, the non-conducting (bad, shaded) k-space volumes viewed from [100] is shown. Each bad slice (volume) contains balls.

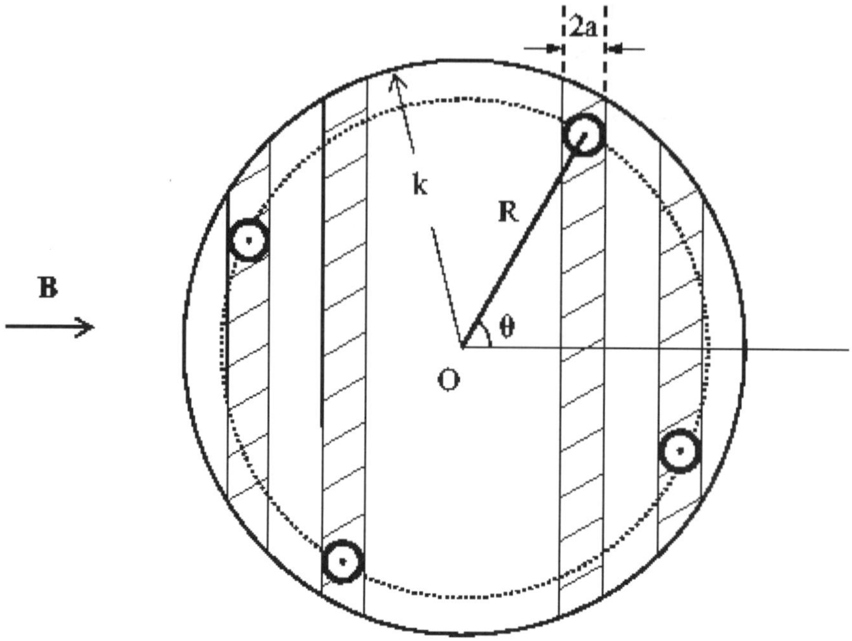

Fig. 3. The electrons circulate perpendicular to the magnetic field **B** in the k-space. If the electron hits the neck of radius a, it will not complete circulation, and it does not contribute to the conduction. The four shaded slices viewed from [100], each a width of $2a$, contain non-conducting electrons.

There are four bad slices here. All centers of the balls in the projected plane perpendicular to [001] lie on the circle of the radius R. This radius R is connected with the Fermi momentum $P_{\mathrm{F}} \equiv \hbar k$ by

$$R = \sqrt{2/3}k. \tag{3}$$

Each bad volume can be calculated, using the following integration formula:

$$\begin{aligned} I(x_0) &= \int_{x_0-a}^{x_0+a} dx\pi(k^2 - x^2) \\ &= \pi\left[2ak^2 - \frac{1}{3}(x_0+a)^3 + \frac{1}{3}(x_0-a)^3\right] \\ &= 2\pi ak^2 - \frac{2}{3}\pi(3x_0^2 a + a^3), \end{aligned} \tag{4}$$

where

$$x_0 = R\cos\theta, \tag{5}$$

is the x-coordinate of the ball center, and θ is the angle shown in Fig. 4. The four centers of the bad balls lie on the circle, separated by $\pi/2$ in θ. We consider the sum

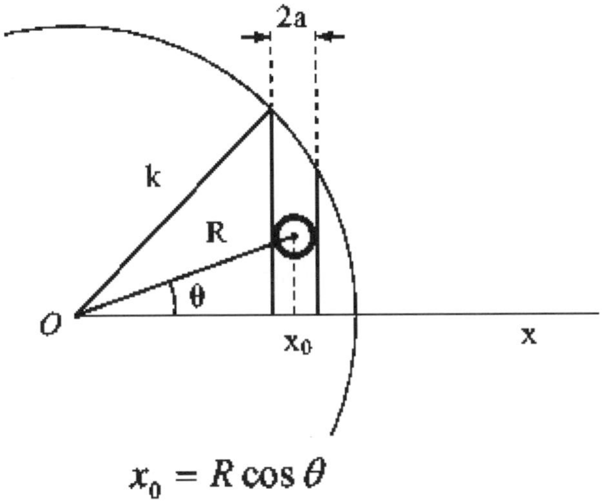

Fig. 4. The center of the bad ball lies on a circle of radius R. Its x-coordinate, x_0, is equal to $R\cos\theta$.

of a pair of the bad volumes associated with the centers at θ and $\theta + \pi/2$. Using (4), $\cos(\theta + \pi/2) = -\sin\theta$, and $\sin^2\theta + \cos^2\theta = 1$, we obtain

$$4\pi ak^2 - 4\pi a^3/3 - 2\pi aR^2. \tag{6}$$

The other pair with the centers at $\theta + \pi$ and $\theta + 3\pi/2$ contributes the same amount. Hence, the total bad volume is

$$V_{\mathrm{C}} = 16\pi ak^2/3 - 8\pi a^3/3, \tag{7}$$

where we used (3). This calculated volume does not depend on the angle θ, supporting the broad minimum (c) in the experiments.

We can calculate the bad volumes for other cases similarly, and obtain

$$V_{\mathrm{A}} = \frac{1}{2}V_{\mathrm{C}} \tag{8}$$

$$V_{\mathrm{E}} = \frac{3}{4}V_{\mathrm{C}}. \tag{9}$$

Going from (c) to (e), the four-slice volume monotonically changes to the three-slice volume, implying that the MR changes smoothly without passing a maximum. Similarly, the four-slice volume changes smoothly to the two-slice volume in going from (c) to (a). Thus, the present model generates no MR maxima. We shall give an explanation for the MR maxima in Sec. 3.

The conducting (good) volume is equal to the total ideal Fermi sphere volume $(4/3)\pi k^3$ minus the bad volume. The conduction electron density n_{c} is given by the ideal density n times the ratio of the good volume over the ideal Fermi sphere volume:

$$n_{\mathrm{c}} = n[(4/3)\pi k^3 - V]/(4/3)\pi k^3. \tag{10}$$

We call the inverse of the magnetoresistivity $\rho(B)$ the magnetoconductivity $\sigma(B)$. Rewriting the $\Delta\rho/\rho_0$ in terms of σ, we obtain

$$\Delta\rho/\rho_0 = \frac{1/\sigma - 1/\sigma_0}{1/\sigma_0} = \frac{\sigma_0}{\sigma} - 1, \tag{11}$$

where $\sigma_0 \equiv 1/\rho_0$ is the zero-field conductivity, which can be calculated through the standard formula:

$$\sigma_0 = \frac{e^2}{m^*} n_0 \tau_0, \tag{12}$$

where m^* is the cyclotron mass, and the suffix 0 denotes the zero-field quantities.

In the experiments, the MR is very large (~ 100) compared with unity. We may omit the -1 in (11). Using (1) and (12), we then obtain

$$\frac{\Delta\rho}{\rho_0} = \frac{M^*}{m^*} \frac{\tau_0}{\tau} \frac{n_0}{n_c}. \tag{13}$$

This expression indicates that the lower the magnetoconduction electron density n_c, the higher is the MR.

3. Discussion

We observe a clear MR maximum near [011] in Fig. 2. Our model calculations in Sec. 2 show a monotonic change in going from the four-slice configuration (c) to the three-slice configuration (e). We propose the following explanation. In going from (c) to (e), an overlap of the bad volumes must occur. This overlap does not occur line-sharp, as assumed in our model. Then, the fluctuations which must occur, generate a dissipation. This causes a resistivity maximum in (d). In going from the four-slice configuration (c) to the two-slice configuration (a), an overlap must also occur too. Since two overlaps occur simultaneously in this case, the fluctuation and the resistance should be greater, generating a MR maximum higher in (b) than in (d).

We see in Fig. 2 that the MR minimum changes from 80 to 340 in going from (e) to (c). This reflects the change in the conduction electron density n_c. We may use these observed values to estimate the ratio of the bad ball diameter a over the Fermi momentum k. Using (7), (9), (10) and (13), we obtain

$$\frac{4\pi k^3/3 - V_C}{4\pi k^3/3 - 3V_C/4} = \frac{340}{80}, \tag{14}$$

and V_C is given in (7). Solving (14) for the ratio a/k, we obtain

$$\frac{a}{k} = 0.40, \tag{15}$$

which is resonable.

We see in Fig. 2 that the MR rises quadratically with the field angle away from the minimum at (e) [and also at (a)]. Our model explains this behavior as follows.

The center of one of the balls lies at $\theta = \pi/2$ for the case (e). We introduce a small angle ϕ:

$$\phi = \theta - \pi/2. \tag{16}$$

Then, we have

$$x_0 = R\cos(\phi + \pi/2) = R\sin\phi \approx R\phi. \tag{17}$$

Using this and (4), we calculate the bad volume near (e), and obtain

$$V = \frac{3}{4}V_c + 2\pi a R^2 \phi^2. \tag{18}$$

Thus, the MR rises quadratically in the deviation angle ϕ, as observed in the experiment.

Our model in Sec. 2 contains a spherical ball. We may consider an ellipsoidal (two parameters) model for the improvement.

Other noble metals such as silver and gold are known to have the Fermi surface with "necks". If the bad balls are greater in relative size (a/k), our theory predicts more prominent MR. Experimental check of this behavior is highly desirable. We suggest that the experiments be done below 1 K, where the phonon scattering is negligible and the MR minima become more visible. Only the minima, and not the maxima, contain important information about the Fermi surface.

Acknowledgments

One of the authors, SF, thankfully acknowledge a travel support from the US Army Research Office.

References

1. See, e.g., N.W. Ashcroft and N.D. Mermin, *Solid State Physics* (Saunders, Philadelphia, 1976), pp. 291–293.
2. J.R Klauder and J.E. Kunzler, in *The Fermi Surface*, eds. Harrison and Webb (Wiley, New York, 1960).
3. S. Fujita, S. Horie, A. Suzuki and D.L. Morabito, *Ind. J. P.A.P.* **44**, 850 (2006); S. Fujita, K. Ito, Y. Kumek and Y. Okamura, *Phys. Rev. B* **70**, 075304 (2004).
4. S.C. Zhang, T.H. Hansson and S. Kivelson, *Phys. Rev. Lett.* **62**, 82 (1989); J.K. Jain, *Phys. Rev. Lett.* **63**, 199 (1989); *Phys. Rev. B* **40**, 8079 (1989); *ibid.* **41**, 7653 (1990).
5. Z.F. Ezawa, *Quantum Hall Effect* (World Scientific, Singapore, 2000); see also eds. R.E. Prange and S.M. Girvin, *Quantum Hall Effect* (Springer-Verlag, New York, 1990).
6. L. Onsager, *Phil. Mag.* **43**, 1006 (1952).
7. A.B. Pippard, *Magnetoresistance in Metals* (Cambridge University Press, Cambridge, UK, 1989), pp. 3–5.

CALCULATION AND INTERPRETATION OF SURFACE-PLASMON-POLARITON FEATURES IN THE REFLECTIVITY OF METALLIC NANOWIRE ARRAYS

PATRICK SCHOLZ,[*] STEPHAN SCHWIEGER,[†] PARINDA VASA,[*,†] and ERICH RUNGE[*]

[*] *Institut für Physik, Technische Universität Ilmenau, 98693 Ilmenau, Germany*

[†] *Institut für Physik, Carl von Ossietzky Universität Oldenburg, 26111 Oldenburg, Germany*

Received 31 July 2008

The far-field reflectivity of metallic nanowire arrays designed to show strong surface-plasmon-polariton (SPP) resonances is studied numerically. The results of calculations in time and frequency space as well as the results of semi-analytic theories using different approximative boundary conditions at the metal surfaces are evaluated and compared. Good agreement between all different methods is obtained in most cases. The SPP-related features are superimposed on a strongly varying background. Combining FDTD simulations, finite element results, and semi-analytical calculations, the microscopic origin of the background contribution is identified. Resonant transmission through sub-wavelength slits leads to pronounced oscillations in the far-field reflectivity as a function of the height of the nanowires.

Keywords: Surface plasmon polaritons; nanowires; Maxwell solvers.

1. Introduction

Surface plasmon polaritons (SPPs) are optical excitations at metal-dielectric interfaces. They are confined to sub-wavelength dimensions and can be guided and manipulated by nano-structuring the interface. SPP-based components have the potential to play a major role for future optical devices. In order to compute SPP properties, one needs reliable, stable and fast numerical methods for solving Maxwell's equations for a given structure and given boundary conditions such as directions, frequencies and amplitudes of the incoming waves. There are two fundamental approaches on how to solve Maxwell's equations: On the one hand, the time-dependent differential equations can be solved directly. This is done within the finite-difference time-domain (FDTD) technique.[1] On the other hand, for the case of external excitation with a harmonic time dependence, Maxwell's equations can be solved in frequency space. Here, often non-rectangular meshes are chosen as done by finite element methods.[2] In addition to these general approaches, semi-analytic approximative theories are often used to investigate SPP properties.[3–7] These methods are limited to special geometries, materials or spectral ranges, but can give additional insight into the microscopic mechanisms of SPP excitation or in the coupling of

SPPs to light[8] or other excitations.[3] The combined use of these conceptually different methods for the investigation of SPPs promises a more complete physical picture, since they all have different strengths and limitations. In order to explore these, we compare in the present work the results of the different methods for a simple test geometry. In particular, we calculate the electromagnetic fields and the reflectivity of an array of metallic nanowires. Such metallic nanowire arrays serve as prototype systems for SPP excitation and show an interesting interplay between far-field radiation, SPP modes, and cavity-like modes in the slits between the wires. Furthermore, this geometry is well suited for fundamental experiments on SPPs.[8,9]

In the next section, we briefly introduce the different theoretical methods. Then, we discuss the physics of our prototype system and compare the results of the different approaches. Based on these findings, we will discuss the applicability as well as the advantages and drawbacks of the different methods.

2. Theories

We will compare and evaluate four different methods. Two of them are general-purpose Maxwell solvers that work in time and frequency space, respectively. Further, two semi-analytic formulations are evaluated that work also in frequency space but use additional approximations. For the time-domain calculations we use the program "FDTD solutions" of Lumerical Solutions, Inc.[10] It allows to calculate the electromagnetic field as a function of time for arbitrary structures and for various beam profiles and source geometries. Fourier transformation allows to calculate frequency-dependent quantities, such as the far-field reflectivity as a function of photon energy. We compute the reflectivity of the structure when illuminated by a long Gaussian pulse (70 fs). The FDTD method cannot deal with general frequency-dependent dielectric functions as needed for metals or excitonic resonances in semiconductors. However, frequency-dependent dielectric functions can often be fitted by, e.g., Lorentzians, which can be accounted for with a FDTD formulation.[1] In our case, Lorentzians allow for a very accurate approximative description of the gold dielectric function in the spectral range of interest.[11] The calculations in frequency space are performed using the finite element formulation provided by the COMSOL RF module.[12] Again, the field distribution of arbitrary structures can be calculated for given boundary conditions. We calculate the far-field reflectivity of the nanowire array for incident plane waves. Obviously, calculations in frequency space can handle arbitrary dielectric functions $\epsilon(\omega)$. This flexibility will be relevant when, e.g., absorbing materials with excitonic as well as band-state absorption are studied.[3]

The considered semi-analytic methods also use a time-harmonic ansatz $\vec{E}$, $\vec{H} \sim e^{-i\omega t}$ for the fields in order to separate the time dependence from the spatial dependence.[4–6] They are designed especially for the considered geometry, i.e., a grating formed by a periodic arrangement of metallic nanowires $[\epsilon = \epsilon_m(\omega)]$ with rectangular cross-section as shown in Fig. 1. The grating period, the slit width, and the wire height are termed a, c, and h, respectively. In the half-spaces above and be-

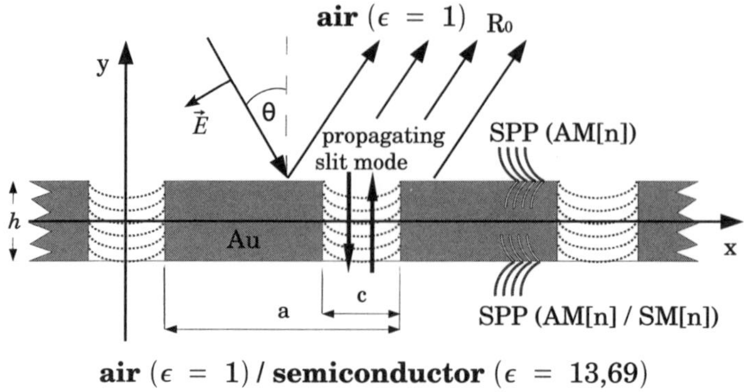

Fig. 1. Sketch of the considered grating structure consisting of a periodic arrangement of gold nanowires, an upper half-space (air with $\epsilon = 1$) and a lower half-space (air or semiconductor with $\epsilon = 13.69$). The grating period, the slit width, and the wire height are termed a, c, and h, respectively. The whole system is illuminated from above by TM-polarized light incident in the x, y-plane under an angle of incidence θ. R_0 is the far-field reflectivity of the $0th$ order. A single propagating slit mode connects the upper and the lower half-space. Surface plasmon polaritons (SPPs) can propagate at an air-metal (AM[n]) or a semiconductor-metal (SM[n]) interface.

low the grating and in the slits are dielectric materials, which are characterized by their dielectric constants ϵ_{abv}, ϵ_{bel}, and ϵ_{slit}, respectively. Here, we restrict ourselves to structures with air above the grating and in the slits ($\epsilon_{abv} = \epsilon_{slit} = 1$). The lower half-space is filled either with air or a semiconductor characterized by the real dielectric constant $\epsilon_{bel} = 13.69$, which is that of GaAs well above the gap.[13] We refer to this geometries as free-standing and semiconductor-supported, respectively. We neglect the semiconductor absorption, because the fields above the grating are only weakly influenced by it. The frequency-dependent gold dielectric function $\epsilon_m(\omega) = \epsilon_{Au}(\omega)$ is taken from Ref. 14. In the following, the y-axis is defined as the film normal, while the x-axis is parallel to the film and perpendicular to the nanowires. The origin is located in the center of a slit (see Fig. 1). A plane wave is incident from above in the x, y-plane under an angle of incidence θ measured against the surface normal. We consider linearly polarized light with the electric field vector in the plane of incidence. We expand the magnetic field ($\| z$) in plane waves

$$H_{abv/bel}(x, y) = \sum_{n=-\infty}^{\infty} \left(A_{n+}^{(\ell)} e^{-\kappa_n^{(\ell)}(y \mp \frac{h}{2})} + A_{n-}^{(\ell)} e^{\kappa_n^{(\ell)}(y \pm \frac{h}{2})} \right) e^{ik_n x} \tag{1}$$

with ℓ standing for "*abv*" or "*bel*" and the upper (lower) sign applying to the fields above (below) the grating. Further,

$$k_n = \frac{\omega}{c_0} \sin\theta + \frac{2\pi}{a} n \text{ and } i\kappa_n^{(\ell)} = \sqrt{\frac{\omega^2}{c_0^2} \epsilon_\ell - k_n^2} \tag{2}$$

are the x and y components of the wave vector, with c_0 being the vacuum velocity

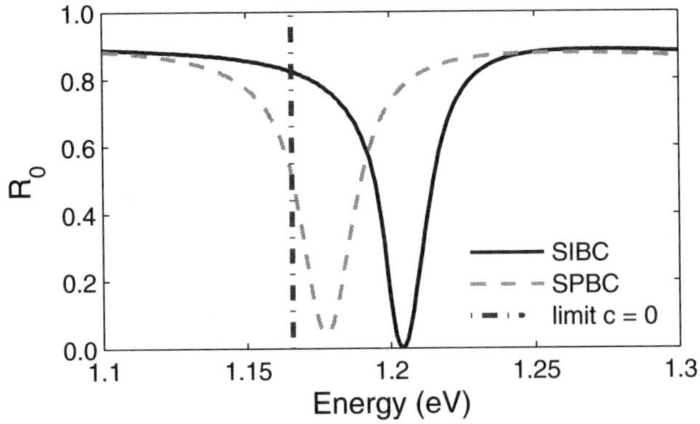

Fig. 2. (Color online) Far-field reflectivity (0*th* order) as function of excitation energy for the semiconductor-supported nanowire array at normal incidence ($\theta = 0°$). Grating parameters: $a =$ 500 nm, $c =$ 10 nm, and $h =$ 80 nm. Comparison of SIBC (solid line), SPBC (dashed line), and analytically obtained SM[-2] resonance position for vanishing slit width (dash-dotted line).

of light. The field in the slits is expanded in slit eigenmodes

$$H_{slit}(x,y) = \sum_{\mu=0}^{\infty} \left(\alpha_{\mu+} e^{-\kappa_\mu^{(s)}(y+\frac{h}{2})} + \alpha_{\mu-} e^{\kappa_\mu^{(s)}(y-\frac{h}{2})} \right) u_\mu(x) \tag{3}$$

with $u_\mu(x) \sim \sin(k_\mu x)$ and $u_\mu(x) \sim \cos(k_\mu x)$ for odd and even slit modes, respectively. The fields $H_{abv/bel}$ and H_{slit} are subject to boundary conditions at the metal-dielectric interface. In Refs. 4–6, the surface impedance boundary condition (SIBC) is used. It reads for the considered geometry

$$\frac{\partial H^{(\ell)}}{\partial \hat{n}} = \eta_\ell \, H^{(\ell)}, \;\; \eta_\ell^{SIBC} = -i\frac{\omega}{c_0}\frac{\epsilon_\ell}{\sqrt{\epsilon_m}}. \tag{4}$$

Here, $\hat{n}$ is a unit vector normal to the interface and pointing outside the metal and ℓ stands for "*abv*", "*bel*" or "*slit*." The SIBC follows from continuity conditions at the metal-dielectric interface and the following fact[15]: For all angles of incidence the attenuated traveling wave inside the metal propagates almost perpendicular to the interface, with $k_{m\perp} \approx (\omega/c_0)\epsilon_m^{1/2}$. This is a consequence of the large norm of the metallic permittivity $|\epsilon_\ell/\epsilon_m| \ll 1$. However, when an SPP is excited the energy is confined to the interface and the penetration depth $i/k_{m\perp}$ is reduced, because $k_{m\perp} \approx (\omega/c_0)\epsilon_m/(\epsilon_m + \epsilon_\ell)^{1/2}$ (see Ref. 16). Therefore, the modified boundary condition

$$\eta_\ell^{SPBC} = -i\,\frac{\omega}{c_0}\frac{\epsilon_\ell}{\sqrt{\epsilon_\ell + \epsilon_m}} \tag{5}$$

was suggested,[3,7] that will be referred to as surface plasmon boundary condition (SPBC) in the following. *A priori*, one expects the SIBC to give better results when no SPP is excited and the SPBC to be superior for the description of SPPs. SIBC and SPBC differ only slightly for small $|\epsilon_\ell/\epsilon_m|$, relevant, e.g., for air-metal

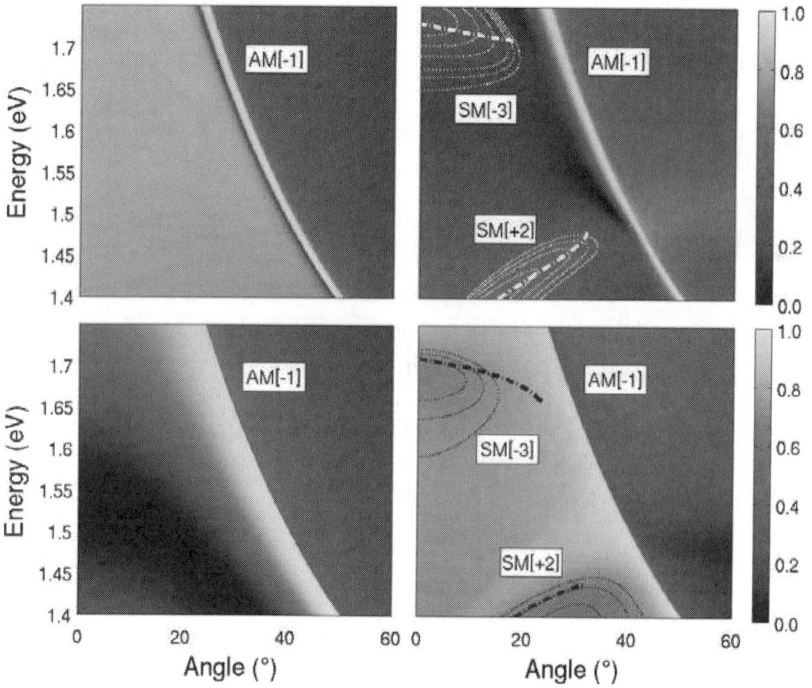

Fig. 3. (Color online) Far-field reflectivity R_0 (color code) as function of excitation energy and angle of incidence θ. Grating parameters: $a = 500$ nm, $c = 140$ nm. Left column: free-standing, right column: semiconductor-supported system, top row: $h = 80$ nm, bottom row: $h = 250$ nm. The structure is illuminated by a plane wave from above. The labeled features are caused by SPPs excited at the air-metal interfaces (AM[n], left) or the air-metal and semiconductor-metal interfaces (AM[n], SM[n], right). The dash-dotted lines mark the resonance positions of the SM[+2] and SM[-3] modes defined by the intensity maximum at the lower grating interface (not shown). The data are calculated using the surface impedance boundary condition (SIBC), see text.

interfaces. Stronger deviation are expected for semiconductor-metal interfaces with $|\epsilon_{sc}/\epsilon_m| \simeq 1/2$.

Using one of these boundary conditions and matching the fields [Eq. (1) and Eq. (3)] at $y = \pm h/2$, all fields and quantities such as the far-field reflectivity or the efficiency of SPP excitation can be calculated. Details are given in Ref. 3.

Figure 2 compares the far-field reflectivity as a function of excitation energy at normal incidence and very narrow slits ($c = 10$ nm) for the two boundary conditions. The minima are caused by excitation of SPPs at the semiconductor-metal interface. For small slit width, the minima are near the well known SPP resonances of a structureless interface[16]

$$E_{spp}^{(n)} = n\frac{c_0}{\hbar}\frac{2\pi}{a}\sqrt{\frac{\epsilon_m + \epsilon_\ell}{\epsilon_m \epsilon_\ell}} \tag{6}$$

($\theta = 0°$, integer n). The SPBC matches the analytic expression better.

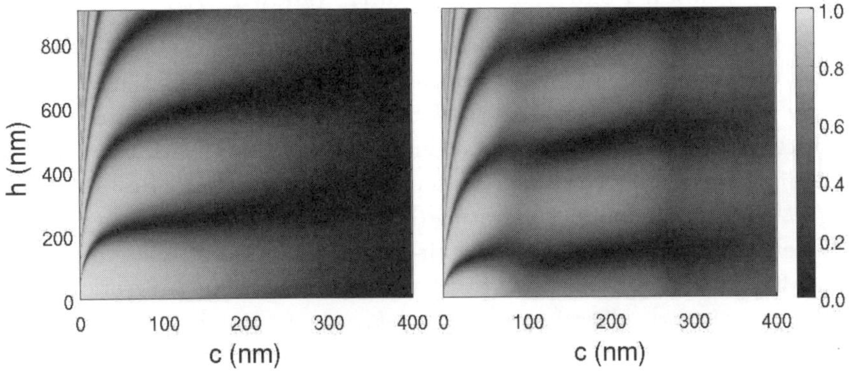

Fig. 4. (Color online) Far-field reflectivity R_0 (color code) of a free-standing (left) and semiconductor-supported (right) nanowire array as a function of wire height h and slit width c for normal incidence ($\theta = 0°$) calculated with the analytical theory using the SIBC. At fixed slit width, R_0 shows oscillations with h. The period of the oscillations is given by the wavelength of the slit mode that propagates perpendicular to the grating in y-direction. Parameters: $a = 500$ nm, excitation wavelength $\lambda_0 = 805$ nm.

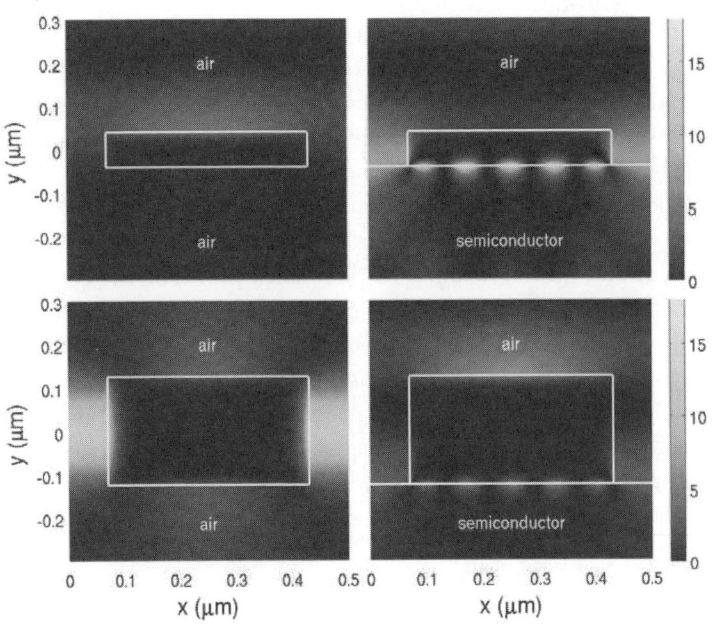

Fig. 5. (Color online) H-field intensity (color code) for the parameters of Fig. 4 and $c = 140$ nm calculated by the time-harmonic finite element method (COMSOL). Top row: $h = 80$ nm, bottom row: $h = 250$ nm. The H-field intensity in the slits is clearly enhanced for the upper right and the lower left panel due to excitation of a slit mode resonance. This leads to large transmission and small reflectivity (see Figs. 3 and 4).

3. Surface Plasmon Polaritons in Metallic Nanowire Arrays

Before we compare the two general-purpose Maxwell solvers (FDTD and COMSOL), we clarify the physical origin of the features to be reproduced by the different calculational methods. Figure 3 shows the far-field reflectivity of 0th order (R_0) as a function of photon energy and angle of incidence calculated with the SIBC for a free-standing grating and a grating on a semiconductor substrate. Results for a thin grating ($h = 80$ nm) and a thicker grating ($h = 250$ nm) are shown. The labeled features are caused by SPPs excited at the air-metal (AM[n]) or at the semiconductor-metal (SM[n]) interface. Due to the translational symmetry in x-direction, an SPP band structure is formed with wave vector $k_{spp} = (\omega/c_0)\sin\theta + 2\pi n/a$ and integer band index n. The angle dependence in Fig. 3 results from the SPP band structure. In all four panels, a very steep and narrow structure can be seen that corresponds to the AM[-1] mode. Additionally, weak, broad and flatter structures are seen for the semiconductor-supported system. They correspond to SPPs excited at the semiconductor-metal interface. Their flatness and large broadening are caused by the large semiconductor refractive index. The SPP resonance positions depend strongly on the angle of incidence and the grating period (not shown), but only weakly on the height of the nanowires.

In order to understand the reflectivity features related to the SPP resonance quantitatively, one has to have a closer look at the background reflectivity that is influenced by the non-trivial spatial structure of the dielectric function. In contrast to the SPP resonance position, the background reflectivity shows an interesting dependence on the height and width of the nanowires (i.e., on the grating thickness and the slit width). For angles θ less than the SPP resonance angle θ_{spp}, we find, e.g., the following: The reflectivity of the free-standing grating is very high for an 80 nm thick film and quite low for a 250 nm thick film. For the supported system this behavior is reversed. The wire-shape dependence of the background reflectivity is shown in more detail in Fig. 4. As expected, reflectivity decreases in general with increasing slit width. However, the oscillatory behavior as a function of wire height may come as a surprise. The oscillation period can be understood using the analytic theories which clearly show that the upper and the lower half-space are connected by a *single* propagating slit mode, only. All other slit modes decay exponentially in y-direction. The wavelength of the propagating mode

$$\lambda_y^{slit} = \frac{2\pi}{\sqrt{\frac{\omega^2}{c_0^2}\epsilon_{slit} + \kappa_{slit}^2}} \tag{7}$$

is determined by the SIBC or SPBC

$$\tanh\left(\frac{c\kappa_{slit}}{\pi}\right) = \frac{\eta_s}{\kappa_{slit}}. \tag{8}$$

The oscillation period in $R_0(h)$ is to a very good approximation $\lambda_y^{slit}/2$. The c-dependence of the reflection minima positions in Fig. 4 is caused by the slit width dependence of λ_y^{slit}. Reflection minima are found when a standing slit mode is

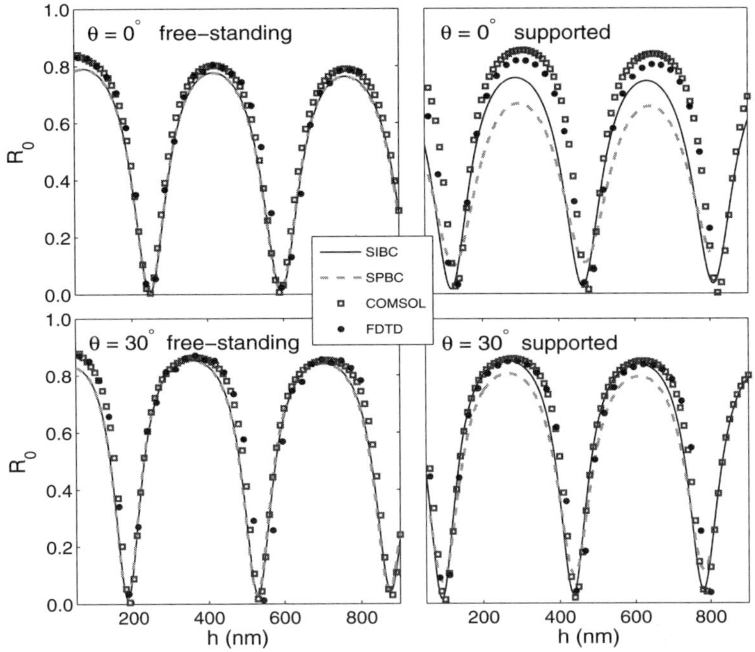

Fig. 6. (Color online) Far-field reflectivity R_0 as a function of wire height at $a = 500$ nm and $c = 140$ nm for a free-standing (left column) and a semiconductor-supported (right column) nanowire array for different angles of incidence (upper row: 0°, lower row: 30°). Comparison of analytical theories (SIBC: solid line, SPBC: dashed line), COMSOL (squares), and FDTD (circles). The excitation wavelength is $\lambda_0 = 805$ nm. All methods agree regarding the oscillation periods and phases. Note, that the oscillation in the semiconductor-supported system is shifted by almost half a period compared to the free-standing system.

excited as demonstrated in Fig. 5, where the (magnetic) field intensity is shown for the four geometries. A large field intensity within the slits is found for parameters corresponding to reflection minima and increased transmission (not shown). The time-dependent FDTD calculations show that these modes decay very slowly. The wavelength λ_y^{slit} of the propagating slit mode and consequently the vertical distance of the minima in Fig. 4 do not depend on the materials above and below the wire array. However, different matching conditions at the latter interface lead to a phase shift of almost half a period between the cases $\epsilon_{bel} = 1$ and $\epsilon_{bel} = 13.69$ (see Figs. 4, 5 and 6). That is why the first resonant transmission is found at 250 nm for the free-standing grating and at 80 nm for the semiconductor-supported system.

4. Comparison of Methods

Having identified the origin of the dominant features in the far-field reflectivity, we are ready to compare the four different methods. Figure 6 shows the far-field reflectivity as a function of metal film thickness h as calculated by FDTD, COMSOL, SIBC and SPBC for two different angles of incidence and two different geometries.

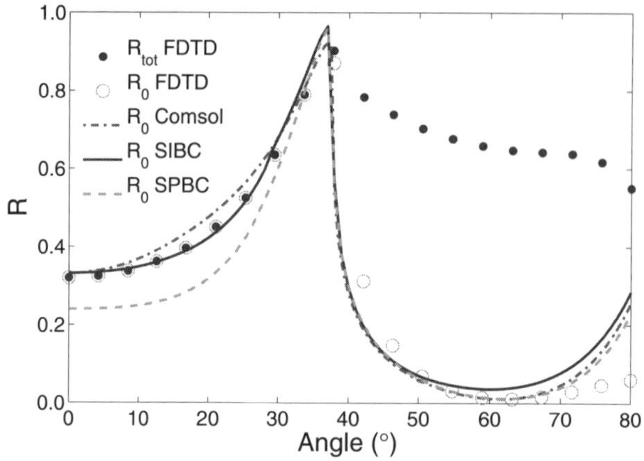

Fig. 7. (Color online) Far-field reflectivity as a function of angle of incidence for a semiconductor-supported nanowire array with $a = 500$ nm, $c = 140$ nm, $h = 162$ nm at excitation wavelength $\lambda_0 = 805$ nm. Dots denote the reflectivity summed over all orders (R_{tot}) calculated by FDTD, open circles are the reflectivity of the $0th$ order only (FDTD). Lines give the $0th$ order reflectivity as calculated by COMSOL (dash-dotted), semi-analytically using the SIBC (solid), and semi-analytically using the SPBC (dashed). SPPs in the AM[-1]-mode are excited between 30° and 40°. For $\theta \gtrsim 37°$ two propagating orders are reflected ($0th$, $-1st$ order), for $\theta \lesssim 37°$ only the $0th$ order is propagating.

The oscillations in $R_0(h)$ resulting from resonant transmission through the slits are clearly to be seen. For the free standing structure (left column), the results agree almost perfectly for $\theta = 0°$, while there are small deviations for $\theta = 30°$. As discussed above, the results for SIBC and SPBC are nearly identical. For the semiconductor-supported array (right column), the deviations between semi-analytical theories and the general-purpose Maxwell solvers are stronger, especially for normal incidence. Indeed, it can be shown that the approximate boundary conditions work best for large ratios between the metal and the dielectric refractive index. There are also slight differences between COMSOL and FDTD at the oscillation maxima. SIBC and SPBC differ clearly, with the SIBC giving much better results. All theories agree almost perfectly with respect to the oscillation period and phase.

Figure 7 compares the angle dependence of the total reflectivity and the reflectivity of the 0th order. The fingerprint of the AM[-1] SPP mode is the pronounced reflectivity maximum at $\theta \approx 36.6°$. Above the resonance the specular reflection R_0 drops rapidly before it increases again at large angles. The total reflectivity R_{tot} is much higher, because scattering into propagating waves with larger κ_n becomes possible. The broad minimum of R_0 near $\theta = 60°$ is well reproduced by all approaches. However, we run into difficulties with Lovalite's FDTD. Their origin is up to now unknown to us, but it is probably related to the fact that for large angles of incidence and periodic boundaries, the incident and reflected beam cross these boundaries very often and even small errors accumulate.

5. Conclusion

The ability of four different approaches to compute far-field reflectivity of a metallic nanowire array was tested. We found SPP excitation and resonant transmission through the sub-wavelength nano slits. Resonant transmission results into an oscillatory behavior of reflectivity as a function of metal film thickness. The results of all theories are in fairly good agreement, especially for free-standing systems. The advantages of FDTD and COMSOL are that no approximations are used to solve Maxwell's equations and that more complicated geometries can be treated. The advantage of the semi-analytic theories is that they give fast results even for large scans over incident wave length, angle of incidence, or structural parameters. Additionally, they can give detailed analytical insight into the physics of the system, because they are formulated with explicitly physical concepts such as the wavelength of the propagating slit mode λ_y^{slit}.

Comparing the two different boundary conditions for the semi-analytic theories, one notices that the SIBC performs better in most cases. The SPBC gives slightly more accurate results when simulating SPPs excited at an interface between a metal and a dielectric with a very high refractive index.

Acknowledgments

The authors thank C. Lienau and D. Zerulla for stimulating discussions. P. Vasa thanks the Alexander von Humboldt Foundation for financial support.

References

1. A. Taflove, S. C. Hagness, *Computational Electrodynamics: the Finite Difference Time-Domain Method*, 2nd ed. (Artech House, Norwood, 2000).
2. J. Jin, *The Finite Element Method in Electrodynamics* (Wiley, New York, 1993).
3. S. Schwieger, P. Vasa, and E. Runge, *Phys. Stat. Sol. B*, doi:10.1002/pssb.200777608.
4. H. Lochbihler and R. A. Depine, *Appl. Optics* **19**, 3459 (1993).
5. H. Lochbihler, *Phys. Rev. B* **50**, 4795 (1994).
6. H. Lochbihler, *Phys. Rev. B* **53**, 10289 (1996).
7. K. G. Lee and Q. H. Park, *Phys. Rev. Lett.* **95**, 103902 (2005).
8. C. Ropers, D. J. Park, G. Stibenz, G. Steinmeyer, J. Kim, D. S. Kim and C. Lienau, *Phys. Rev. Lett.* **94**, 113901 (2005).
9. S. Rehwald, M. Berndt, F. Katzenberg, S. Schwieger, E. Runge, K. Schierbaum and D. Zerulla, *Phys. Rev. B.* **76**, 085420 (2007).
10. FDTD Solutions, Release 5.1, Lumerical Solution, Inc., Vancouver, BC, Canada.
11. A. Vial, A.-S. Grimault, D. Macias, D. Barchiesi and M. L. de la Capelle, *Phys. Rev. B.* **71**, 085416 (2005).
12. COMSOL 3.3, RF Module Reference guide (2006).
13. D. E. Aspness, in *Properties of Gallium Arsenide*, 2nd ed., (The Institution of Electrical Engineers, London, 1990).
14. P. B. Johnson and R. W. Christy, *Phys. Rev. B* **6**, 4370 (1972).
15. L. D. Landau and E. M. Lifshitz, *Electrodynamics of Continuous Media*, 2nd ed. (Pergamon Press, Oxford, 1984), p. 300.
16. H. Raether, *Surface Plasmons* (Springer Verlag, Berlin, 1988), p. 5.

THE SURPRISING PHENOMENON OF LEVEL MERGING IN FINITE FERMI SYSTEMS

JOHN W. CLARK*

Institut für Theoretishe Physik, Johannes Kepler University
A-4040 Linz A-4040, Austria
jwc@wustl.edu

VICTOR A. KHODEL†

Russian Research Centre Kurchatov Institute
Moscow 123182, Russia
khodel@pretty.mbslab.kiae.ru

HAOCHEN LI‡

Department of Physics, Washington University
St. Louis, MO 63130 USA
haochenli@gmail.com

MIKHAIL V. ZVEREV

Russian Research Centre Kurchatov Institute
Moscow 123182, Russia
zverev@pretty.mbslab.kiae.ru

Received 31 July 2008

When applied to a finite Fermi system having a degenerate single-particle spectrum, the Landau-Migdal Fermi-liquid approach leaves room for the possibility that different single-particle energy levels merge with one another. It will be argued that the opportunity for this behavior exists over a wide range of strongly interacting quantum many-body systems. An inherent feature of the mergence phenomenon is the presence of nonintegral quasiparticle occupation numbers, which implies a radical modification of the standard quasiparticle picture. Consequences of this alteration are surveyed for nuclear, atomic, and solid-state systems.

Keywords: Finite Fermi systems; quasiparticle picture; non-Fermi-liquid behavior; Landau theory.

*Permanent address: Department of Physics & McDonnell Center for the Space Sciences, Washington University, St. Louis, M0 63130 USA

†Secondary affiliations: Department of Physics & McDonnell Center for the Space Sciences, Washington University, St. Louis, MO 63130 USA

‡Current address: Norris Cotton Cancer Center, Dartmouth Medical School, Lebanon, NH 03756 USA

1. Introduction

Lev Landau's Fermi Liquid (FL) theory, introduced in the mid-1950s,[1] is one of the foundation stones of quantum many-body theory, and of our understanding of the thermodynamic and transport properties of macroscopic condensed-matter systems at low temperature. In its essence, FL theory is a description of a system of N interacting fermions in terms of an equal number of distinct, whole, long-lived, weakly interacting quasiparticles. Adapted to finite Fermi systems in the mid-1960s by Arkady Migdal and his associates,[4] who focused on nuclei, the FL theory has provided a deeper understanding of both nuclear and atomic shell models. We point out two salient implications of the Fermi Liquid theory of finite Fermi systems in nuclear and atomic physics:

(i) The total angular momenta of the ground states of odd-A nuclei must be carried by a single quasiparticle.

(ii) "The electronic configurations of ions of elements of the periodic table must repeat those of preceding atoms"[2] (rule for filling electronic orbitals of atoms).

Yet the standard Landau FL theory is known to fail under certain circumstances. In condensed-matter systems of macroscopic dimensions, diverse non-Fermi-liquid (NFL) behavior is manifested near a putative quantum critical point in high-T_c ceramic compounds, heavy-fermion metals, the 2D electron gas, and liquid ^{3}He films. The source of this NFL behavior is a subject of heated debate in condensed-matter physics.

What about *finite* Fermi systems?

Although finite Fermi systems are not subject to the damping processes sometimes identified as a source of non-Fermi-liquid behavior in macroscopic systems,[3] they *also* exhibit violations of standard FL theory. We may cite two prominent examples:

(i) In nuclear physics, the total angular momenta of the ground states of many odd-A nuclides in the transition region cannot be attributed to just one single-particle (sp) state.

(ii) In atomic physics, the electronic configurations of elements that do not belong to the principal groups of the periodic table depart from those expected from FL theory.

In this paper, we will demonstrate a particular mechanism by which NFL behavior may emerge in finite Fermi systems. In Sec. 2, we frame the general argument in concrete terms, by considering nuclei with closed shells for which the sp spectrum is degenerate with respect to the magnetic quantum number m. We show that as particles enter the open shell, NFL behavior can arise *within the Landau approach*[1] *itself*, due to the *merging* of a sp level being filled with the nearest empty sp level. This phenomenon stems from the fact that the sp energies experience shifts when the occupation of the lowest unfilled sp level is changed, because of the interaction between the quasiparticles. We find that the energetic distance between the sp level being filled and the nearest empty level shrinks progressively as the former level is filled, leading to a crossing of the levels in cases where standard FL holds. But we find that under plausibly realizable conditions, the levels do not cross but instead they merge. A primary condition for merging to occur is that the Landau-Migdal interaction function f is repulsive in coordinate space, which is true in the nuclear interior and for the electron-electron (e-e) interaction. In Sec. 3 we present and interpret an illustrative numerical example. In Secs. 4 and 5, we discuss some implications of the level mergence phenomenon for nuclear physics and atomic physics, respectively. Section 6 is devoted to concluding remarks. These include (i) a possible explanation of certain NFL behavior exhibited by heavy-fermion metals, based on the level-merging mechanism revealed by our study, and (ii) a comparison of the phenomenon of level mergence with that of fermion condensation in infinitely extended Fermi systems.

2. How Merging of Single-Particle Levels Can Arise

We begin by recalling some key formulas of Landau theory written for a homogeneous isotropic system, which will be adapted to the case of a finite Fermi system. In Landau's quasiparticle (qp) picture, the total energy E of the interacting system is viewed as a functional of the quasiparticle momentum distribution $n_{\mathbf{p}\sigma}$. The single-quasiparticle energy (also called just the sp energy) is identified as

$$\epsilon_{\mathbf{p}\sigma} = \frac{\delta E[n_{\mathbf{p}\sigma}]}{\delta n_{\mathbf{p}\sigma}} \, . \tag{1}$$

Importantly, the qp energy $\epsilon_{\mathbf{p}\sigma}$ changes under addition of another qp of momentum/spin $(\mathbf{p}', \sigma')$, which changes the original qp distribution by $\delta n_{\mathbf{p}'\sigma'}$. Thus

$$\delta\epsilon_{\mathbf{p}\sigma} = \sum_{\mathbf{p}'\sigma'} f_{\mathbf{p}\sigma,\mathbf{p}'\sigma'} \delta n_{\mathbf{p}'\sigma'} \, , \tag{2}$$

where the qp interaction $f_{\mathbf{p}\sigma,\mathbf{p}'\sigma'}$ producing the energy shift is identified with the second variation of the total energy with respect to the qp distribution,

$$f_{\mathbf{p}\sigma,\mathbf{p}'\sigma'} = \frac{\delta^2 E}{\delta n_{\mathbf{p}\sigma} \delta n_{\mathbf{p}'\sigma'}} \, . \tag{3}$$

The generalization of Eq. (2) to the geometry of the finite system will provide the basis for the analysis below.

To illustrate the merging phenomenon, we analyze the behavior of a model system involving two neutron levels in an open shell of a spherical even-even nucleus of mass number A and radius $R = r_0 A^{1/3}$. The two levels are denoted by 0 and +, in order of increasing energy. The sp energies ε_λ and wave functions $\varphi_\lambda(\mathbf{r}) = R_{nl}(r)\Phi_{jlm}(\mathbf{n})$ are solutions of the equation

$$\left[\frac{\mathbf{p}^2}{2M} + \Sigma(\mathbf{r}, \mathbf{p}) \right] \varphi_\lambda(\mathbf{r}) = \epsilon_\lambda \varphi_\lambda(\mathbf{r}) \, , \tag{4}$$

where Σ stands for the self-energy. We note that a spherical nucleus with even numbers of neutrons and protons has zero total angular momentum in its ground state due to pairing correlations, and the energies ϵ_λ are independent of the magnetic quantum number m. In our treatment, self-interaction corrections will be neglected, and only the major, spin- and momentum-independent component $V(r)$ of the self-energy $\Sigma(\mathbf{r}, \mathbf{p})$ will be retained. It is assumed that the orbital angular momentum quantum numbers l_0 and l_+ of the levels 0 and + are different and both are large compared to unity.

The analysis proceeds by following the variation of the distance between the levels 0 and +, as $N \gg 1$ neutrons are added to level 0, changing the density $\rho(r)$ by $\delta\rho(r) = NR^2_{n^{(0)}l^{(0)}}(r)/4\pi$. We know from the finite-system analog of Eq. (2) that addition of particles to the system, producing an increase of density, is accompanied by a shift in the single-particle energy, hence a shift in the self-energy. Retaining only a dominant, $\delta(\mathbf{r})$-like portion of the interaction function f, the counterpart of Eq. (2) for our model system simplifies to[4,5]

$$\delta V(r) = f[\rho(r)]\,\delta\rho(r)\,. \tag{5}$$

The four interaction matrix elements involved in the model, having the form

$$f_{kk'} = \int R_k^2(r) f\left[\rho(r)\right] R_{k'}^2(r)\frac{r^2 dr}{4\pi} \tag{6}$$

with $k, k' = \{0, +\}$, are assigned values $f_{00} = u$, $f_{++} = v$, and $f_{0+} = f_{+0} = w$. In a semiclassical approximation where $R_k(r) \sim r^{-1}R^{-1/2}\cos\int p_k(r)dr$, one obtains the estimates $u \simeq v \simeq 3w/2$.

At this point we introduce a convenient notation. Let D be the initial energetic distance $\epsilon_+(0) - \epsilon_0(0)$ between levels + and − (before any neutrons are added), and define dimensionless matrix elements by $U = u(2j_0 + 1)/D$ and $W = w(2j_0 + 1)/D$. We define occupation numbers $n_k = N_k/(2j_k + 1)$ for the two levels that range between 0 and 1, where N_k is the actual number of neutrons in level k and $2j_k + 1$ is its multiplicity.

With these conventions and under the above assumptions, the dimensionless energy shifts

$$\xi_k(N) = \left[\epsilon_k(N) - \epsilon_k(0)\right]/D \qquad (k = 0, +) \tag{7}$$

are given by

$$\xi_0(N) = n_0 U \quad \text{ and } \quad \xi_+(N) = n_0 W \tag{8}$$

for simple filling of level 0 in the open shell. (We neglect corrections having little effect on the results since they are almost independent of the sp quantum numbers.) These results tell us that under addition of N fermions (neutrons, in this example), the dimensionless distance

$$d(N) = \left[\epsilon_+(N) - \epsilon_0(N)\right]/D = 1 + \xi_+(N) - \xi_0(N) \tag{9}$$

changes sign when N reaches $N_c = D/(u - w)$. Clearly, this occurs before filling of the level 0 is complete if the distance D is sufficiently small. It is likewise clear that for N above the critical value N_c, the relations

$$\xi_0(N) = n_0 U \quad \text{and} \quad \xi_+(N) = n_0 W \tag{10}$$

no longer apply, since the pattern of orbital filling must change from simple filling of the open shell.

In the standard FL picture, which allows only for crossing of sp levels, all the added $N > N_c$ quasiparticles must migrate to the empty sp level +, which however entails a change of the relevant sp energies – so the story is not over. In turn, the changes induced by migration from level 0 to level + would cause the N fermions to abandon level + and return to level 0, and so on in a 2-cycle in the sense of Poincare mapping, as long as one insists, as required in standard FL theory (or the shell model), that *one and only one* level can be partially occupied. Speaking colloquially, the system is driven back and forth between a rock and a hard place.

In attempting to determine the actual outcome chosen by nature when $N > N_c$, a crucial factor is the sign of the difference

$$\epsilon_r(N) = \delta\epsilon_+(N) - \delta\epsilon_0(N) \tag{11}$$

of the shifts of the sp energies ϵ_+ and ϵ_0 due to total migration of the quasiparticles from level 0 into level +. With the aid of Eq. (5), it is found that

$$\epsilon_r(N) = N(u - 2w + v)\,. \tag{12}$$

In our model, $\epsilon_r(N) \simeq 2Nu/3$ is positive. This guarantees that the level + remains *above* the level 0 through the migration process.

The standard FL theory must then encounter a catastrophe and present a conundrum. The quasiparticles must leave level 0, yet their total migration to level + is prohibited. To break the deadlock *both* levels must be occupied, thus departing from the restriction imposed in standard FL theory (or Hartree-Fock theory).

Such dual partial occupation is possible only if the sp energies ϵ_0 and ϵ_+ coincide with the chemical potential μ. If they do, then at $N > N_c$ we must have

$$\epsilon_0(0) + N_0 u + N_+ w = \epsilon_+(0) + N_0 w + N_+ v\,, \tag{13}$$

which has the solution

$$N_+ = (N - N_c)(u - w)(u + v - 2w)^{-1}\,. \tag{14}$$

We emphasize that a resolution of the dilemma has been found within the Landau framework itself. To understand the consistency of this resolution, recall that the occupation numbers of Landau quasiparticles are given by[1]

$$n_\lambda(T) = [1 + e^{(\epsilon_\lambda - \mu)/T}]^{-1}\,. \tag{15}$$

At zero temperature, this restricts qp occupation numbers to 0 and 1, but *only* for those sp levels having energies ϵ_λ different from μ; otherwise the index of the exponent is indeterminate. The proposed merging of two or more sp levels drives the levels exactly to the Fermi surface. Solution of the equations of merging eliminates the remaining uncertainty at the cost of introducing fractional occupation numbers n_λ, violating what would appear to be an elementary truth of FL theory, but in fact preserving its consistency.

In the presence of merging, the Green function has the familiar form

$$G(\mathbf{r},\mathbf{r}',\epsilon) = \sum_{k=0,+} \left(\frac{1-n_k}{\epsilon - \epsilon_k + i\delta} + \frac{n_k}{\epsilon - \epsilon_k - i\delta} \right) \varphi_k(\mathbf{r})\, \varphi_k^*(\mathbf{r}')\,, \tag{16}$$

but the occupation numbers n_k of the merging sp levels become fractional.

3. An Illustrative Numerical Example

Results from numerical calculations are shown in Fig. 1, which is made of two columns corresponding to two sets of parameters of the model, with each column featuring two panels. The upper panels contain plots of the dimensionless ratio

$$d(x) = \left[\epsilon_+(x) - \epsilon_0(x)\right]/D \tag{17}$$

versus $x = N/(2j_0 + 2j_+ + 2)$, which runs from 0 to 1. The lower panels give the occupation numbers $n_+(x)$ and $n_0(x)$, which, in the range of x where $d(x) = 0$, behave as

$$\begin{aligned} n_+(x) &= \frac{1}{2}\left[x(1+r) - \frac{r}{(U-W)}\right], \\ n_0(x) &= \frac{1}{2}\left[\frac{x(1+r)}{r} + \frac{1}{(U-W)}\right], \end{aligned} \tag{18}$$

with $r = (2j_0 + 1)/(2j_+ + 1)$.

There are three distinct regimes in the variable x. In two of them, $d \neq 0$ and the standard FL picture holds. It fails in the third region, where $d = 0$ and integration of the Green function over ϵ yields the density for the added quasiparticles,

$$\rho(\mathbf{r}) = \sum_{k=0,+} n_k \varphi_k(\mathbf{r}) \varphi_k^*(\mathbf{r})\,. \tag{19}$$

This density cannot be attributed to a particular sp level. Well-defined sp excitations in the familiar Landau sense *no longer exist.*

The two sp levels remain merged until one of them is completely filled. If the level 0 fills first, the episode of merging is ended by repulsion of the two levels, as if they possess the *same symmetry* — in spite of the fact that in the open shell, they always have *different* symmetries. In the case where level + becomes fully occupied before level 0, the distance $d(x)$ becomes negative, and the two levels just cross each other at this point.

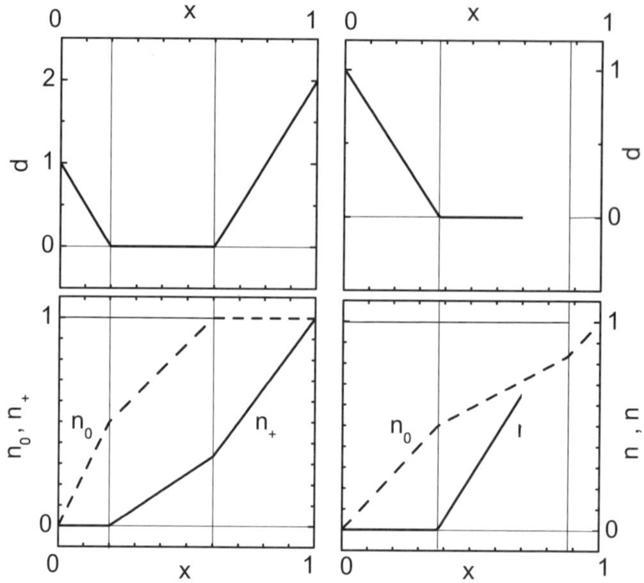

Fig. 1. Dimensionless distance $d = (\epsilon_+ - \epsilon_0)/D$ between levels + and 0 versus the ratio $x = N/(2j_0 + 2j_+ + 2)$. Lower panels: Occupation numbers n_k for levels 0 and +. Input parameters: $U = V = 3, W = 1$. For the left column, the ratio $r \equiv (2j_0 + 1)/(2j_+ + 1) = 2/3$; for the right, $r = 3$.

4. Implications of Level Mergence in Nuclear Physics

In nuclear physics, the sp level degeneracy (whether initially present or due to merging) is lifted when pairing correlations are explicitly involved.[4–6] We note that realistic pairing forces affecting nucleons added outside a closed shell are rather weak, with the result that the nuclear pairing energy becomes part of a shell correction to the liquid-drop mass formula.

To study the interplay of pairing and the mechanism that promotes level-merging, we have carried out a BCS treatment of pairing within the two-level model developed above.[7] In this treatment, the pairing gap plays the role of $D_{\rm min}$, the minimum value of the separation D between levels 0 and +. The inclusion of pairing correlations does indeed lift the degeneracy of the sp levels, resulting in a shrinkage of D. However, the value of the lowest of the energies E_k of the Bogoliubov quasiparticles remains markedly less than D. In Fig. 2 we show the results for the gap Δ and quasiparticle energies E_k in two cases: when the shrinkage effect is taken into account and when it is not. With shrinkage in play, the dip is filled in between two peaks of the gap plotted as a function of $x = N/(2j_0 + 2j_+ + 2)$.

According to this model study, the introduction of merging leads to roughly a doubling of the pairing gap, significantly increasing the pairing energy. A more realistic calculation is needed to determine whether or not this effect can give rise to a new minimum in the ground-state energy functional for superheavy nuclei.

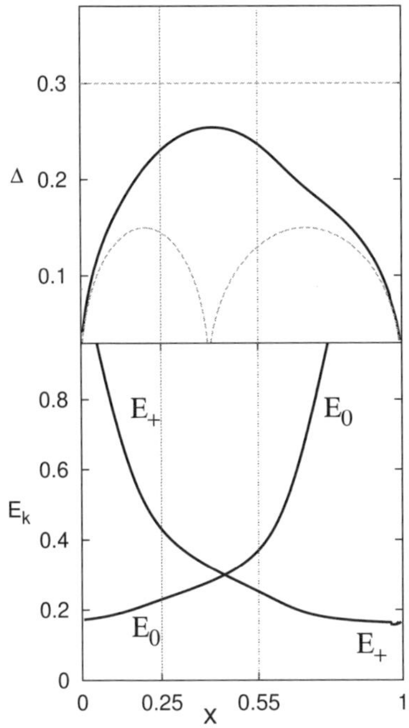

Fig. 2. Top panels: Pairing gap Δ (in units of D) plotted versus $x = N/(2j_0 + 2j_+ + 2)$, both accounting for the shrinkage of the interlevel distance (solid line) and neglecting it (dashed line). Bottom panels: Energies $E_k = \Delta/2\,[n_k(1-n_k)]^{1/2}$ of Bogoliubov quasiparticles. Pairing constant: $g = 0.3$. Other input parameters: $U = 4.0$, $W = 2.4$, $r = (2j_0+1)/(2j_+ + 1) = 2/3$. (Results are similar for $r = 3$.)

We now turn to issues of symmetry violation, which may have consequences for atoms and atomic clusters, as well as nuclei. In nuclei, the members of pairs of sp levels with quantum numbers n, l and $n \pm 1, l \pm 2$ lie quite close to each other. As the lower of the two levels is being filled, the distance from its neighbor decreases, resulting in an increase of the nuclear quadrupole moment. If merging occurs, spherical symmetry is broken. This mechanism is presumably responsible for the occurrence of islands of nuclear deformation beyond the "mainland" of rare-earth elements.

Level mergence can also enhance parity violation effects, both in nuclei and atoms.

5. Implications of Level Mergence in Atomic Physics

Two features of the problem complicate the analysis of merging electron sp levels in atoms. For one thing, the sp energies ϵ_{njlm} do exhibit a dependence on the magnetic quantum number m, due to the absence of pairing. The attendant difficulties can be reduced by tracking the center-of-gravity energies $\bar{\epsilon}_k = \sum_m \epsilon_{km}/(2j_k+1)$ of the levels,[8] rather than individual m-levels or a band. (One can argue that small correlation components of the effective e-e interaction may significantly affect every sp level, yet affect the sum over m only weakly.) Then one only has to consider the spherically symmetric part of $\delta\rho(\mathbf{r})$, much as in the nuclear problem. The second complication is that the self-energy Σ acquires a nonlocal character due to the long range of the Coulomb interaction between electrons.

In other respects, quantitative description of merging sp levels is simpler in atoms than in the nuclear case. The spacing parameter $r_s = r_0/a_B$ (ratio of volume per electron to Bohr radius) is less than unity. The correlation contributions to the e-e interaction function must therefore be rather small compared to exchange.[9] It is then reasonable to suppose that the e-e version of the quasiparticle interaction f takes the Hartree-Fock form. Repeating the operations that led to the positivity of $\epsilon_r(N) = \delta\epsilon_+(N) - \delta\epsilon_0(N)$ as a necessary condition for level mergence, we obtain the corresponding condition

$$f_{nl}^{nl} + f_{n'l'}^{n'l'} - 2f_{nl}^{n'l'} > 0 \tag{20}$$

for the merging of two electron sp levels with quantum numbers n, l and n', l'. Constructing the f matrix elements from the bare e-e interaction, and making semiclassical estimates similar to those applied for nuclei, the nondiagonal matrix elements are found to be of much smaller size than the diagonal ones. On this basis, the necessary condition for merging of sp levels is met in atoms.

It is instructive to compare estimated values of the critical parameter $N_c = D/(u - w)$ for nuclei and atoms, recalling that $\epsilon_+(N) - \epsilon_0(N)$ changes sign at $N = N_c$.

(i) *Nuclei.* The nucleon-nucleon interaction is characterized by the dimensionless constant[4] $F_{NN} = \mathcal{F}_{NN} p_F M/\pi^2 \simeq 1$. Estimating the corresponding $f_{kk'}$ matrix elements, one finds $u \simeq \epsilon_F^0/A$, where ϵ_F^0 is the Fermi sp energy. On the other hand, the spacing D of sp levels in the open shell of a spherical nucleus with a closed shell is of order $\epsilon_F^0/A^{2/3}$, so in this case the critical particle number $N_c = D/(u - w)$ must be of order $A^{1/3}$. Therefore not every pair of nuclear sp levels adjacent to the Fermi surface is susceptible to merging.

(ii) *Heavy atoms.* The diagonal matrix elements of the e-e interaction in the open shell are of order several eV. This is greatly enhanced in the event of f-orbital collapse, as occurs in rare-earth and transuranic elements.[2,12–14] (Note: f-orbital collapse refers to a rapid drop in the energy and size of a d or f atomic orbital in response the change in the effective potential (which includes the centrifugal barrier) under increase of atomic number Z)). The separation D of sp levels adjacent to the Fermi surface in atoms with closed electron shells is some 1–2 eV as well, so the critical number $N_c = D/(u-w)$ in atoms is *of order of unity.* One may infer that in elements with nuclear charge $Z > 20$, the sp Green function has the level-merged form — except for elements of principal groups, where the standard FL picture still holds.

Significantly, the estimation made for the case of atoms indicates that when N substantially exceeds N_c, as in the case of many rare-earth and transuranic atoms, merging of sp levels triggered by collapse of the f orbitals unavoidably involves most of the sp levels in the open shell. Fractional occupation becomes inevitable, with the consequence that these elements lose their chemical individuality, a well-known property of the sequence of rare-earth elements. It is possible to test these inferences by means of precise measurements of the difference $\sigma(Z+1) - \sigma(Z)$ between cross sections for elastic scattering of charged particles by rare-earth-metal or transuranic atoms with atomic numbers differing by unity. In Maria Goeppert-Mayer's theory,[12] this difference is expressed directly in terms of a single $4f$ wave function, whereas if merging occurs, the density change involves all the merging levels.

6. Conclusions

Based on the results flowing from our study of the two-level model, one expects that in finite Fermi systems devoid of pairing, typified by atoms, the centers of merged sp levels "get stuck" at the Fermi surface. We observe that this could provide a simple mechanism for pinning narrow bands in solids to the Fermi surface, as is seen in certain heavy-ion metals.

To exemplify this possibility, let us consider a model in which the electron sp spectrum, calculated in local-density approximation (LDA), is exhausted by (i) a wide band (w-band) that disperses through the Fermi surface, and (ii) a narrow band (n-band) whose center is located a distance D_n below the Fermi surface. To simplify the analysis, we suppose that only the diagonal matrix elements f_{nn} of the interaction function referring to the n-band are significant and neglect their momentum dependence. The shift $\delta\epsilon_n$ in the location of the (center of the) n-band caused by switching on the intraband correlations is then given schematically by $\delta\epsilon_n = f_{nn}\delta_n$, where δ_n is the initial density of the n-band. If this shift is less than the separation D_n of the band center from the Fermi surface, the filling of the n-band stays the same. However, if $\delta\epsilon_n$ exceeds D_n, then the standard FL scenario requires the n-band to be completely emptied. But if that happens, $\delta\epsilon_n$ must vanish, and the lost quasiparticles must return to the n-band. As we have learned, this quandary can be expunged if only a fraction of the particles leave the n-band, in just the right proportion to equalize the band chemical potentials. Thus, there is a feedback mechanism that positions the narrow band exactly at the Fermi surface, resolving a long-standing problem in the LDA scheme.

This same analysis reveals an opportunity for the onset of *fermion condensation* in the condensed-matter systems involved. Fermion condensation, primarily explored in homogeneous media, is a rearrangement of single-particle degrees of freedom leading to wholesale mergence of sp levels – indeed, the pinning of a continuum of sp states to the chemical potential. It has been proposed that such a rearrangement may be associated with the flattening of the sp spectrum that has been observed in strongly correlated Fermi systems displaying NFL behavior, and is triggered by the vanishing of the quasiparticle group velocity. In the above model, the bare n-band group velocity, proportional to the bandwith $W^{(n)}$, is rather small, whereas the matrix elements f_{nn} do not contain this suppressive factor. Accordingly, the group velocity might change its sign when the intraband interactions are taken into account, opening the way to fermion condensation.

Although the two phenomena show clear similarities, there is a crucial difference between the conditions for fermion condensation in homogeneous Fermi fluids and level mergence in finite Fermi systems with degenerate sp levels. In the former, the presence of a significant velocity-dependent component in the interaction function f is required to produce fermion condensation, while in the latter, sp levels can merge even if f is velocity-independent. There is a simple reason for this difference. In the homogeneous case, the quantities u, v, and w are equal to each other, so the energy gain due to the rearrangement (cf. Ref. Eq. (12)) vanishes when velocity-dependent forces are absent.

To summarize: exploration of the phenomenon of mergence of single-particle levels in finite Fermi systems reveals that if the merging of two or more sp levels occurs, the ground state experiences a rearrangement that introduces a multitude of quasiparticle terms, endowing it with a more complex character, as in the comparison of a chorus with a dominant soloist.

Acknowledgments

This research was supported in part by the McDonnell Center for the Space Sciences. Travel support from the Army Research Office through a grant to Southern Illinois University is gratefully acknowledged.

References

1. L. D. Landau, *JETP* **30**, 1058 (1956); *Sov. Phys. JETP* **3**, 920 (1957).
2. L. D. Landau and E. M. Lifshitz, *Quantum Mechanics, Non-Relativistic Theory* (Pergamon, New York, 1965).
3. D. Pines and P. Nozières, *Theory of Quantum Liquids*, Vol. 1 (W. A. Benjamin, New York-Amsterdam, 1966).
4. A. B. Migdal, *Theory of Finite Fermi Systems and Applications to Atomic Nuclei* (Wiley, New York, 1967).
5. P. Ring and P. Schuck, *The Nuclear Many-Body Problem* (Springer-Verlag, Berlin, 1980).
6. S. T. Belyaev, *Mat. Fys. Medd. Dan.* **31**, No. 11 (1959).
7. V. A. Khodel, J. W. Clark, H. Li and M. V. Zverev, *JETP Lett.* **84**, 588 (2006).
8. J. C. Slater, J. B. Mann, T. M. Wilson and J. H. Wood, *Phys. Rev.* **184**, 672 (1969).
9. M. Ya. Amusia and V. R. Shaginyan, *J. Phys. B* **25**, L345 (1992), and references therein.
10. V. A. Khodel and V. R. Shaginyan, *JETP Lett.* **51**, 553 (1990); *Condensed Matter Theories* **12**, 221 (1997).
11. V. A. Khodel, V. V. Khodel and V. R. Shaginyan, *Phys. Rep.* **249**, 1 (1994); V. A. Khodel, M. V. Zverev and V. M. Yakovenko, *Phys. Rev. Lett.* **95**, 236402 (2005).

12. M. Goeppert-Mayer, *Phys. Rev.* **60**, 184 (1941).
13. D. G. Griffin and K. L. Andrew, *Phys. Rev.* **177**, 62 (1969).
14. I. M. Band *et al.*, *Sov. Phys. JETP* **67**, 1561 (1988).

THE EFFECTIVE MASS OF A CHARGED CARRIER IN A NONPOLAR LIQUID: APPLICATIONS TO SUPERFLUID HELIUM

ANDREI VARLAMOV

INFM-CNR, COHERENTIA, via del Politecnico, 1, I-00133, Rome, Italy
varlamov@ing.uniroma2.it

IOULIA CHIKINA

DRECAM/SCM/LIONS CEA - Saclay, 91191 Gif-sur-Yvette Cedex, France

VALERIY SHIKIN

ISSP, RAS, Chernogolovka, Moscow District, 142432 Russia

Received 31 July 2008

The problem of a correct definition of the charged carrier effective mass in superfluid helium is revised. It is shown that the effective mass of such a quasi-particle can be introduced without Atkins's idea about the solidification of liquid He^4 in the close vicinity of an ion (the so-called "snowball" model). Moreover, in addition to generalization of the Atkins's model, the charged carrier effective mass formation is considered within the framework of the two-fluid scenario. The physical reasons of the normal fluid contribution divergency and the way of corresponding regularization procedure are discussed. Agreement between the theory and the available experimental data is found in a wide range of temperatures.

Keywords: Associated mass; viscous flow; two-fluid model.

1. Introduction

The ion-dipole interaction between the inserted charged particle and the induced electric dipoles of surrounding atoms is one of the many interesting phenomena which take place in a non-polar liquid. It is this interaction which is generally responsible for the different solvation phenomena there,[1] while in non-polar cryogenics liquids (like He, Ne, Ar, etc.) it leads to the so-called "snowball effect". The latter consists of the formation of a non-uniformity $\delta\rho(r)$ in a density of a liquid around the inserted charged particle.

The ion-dipole interaction $U_{i-d}(r)$ between the inserted charged particle and solvent atoms in its simplest form can be written as

$$U_{i-d}(r) = -\frac{\alpha}{2}\frac{e^2}{r^4}, \tag{1}$$

where α is the polarization of a solvent atom and r is the distance to the charged particle. The presence of attraction potential (1) in equilibrium has to be compensated by the growth of the solvent density in the direction of the charged particle. The latter can be estimated, using the requirement of the chemical potential constancy:

$$\delta\mu(r) = v\delta P(r) - \frac{\alpha}{2}\frac{e^2}{r^4} = 0,$$

$$\delta\rho(r) = \left(\frac{\partial\rho}{\partial P}\right)\delta P(r) \equiv \frac{1}{s^2}\delta P(r) \tag{2}$$

where $\rho(P)$ is the local liquid density, $\delta P(r)$ is the additional pressure that appears around the ion due to the polarization of media, v is the volume of individual solvent atom and s is the sound velocity.

The distance $R^{(s)}$ at which the total local pressure $P(r) = P_0 + \delta P(r)$ reaches the value of a solvent solidification pressure $P^{(s)}$, corresponds to the radius of the rigid sphere which is called a "snowball". In the case of positive ions in liquid helium, where the inter-atomic distance $\mathfrak{a}_{(He^4)} \approx 3$ Å and the solidification pressure $P^{(s)}_{(He^4)} \approx 25$ atm, the snowball radius for the normal external pressure $P_0 = 1$ atm is estimated as $R^{(s)}_{(He^4)} \geq 7$ Å.

The so-called Atkins's snowball model described above[2] is quite transparent and was found useful in various qualitative predictions. In particular, it provides by the natural definition and estimation for the effective mass M of such a quasi-particle as the sum of three contributions: the solid body nucleus mass $M^{(s)}$ (Ref. 3)

$$M^{(s)}(R_s) = \frac{4\pi}{3}\rho_s\left[R^{(s)}\right]^3, \tag{3}$$

$$M^{(s)}_A\left(R^{(s)}_{(He^4)}\right) \simeq 32m_{(He^4)},$$

(ρ_s is the density of the solid phase in the absence of perturbation at normal pressure P_0), mass $M^{(ind)}$caused by the induced density perturbation $\delta\rho(r)$

$$M^{(\mathrm{ind})}\left(R^{(s)}\right) = 4\pi\int_{R^{(s)}}^{\infty} r^2\delta\rho(r)\,dr; \tag{4}$$

$$M^{(\mathrm{ind})}_A\left(R^{(s)}_{(He^4)}\right) \simeq 28m_{(He^4)}$$

and the associated mass, related to the occurrence of the velocity distribution around the sphere moving in an ideal liquid (Ref. 4)

$$M^{(\mathrm{ass})(0)}\left(R^{(s)}\right) = \frac{2}{3}\pi\rho_l\left[R^{(s)}\right]^3, \tag{5}$$

$$M^{(\mathrm{ass})(0)}_A\left(R^{(s)}_{(He^4)}\right) \simeq 16m_{(He^4)}$$

(ρ_l is the density of the liquid phase in the absence of perturbation at normal pressure P_0).

When calculated this way, the snowball effective mass turns out to be $M \gtrsim 75m_{(He^4)}$, value, which correctly determines the scale of the positive ion mass renormalization in liquid helium,[5–9] although numerically it considerably exceeds the values $M \lesssim 30m_{(He^4)}$, observed within the low temperature limit.

This difficulty is hard to overcome within the frame of the snowball model. Indeed, the solidification pressure is reliably measured. The definition of $R^{(s)}$ by means of the local equilibrium condition (2) is well substantiated. Thus, possibilities of the Atkins's two phase model are exhausted and further improvement of the accuracy of evaluations (3)–(5) is impossible. Moreover, the value of $R^{(s)}_{(He^4)}$ turns out to be comparable to the width of the transition region between the solid and liquid phases, and the very existence of such a small solid phase nucleus itself can be called in question.

The aim of this paper is to extend the Atkins's snowball model in a way to exclude from it the solid phase nucleus altogether, keeping in mind only the model cornerstone, i.e., the existence of the density perturbation (2), which indeed is the unique origin of the abnormally large effective mass of the charge carrier. We will refer to such generalization as the "*cloud*" model. Due to such approach, the parameter $R^{(s)}$, which in the Atkins's model strongly affects the values of all three contributions to the effective mass of the charge carrier, does not appear at all.

Within the framework of the cloud model, the first contribution to the effective mass is absent, since one does not assume existence of the solid nucleus as such. The contribution of the static consolidation remains in its form (4), but the lower limit of integration in it requires a numerical estimate, what will be done below. As for the third contribution to the effective mass, it has to be revised completely. The matter of fact is that its strong temperature dependence, which has been systematically observed experimentally (see Refs. 6, 9), cannot be explained within the "ideal" liquid flow picture. This is why we propose below the hydrodynamics scenario of the associated mass $M^{(\mathrm{ass})}$ formation that takes into account also the viscous part of this problem. In spite of being considered as a hydrodynamic associated mass of the Atkins solid phase nucleus moving in helium (see (5)), the cloud associated mass $M^{(\mathrm{ass})}$ will be identified below with the kinetic energy of the normal and the superfluid velocity fields appearing in helium in the process of such a motion. The non-zero viscosity results in the motion of a spacious domain of viscous liquid around the moving particle. Its size grows dramatically with the decrease of the dragging velocity and, consequently, increases strikingly the effective associated mass. Nevertheless, the arising divergency can be cut off either by accounting for the non-linear effects in stationary flow, or using the *ac* exiting field.

2. Viscous Part of Associated Mass

Let us now discuss the viscous part of associated mass within the cloud model approximation. Here, in order to find the velocity field related to motion in liquid of the density perturbation $\delta\rho(r)$, we have to start from the Navier-Stokes equation. It is important for further consideration that far from the ion, moving in the viscous liquid, the velocity field $\mathbf{v}_n(\mathbf{r})$ coincides with that one of the rigid sphere.

Assuming that there is an analogy between the rigid sphere and density compression (cloud) motion in a viscous liquid at large distances, we will proceed with the evaluation of the cloud normal mass component in the approximation of a rigid sphere of the radius $R_{(\mathrm{n})}$. This simplification will permit us to take into consideration the nonlinear and non-stationary effects of Navier-Stokes hydrodynamics for the velocity field $\mathbf{v}_n(\mathbf{r},\omega)$ which are necessary to avoid the nonphysical divergencies of the linear theory and specifics of the associated mass temperature behavior observed in experiment at finite frequencies.

2.1. *Stationary motion*

The distribution of velocities in viscous liquid occurring in process of a rigid sphere stationary motion in it is described by the Navier-Stokes equation. For small Reynolds numbers $\mathrm{Re} = \rho_n V R_{(\mathrm{n})}/\eta \ll 1$ and distances $R_{(\mathrm{n})} < r < R_{(\mathrm{n})}/\mathrm{Re}$ the gradient term $(\mathbf{v}\cdot\nabla)\,\mathbf{v}$ in it can be ignored and the solution of such linearized equation in zero approximation turns out to be independent on the liquid viscosity η. In the center of mass frame it has the form (see Refs. 4, 13):

$$v_r(r,\theta) = -V\cos\theta\left(\frac{3R_{(\mathrm{n})}}{r} - \frac{R^3_{(\mathrm{n})}}{2r^3}\right), \tag{6}$$

$$v_\theta(r,\theta) = V\sin\theta\left(\frac{3R_{(\mathrm{n})}}{4r} + \frac{R^3_{(\mathrm{n})}}{4r^3}\right). \tag{7}$$

Here θ is the polar angle counted from the x-axes, which coincides with the ion velocity direction.

One can easily see, that corresponding contribution to the kinetic energy diverges at the upper limit of integration. To regularize this divergence it is necessary to use at large distances $r \gtrsim l_\eta = R_{(\mathrm{n})}/\mathrm{Re}$ ($l_\eta = \eta/\rho_n V$ is the characteristic viscous length) more precise, so-called Oseen, solution of the Navier-Stokes equation which was obtained with the gradient term in it.[4] This solution shows that almost for all angles, besides the domain restricted by the narrow paraboloid $\theta(x) = \pi - \sqrt{\eta/\rho_n V|x|}$ behind the snowball (so-called "laminar trace") at large distances the velocity field decays exponentially. In latter the velocity decays by power law[4] and it gives logarithmically large contribution with respect to other domain of disturbed viscous liquid. Such specification allows to cut off formal divergency of the kinetic energy and to find with the logarithmic accuracy the value of the normal component of associated mass for the stationary moving in viscous liquid charge carrier:

$$\begin{aligned} M^{(\mathrm{ass})}_{n(\mathrm{st})} &\sim M_0^{(\mathrm{ass})}\left(\rho_n, R_{(\mathrm{n})}\right)\left(\frac{l_\eta}{R_{(\mathrm{n})}}\right)\ln\frac{l_\eta}{R_{(\mathrm{n})}} \\ &= R^2_{(\mathrm{n})}\frac{\eta}{V}\ln\frac{\eta}{\rho V R_{(\mathrm{n})}}. \end{aligned} \tag{8}$$

Here the associated mass $M_0^{(\mathrm{ass})}\left(\rho_n, R_{(\mathrm{n})}\right)$ is defined by Eq. (5) with radius $R_{(\mathrm{n})}$ and density ρ_n. Since in our assumptions $l_\eta \gg R_{(\mathrm{n})}$ the value of $M^{(\mathrm{ass})}_{n(\mathrm{st})}$ turns out much larger than the value of the associated mass in the ideal liquid. Moreover, when velocity $V \to 0$ such definition, in spite of the performed above regularization procedure, fails since Eq. (8) diverges.

2.2. *Oscillatory regime*

More suitable is the mass definition using ac exciting field. However, this way is problematic for the unambiguous measurement of the associated effective mass. The point is that in the course of oscillations the quasi-particle passes through the points where its velocity becomes zero. Therefore, the associated mass is not constant throughout the oscillation period changing from zero to its maximum value depending on the oscillation frequency. This sort of dynamics should be discussed separately. Its details which are equally relevant to both normal and superfluid parts of the problem are most conveniently discussed using the known results for the motion of a sphere in viscous liquid. Let us start with the definition of the force $F(t)$ acting on a sphere of radius $R_{(\mathrm{n})}$ which moves with the time-dependent velocity $V(t)$ (see Refs. 4, 13):

$$\begin{aligned} F(\omega) &= 6\pi\eta R_{(\mathrm{n})}\left(1+\tfrac{R_{(\mathrm{n})}}{\delta(\omega)}\right)V(\omega) \\ &+3\pi R_{(\mathrm{n})}^2\sqrt{\tfrac{2\eta\rho_{\mathrm{n}}}{\omega}}\left(1+\tfrac{2R_{(\mathrm{n})}}{9\delta(\omega)}\right)i\omega V(\omega), \end{aligned} \tag{9}$$

where $\delta(\omega) = (2\eta/\rho_n\omega)^{1/2}$ is so-called dynamic penetration depth. The definition (9) is correct, if $x(t) \ll R_{(\mathrm{n})}$.

It is natural to identify the coefficient in front of the Fourier transform of acceleration $i\omega V(\omega)$ with the effective dynamic associated mass, what gives:

$$M_{n(\mathrm{dyn})}^{(\mathrm{ass})}\left(\omega,\rho_n,R_{(\mathrm{n})}\right) = M_0^{(\mathrm{ass})}\left(\rho_n,R_{(\mathrm{n})}\right)\left(1+\frac{9}{2}\frac{\delta(\omega)}{R_{(\mathrm{n})}}\right). \tag{10}$$

One can see that for high frequencies ($\delta(\omega) \ll R_{(\mathrm{n})}$) the dynamic associated mass coincides with that one of a sphere moving stationary in an ideal liquid, while when $\omega \to 0$, $M_{n(\mathrm{dyn})}^{(\mathrm{ass})}(\omega)$ diverges as $\omega^{-1/2}$. As we have already seen above, this formal divergence is related to fall down of the linear approximation in the Navier-Stokes equation assumed in derivation of Eq. (9). It is clear that the definition (10) is valid for high enough frequencies, until $\delta(\omega) \lesssim l_\eta$ (i.e., $\omega \gtrsim \widetilde{\omega} = \tau^{-1} = \eta/\rho_n R_{(\mathrm{n})}^2$).When ω becomes lower than $\widetilde{\omega}$, the penetration depth $\delta(\omega)$ in Eq. (10) has to be substituted by l_η and up to the accuracy of $\ln l_\eta/R_{(\mathrm{n})}$ the dynamic definition Eq.(10) matches the static one (see Eq. (8)).

The alternative possibility to formulate the beginning of non-linear regime in $M_n(\omega)$ behavior is the reorganization of requirement $l_\eta \sim R_{(\mathrm{n})}$, following from Eq. (8), to the analogous one: $\omega\tau \sim 1$. In the region $\omega < \tau^{-1}$ frequency dependence of mass $M_{n(\mathrm{dyn})}^{(\mathrm{ass})}(\omega)$ is saturated and ω in Eq. (10) has to be substituted by τ^{-1}.

Thus, the following qualitative picture arises. In the stationary regime the moving sphere disturbs an enormous domain in viscous liquid. This perturbation results in the large effective mass (8). When going to the oscillatory regime the viscous "coat" and especially the "laminar trace" do not keep pace with the charge in its oscillations; the perturbed domain shrinks with the growing frequency thus reducing the value of effective mass $M_{n(\mathrm{dyn})}^{(\mathrm{ass})}(\omega)$.

3. Discussion

The main conclusion based on the above considerations consists in the statement that in oscillatory experiments the dynamic associated mass is the function both of temperature and frequency: $M^{(\mathrm{ass})}_{(\mathrm{dyn})} = M^{(\mathrm{ass})}_{(\mathrm{dyn})}(\omega, T)$.

With respect to the snowball effective mass temperature dependence, all the authors of experimental investigations[6,9] indeed, denote its existence. For instance, in the case of positive charge carriers in liquid He^4 growth of the snowball effective mass has been observed in the interval from $50\,mK$ up to $2\,K$, with the average rate of $28\,m_{He^4}/K$. Nevertheless, none of the authors indicates the existence of the frequency dependence of $M^{(\mathrm{ass})}_{(\mathrm{dyn})}$.

Let us discuss the results of Ref. 6. In this paper the authors measured real and imaginary parts of the *ac* ± ions mobility at finite frequencies side by side with the value of the ions *dc* mobility. In assumption of the frequency independence of the snowball effective mass, the latter could be extracted from these data in approximation of the Drude-type equation of motion. The results of such analysis of experimental data[6] in the case of positive ions are presented in Fig. 1 by white circles. The analogous measurements performed for the negative charge carriers (electron bubbles) do not show any temperature dependence of the effective mass (see Fig. 5 of Ref. 6).

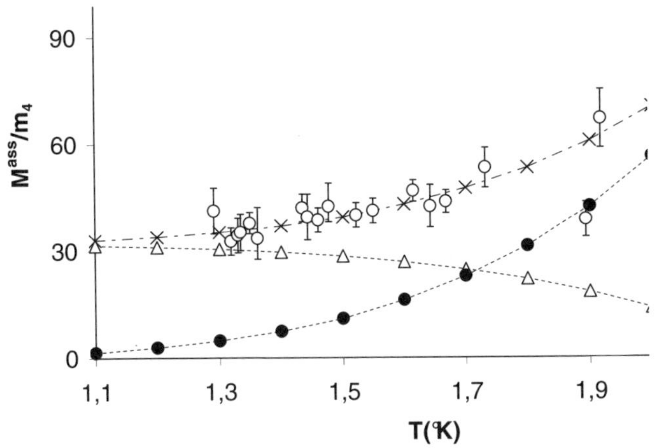

Fig. 1. The temperature dependencies of $M_s^{(\mathrm{ass})}(T)$ (triangles), $M_n^{(\mathrm{ass})}(T)$ (black dots) and their sum $M_s^{(\mathrm{ass})}(T) + M_n^{(\mathrm{ass})}(T)$ (crosses) are presented separately. The dependence $\rho_n(T)$ is taken from Ref. 15. The experimental data (open circles with error bars) are taken from Ref. 6.

4. Conclusion

In conclusion, it should be emphasized that the observed temperature (actually, frequency) dependence of the snowball effective mass is interesting not only in itself. It also turns out to be a sensitive indicator revealing qualitative difference in the velocity distribution field around the moving sphere in either viscous or ideal regime. This difference has been known for a long time. However, it did not attract much attention since in the applications (calculation of the Stokes drag force) slow decrease (of the $1/r$ type) of the velocity field does not result in any divergence in its real part. The situation is quite different in the calculation of the imaginary part, and the above insight into this field is one of the important conclusions of this paper.

References

1. B.E. Conway, in *Physical Chemistry: An Advanced Treatise*, eds. H. Eyrung, D. Henderson and W. Yost, Academic, New York, (1970), Vol. IXA; H.L. Friedman and C.V. Krishnan in *Water a Comprehensive Treatise*, ed. F. Franks, Plenum, New York, (1973), Vol. 3.
2. K. Atkins, *Phys. Rev.* **116**, 1339 (1959).
3. I.M. Khalatnikov, *An Introduction to the Theory of Superfluidity* (Perseus Books Group, USA, 2000).
4. L.D. Landau and E.M. Lifshitz, *Fluid Mechanics*, 2nd edn. (Pergamon, London, 1987).
5. J. Poitrenaud and F.I.B. Williams, *Phys. Rev. Lett.* **29**, 1230, (1972).
6. A. Dahm and T. Sanders, *J. of Low Temp.Phys.* **2**, 199 (1970).
7. M. L. Ott-Rowland *et al.*, *Phys. Rev. Lett.* **49**, 1708 (1982).
8. C. Mellor, C. Muirhead, J. Travers and W. Vinen, *J. Phys. C*, **21**, 325 (1988).
9. C. Mellor, C. Muirhead, J. Travers and W. Vinen, *Surface Sci.* **196**, 33 (1988).
10. B. Esel'son, Yu. Kovdria and V. Shikin, *Soviet JETP* **32**, 48, (1970).
11. R. Bowley and J. Lekner, *J. Phys. C*, **3**, L127 (1970).
12. M. Kushnir, J. Ketterson and P. Roach, *Phys. Rev.* **A 6**, 341 (1972).
13. H. Lamb, *Hydrodynamics*, 6th edn. (Dover, New York, 1932), or 6th edition (1993).
14. T. O'Malley *Phys. Rev.* **130** 1020 (1963).
15. S. Putterman, *Superfluid Hydrodynamics* (North Holland Pub. Company, Amsterdam, 1974).

DEPENDENCE OF NON-ABELIAN MATRIX BERRY PHASE OF A SEMICONDUCTOR QUANTUM DOT ON GEOMETRIC PROPERTIES OF ADIABATIC PATH

S. C. KIM, N. Y. HWANG, P. S. PARK, Y. J. KIM, C. J. LEE and S.-R. ERIC YANG*

Physics Department, Korea University, Seoul, Korea

A matrix Berry phase can be generated and detected by all electric means in II-VI and III-V n-type semiconductor quantum dots by changing the shape of the confinement potential. This follows from general symmetry considerations in the presence of spin-orbit coupling terms. We explain these results and discuss how the matrix Berry phase depends on geometric properties of adiabatic paths. We suggest how the matrix Berry phase may be detected in transport measurements.

1. Introduction

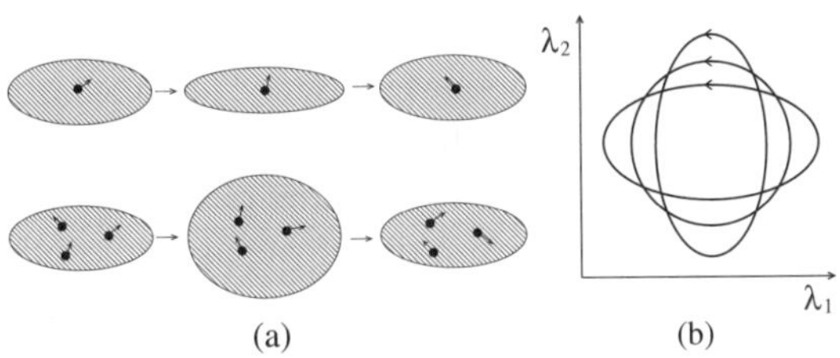

Fig. 1. (a) Value of electron spins may not return to the initial value after a cyclic adiabatic change of the shape of the electric confinement potential of a quantum dot. (b) Adiabatic parameters λ_1 and λ_2 provides a means to control the shape of the dot electrically. Several adiabatic paths are shown.

Single and few electron control in semiconductor quantum dots by electric means would be valuable for spintronics, quantum information, and spin qubits.[1–3] Adiabatic time evolution of degenerate eigenstates of a quantum system provides a means for controlling individual quantum states through the generation of non-Abelian matrix Berry phases.[4–6] This method can be used to perform universal quantum computation.[7] There are several semiconductor nanosystems that exhibit matrix

*Corresponding author, eyang@venus.korea.ac.kr.

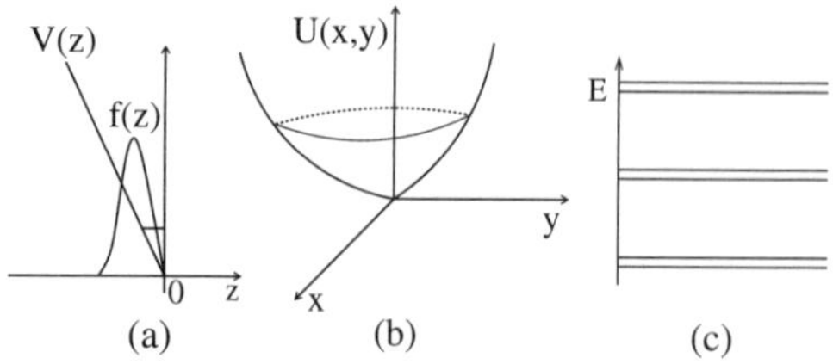

Fig. 2. Basic mechanism of adiabatic control is based on non-trivial degeneracy of II-VI and III-V semiconductors. (a) An electric field along the z-axis quantizes the electronic motion in a triangular potential along the axis. An adiabatic change can be induced by changing the magnitude of the electric field. (b) The two-dimensional electronic motion is quantized in a distorted parabolic potential. An adiabatic change can be induced by changing the magnitude of the distortion potential. (c)In the presence of the spin-orbit terms each discrete eigenstate of a semiconductor quantum dot has a double degeneracy due to time reversal symmetry in the absence of a magnetic field.

Berry phases: they include excitons,[8] CdSe nanocrystals,[9] and acceptor states of p-type semiconductors.[10] It is desirable to generate and detect matrix Berry phases by *all electric means* in semiconductor nanosystems. Recently, we have demonstrated theoretically that it is possible to control *electrically* electron spins of II-V and III-V n-type semiconductor quantum dots[11,12] and rings[13] by changing the shape of the electric confinement potential (see Fig. 1) The mechanism is based on spin orbit couplings which generate non-Abelian vector potentials. Its main ingredients are the following general symmetry principles: [a] Time-reversal symmetry, which leads to double degeneracy.[11,14] As depicted schematically in Fig. 2. [b] Non-Abelian U(2) gauge theory. [c] Adiabatic changes that break the parity symmetry of the electric confinement potential of the dot. The matrix Berry phase can be also generated in the presence of several electrons and the effect of many body correlations can be included in a compact way.[12,15] The matrix Berry phase can be present even in the absence of the spin-orbit terms. An example of this is semiconductor quantum dot pumps,[16,17] which can be understood as a manifestation of a matrix Berry phase.[18]

Let us give a brief explanation of matrix Berry phases. The electron state at time t is given by

$$|\Psi(t)\rangle = c_1(t)|\psi_1(t)\rangle + c_2(t)|\psi_2(t)\rangle, \tag{1}$$

where $|\psi_i(t)\rangle$ are the instantaneous basis states satisfying $H(t)\psi_i(t) = E(t)\psi_i(t)$ for $i = 1, 2$ ($H(t)$ is the Hamiltonian with the eigenenergy $E(t)$ at time t). The matrix Berry phase after the first cycle $\Phi_C(1)$ connects $(c_1(T), c_2(T))$ to $(c_1(0), c_2(0))$, where T is the period of the cycle. It is therefore a unitary time evolution operator connecting the states at $t = 0$ and T. It is given by

$$\begin{aligned}\Phi_C(1) &= e^{i\sum_p A_p(t_n)d\lambda_p} \ldots . e^{i\sum_p A_p(t_1)d\lambda_p} \\ &= Pe^{i\oint_C \sum_p A_p d\lambda_p},\end{aligned} \tag{2}$$

where P orders time slices as $t_1 < \cdots < t_n < T$. In Eq. (2) the matrix vector potentials A_p

$$(A_p)_{i,j} = i\langle\psi_i|\frac{\partial\psi_j}{\partial\lambda_p}\rangle, \quad p = 1, 2 \tag{3}$$

are integrated along the path C in the parameter space in the order of increasing time. The path C is parameterized in time as $(\lambda_1(t), \lambda_2(t))$. The matrix Berry phase is independent of the functions $\lambda_p(t)$ as long as they describe the same path C.

In applying matrix Berry phases to II-VI and III-V semiconductor quantum dots there are several issues that need to be addressed. One issue is whether the line integral along C can be *generally* converted into an area integration over A in Eq. (2). We will investigate the applicability of Stokes' theorem in the case of Rashba and Dresselhaus terms. For this purpose it is useful to consider the field strength 2-form

$$F = \sum_{\mu\nu} F_{\mu\nu} \mathrm{d}\lambda_\mu \wedge \mathrm{d}\lambda_\nu, \tag{4}$$

where

$$F_{\mu\nu} = \partial_\mu A_\nu - \partial_\nu A_\mu + [A_\mu, A_\nu] \tag{5}$$

with $\partial_\mu A_\nu = \frac{\partial A_\nu}{\partial \lambda_\mu}$. For a small area A it can be shown[5] that the matrix Berry phase is

$$\Phi_C(1) = \exp(iAF_{12}). \tag{6}$$

For an arbitrarily large A there are two cases: [1] If $[A_\mu, A_\nu] = 0$, the curvature reduces to $F_{\mu\nu} = \partial_\mu A_\nu - \partial_\nu A_\mu$ and one can use Stokes' theorem to convert the line integral along C to an area integration over A:

$$\Phi_C(1) = \exp(i\frac{1}{2}\int_A F_{\mu\nu} \mathrm{d}\lambda_\mu \wedge \mathrm{d}\lambda_\nu). \tag{7}$$

[2] When $[A_\mu, A_\nu] \neq 0$ it is not possible to convert the line integral into an area integration.

In semiconductor quantum dots with spin orbit coupling terms the matrix Berry phase is a unitary 2×2 matrix and it may be thought of as a spin rotation matrix.[19] It has the general form of

$$\Phi_C(n) = e^{\frac{i}{2}(2\alpha n)\vec{m}\cdot\vec{\sigma}} = \cos(\alpha n)I + i\sin(\alpha n)\vec{m}\cdot\vec{\sigma}, \tag{8}$$

where $\vec{\sigma}$ are Pauli spin matrices, I is the identity matrix, α is the angle of rotation, and n is the number of periodic adiabatic cycles. This expression is valid *only* at time $t = nT$, and it should be stressed that in the time interval $(n-1)T < t < nT$ the probability amplitudes $c_1(t)$ and $c_2(t)$ exhibit a much more complicated behavior, see Fig. 6. A spin rotation matrix is characterized by the direction and angle of rotation. It is unclear how the geometric properties of the adiabatic path are reflected on them. We show in this paper how the information of the path C can be encoded in the direction and angle of the rotation. Another issue is whether the matrix Berry

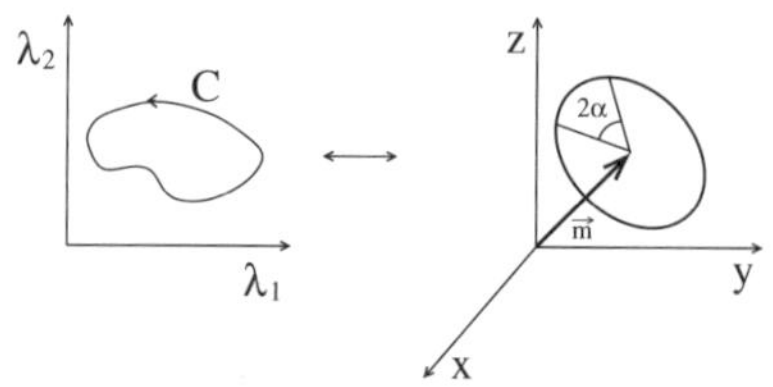

Fig. 3. The geometric information about the closed curve in the parameter space is encoded into the rotation axis $\vec{m}$ and angle 2α. When only the Rashba term is present $\vec{m}$ is independent of path C.

phase can be detected in a transport measurement, which would complement the detection of matrix Berry phases by infrared optical measurements.[11]

Let us now give a brief summary of the main results of this paper. The information about the *closed* curve in the parameter space is encoded into the rotation axis represented by the unit vector

$$\vec{m} = (\mathrm{Re}(\beta), -\mathrm{Im}(\beta), \sqrt{1-|\beta|^2}) \tag{9}$$

and the angle of rotation α, see Fig. 3. The constants α and β are real and complex numbers, respectively. This result is even valid for many electron systems with correlations as long as double degeneracy is present.

In addition when only the Rashba term is present we find $\beta = 1$ and $\vec{m} = (1, 0, 0)$, and all the geometric information of the path is contained in the rotation angle α. We have derived an analytical expression for α. Furthermore, remarkably $[A_\mu, A_\nu] = 0$ and Stokes' theorem can be applied. When both Rashba and Dresselhaus terms are present each path has different $\vec{m}$ and α. We have investigated how they depend on the shape of each path. Moreover, we find that $[A_\mu, A_\nu] \neq 0$ and that Stokes' theorem cannot be applied. We propose how the presence of the matrix Berry phase can be measured in transport experiments by changing the electric confinement potential suddenly. This is shown schematically in Fig. 7.

2. Semiconductor Quantum Dots with the Rashba and Dresselhaus Spin Orbit Couplings

2.1. *Model*

The Hamiltonian is $H = -\frac{\hbar^2\nabla^2}{2m^*} + U(x,y) + V(z) + H_\mathrm{R} + H_\mathrm{D}$. We take the two-dimensional potential to be $U(x,y) = \frac{1}{2}m^*\omega_x^2 x^2 + \frac{1}{2}m^*\omega_y^2 y^2 + \epsilon' y$, where the term $\epsilon' y$ represents a distortion of the two-dimensional harmonic potential. The strengths of the harmonic potentials are denoted by ω_x and ω_y, and the effective electron mass is denoted by m^*. The characteristic lengths scales along x- and y-axis are $R_{x,y} = \sqrt{\frac{\hbar}{m^*\omega_{x,y}}}$. In our work the triangular potential $V(z)$ is sufficiently strong and only the lowest energy subband is included. The characteristic length scale along the z-axis is $R_z = 1/\sqrt{0.8(2m^*eE/\hbar^2)^{2/3}}$, where E is the Rashba electric field applied along the z-axis. The Rashba spin orbit term[20,21] is

$$H_{\rm R} = c_{\rm R}\left(\sigma_x k_y - \sigma_y k_x\right), \tag{10}$$

and the Dresselhaus spin orbit term[22] is

$$H_{\rm D} = c_{\rm D}\left(\left(\sigma_x k_x\left(k_y^2 - k_z^2\right)\right) + \left(\sigma_y k_y\left(k_z^2 - k_x^2\right)\right)\right). \tag{11}$$

Note that the Rashba spin orbit constant c_R *depends on the external electric field* E. Here $k_{x,y}$ are momentum operators ($k_x = \frac{1}{i}\frac{d}{dx}$ and similarly with k_y). The Hamiltonian matrix is represented in the basis states $|mn\sigma\rangle$ of the harmonic oscillator states in the xy plane with spin component σ. The subband wavefunction $f(z)$ is suppressed in the notation $|mn\sigma\rangle$. Let us take $\omega_x = 2\omega_y$ (Other values of ω_x can also be chosen). Then the lowest eigenenergy state of H is $|mn\rangle = |00\rangle$ with the energy $E_0 = \frac{3}{2}\hbar\omega_y$ and the next lowest eigenenergy state is $|01\rangle$ with the energy $E_1 = \frac{5}{3}E_0$. The typical value of the energy spacing between the quantum dot levels, E_0, is of order $1 - 10meV$. It can be several times larger in self-assembled dots. The energy scale of the Rashba term is $E_R = c_R/R \sim 0.01 - 10meV$, where the length scale $R \sim 100$ Å is the lateral dimension of the quantum dot. The energy scale of the distortion potential is $E_p = \langle 0|\epsilon' y|1\rangle$. Its magnitude is of order $1 - 10meV$, depending on the electric field applied along the y-axis. The energy scale of the Dresselhaus term is $E_D = c_D/R^3$, and it can be larger or smaller than the Rashba term, depending on the material.[20] Here $E_D = ic_D\langle 0|k_y|1\rangle(\langle f(z)|k_z^2|f(z)\rangle - \langle 0|k_x^2|0\rangle) = c_D/R_y R_z^2 - c_D/R_y R_x^2$. It originates from the second term of the Dresselhaus term (the first term in the Dresselhaus term is zero in our model since $\langle 0|k_x|0\rangle = 0$).

We will use a truncated version of the Hamiltonian matrix, which makes it possible to write the eigenstates as a linear combination of four basis states made out of $|mn\rangle$ and spin degree of freedom:

$$\begin{aligned} |\psi\rangle = \quad & c_{0,0,\uparrow}|0,0,\uparrow\rangle + c_{0,1,\uparrow}|0,1,\uparrow\rangle \\ & + c_{0,0,\downarrow}|0,0,\downarrow\rangle + c_{0,1,\downarrow}|0,1,\downarrow\rangle. \end{aligned} \tag{12}$$

The advantage of this truncated model is that it is *exactly solvable*. In this paper we choose the following eigenstates of the lowest energy shell as the instantaneous basis vectors:

$$\begin{aligned} |\psi_1\rangle &= \frac{1}{N_1}\begin{pmatrix} 3E_p \\ E_0 - \sqrt{E_0^2 + 9(E_D^2 + E_p^2 + E_R^2)} \\ 3(E_D - iE_R) \\ 0 \end{pmatrix}, \\ |\psi_2\rangle &= |\overline{\psi}_1\rangle. \end{aligned} \tag{13}$$

Here $|\overline{\psi}_1\rangle$ is the time reversed state of $|\psi_1\rangle$, and N_1 is the normalization factor. In the following the non-Abelian vector potentials are evaluated with respect to these basis vectors.

In the following we choose the adiabatic parameters as the Rashba energy and the distortion energy

$$\lambda_1 = E_R = c_R/(\sqrt{2}R_y)$$
$$\lambda_2 = E_p = \langle 0|\epsilon' y|1\rangle = \frac{\epsilon'}{\sqrt{2}}R_y. \tag{14}$$

The cyclic adiabatic path is given by

$$E_R(t) = E_{R,c} + \Delta E_R \cos(\omega t) \quad E_p(t) = E_{p,c} + \Delta E_p \sin(\omega t), \tag{15}$$

where $\omega = 2\pi/T$. This path is an ellipse with the center $(E_{R,c}, E_{p,c})$.

2.2. *Geometric dependence*

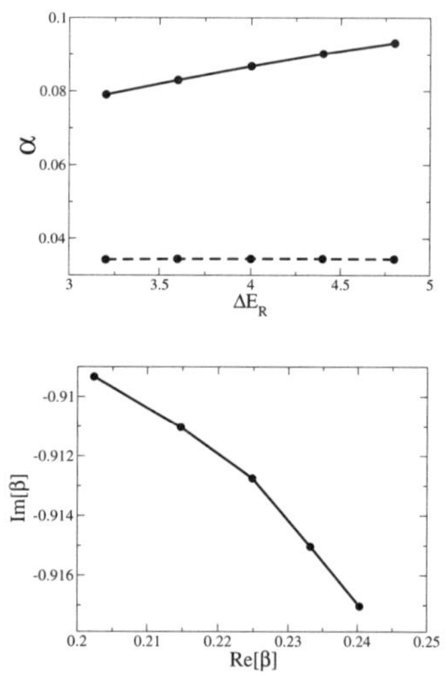

Fig. 4. For elliptic paths with fixed area with ΔE_R changing from $3.2E_0$ to $4.8E_0$. In the presence of both of the Dresselhaus and Rashba terms (solid line). Only the Rashba term present (dashed). How β changes with semi-axis ΔE_R in the presence of both of the Dresselhaus and Rashba terms.

When only the Rashba term is present the structure of the non-Abelian vector potentials is remarkably simple. The orthonormalization $\langle\psi_i|\psi_j\rangle = \delta_{ij}$ gives that the diagonal matrix elements $(A_p)_{i,i}$ are real and that the off-diagonal elements satisfy $(A_p)_{i,j} = (A_p)^*_{j,i}$. We find non-Abelian vector potentials A_1 and A_2

$$A_1 = -\frac{E_p\left(1+\frac{1}{\sqrt{9E_p^2+9E_R^2+1}}\right)}{2\left(E_p^2+E_R^2\right)}\sigma_x, \tag{16}$$

$$A_2 = \frac{E_R\left(1+\frac{1}{\sqrt{9E_p^2+9E_R^2+1}}\right)}{2\left(E_p^2+E_R^2\right)}\sigma_x. \tag{17}$$

(Note that $\langle f(z)|\frac{\partial}{\partial E_R}|f(z)\rangle = 0$, where $f(z)$ is the subband wavefunction along the z-axis, see Fig. 2). The diagonal elements $(A_p)_{ii}$ are zero and the off-diagonal elements A_{ij} are real. Since A_1 and A_2 are proportional to σ_x it follows from Eqs.(8) and (9) that the parameter $\beta = 1$. The matrix Berry phase is given by

$$\Phi_C(n) = \begin{pmatrix} \cos(\alpha n) & i\sin(\alpha n) \\ i\sin(\alpha n) & \cos(\alpha n) \end{pmatrix}. \tag{18}$$

In addition, one can show that $[A_1, A_2] = 0$, which implies that Stokes' theorem, Eq. (7), is valid. The field strength has a simple form $F_{12} = f\sigma_x/E_0^2$, where

$$f = -\frac{9}{2\left(9x^2 + 9y^2 + 1\right)^{3/2}} \tag{19}$$

with $x = E_p/E_0$ and $y = E_R/E_0$. From this we can compute the rotation angle

$$\alpha = \int_A f dxdy, \tag{20}$$

Eq. (19) implies that $|\alpha| < 0.7854$. Figure 4 shows numerical results for α when only the Rashba term is present. We have *verified* that these numerical results agree well with the results obtained from the equation Eq. (20).

When both the Rashba and Dresselhaus terms are present Stokes' theorem is not applicable. We have investigated how $\vec{m}$ and α depend on the shape of the paths. They can be computed numerically by solving the time dependent Schrödinger equation

$$i\hbar\dot{c}_i = -\sum_j A_{ij}c_j. \tag{21}$$

The matrix elements A_{ij} are given by $A_{ij} = \hbar\sum_p (A_p)_{i,j}\frac{d\lambda_p}{dt}$, where the sum over p in A_{ij} is meant to be the sum over λ_p. We have considered elliptic paths of Eq. (15) with the area $\pi\Delta E_R\Delta E_p = 16\pi E_0^2$ with $E_{R,c} = 5E_0$ and $E_{p,c} = 5E_0$. The results are shown in Fig. 4. (Note that the dependence of E_D on the adiabatic parameter E_R is not well-known, but the essential physics does not depend on the exact functional form, as discussed in Ref. 11).

2.3. *Time dependence*

Let us now investigate the time evolution of the system in the presence of the Rashba term. The time-dependent Schrödinger equation is solved using the Runge-Kutta method. For the parameters $E_{R,c} = 2E_0$, $E_{p,c} = E_0$, $\Delta E_R = 1.9E_0$, $\Delta E_p = 0.9E_0$, and $\hbar\omega_1 = 0.1E_0$ we calculate $|c_1(t)|^2$ and $|c_2(t)|^2$, see Fig. 5. We can fit the data to $|c_1(nT)|^2 = \cos^2(\alpha n)$, $|c_2(nT)|^2 = \sin^2(\alpha n)$, with $\alpha \approx 0.182$. In this case the matrix Berry phase takes a simple form with $\beta = 1$. We have *tested* our numerical method as follows: For two different periods T_1 and T_2 we find the matrix Berry phases are the same, i.e., $c_2(T_1) = c_2(T_2)$ and $c_1(T_1) = c_1(T_2)$. This is consistent with the fact that the matrix Berry phase is a geometric effect independent of how

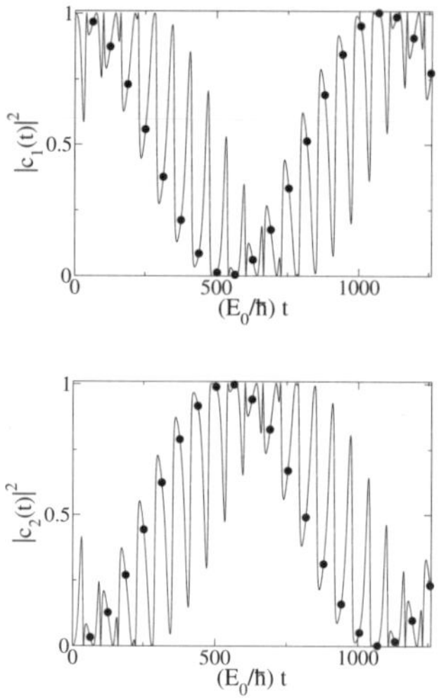

Fig. 5. Probabilities as a function of time in the presence of the Rashba spin orbit coupling. Black dots are fitted values at $t = nT$ using the analytical expression for the matrix Berry phase, Eq. (18).

the path is parameterized. Let us now include the Dresselhaus spin orbit term as described in Ref. 11. In this case the matrix Berry phase is more complicated since $\beta \neq 1$. The diagonal elements A_{ii} are non-zero and the off-diagonal elements A_{ij} are complex numbers. Using the same parameters $E_{R,c}$, $E_{p,c}$, ΔE_R, ΔE_p, and ω as in the previous calculation, we have computed $|c_1(t)|^2$ and $|c_2(t)|^2$, which are displayed in Fig. 6. These results can be fitted with the parameters $\alpha \approx -0.477$ and $\beta \approx -0.190 + i0.965$. In both cases the numerical results are consistent with those of the analytical expression at $t = nT$, given by Eq. (8). In the time interval $(n-1)T < t < nT$ the coefficients $c_{1,2}(t)$ display a complicated behavior. It should be stressed that even when the adiabatic path is *not* closed the actual state $\Psi(t)$ is given by a linear combination of the instantaneous eigenstates $\psi_1(t)$ and $\psi_2(t)$

$$|\Psi(t)\rangle = c_1(t)|\psi_1(t)\rangle + c_2(t)|\psi_2(t)\rangle. \tag{22}$$

In this case a matrix phase of the form

$$\Phi_C = Pe^{i\int_C \sum_p A_p d\lambda_p}, \tag{23}$$

will appear, where the path C is unclosed.

Experimental investigations of matrix Berry phases in II-VI and III-V semiconductor quantum dots would be most interesting. Optical dipole measurements[23] were proposed before.[11] Here we propose a transport measurement. In order to measure the matrix Berry phase the following set of manipulations can be executed:

(1) Prepare an initial state by applying a small magnetic field and taking the zero field limit, as explained in detail in Ref. 11.
(2) Perform an adiabatic cycle by changing E_R and E_p that defines the shape of the electric confinement potential, see Fig. 7(a).

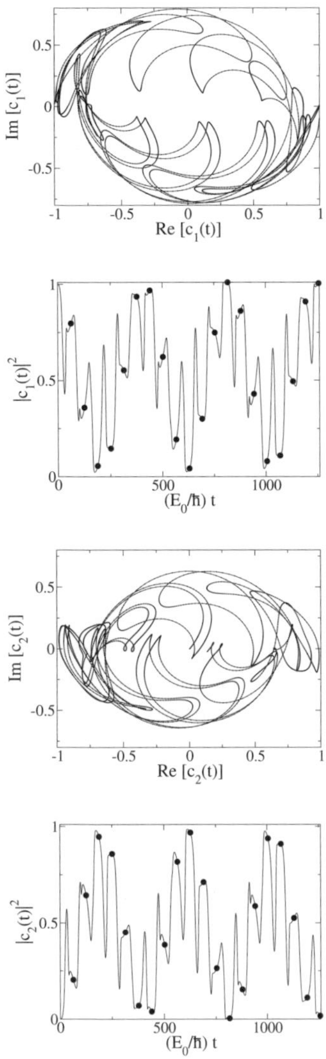

Fig. 6. Probability $c_1(t)$ when Dresselhaus spin orbit interaction is included. Black dots are fitted values using the analytical expression for the matrix Berry phase, Eq. (8). The initial state is $c_1(0) = 1$ and $c_2(0) = 0$.

(3) Change suddenly the electric field along the z-axis so that the electron can tunnel out of the dot, see Fig. 7(b).
(4) Measure the spin dependence of the tunneling current.[2]
(5) Repeat this set of procedures many times.

After this set of procedures the entire scheme is repeated one more time except step 2. If a matrix Berry phase exists it will show up as a difference between, for example, spin-up currents of these two schemes.

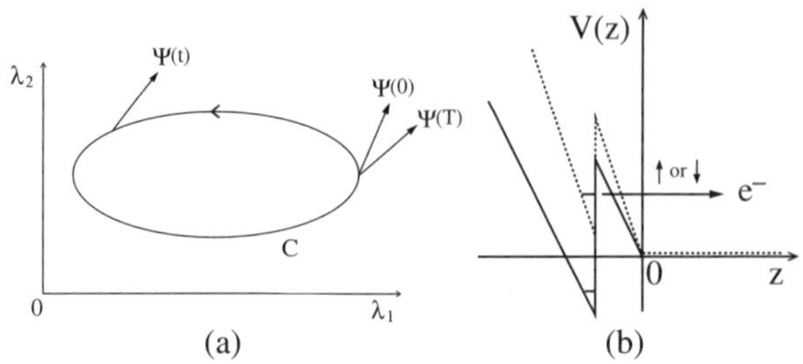

Fig. 7. (a) A matrix Berry phase is generated when the adiabatic parameters λ_1 and λ_2 go through a cyclic change. (b) After a matrix Berry phase is generated the Rashba electric field is suddenly increased so that an electron can tunnel out the dot.

3. Discussion and Conclusion

We have investigated matrix Berry phases of II-VI and III-V n-type semiconductor dots with spin orbit coupling terms. We have shown that the geometric information of a closed adiabatic path can be encoded into the matrix Berry phase via the rotation axis and angle of the spin half matrix. This result is applicable even for strongly correlated states, as long as the states are doubly degenerate. In our truncated model for the Hamiltonian the matrix Berry phase can be found *exactly* when the Rashba term is present. In this case the Stokes' theorem is applicable and the relevant field strength can be calculated analytically. Our truncated model can be applied to self-assembled quantum dots,[24] which have large energy shell differences. In these dots exchange and correlation effects can be neglected,[12] and only the electron in the last occupied energy shell is relevant. On the other hand, in gated quantum dots[25] with smaller energy differences between the shells correlation effects may be relevant. A computational scheme to calculate the matrix Berry phase in such a system is developed in Ref. 12 and 15.

Acknowledgments

This work was supported by Grant No. R01-2005-000-10352-0 from the Basic Research Program of the Korea Science and Engineering Foundation and by Quantum Functional Semiconductor Research Center (QSRC) at Dongguk University of the Korea Science and Engineering Foundation. In addition this work was supported by the Second Brain Korea 21 Project and by Grant No. C00275(I00410) from Korea Research Foundation.

References

1. D. D. Awschalom, D. Loss and N. Samarth, *Semiconductor Spintronics and Quantum* Computation (Springer, Berlin, 2002).
2. S. Datta and B. Das, *Appl. Phys. Lett.* **56**, 665 (1990).
3. D. Loss and D.P. DiVincenzo, *Phys. Rev. A* **57**, 120 (1998); S.D. Lee, S.J. Kim, Y.B.

Cho, J.B. Choi, Sooa Park, S.-R. Eric Yang, S.J. Lee and T.H. Zyung, *Appl. Phys. Lett.* **89**, 023111 (2006).
4. F. Wilczek and A. Zee, *Phys. Rev. Lett.* **52**, 2111 (1984).
5. A. Shapere and F. Wilczek (eds.), *Geometric Phases in Physics*, (World Scientific, Singapore, 1989).
6. A. Bohm, A. Mostafazadeh, H. Koizumi, Q.Niu and J. Zwanziger, *The Goemetric Phase in Quantum Systems* (Springer-Verlag, Berlin 2003).
7. P. Zanardi and M. Rasetti, *Phys. Lett.* **264**, 94 (1999).
8. P. Solinas, P. Zanardi, N. Zanghi and F. Rossi, *Phys. Rev. A* **67**, 062315 (2003).
9. Yu. A. Serebrennikov, *Phys. Rev. B* **70**, 064422 (2004).
10. B. A. Bernevig and S.-C. Zhang, *Phys. Rev. B* **71**, 035303 (2005).
11. S.-R. Eric Yang and N.Y. Hwang, *Phys. Rev. B* **73**, 125330 (2006).
12. S.-R. Eric Yang, *Phys. Rev. B* **75**, 245328 (2007).
13. S.-R. Eric Yang, *Phys. Rev. B* **74**, 075315 (2006).
14. M. Valin-Rodriquez, A. Puente and L. Serra, *Phys. Rev. B* **69**, 085306 (2004).
15. S.C. Kim, Y.J. Kim, P.S. Park, N.Y. Hwang and S.-R. Eric Yang submitted to *J. Phys. C.*
16. D.J. Thouless, *Phys. Rev. B* **27**, 6083 (1983).
17. P. W. Brouwer, *Phys. Rev. B* **58**, R10135 (1998).
18. N.Y. Hwang, S.C. Kim, P.S. Park and S.-R. Eric Yang, *Solid State Comm.* **145**, 515 (2008).
19. M.A. Nielson and I.L. Chung, *Quantum Computation and Quantum Information* (Cambridge, University Press, 2000).
20. E.I. Rashba, *Physica E* **34**, 31 (2006).
21. J. Nitta, T. Akazaki, H. Takayana and T. Enoki, *Phys. Rev. Lett.* **78**, 1335 (1997).
22. G. Dresselhaus, *Phys. Rev.* **100**, 580 (1955).
23. D. Heitmann and J. P. Kotthaus, *Phys. Today* **56** (6), 56 (1993).
24. P. M. Petroff, A. Lorke and A. Imamoglu, *Phys. Today* **54** (5), 46 (2001).
25. M.A. Kastner, *Rev. Mod. Phys.* **64**, 849 (1992).

PHYSICS OF THE MIND: OPINION DYNAMICS AND DECISION MAKING PROCESSES BASED ON A BINARY NETWORK MODEL

F. V. KUSMARTSEV and KARL E. KÜRTEN*

Department of Physics, Loughborough University, LE11 3TU, UK

**Faculty of Physics, University of Vienna, 5, Boltzmanngasse, A-1090 Vienna, Austria*

We propose a new theory of the human mind. The formation of human mind is considered as a collective process of the mutual interaction of people via exchange of opinions and formation of collective decisions. We investigate the associated dynamical processes of the decision making when people are put in different conditions including risk situations in natural catastrophes when the decision must be made very fast or at national elections. We also investigate conditions at which the fast formation of opinion is arising as a result of open discussions or public vote. Under a risk condition the system is very close to chaos and therefore the opinion formation is related to the order disorder transition. We study dramatic changes which may happen with societies which in physical terms may be considered as phase transitions from ordered to chaotic behavior. Our results are applicable to changes which are arising in various social networks as well as in opinion formation arising as a result of open discussions. One focus of this study is the determination of critical parameters, which influence a formation of stable mind, public opinion and where the society is placed "at the edge of chaos". We show that social networks have both, the necessary stability and the potential for evolutionary improvements or self-destruction. We also show that the time needed for a discussion to take a proper decision depends crucially on the nature of the interactions between the entities as well as on the topology of the social networks.

1. Introduction

The behavior of any human being is obeying a set of rules. The rules are dictated by society, education, church and family. The formation of his mind is somehow associated with the rules which he is using through his whole life. The society is playing a very important role in this process. In other words the human is a product of the society interaction, which is existing and supporting via social links. Any social group is associated with a social network, which are formed from social links where the nodes are humans. The structure and dynamical evolution within such a network has a crucial influence on our life. Social networks are formed spontaneously and their lifetime depends on many different conditions such as the stability of the state, economy and society. They consist of nodes associated with individual human beings and with social links which are formed due to the human interactions. Among a large variety of different network topologies the social network have preferentially one type of topology known as the scale free one.[1–3] The nature of this topology is related to the creation of nodes with different power that in the language of the social psychology means that all people are different. There are more or less attractive people or popular people. The nodes associated with these people have many links and named hubs. This variety of nodes leads to a formation of the scale free distribution of links attached to the nodes so-called a scale free distribution or scale free network topologies. Scale-free network topologies have been found almost everywhere in the real world. Especially this is relevant to dynamical networks which are developing and expanding in the time. Many networks expand through the addition of nodes to an already existing network, and those nodes attach preferentially to nodes which are already well connected, i.e., to very popular nodes. The example is Pareto law in economics known in popular words as the rich gets richer. When this is the case, a scale-free network naturally arises. In fact, the scale-free network is a very specific kind of network in which the distribution of connectivity or links is extremely uneven: some nodes act as "very connected" hubs, whereas most of the nodes are rather sparsely connected. Examples of networks with such a power-law distribution are computer networks, Internet and the world wide web. Such systems react significantly different from randomly connected networks in the presence of perturbations and therefore have drastically different stability properties. If nodes fail randomly, scale-free networks behave much better than randomly connected networks, because random failures are unlikely to harm an important hub. However, if the failure of nodes is not random, scale-free networks can fail catastrophically. The properties of the networks, of course, depend not only on the topology. The important role is played by the properties of the individual agents or units and how they interact with each other. In the present work we would like to introduce a general model of social networks which takes all these manifold sides of social networks into account.

2. A Model of Discrete Opinions

Let us introduce a most general model for a social network.[4–6] Such a network is made from individual multifunctional units. Each of these individuals has rules which allow him to respond to signals from other units connected to this chosen one. Thus, such rules control the everyday behavior of the individual. As a result there arises a collective behavior of such individuals, which is completely based on the rules that the individuals are using. The minimal requirement for the model of such social network is that the network should include N such participants. Each participant has a spectrum of activities which are realized via rules controlling these activities. For example, some person (here a node of the network) has an opinion about some issue, for example, which candidate to choose in presidential elections. In many such cases an individual must take a simple decision, yes or no. Such situation arises very often in a group of individuals who are sitting around a table and discuss various management and business issues. After this discussion eventually the group should come to a consensus which is expressed in the form of mutual agreement, YES, or NO. Such consensus depends on yes or no decision of

Fig. 1. Connectivity structure: Normal, Poissonian distribution. In this network all nodes are connected randomly, there are no preferential nodes and the majority of nodes have the same number of links.

each individual will make and the strength of the node. The opinion of participant, say of the ith individual unit, may be described by the state variables σ_i only capable to take the value $+1$ or -1 corresponding to the states of the two opinions YES and NO, respectively. We assume that each node, i, is influenced by K_i other nodes of the network with $1 \leq K_i \leq N$. In the simplest case the strength of all nodes are the same. When $K_i = N$ we have the "cave world" where everybody is bound with everybody, while in modern society the connectioon between people is very frustrated. The formation of human mind in the cave or in the modern human society or, for example, an opinion whom to give your vote in the presidential election is formed with the time. In many cases the formation of an opinion requires some discussion where the individuals may exchange their views to form some firm opinion. To describe this discussion process and how this progresses we split time in discrete time steps. At each time step the opinion of any participant may be changed. At each step we update the opinion of all participants in parallel. Of course this is a simplification, in reality it may happen more in a sequence or randomly, according to some stochastic process. It is very convenient to describe this group as connected individuals as a network of connected nodes. The dynamical rules governing the deterministic time evolution of each node of the social network in discrete time steps, that is, whether the node, i, will say YES or NO at the next time step, may

Fig. 2. Connectivity structure: Powerlaw distribution of links attached to individual nodes. In this network all nodes are not randomly connected, there are preferential nodes, to which the majority links are attached.

be described by the linear threshold rule

$$\sigma_i(t+1) = \operatorname{sgn}\left(\sum_{(j)} c_{ij}\sigma_j(t) - h_i\right) \qquad i = 1, ..., N \quad . \tag{1}$$

Here, the state of the ith node at the time $t+1$ depends only on the value of its K_i inputs variables at the previous time step at time, t. The bracket term specifying the net internal stimulus felt by the ith node, is given in terms of a weighted sum over the input states of those nodes which influence with the node, i. The quantity h_i represents a threshold for node or unit i, that is, a resistance of the unit i to change the opinion under the influence of the connected nodes. The elements c_{ij} are usually not symmetric with respect to interchange of the subscripts i and j. This coefficient defines the strength with which the individuals j acts on the individual i. When we exchange the indices then the element c_{ji} will define the strength with which the individuals i acts on the individual j. Obviously, the action from the individual i on j is not equal to the action from the individual j on i. So the third law of Newton is not valid for social networks. The similar situation arises for neural networks where the value of c_{ij} takes a positive (negative) value, if unit j has an excitatory (inhibitory) effect on unit i; its magnitude quantifies the efficacy or interaction strength of unit j upon the target unit i. It is zero, if the unit j does not influence the unit i. If there is a simple majority rule, where the node i is taken the state which is associated with the majority states from the connected nodes, then all coefficients $c_{ij} = 1$ with threshold $h = 0$.[7] In the opposite case of the minority rule, when the ith individual takes an opinion opposite to the opinion expressed by the majority people connected to him then all coefficients $c_{ij} = -1$. It is now important to make a few remarks about the dynamical nature of the network evolution. Due to the binary nature of the variables there exist only 2^N different configurations in phase-space. During the evolution the system may go either through the same configuration around a close trajectory in the phase space, then the evolution will be inevitably periodical or it may have a fixed point. Therefore, the network dynamics driven by Eq. (1) will inevitably evolve through a transient phase to an attractor, which is either a limit cycle or a fixed point, characterized by a peridic change of the opinions of arbitrary agents with the time. Generally, the long-term behavior of such networks is considered to determine their usefulness in a prediction societies behavior, in both the natural and artificial contexts: Some aspects of cycling and fixed point phenomena, often interpreted as the response of the network to stimuli expressed as initial conditions, are of special interest. In our model the time evolution of the opinion vector $\overline{\sigma}$, where each participant expresses his opinion, describes the ongoing discussion in a series of time steps, which — due to the deterministic nature of the dynamics — might be periodic or reach a fixed point. The length of the period depends on arguments and the influence of the participants and the threshold associated with the confidence of the chosen ith participant. Our model expressed by Eq. (1) effectively describes the evolution of the social network and is summarizing the discussion taking into account individual opinions σ_i of all participants at each time step and the influence of the opinions from their neighbors at each time step of the network evolution. Therefore, it may give an important result both on the short- and long-terms of the state of the social network. Similar mathematical models have been used in the past to mimic the time evolution of neural network. The present situation differs by the complexity of the social network which consists of the group of individuals, where each has very different rules of behavior, different strength of interaction, which depends on the particular other unit and which depends on the confidence of the ith individual described by a set of variables h_i. Such variables may have different distributions, depending on society and the environment the individuals live. In spite of such differences we expect that dynamical behavior leading to a formation of limit cycles will be similar to the Kauffman's evolutionary model for cell differentiation[8] where the total number of cyclic modes scales with the highly limited number of different cell types in living organisms or to neural network models where the limit cycles can be considered as model analogues of content-addressable memories.

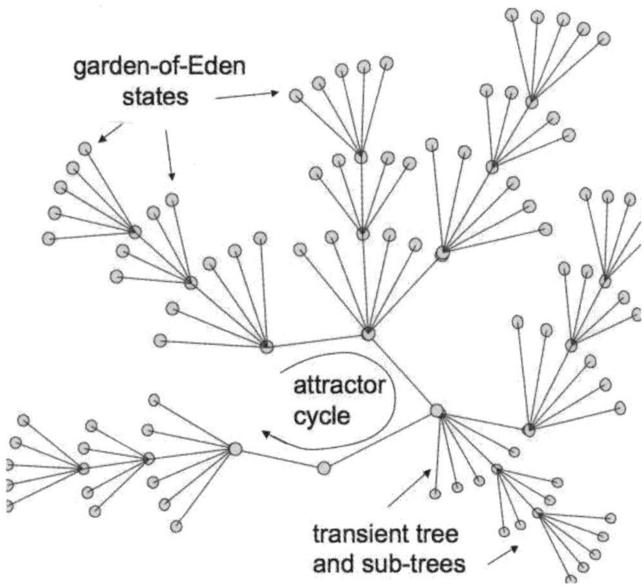

Fig. 3. Schematic representation of the possible evolutions of social networks presented as the garden of Eden states: the Eden tree. This tree includes the limit cycle, which is just in a center of the diagram, then transient trees and sub-trees, which are ends by fixed points and other attractors.

3. Dynamical Evolution of a Randomly Connected Social Network

Let us consider a specific model where (a) there is a random connectivity between the people and they are (b) sparsely connected ($K \sim \log(N)$). Such social network arises very often on a party where all people invited do not know each other like at the introduction's reception on the first day of holiday break. On that party the people begin to meet each other with probability distribution P_k, and when they got some friends, they are connected. The number of connections depends how long the reception lasts and they are usually connected sparsely for all sorts of reasons. We may also assume that (c) there are random weights c_{ij} with probability distribution $\rho(c_{ij})$. Under these assumptions the time evolution of the normalized Hamming distance representing the fraction of bits being different in two initial configurations $\sigma^{(1)}(t)$ and $\sigma^{(2)}(t)$

$$d(t) = \frac{1}{2N} \sum_{\nu=1}^{N} |\sigma_\nu^{(1)}(t) - \sigma_\nu^{(2)}(t)| \tag{2}$$

can be derived exactly.[9–11] This distance may describe how gossips spreads around this small community. In our context the concept of a perturbation is opinion changes in one or a few agents. In order to evaluate the effects we consider an identical replica of the system and compare the time evolution of the two replicas with slightly different initial conditions, where some agent starts with different opinions which happen due to an origination of some gossip. The so-called Hamming distance defined by Eq. (2), then tells us how many agents have changed their mind at each time stepe of the time evolution. One can also study how the perturbations spreads over the system as a function of time. These damage spreading processes have long been applied originally introduced by Kauffman,[8] who intended to study the reaction of gene regulatory networks due to external influences.[1] The evolution leads to a formation of two stationary states. The first one is the ordered state, where an infinitesimally small initial distance $d(t = 0)$ either remains confined or eventually disappears in the large time limit. This happens when nobody buys or has no interest in this gossip and either it is accepted as an information or vanishes with the time. The second stationary state of the small community may be termed as chaotic phase: there an infinitesimally small initial distance $d(t = 0)$ evolves to a finite distance. For a symmetric distribution $\rho(c_{ij}) = \rho(-c_{ij})$, the time evolution of the individual Hamming distance $d_k(t)$ can be written as

$$d_k(t+1) = 1 - \sum_{j=0}^{k} \binom{k}{j} I_{kj} d^j(t)(1-d(t))^{k-j} \tag{3}$$

where the probability integral

$$I_{kj} = \int \cdots \int dx_1 \cdots dx_k \rho(x_1) \cdots \rho(x_k) \theta\{(x_{j+1} + \cdots + x_k + h)^2 - (x_1 + \cdots + x_j)^2\} \tag{4}$$

specifies the probability that a sign reversal of j randomly chosen input variables σ_j will *not* affect the output state of the system. The sensitivity integrals I_{kj} depend on the number of input variables K, on the probability distribution of the synaptic weights $\rho(c_{ij})$, and on the threshold h. The dynamical long-term behavior of the system then depends on whether or not the natural fixed point $d \equiv 0$ is attractive. One may easily estimate the necessary condition for stability, where the Hamming-distance vanishes:

$$\frac{\partial d(t+1)}{\partial d(t)}|_{d\equiv 0} = k(1 - I_{k1}) = 1. \tag{5}$$

The stability only depends on the one spin flip integral I_{k1}

$$I_{k1} = \int \ldots \int dx_1 \ldots dx_k \rho(x_1) \ldots \rho(x_k) \theta\{(x_2 + \ldots + x_k + h)^2 - x_1^2\}. \tag{6}$$

Provided that the normalized distribution of the connectivity P_k is known, the generalized Hamming distance $d(t)$ is a linear combination of the indiviual distances $d_k(t)$:

$$d(t) = \sum_{k=0}^{K} P_k d_k(t) = 1 - \sum_{k=0}^{K} \sum_{j=0}^{k} P_k \binom{k}{j} I_{kj} d^j(t)(1-d(t))^{k-j} \tag{7}$$

with the generalized stability condition

$$\sum_{k=1}^{K} P_k k(1 - I_{k1}) = 1. \tag{8}$$

4. Applications to Stability of Social Networks: Communities with Poisson- and Dirac-Versus Power-Law Probability Distributions

For the connectivity structure of the exemplary typical human communities we choose a Poisson- or power-law probability distribution

$$P(k) = \frac{\alpha^k}{k!} / \sum_{j=0}^{K} \frac{\alpha^j}{j!} \tag{9}$$

or

$$P(k) = \frac{1}{(k+1)^\gamma} / \sum_{j=0}^{K} \frac{1}{(j+1)^\gamma}, \tag{10}$$

respectively. The other aspect is the strength of interactions between the individuals which is described by the couplings c_{ij}. To describe two contrast communities where the first one associated with a society where all people are cloned and therefore identical. In this case the strength of their mutual interaction is identical. Such distribution of interactions is modeled by bivalent Dirac-delta functions $\delta(x)$ via

$$\rho(c_{ij}) = \frac{1}{2}[\delta(c_{ij} - 1) + \delta(c_{ij} + 1)]. \tag{11}$$

The second example is most relevant to a majority of people's communities, where the constant of interactions is distributed according to a power-law

$$\rho(c_{ij}) = \frac{1}{2} \frac{(1-\alpha)}{|c_{ij}|^\alpha} \tag{12}$$

with the power index equal to α. The corresponding single spin flip integrals Eq. (6) for threshold zero, i.e., $h = 0$, and the Dirac distribution, Eq. (11) are analytic. We have here that[3]

$$I_{k1} = 1 - \frac{1}{2^k} \binom{k}{[\frac{k-1}{2}]}. \tag{13}$$

The square brackets in the binomial coefficient denote floor function. For the power-law distribution Eq. (12) the integrals have to be calculated numerically with suitable multi-dimensionional Monte Carlo methods.[3] Inserting the connectivity distributions $P(k)$ together with the corresponding stability integrals I_{k1} into Eq. (8) we can solve the non-linear equation for the parameters γ or α in order to get the critical parameter values γ_c and α_c, respectively. Note that for the connectivity the

Table 1. Critical average connectivity $< K >$ for two probability distributions for the network architecture and the coupling coefficients, respectively. The maximal in-degree is $K = 30$.

TOPOLOGY/COUPLINGS	Dirac	Power-law ($\alpha = \frac{2}{3}$)
Poisson	1.85	2.63
Power-law ($\gamma = 1.5$)	3.44	5.29

Poisson- as well as the power-law distributions allow that a fraction of units specified by $P(k=0)$ does not receive any stimulus from the whole network. These are outsides of the society, i.e., they are decoupled from the rest of the network and therefore they have vanishing influence on the opinion formation. Considering that the stability of the original Kauffman model[8] for gene regulation was limited to an average connectivity $< K > \quad \approx \quad 2$, the results manifest a clear advantage for the power-law distributions in promoting dynamically stable networks, earlier reported in the work by Aldana and Cluzel.[2]

5. Comparison of Opinion Formation in Societies having a Connectivity Architecture with Uniform and Power-Law Distributions

Let us first consider the limiting power-law connectivity distribution $P(k)$ defined by Eq. (10) for $\gamma = 0$ which represents a uniform distribution, where each node has the same chance to have k incoming connections with $0 \leq k \leq N$. This is of course a very artificial society where each member of the society may have any number of acquaintances or friends with equal probability. This is the highly connected society since in this society the number of hubs are the same as the number of weakly connected nodes. Therefore, betweenness of nodes is high. That means that each individual obtain information from different agents at the same time, which are contradictory to each other. Therefore in this society many hubs may compete with each other and may have an opposite influence on the formation of the individual opinions. Obviously, this results in chaotic fluctuations of the individual opinions. Figure 4 (left) depicts the time evolution of opinions in such a society. At each moment of time the opinions of individual members are presented on the horizontal line

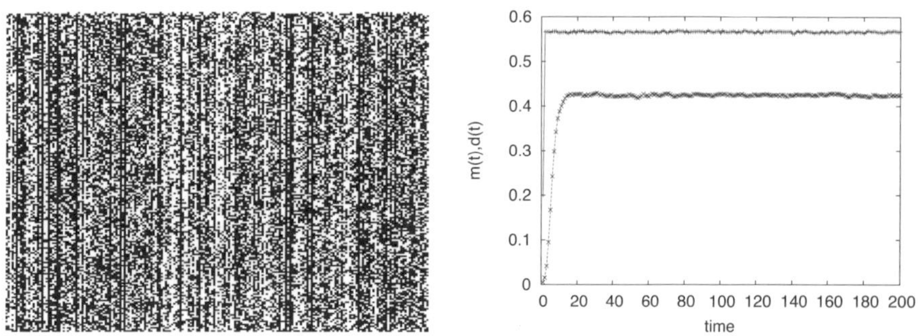

Fig. 4. (left) The evolution of the opinion of the individual members of the society, i.e., what each individual thinks. The black square corresponds to YES, while the white square corresponds to NO. We distribute the individuals(squares) along the horizontal line and present here the space-time plot, where time is changing along the vertical axes, from the top down; (right) The time evolution of the opinion of the whole society, i.e., what thinks the majority, which we call as the "magnetization" of the society and denote as $m(t)$. The function $d(t)$ shows the evolution of the opinion changes of the society, i.e., a fraction individuals who change their opinion at each time step.

consisting of the small black and white squares. The black square corresponds to YES, while the white square corresponds to NO. The initial starting configuration of opinions across the society corresponds to the uppermost line. The time changes from the top down. From the evolution pattern made of YES- and NO- (or black- and white-)squares presented in Fig. 4 (left) we may conclude that in such society the opinion evolution of an individual member, which corresponds to the vertical line, fluctuates chaotically as we have expected. This observation is confirmed by an calculation of the asymptotic value of the distance, $d^* \approx 0.42$. That indicates that any fluctuation or a gossip arising in such society strongly influences the opinion of individual members creating numerous discussions which are never ended. One may call such society as socially unstable respect to any external perturbation or information. And it does not matter if this information right or false. However, in spite of the fact that each member of such a society changes his opinion chaotically, it is interesting to know the opinion of the majority. One may investigate here the evolution of the opinion of the whole society, i.e., what the majority thinks, YES or NO and how large is this majority. To quantify this property it is convenient to introduce the value of an average opinion of the society at some chosen time t. In the language of the spin system this average opinion will correspond to the normalized "magnetization" of the society $m(t) = \sum_i^N \sigma_i(t)/N$. Figure 4 (right) shows the corresponding time evolution of the society magnetization $m(t)$ and the distance $d(t)$. Note that the asymptotic value for the magnetization is reached after the first time step. That is, in spite that the opinion of individual members changes randomly,

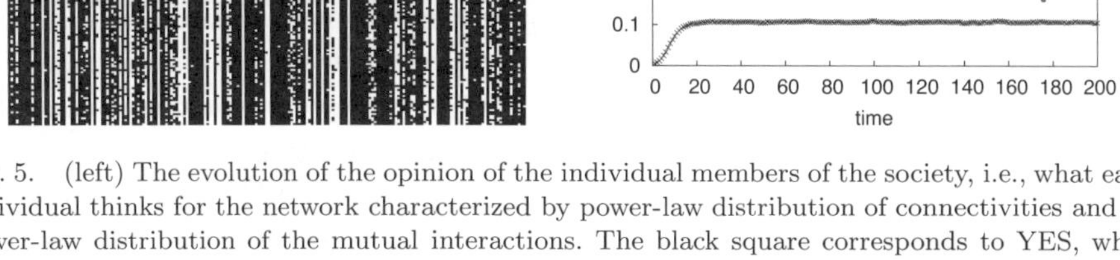

Fig. 5. (left) The evolution of the opinion of the individual members of the society, i.e., what each individual thinks for the network characterized by power-law distribution of connectivities and by power-law distribution of the mutual interactions. The black square corresponds to YES, while the white square corresponds to NO. We distribute the individuals(squares) along the horizontal line and present here the space-time plot, where time is changing along the vertical axes, from the top down; (right) The time evolution of the opinion of the whole society, i.e., what the majority thinks, which we call as the "magnetization" of the society and denote as $m(t)$. The function $d(t)$ shows the evolution of the opinion changes of the society, i.e., a fraction individuals who change their opinion at each time step. The space-time portrait made of the black and white squares is arising here at the conditions close the the "phase transition" between two phases of society, opinion ordered or chaotic. In other words these conditions are at the edge of chaos where $d = 0.05$. For the scale free network the power index $\gamma = 2.5$.

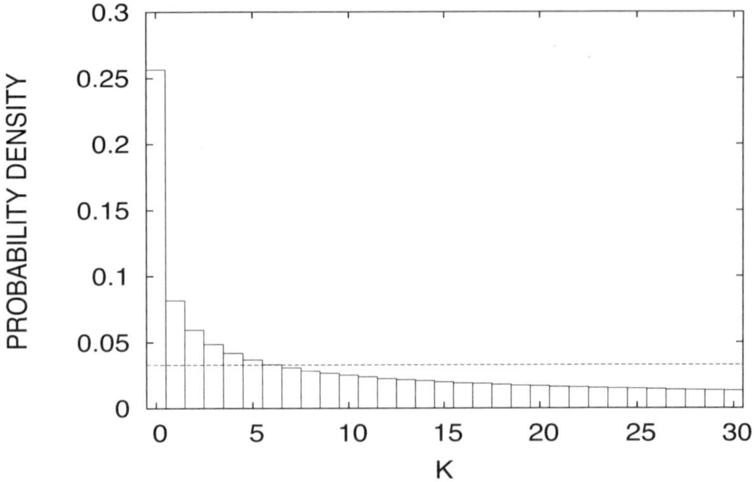

Fig. 6. Probability density for the power-law distribution with $\gamma = 1.5$ (boxes) and corresponding uniform distribution (dotted line) with $\gamma = 0$ for a maximal value of $K = 30$.

the majority of the society came to a consensus with $m^* \approx 0.57$ very fast. The fluctuations around this asymptotic value can be shown to be Gaussian, the amplitudes depend on the system size N. The distance $d(t)$ first increases exponentially[11] and approches monotonically its asymptotic value $d^* \approx 0.42$. With increasing power-law control parameter γ the social stability can be strongly enhanced. However eventually at very large values of γ it turns out to be counterproductive, since for larger values of γ the network will have less and less hubs and more people will have fewer links. This means that the individual members of the society become more and more disconnected. The situation opposite to the chaotic regime described above is the ordered society where each member has a very firm opinion, which is not changed within time. Such an opinion is formed very fast and the society is well ordered. Changing the topological structure of the social network may cause drastic changes reminiscent of a phase transition arising in many physical systems. Let us investigate what happens if we are in the vicinity of such a phase transition. Figures 5(a)–(b) depict the corresponding situation close to a transition, where the state of the society may be changed. In this case the society is close to the edge of chaos (supercritical state)[8,12] with the finite distance $d^* \approx 0.11$. Here the power in the scale free distribution of connectivities is $\gamma = 1.5$. We observe that a big portion of the individuals in such a society do not change their mind; however there is still a small portion of individuals which show moderate cycling behavior, i.e., their opinions are periodically changing. Figure 6 depicts the power-law density distribution (boxes) for $\gamma = 1.5$ and the uniform distribution (presented by the dotted line) for $\gamma = 0$. We observe here that about 25% of the agents do not receive any stimulus ($P(k = 0) \approx 0.25$), although they might have outgoing connections and stimulate other units. There the society state is frozen and uniquely specified by the initial

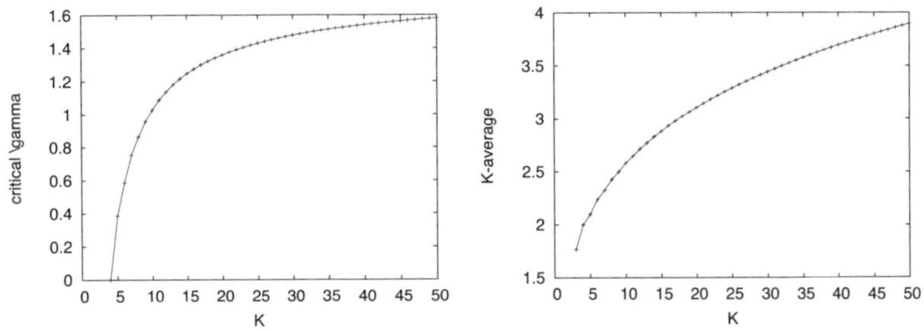

Fig. 7. Critical scale free parameter γ_c (left) and the mean connectivity $< K >$ (right) as a function of the maximal in-degree K.

condition or its history. The most important factor stabilizing this society is the existence of a few hubs which have many connections and therewith may have a coherent influence on many individuals pushing the majority to form a firm opinion. However, when the proportion of hubs increases, this leads to a chaotic instability, where society may be transformed into a chaotic phase. Figure 7 (left) shows the critical control parameter γ_c which specifies the power-law distribution as a function of K. For $K = 4$ we have $\gamma_c = 0$ such that already the uniform distribution stabilizes the system. It is not surprising because here the majority of individuals will have only a few connections anyway. With increasing values of γ the distribution deviates more and more from the uniform and the critical value of γ_c increases logarithmically. The corresponding mean connectivity $< K >$ of the network depicted in Fig. 7 (right) seems to increase according to a power-law $\approx K^{0.22}$.

6. Conclusion

In summary we have proposed here to study social networks with the use of a binary network model. Similar models have been used before for various biological systems.[8–11] Special attention has been given to the issue of the public opinion formation or a presidential elections. It was shown that the stability of the public opinion as well as opinions of individual members of human society depend strongly on the topology of the network architecture and on the strength of mutual interactions between individuals. In the framework of binary network models we have shown that the psychological stability of the society associated with the formation of firm opinions of the individual members of the society and therewith a firm public opinion depends crucially on the probability distribution of the connectivity structure as well as on the probability distribution of the interaction weights. The importance of the hubs for the stability of the social network of opinions and a formation of a public opinion has been also discovered. The existence of a small portion of hubs stabilizes the formation of a stable public opinion and of a stable opinion of individual members; however when the proportion of hubs increases, they begin to destabilize the formation of stable opinion of individual members that leads to a chaotic fluctuation of a public opinion. Due to the existence of the hubs and their synchronous influence on many individuals the widely discussed scale free distributions of the social connections clearly outperform the conventional choices such as the random or Poisson distribution. For social networks which are intimately scale free networks the mean connectivity of the systems, which was limited to $< K >= 2$, in the original models,[8,9] could be largely shifted to higher values of $< K >$ giving rise to more realistic models of biological and social networks, where the nodes are usually governed my more than two incoming connections on average.

Acknowledgments

The authors thank J.W. Clark, L. Raeymakers and D. Stauffer for numerous useful discussions about biological and social implications. This work has been supported by the European Science Foundation (ESF) in the framework of the network program: "Arrays of Quantum Dots and Josephson Junctions".

References

1. A.-L. Barabási and R. Albert, *Science* **286**, 509 (1999).
2. M. Aldana and P. Cluzel, *Proc. Natl. Acad. Sci. (USA)* **100**, 8710 (2002).
3. K. E. Kürten an J. W. Clark, *Phys. Rev. E* **77**, 1 (2008).
4. C. Castellano, S. Fortunato and V. Loreto, Statistical physics of social dynamics, arXiv:0710.3256v1 (2007).
5. D. Stauffer Opinion dynamics and sociophysics, arXiv:0705.0891 (2008).
6. S. Galam, *J. Math. Psychol.* **30**, 426 (1986).
7. S. Galam, *Physica A* **33**, 453 (2003).
8. S. A. Kauffman, *Origins of Order: Self-Organization and Selection in Evolution* (Oxford University Press, Oxford, 1993).
9. K. E. Kürten, *Phys. Lett.* **A129**, 157 (1988).
10. B. Derrida, *J. Phys. A* **20**, L721 (1987).
11. K. E. Kürten, *J. Phys. A: Math. Gen.* **21**, L615 (1988).
12. K. E. Kürten and H. Beer, *J. Stat. Phys.* **87**, 929 (1997).

NANOLAYERED MAX PHASES FROM *ab initio* CALCULATIONS

W. LUO, C. M. FANG and R. AHUJA*

Condensed Matter Theory Group, Department of Physics and Materials Physics, Uppsala University, Box 530, SE-751 21 Uppsala, Sweden

rajeev@fysik.uu.se

Received 31 July 2008

The advancement in new materials processing and fabrication techniques has made it possible to better control the atomistic level of structures in a way, which was not feasible only a decade ago. If one can couple this atomic control with a good understanding of the relationship between structure and properties, this will in the future lead to a significant contribution to the synthesizing of tailor-made materials. In this paper we have focused on, the structurally related nanolayered ternary compounds $M_{N+1}AX_N$, (MAX) where N = 1, 2 or 3, M is an early transition metal, A is an A-group (mostly IIIA and IVA) element, and X is either C and/or N, which has attracted increasing interest owing to their unique properties. The general relations between the electronic structure and materials properties of MAX phases have been elaborated based on *ab initio* calculations.

Keywords: High pressure; electronic structure; total energy; MAX phases.

In recent years, the structurally related layered ternary compounds $M_{N+1}AX_N$, where N = 1, 2 or 3, M is an early transition metal, A is an A-group (mostly IIIA and IVA) element, and X is either C and/or N, attract increasing interest owing to their unique properties.[1–10] These ternary carbides and nitrides combine unusual properties of both metals and ceramics. Like metals, they are good thermal and electrical conductors with electrical and thermal conductivities ranging from 0.5 to 14×10^6 $\Omega^{-1}m^{-1}$, and from 10 to 40 W/m·K, respectively. They are relatively soft with Vickers hardness of about 2–5 GPa. Like ceramics, they are elastically stiff, some of them like Ti_3SiC_2, Ti_3AlC_2 and Ti_4AlN_3 also exhibit excellent high temperature mechanical properties. They are resistant to thermal shock and unusually damage tolerant, and exhibit excellent corrosion resistance. Above all, unlike conventional carbides or nitrides, they can be machined by conventional tools without lubricant, which is of great technological importance for the application of the $M_{N+1}AX_N$ phases. These excellent properties mentioned above make the $M_{N+1}AX_N$ phases another family of technically important materials.

*Corresponding author.

To date, the most extensively studied $M_{N+1}AX_N$ phase is Ti_3SiC_2, which is demonstrated to have promising applications in the following areas owing to its unique properties. Firstly, Ti_3SiC_2 and possibly some of the other ternaries are good candidates for high temperature applications. The density of Ti_3SiC_2 is roughly half that of the currently used Ni-based superalloys, but double their stiffness and possess very decent high temperature mechanical properties, which make them potential candidates to substitute for Ni-based superalloys. Secondly, they can be used as machinable ceramics. The conventional machinable ceramics require a sintering step after machining which usually results in about 2% shrinkage. The advantage of using Ti_3SiC_2 and other related ternary compounds as machinable ceramics is that the machining is final, i.e., they do not need the fired state. Thirdly, they can be used as kiln furniture. The good oxidation resistance and machinability, relatively low cost of raw materials and excellent thermal shock and chemical resistance make this application very possible. Finally they can be used as heat exchanges. The excellent thermal conductivity together with its chemical stability, good machinability and resistance to thermal shock make this application obvious.

Although significant progress has been made in the last few years in understanding these layered ternary compounds, there are several issues that have either been glossed over or ignored, amongst them, especially the theoretical investigations on these compounds. For example, what about the electronic structure and bonding properties of these $M_{N+1}AX_N$ phases? Since there are roughly 50 $M_{N+1}AX_N$ phases, experimental study alone will not be efficient from the time and economic point of view, hence theoretical calculation is necessary. These compounds share the same characteristic that they do not melt but decompose into MX and A element at high temperatures, so what about their phase stability and transformation mechanism? These works cannot be done by experiments or alone. Therefore, systematic theoretical work is very necessary and instructive to investigate these ternary compounds by First Principles.

We present results of first-principles calculations of the (novel) ternary transition metal carbides, the M_3SiC_2 series (M = early transition metals). New structures of the series were predicted for future experimentalists. The relationship of the mechanical properties between the ternary carbides M_3SiC_2 and their corresponding binary carbides MC is found to obey the 3/4 rule. Mechanism of this rule is analyzed. The present calculations are based on the density functional theory (DFT) within both local density approximation (LDA) and generalized gradient approximation (GGA) which were performed using VASP code.[11,12] The projector augmented wave (PAW) potentials are employed. The relaxation convergence for ions was 1×10^{-4} eV and that for electrons was 1×10^{-5} eV.[13] The conjugate gradient optimization convergence of the wave functions, the reciprocal-space integration with a Monkhorst-Pack scheme,[14] the k-points grid of $7 \times 7 \times 7$ and the tetrahedron method with Blchl corrections[15] for the energy were used. The equilibrium volume, the bulk modulus and its pressure derivative (B′) were derived by fitting the calculated data to the third-order Birch-Murnaghan's equations of states (B-M EOS).[16]

Most of the binary early transition metal carbides have the NaCl-type structure except WC which has a hexagonal-type structure.[17–19] The calculated lattice parameter and bulk modulus for these binary carbides are in good agreement with the experimental data and former theoretical calculations.[17–20] Table 1 lists the calculated results (lattice parameters, volumes, internal structural parameters, c/a ratio and bulk modulus) for the M_3SiC_2 compounds. As shown in Table 1, the optimized lattice parameters for Ti_3SiC_2 are in good agreement with the experimental values,[1,7] as well as other calculations,[20,21] which as a test of our computational technique. Note that the calculated bulk modulus depends to some extent on the computational method (LDA or GGA, VASP, FP-LMTO or FPLAPW), with a scatter in the calculated data of approximately 20%. Since the other MAX phases are predicted, no experimental data are available to compare these calculations with.

Table 1. The calculated results for the (novel) M_3SiC_2 (M = a transition metal) compounds using Density Functional Theory (DFT). M_3SiC_2 with Space group $P6_3/mmc$ (no. 194 in the Table), M1 at 2a (0 0 0), M2 at 4f (1/3 2/3 z_{M2}), Si at 2b (0 0 1/4) and C at 4f (1/3 2/3 z_C). The experimental data for Ti_3SiC_2 are included for comparison.

	c/a	z_{M2}	z_C	$V(a^3)$	B (GPa)
Ti_3SiC_2	5.7823	0.1340	0.9279	146.80	187
	5.7811[7]	0.135	0.932[7]	143.50[7]	190[2,6]
					206[2,7]
V_3SiC_2	6.0150	0.1318	0.9295	129.06	219
Cr_3SiC_2	5.8639	0.1307	0.9281	121.07	234
Zr_3SiC_2	5.5848	0.1399	0.9256	183.94	173
Nb_3SiC_2	5.9557	0.1358	0.9288	161.53	201
Mo_3SiC_2	5.8522	0.1377	0.9277	147.96	236
Hf_3SiC_2	5.6174	0.1377	0.9262	176.26	185
Ta_3SiC_2	6.0181	0.1358	0.9295	158.40	231
W_3SiC_2	5.9563	0.1347	0.9287	147.79	273

The Zr_3SiC_2 is predicted to have the largest volume in the series M_3SiC_2 compounds (Table 1). Further, the MAX compounds with M = V, Nb and Ta, have larger c/a ratios than the others. The volumes of the M_3SiC_2 compounds decrease for M metals in the same row in the periodic table with increasing the atomic number, while the mechanical properties show a different behavior. The bulk modulus increases with increasing atomic number. The same type of behavior was found for the M_2SiC series.[10] The calculated bulk modulus for Ti_3SiC_2 is about 187 GPa, in good agreement with the experimental data (190 ± 10 GPa or 206 GPa),[1,7,22,23] and the previous theoretical value (204 GPa).[24] Table 1 also shows that the bulk modulus of the M_3SiC_2 systems (M = early *3d* metal) increases from M = Ti (187 GPa) to M = V (219 GPa) to M = Cr (234 GPa), also in line with the behavior of the M_2SiC series.[10] As shown in Fig. 1, the value of the bulk modulus of a ternary

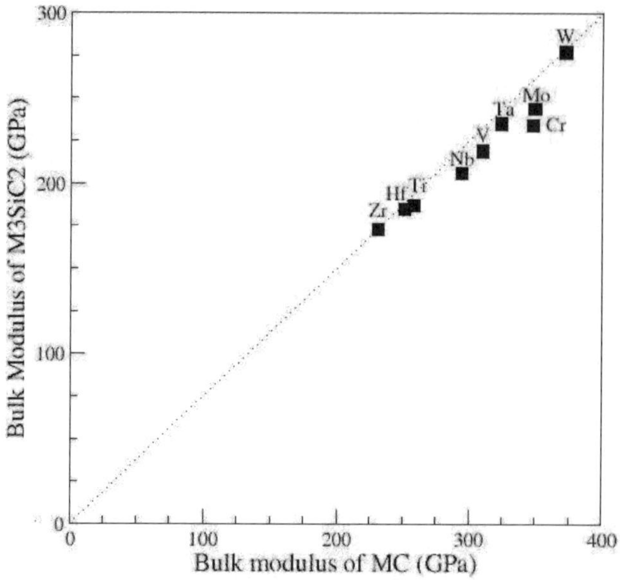

Fig. 1. The bulk moduli of M_3SiC_2 versus the bulk moduli of the corresponding MC, where M is a transition metal. The dotted line corresponds to the 3/4 rule.

M_3SiC_2 compound is approximately 3/4 of that of the corresponding binary transition metal carbide. Two ternary 4*d* metal (M = Nb and Ta) compounds have their bulk modulus somewhat under the 3/4 line. We also noted that Cr_3SiC_2 is the only magnetic compound in this series. The magnetic moment is calculated to be about 0.45 μB/Cr-atom at the equilibrium volume, and we find that the magnetic moment decreases slightly with pressure. The calculations show that the novel CrC compound with the NaCl-type structure does not order magnetically. The calculated bulk modulus of the Cr_3SiC_2 phase deviates most from the 3/4 line (see Fig. 1), which is likely to be caused by the magnetism in this system.

In order to analyze the origin of the 3/4 rule we proceed with a simple model where the energy of the bonds between Ti and C planes is written as $E = K_1(c-c_0)^2$, where c_0 is the equilibrium distance between Ti and C planes and K_1 is a force constant. The energy of the binding between Ti and Si planes is written in a similar way as $E = K_2(c - c_0)^2$, no other contributions are considered since as pointed out by Ahuja *et al.* these systems are rather isotropic when it comes to the bulk modulus and for this reason an analysis based on interactions in one dimension suffices. A conventional unit cell of, e.g., Ti_3SiC_2 contains 8 Ti-C bonds and 4 Ti–Si bonds, which gives $E_{MAX} = 8K_1(c-c_0)^2+4K_2(c-c_0)^2$. A similar expression can be obtained for TiC by replacing the 4 Ti-Si bonds with Ti-C bonds and we obtain in this case $E_{TiC} = 12K_1(c - c_0)^2$. With this simple model the bulk modulus can be calculated analytically and we obtain a ration between the bulk modulus of the MAX phase to that of TiC as, $B_{MAX}/B_{TiC} = 2/3 + K_2/3K_1$. We can now identify a criterion between the force constants, K_1 and K_2, for when the ratio B_{MAX}/B_{TiC} should equal 3/4, and this criterion is $K_1 = 4\ K_2$, in other words the Ti-C binds should have a force constant four times stronger than that of the Ti-Si bonds. In summary, using a first-principles method, we have predicted the structure and mechanical properties of a series of new M_3SiC_2 phases. The bulk moduli of these ternary carbides are approximately 3/4 of those of their corresponding binary carbides. Our predictions for new MAX phases will hopefully motivate further experimental work on these new MAX phases.

Acknowledgments

We acknowledge support from the Carl Tryggers Stiftelse (CTS) and useful discussion with Dr. C.M. Fang.

References

1. M.W. Barsoum, *Prog. Solid St. Chem.* **28**, 210 (2000).
2. J.E. Spanier, S. Gupta, M. Amer and M. W. Barsoum, *Phys. Rev. B* **71**, 012103 (2005).
3. M.W. Barsoum and L. Farber, *Science* **284**, 937 (1999).
4. M.W. Barsoum, T. Zhen, S. R. Kalidindi, M. Radovic and A. Murugaiak, *Nature Materials* **2**, 107 (2003).
5. T. El-Raghy, A. Zavaliangos, M.W. Barsoum and S.R. Kalidindi, *J. Am. Ceram. Soc.* **80**, 513 (1997).
6. C.J. Gilbert, D.R. Bloyer, M.W. Barsoum, T. EI-Raghy, A.P. Tomsia and R.O. Ritchie, *Scripta Mater.* **42**, 761 (2000).
7. M.W. Barsoum and T. El-Raghy, *J. Amer. Ceram. Soc.* **79**, 1953 (1996).
8. H. Yoo, M.W. Barsoum and T. El-Raghy, *Nature* **407**, 581 (2000).
9. S.E. Lofland, J.D. Hettinger, K. Harrell, P. Finkel, S. Gupta, M.W. Barsoum and G. Hug, *Appl. Phys. Lett.* **84**, 508 (2004).
10. Z.M. Sun, R. Ahuja, S. Li and J.M. Schneider, *Appl. Phys. Lett.* **83**, 899 (2003).
11. G. Kresse and J. Hafner, *Phys. Rev. B* **48**, 13115 (1993).
12. G. Kresse and J. Hafner, *Phys. Rev. B* **49**, 14251 (1994).
13. J.P. Perdew and Y. Wang, *Phys. Rev. B* **45**, 13244 (1992).
14. H.J. Monkhorst and J.D. Pack, *Phys. Rev. B* **13**, 5188 (1976).
15. P. E. Blöchl, *Phys. Rev. B* **50**, 17953 (1994).
16. F. Birch, *J. Geophys. Res.* **83**, 1257 (1978).
17. R.W.G. Wyckoff, *Crystal Structures* (John Wiley & Sons, New York, London, Sydney, 1967).
18. J.C. Grossman, A. Mizel, M. Cote, M.L. Cohen and S.G. Louie, *Phys. Rev. B* **60**, 6343 (1999).
19. J. Haglund, G. Grimvall, T. Jarlborg and A.F. Guillermet, *Phys. Rev. B* **43**, 14400 (1991).
20. R. Ahuja, O. Eriksson, J.M. Wills and B. Johansson, *Phys. Rev. B* **53**, 3072 (1996).
21. J.Y. Wand and Y.C. Zhou, *J. Phys.: Condens. Matter* **15**, 1983 (2003).
22. P. Finkel, M.W. Barsoum and T. El-Raghy, *J. Appl. Phys.* **85**, 7123 (1999).
23. A. Onedera, H. Hirano and T. Yuasa, *Appl. Phys. Lett.* **74**, 3782 (1999).
24. R. Ahuja, J.M. Wills and B. Johansson, *Appl. Phys. Lett.* **76**, 2226 (2000).

EFFECT OF DISORDER ON THE INTERACTING FERMI GASES IN A ONE-DIMENSIONAL OPTICAL LATTICE

GAO XIANLONG

Department of Physics, Zhejiang Normal University, Jinhua, Zhejiang Province, 321004, China

M. POLINI and M. P. TOSI

NEST-CNR-INFM and Scuola Normale Superiore, I-56126 Pisa, Italy

B. TANATAR*

Department of Physics, Bilkent University, Ankara, 06800, Turkey

** tanatar@fen.bilkent.edu.tr*

Received 31 July 2008

Interacting two-component Fermi gases loaded in a one-dimensional (1D) lattice and subjected to a harmonic trapping potential exhibit interesting compound phases in which fluid regions coexist with local Mott-insulator and/or band-insulator regions. Motivated by experiments on cold atoms inside disordered optical lattices, we present a theoretical study of the effects of a correlated random potential on these ground-state phases. We employ a lattice version of density-functional theory within the local-density approximation to determine the density distribution of fermions in these phases. The exchange-correlation potential is obtained from the Lieb-Wu exact solution of Fermi-Hubbard model. On-site disorder (with and without Gaussian correlations) and harmonic trap are treated as external potentials. We find that disorder has two main effects: (i) it destroys the local insulating regions if it is sufficiently strong compared with the on-site atom-atom repulsion, and (ii) it induces an anomaly in the inverse compressibility at low density from quenching of percolation. For sufficiently large disorder correlation length the enhancement in the inverse compressibility diminishes.

Keywords: Fermi-Hubbard model; optical lattices; disorder.

1. Introduction

Disorder and interaction effects in condensed matter systems have a long and rich history. The interplay between them has been the subject of continuing interest. The notable examples range from two-dimensional electron systems with long-ranged Coulomb interactions[1] to liquid ^{4}He absorbed in various substances such as aerogel and vycor[2] and granular superconductors.[3] In the former system experimental and theoretical investigations[4] reveal the crucial role played by disorder as the two-dimensional electron system undergoes a metal-insulator transition. Furthermore, thermodynamic quantities carry the signature of phase transition[5,6] along with transport properties.

In recent years, cold atomic systems are being used to investigate the interplay between single-particle randomness and interaction effects with the help of optical lattices allowing access to strong coupling regimes through the depression of kinetic energy.[7–9] Furthermore, the on-site interaction can be tuned either indirectly by changing the strength of the lasers that create the optical lattice or directly by means of a Feshbach resonance.[12]

After the initial observation of superfluid to Mott insulator transition using bosonic atoms in an optical lattice[10] and creation of disorder potentials,[11] studies on fermionic atoms[12,13] are beginning to be explored. A couple of forthcoming reviews[14] encompass many aspects of the theory of ultracold Fermi gases.

In this work we study the interaction and disorder effects on a one-dimensional, two-component Fermi gas trapped in a harmonic confinement potential and an optical lattice. We use the lattice version of density functional theory[15] taking advantage of the exact solution of the one-dimensional Fermi-Hubbard model.[16] Ground-state calculations in the absence of disorder have been performed within a number of numerical techniques[17,18] to identify various phases. In the present approach, the harmonic confinement potential and the disorder potential are treated as part of the Kohn-Sham potential within a local-density approximation which has an exact representation for the uniform case.

Generalizing our earlier work[19] on the effects of uncorrelated disorder, we here study correlated disorder which depends on the correlation length as a parameter. The site occupation profiles and thermodynamic stiffness are determined to study the effects of disorder. Our motivation for considering correlated disorder comes from the experiments[7] on Bose-Einstein condensates in optical speckle potentials where it was noted that the correlation length is several times longer than the lattice spacing.

The paper is organized as follows. In Sec. 2 we briefly outline the Fermi-Hubbard model in a harmonic potential and disorder. We then give the main ingredients of our density-functional theory approach. Section 3 presents our results on site occupations in the presence of correlated disorder. We conclude with a brief summary in Sec. 4.

2. Theory and Model

We consider a two-component Fermi gas with N atoms constrained to move under harmonic confinement of strength V_2 inside a disordered $1D$ optical lattice with unit lattice constant and L lattice sites $i \in [1, L]$. The system is described by a single-band Hubbard Hamiltonian,

$$\hat{\mathcal{H}} = -t \sum_{i=1}^{L-1} \sum_{\sigma} (\hat{c}_{i\sigma}^{\dagger} \hat{c}_{i+1\sigma} + \text{H.c.}) + U \sum_{i=1}^{L} \hat{n}_{i\uparrow} \hat{n}_{i\downarrow}$$

$$+ V_2 \sum_{i=1}^{L} (i - L/2)^2 \hat{n}_i + \sum_{i=1}^{L} \varepsilon_i \hat{n}_i \,. \tag{1}$$

Here $t_{ij} = t > 0$ if i, j are nearest sites and zero otherwise, $\sigma = \uparrow, \downarrow$ is a pseudospin-1/2 label for two internal hyperfine states, $\hat{n}_{i\sigma} = \hat{c}^{\dagger}_{i\sigma}\hat{c}_{i\sigma}$ is the pseudospin-resolved site occupation operator, and $\hat{n}_i = \sum_\sigma \hat{n}_{i\sigma}$. The effect of disorder is simulated by the last term in Eq. (1). The above Hamiltonian without the confining potential and disorder term has been solved exactly using the Bethe Ansatz technique by Lieb and Wu.[16] The exact solution provides us with the ground-state energy at any coupling strength $u = U/t$ and filling $n = N/L$. In the noninteracting and unconfined limit (i.e., without the U and trap terms) one recovers the Anderson model[20] studied intensively for the Anderson localization problem. The successful creation of optical lattices to study cold atomic systems has led to renewed interest in these model systems, since the atomic systems offer a wide range of control on the various parameters.

In this work we generalize our previous study[19] on uncorrelated noise to Gaussian correlated disorder, defined as

$$\varepsilon_i = \varepsilon_i(\xi) = \frac{1}{\sqrt{2\pi\xi}} \sum_{j=1}^{L} \exp\left[-\frac{(i-j)^2}{2\xi}\right] \mathcal{W}_j \,, \tag{2}$$

where ξ is the correlation length (in units of the lattice spacing, which is set to unity throughout this work) and $\mathcal{W}_j$ is randomly chosen at each site j from a uniform distribution in the range $[-W/2, W/2]$. Because of the following mathematical identity,

$$\lim_{\xi \to 0} \frac{1}{\sqrt{2\pi\xi}} \exp\left[-\frac{(i-j)^2}{2\xi}\right] = \delta_{ij} \,, \tag{3}$$

Eq. (2) tends smoothly to uncorrelated (*i.e.* white) noise in the limit $\xi \to 0$. We illustrate typical behavior of ε_i over a lattice of $L = 200$ sites in Fig. 1 for some values of ξ. As the correlation length increases the correlated disorder potential becomes smoother with amplitude smaller than that in the uncorrelated case.

A particular set of values ε_i is a realization of disorder. Each realization defines an external potential $\mathcal{V}_i = V_2(i - N_s/2)^2 + \varepsilon_i$ which is the sum of a harmonic trap potential and disorder. the site-occupation functional theory[15] (SOFT) to determine the site occupation $n_i = \langle\Psi|\hat{n}_i|\Psi\rangle$ where Ψ is the ground-state of $\mathcal{H}$ for this particular disorder realization. The site occupation $\mathcal{N}_i$ is obtained by the disorder ensemble average, $\mathcal{N}_i = \langle\langle n_i \rangle\rangle_{\text{dis}}$. SOFT is the discrete or lattice version of the density functional theory which has been successfully applied to the present system in the absence[18] and presence[19] of uncorrelated disorder. In the clean limit the local-density approximation which we adopt has been shown to be reliable through extensive comparisons with accurate quantum Monte Carlo calculations.[18] Local-density approximation based density-functional schemes for disordered systems have been employed to study the low-density compressibility anomaly in the two-dimensional

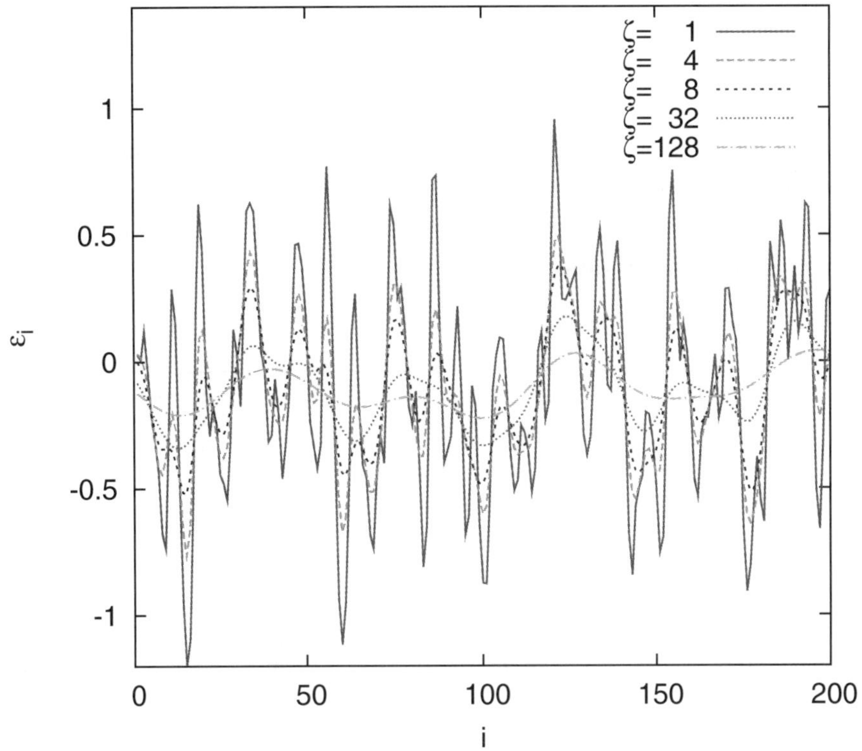

Fig. 1. (Color online) The correlated disorder potential ε_i as a function of the site position for a lattice of $L = 200$ sites. The disorder strength is $W/t = 3$. Different values of ξ are indicated in the legend.

metal-insulator transition (see Refs. 6, 21) and the statistical properties of $2D$ disordered quantum dots.[22]

The total energy is a unique functional of n_i which can be written as

$$\mathcal{E}[n] = \sum_i \epsilon(n, u) + \sum_i \mathcal{V}_i(z_i) n(z_i) , \tag{4}$$

where $\epsilon(n, u)$ is the ground-state energy of the Hubbard Hamiltonian as obtained by Lieb and Wu.[16] The Euler-Lagrange equation that follows from the above energy density is

$$\frac{\partial \epsilon}{\partial n} + v_{\mathrm{KS}}(z_i) = \text{constant} , \tag{5}$$

where the Kohn-Sham potential is

$$v_{\mathrm{KS}} = \frac{1}{2} U n + v_{\mathrm{xc}}(z_i) + \mathcal{V}_i(z_i) . \tag{6}$$

Within the local-density approximation, exchange-correlation potential is approximated by

$$v_{\mathrm{xc}} = \frac{\partial}{\partial n}\left[\epsilon(n,u) - \epsilon(n,0) - \frac{1}{4}Un^2\right] . \tag{7}$$

The above set of equations allow us to obtain the site occupations n_i and their disorder averages $\mathcal{N}_i$.

3. Results and Discussion

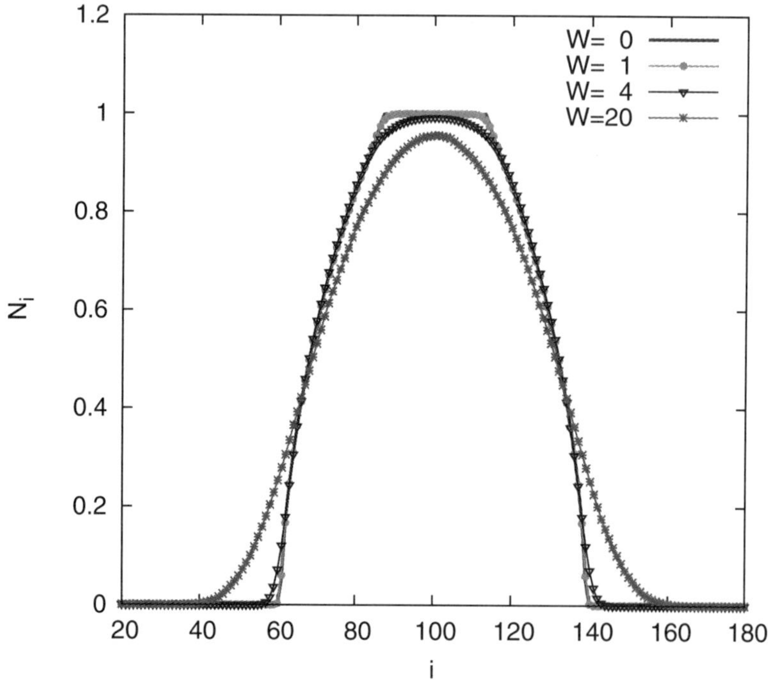

Fig. 2. (Color online) Site occupation $\mathcal{N}_i$ as a function of site position i for $N = 60$ fermions with $u = 4$, $\xi = 4$, and $V_2/t = 2.5 \times 10^{-3}$ in a lattice with $L = 200$ sites. The four curves have been calculated for different values of disorder strength: $W/t = 0$ (solid line), $W/t = 1$ (circles), $W/t = 4$ (triangles), and $W/t = 20$ (squares).

Before we describe the effects of disorder on the trapped lattice fermions, we outline the various ground-state phases of a clean system as obtained by previous numerical studies.[18] There are altogether five phases ($\mathcal{A}\ldots\mathcal{E}$) controlled by the interaction strength U/t, number of fermions N, and number of lattice sites L. Phase $\mathcal{A}$ is a fluid with $0 < n_i < 2$. In phase $\mathcal{B}$ a Mott insulated occupies the central region of the trap with $n_i = 1$. In phase $\mathcal{C}$ a fluid with $1 < n_i < 2$ is embedded in the

Mott plateau. Phase $\mathcal{D}$ is a band insulator with $n_i = 2$ surrounded by fluid edges and embedded in the Mott plateau. Finally, in phase E a band insulator in the central region of the trap coexists with fluid edges. The sketch of site occupations in these phases was given in Ref. 19.

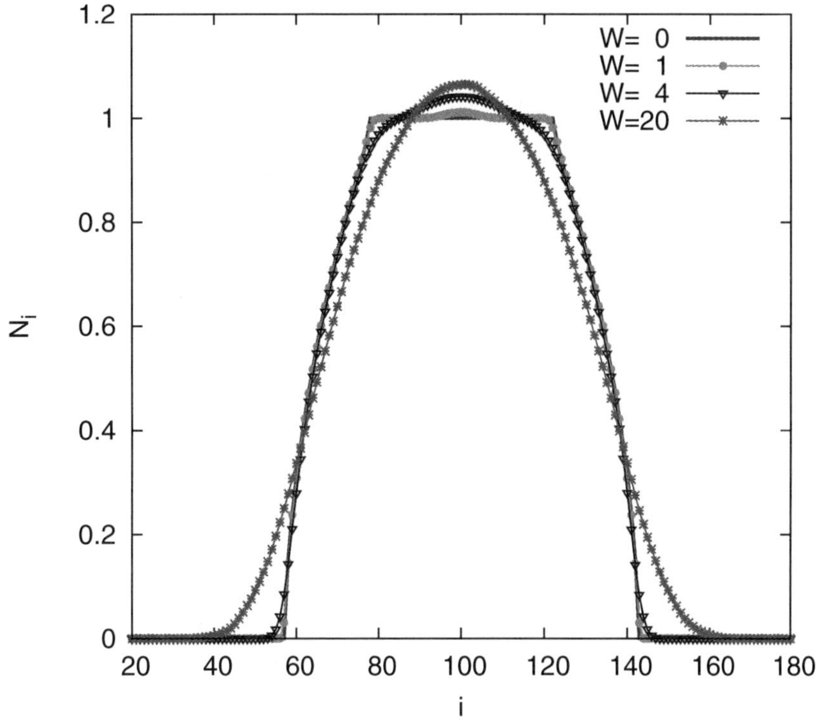

Fig. 3. (Color online) Site occupation $\mathcal{N}_i$ as a function of site position i for $N = 70$ fermions with $u = 4$, $\xi = 4$, and $V_2/t = 2.5 \times 10^{-3}$ in a lattice with $L = 200$ sites. The four curves have been calculated for different values of disorder strength: $W/t = 0$ (solid line), $W/t = 1$ (circles), $W/t = 4$ (triangles), and $W/t = 20$ (squares).

In Fig. 2 we show the disorder-averaged site occupation $\mathcal{N}_i$ for a system of fermions with $N = 60$ in an optical lattice with $L = 200$ sites. The interaction strength is $u = 4$ and the trap potential is $V_2/t = 1.5 \times 10^{-3}$. Here we fixed the disorder correlation length to be $\xi = 4$ and varied the disorder strength W/t. The clean system is in phase $\mathcal{B}$. As in the case of uncorrelated disorder[19] we find that with increasing W/t the Mott insulating region is depleted and the site occupation profile $\mathcal{N}_i$ broadens. However, the rate of depletion and broadening are smaller than the uncorrelated disorder. In other words, the Mott insulating region is more stable against the formation of a disordered fluid phase in the presence of correlated disorder.

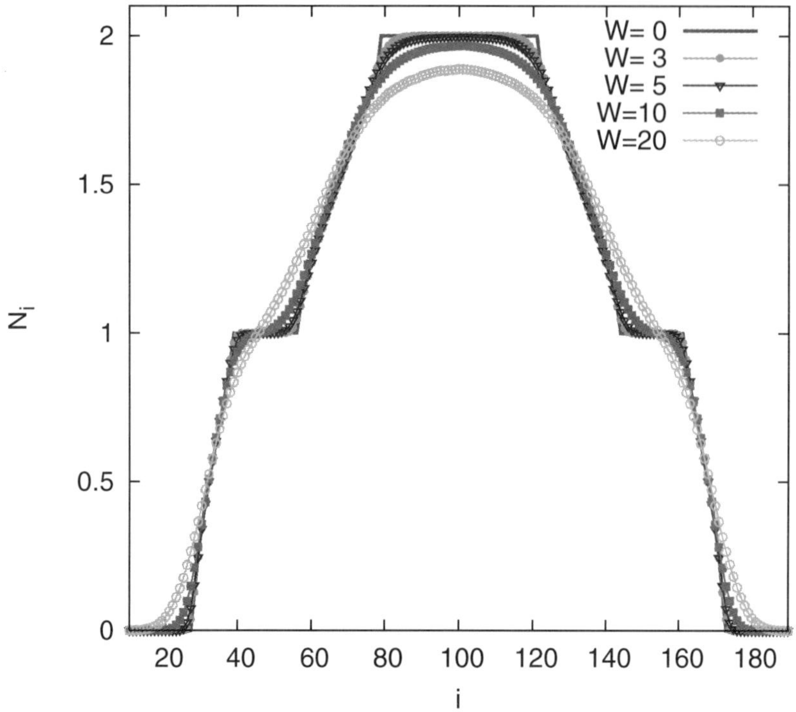

Fig. 4. (Color online) Site occupation $\mathcal{N}_i$ as a function of site position i for $N = 200$ fermions with $u = 8$, $\xi = 4$, and $V_2/t = 2.5 \times 10^{-3}$ in a lattice with $L = 200$ sites. The five curves have been calculated for different values of disorder strength: $W/t = 0$ (solid line), $W/t = 3$ (solid circles), $W/t = 5$ (triangles), $W/t = 10$ (squares), and $W/t = 20$ (empty circles).

In Fig. 3 we display the disorder averaged site occupation $\mathcal{N}_i$ for a system with $N = 70$ atoms keeping the rest of the parameters the same as in Fig. 2. $N = 70$ is the critical number of atoms at which the phase transition $\mathcal{B} \to \mathcal{C}$ occurs in the clean limit. For weak uncorrelated disorder a fluid phase with $\mathcal{N}_i > 1$ is induced at the center of the trap. Essentially the same behavior is observed for correlated disorder ($\xi = 4$). Sufficiently strong W/t destroys the Mott plateau completely and the system becomes a disordered fluid with $\mathcal{N}_i < 1$. We find that the critical disorder strength for this to happen is much larger for correlated disorder than the corresponding value of W/t for uncorrelated disorder.

In Fig. 4 we show the site occupation for a strongly interacting system ($U/t = 8$) with correlated disorder ($\xi = 4$). The disorder free system is in the $\mathcal{D}$ phase discussed above. With increasing W/t the band insulating region at the center of the trap is depleted. At the same time, the Mott insulating regions are also slowly destroyed. To see more clearly the rate at which the band and Mott insulating regions are destroyed, we plot in Fig. 5 the number of consecutive sites N_{Mott} and N_{Band} such

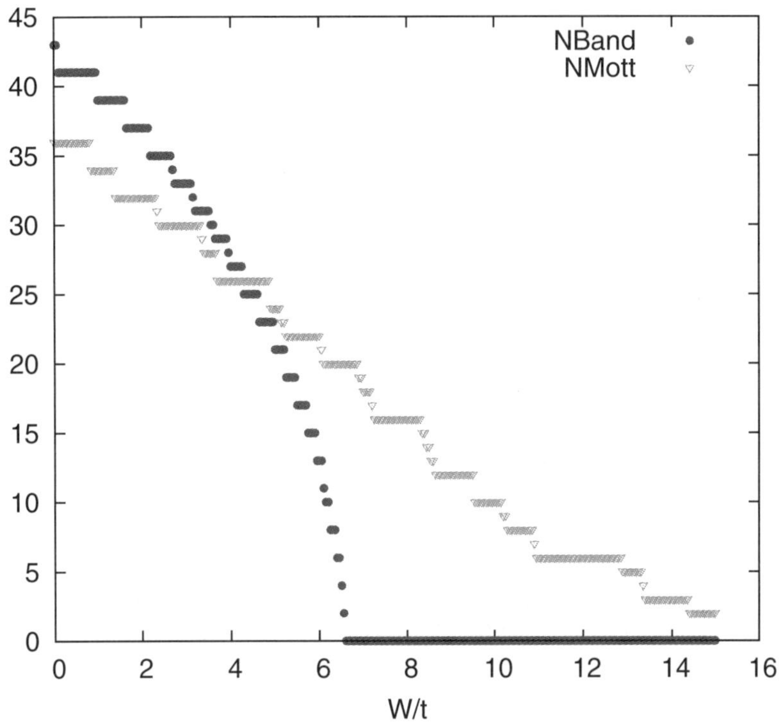

Fig. 5. (Color online) The number of consecutive sites $N_{\rm Mott}$ and $N_{\rm Band}$ such that $|\mathcal{N}_i - 1| < 0$ and $|\mathcal{N}_i - 2| < 0$, respectively, as a function of W/t. The parameters are the same as in Fig. 4.

that $|\mathcal{N}_i - 1| < 0$ and $|\mathcal{N}_i - 2| < 0$, respectively, as a function of W/t. Comparison with the same calculation in the uncorrelated disorder case[19] reveals the increased stability range of insulating regions for correlated disorder.

The effect of disorder on 1D fermions and in particular transitions between different phases can also be assessed through the thermodynamic properties. For this purpose we calculate the inverse compressibility defined as $\kappa^{-1} = \langle\langle N^2 \delta\mu/\delta N \rangle\rangle_{\rm dis}$ where μ is the chemical potential. In Fig. 6 we show κ^{-1} as a function of N for various values of the disorder correlation length ξ. The phase transitions can be identified as sharp kinks in a clean system ($W/t = 0$). Uncorrelated disorder has two main effects on κ^{-1} as discussed in our previous work.[19] Firstly, the sharp features indicating phase transitions are smoothed out. Secondly, a large enhancement of κ^{-1} at low density is observed. This is reminiscent of a similar behavior found in 2D electron systems. As N decreases the atoms mostly occupy the deepest valleys in the disorder landscape, thus the high density regions in the system tend to become disconnected. For a given interaction strength u and low N the system stiffens as disorder grows. In the present situation we observe that with increasing correlation length ξ the low density enhancement in κ^{-1} is diminished. The basic reason for

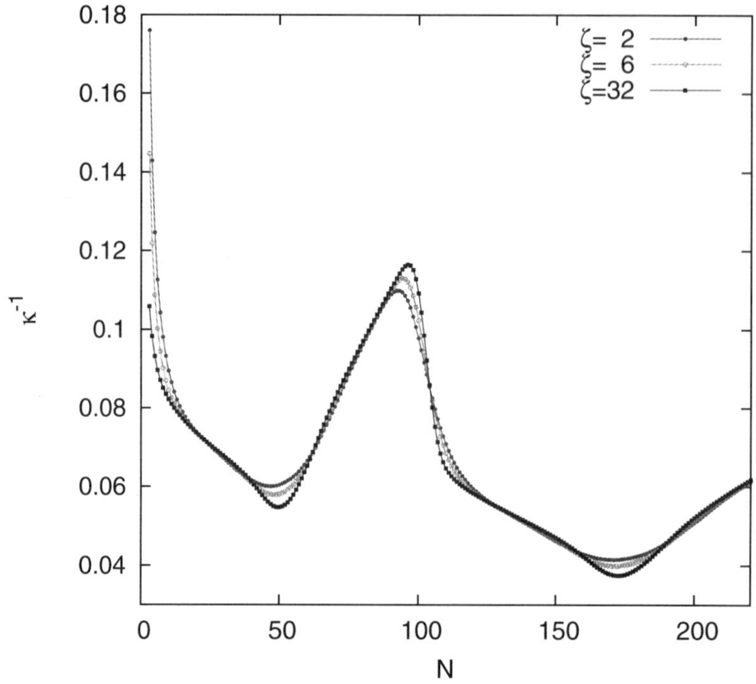

Fig. 6. (Color online) The inverse compressibility κ^{-1} (in units of t) in the presence of correlated disorder as a function of N for $V_2/t = 2.5 \times 10^{-3}$, $W/t = 5$, $u = 8$, and $L = 200$ lattice sites. The correlation length $\xi = 0$ (circles), $\xi = 6$ (squares), and $\xi = 32$ (triangles).

this behavior is that increasing ξ makes the disorder potential more smooth and more shallow compared to the $\xi = 0$ case. Therefore the atoms are less localized in a correlated disorder potential.

4. Concluding Remarks

We have studied the one-dimensional Fermi-Hubbard model in a harmonic confinement potential and in the presence of correlated disorder. This is believed to represent cold fermionic atoms in an optical lattice created by standing laser waves. Our numerical calculations of the site occupations are based on a lattice version of density functional theory in which we make use of the exact solution of the one-dimensional Hubbard model to treat the exchange-correlation effects. Disorder affects the ground-state phases of the interacting Fermi gases confined in a harmonic potential and an optical lattice. The insulating regions appear to be stable against both uncorrelated and correlated disorder. The anomalous enhancement of the stiffness observed at low density for uncorrelated disorder decreases with increasing correlation length. We hope that our results on disorder effects will stimulate experimental investigations with cold Fermi atoms in the future.

Acknowledgments

G. X. was supported by NSF of China under Grant No. 10704066. B. T. acknowledges the support by TUBITAK (No. 106T052), TUBA and a travel grant from Julian Swchiwinger Foundation.

References

1. See for instance, E. Abrahams, S.V. Kravchenko and M.P. Sarachik, *Rev. Mod. Phys.* **73**, 251 (2001); B.L. Altshuler, D.L. Maslov and V.M. Pudalov, *Physica E* **9**, 209 (2001).
2. J.D. Reppy, *J. Low Temp. Phys.* **87**, 205 (1992); B. Fåk, O. Plantevin, H.R. Glyde and N. Mulders, *Phys. Rev. Lett.* **85**, 3886 (2000), and references therein.
3. R. Fazio and H. van der Zant, *Phys. Rep.* **355**, 235 (2001).
4. A. Punnoose and A.M. Finkel'stein, *Science* **310**, 289 (2005); A.A. Shashkin, S. Anissimova, M.R. Sakr, S.V. Kravchenko, V.T. Dolgopolov and T.M. Klapwijk, *Phys. Rev. Lett.* **96**, 036403 (2006).
5. J.P. Eisenstein, L.N. Pfeiffer and K.W. West, *Phys. Rev. Lett.* **68**, 674 (1992); S. Ilani, A. Yacoby, D. Mahalu and H. Shtrikman, *ibid.* **84**, 3133 (2000); S.C. Dultz and H.W. Jiang, *ibid.* **84**, 4689 (2000).
6. B. Tanatar, A.L. Subaşı, K. Esfarjani and S.M. Fazeli, *Int. J. Mod. Phys. B* **21**, 2134 (2007).
7. J.E. Lye, L. Fallani, M. Modugno, D.S. Wiersma, C. Fort and M. Inguscio, *Phys. Rev. Lett.* **95**, 070401 (2005); D. Clément, A.F. Varón, M. Hugbart, J.A. Retter, P. Bouyer, L. Sanchez-Palencia, D.M. Gangardt, G.V. Shlyapnikov and A. Aspect, *ibid.* **95**, 170409 (2005); C. Fort, L. Fallani, V. Guarrera, J.E. Lye, M. Modugno, D.S. Wiersma and M. Inguscio, *ibid.* **95**, 170410 (2005); for a review see V. Ahufinger, L. Sanchez-Palencia, A. Kantian, A. Sanpera and M. Lewenstein, *Phys. Rev. A* **72**, 063616 (2005).
8. J.I. Cirac and P. Zoller, *Science* **301**, 176 (2003).
9. D. Clement, P. Bouyer, A. Aspect and L. Sanchez-Palencia, *Phys. Rev. A* **77**, 033631 (2008); Y.P. Chen *et al.*, e-print arXiv:0710.5187, to be published in *Phys. Rev. A*.
10. M. Greiner, O. Mandel, T. Esslinger, T.W. Hänsch and I. Bloch, *Nature* **415**, 39 (2002).
11. T. Schulte, S. Drenkelforth, J. Kruse, W. Ertmer, J. Arlt, K. Sacha, J. Zakrzewski and M. Lewenstein, *Phys. Rev. Lett.* **95**, 170411 (2005).
12. H. Moritz, T. Stöferle, K. Günter, M. Köhl and T. Esslinger, *Phys. Rev. Lett.* **94**, 210401 (2005).
13. W. Hofstetter, C.I. Cirac, P. Zoller, E. Demmler and M.D. Lukin, *Phys. Rev. Lett.* **89**, 220407 (2002); S. Trebst, U. Schollwök, M. Troyer and P. Zoller, *Phys. Rev. Lett.* **96**, 250402 (2006).
14. I. Bloch, J. Dalibard and W. Zwerger, arXiv:0704.3011v2, to be published in *Rev. Mod. Phys.*; L. Giorgini, L.P. Pitaevskii and S. Stringari, e-print arXiv:0706.3360v2, to be published in *Rev. Mod. Phys.*
15. K. Schönhammer, O. Gunnarsson and R.M. Noack, *Phys. Rev. B* **52**, 2504 (1995); N.A. Lima, M.F. Silva, L.N. Oliveira and K. Capelle, *Phys. Rev. Lett.* **90**, 146402 (2003).
16. E.H. Lieb and F.Y. Wu, *Phys. Rev. Lett.* **20**, 1445 (1968).
17. M. Rigol, A. Muramatsu, G.G. Batrouni and R.T. Scalettar, *Phys. Rev. Lett.* **91**, 130403 (2003); M. Rigol and A. Muramatsu, *Phys. Rev. A* **69**, 053612 (2004); *Opt.*

Commun. **243**, 33 (2004); X.-J. Liu, P.D. Drummond and H. Hu, *Phys. Rev. Lett.* **94**, 136406 (2005); V.L. Campo, Jr. and K. Capelle, *Phys. Rev. Lett.* **72**, 061602 (2005).
18. G. Xianlong, M. Polini, M.P. Tosi, V.L. Campo, Jr., K. Capelle and M. Rigol, *Phys. Rev. B* **73**, 165120 (2006).
19. G. Xianlong, M. Polini, B. Tanatar, and M.P. Tosi, *Phys. Rev. B* **73**, 161103 (2006).
20. P.W. Anderson, *Phys. Rev.* **109**, 1492 (1958).
21. J. Shi and X.C. Xie, *Phys. Rev. Lett.* **88**, 086401 (2002).
22. K. Hirose, F. Zhou and N.S. Wingreen, *Phys. Rev. B* **63**, 075301 (2001); K. Hirose and N.S. Wingreen, *ibid.* **65**, 193305 (2002); H. Jiang, D. Ullmo, W. Yang and H.U. Baranger, *ibid.* **69**, 235326 (2004); E. Räsänen and M. Aichinger, *ibid.* **72**, 045352 (2005).

A NEW LOOK AT SUPER HEAVY NUCLEI

G. S. ANAGNOSTATOS

Institute of Nuclear Physics, National Center for Scientific Research "Demokritos"
Aghia Paraskevi, Attiki, 15310 Greece

Received 31 July 2008

A new look at super heavy nuclei is attempted by employing the Isomorphic shell Model or, in different wording, the Multiharmonic Shell Model. After ^{208}Pb the same magic numbers, like the conventional Shell Model, are predicted, i.e., at Z = 126 and N = 184. Good results for charge radii and binding energies for a sample of nuclei are also presented.

1. Introduction

The novelty of the model employed, i.e., the Isomorphic Shell Model or, in different wording, the Multiharmonic Shell Model, is that it takes advantage of the fermionic nature of neutrons and protons and their antisymmetric wave function. Specifically, it takes advantage of the well known property (experimentally and theoretically) that fermions behave as if a repulsive force (of unknown nature) is acting among them.[1] This fact separates their most probable positions at the maximum. As a result of this property of fermions the nuclear many-body problem is shifted to the problem of finding the equilibria of repulsive particles on a sphere, like the sphere of a nuclear shell. This problem, however, was solved in 1957 by J. Leech,[2] who showed that equilibrium of repulsive particles on a sphere is obtained only for specific numbers of particles and when these particles are arranged at the vertices, or middles of faces, or middles of edges, or simultaneously at these characteristic points of regular polyhedra or their derivative polyhedra. Such polyhedra in the frame of the model stand for the average forms of nuclear shells as it has been verified by many nuclear properties.

In the first two rows of Fig. 1 the regular and semi-regular polyhedra together with their names are shown, while in the third row their relative orientation to obtain the maximum relative symmetry is demonstrated together with derivative polyhedra resulting from this orientation as shown with thinner lines.

It is very important to emphasize at this point that the Leech polyhedra, besides their equilibrium properties mentioned earlier, inherently possess in their structure the quantization of orbital angular momentum.[3–6] Figure 2 shows the well-known drawing for quantization of the orbital angular momentum in the case $\ell = 2$ and the formula defining the relevant characteristic angle between the vector of the

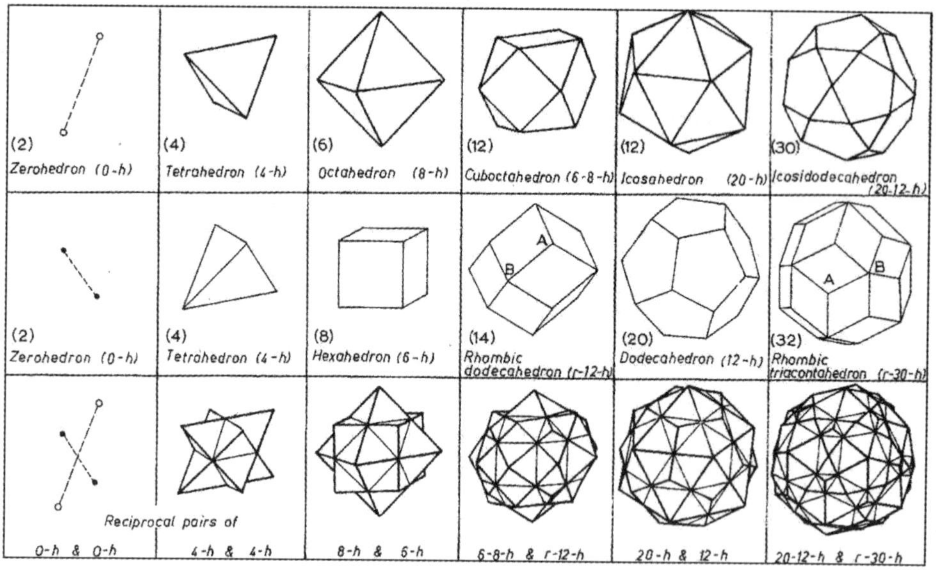

Fig. 1. Leech-type equilibrium polyhedra.

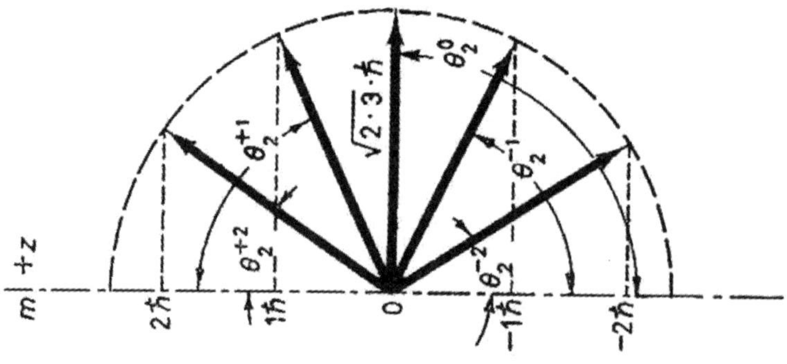

$$\theta_l^m = \cos^{-1}(m/\sqrt{l(l+1)})\,.$$

Fig. 2. Quantization of orbital angular momentum.

orbital angular momentum $sqrt[\ell(\ell+1)]$ and its projection m. Figure 3 shows this quantization for all m with $\ell = 1-6$ in relation to Leech polyhedra with respect

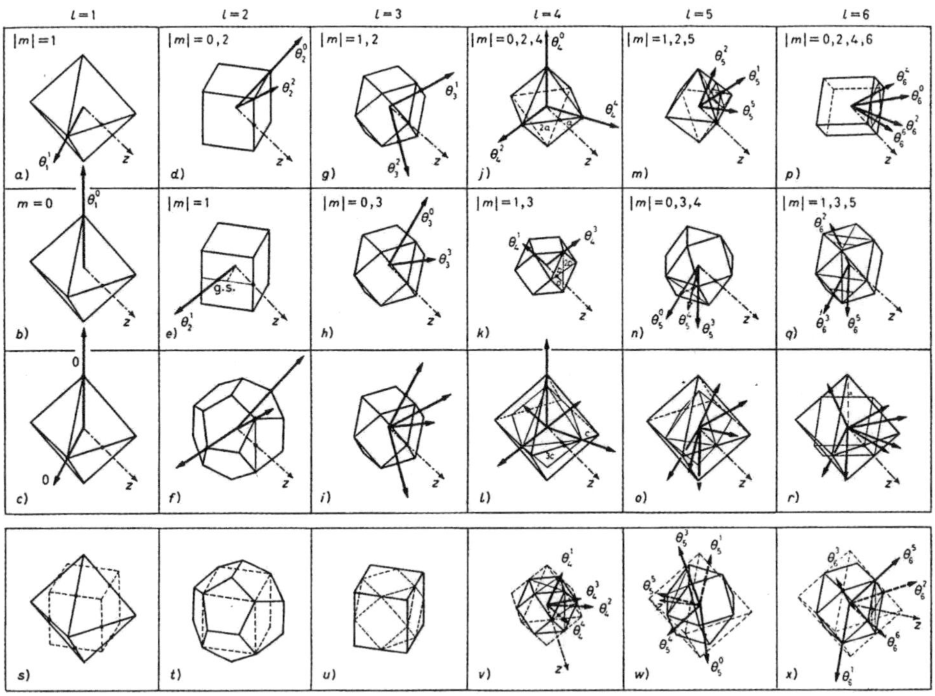

Fig. 3. Angular-momentum quantization of direction vectors in relation to the polyhedra shown.

to a quantization axis common for all polyhedra involved in the nuclear structure supported by this model.[6]

The first requirement, of course, of a nuclear structure model related to polyhedra is the verification of magic and semi-magic numbers. Indeed, these numbers are very successfully verified, as we will present later. Other important requirements of the above polyhedral structure of the nuclear shells are the reproduction of the nuclear sizes and binding energies in the whole periodic table. These requirements are also fulfilled here.

From the beginning of the application of this model, we have to decide if we will follow a semiclassical or a completely quantum mechanical approach. For the first approach we use a two-body potential[7] and the work takes place in the part of the model called the Isomorphic Shell Model.[8] For the second approach we use a harmonic oscillator as central potential for each shell (not for the whole nucleus) and the work takes place in the part of the model called the Multiharmonic Shell Model.[9] Now, both parts of the model are in a much more advanced stage of their development than in their initial appearance.

In the present work we will follow the multiharmonic part of the model and the $\hbar\omega$ for each harmonic oscillator potential will be determined by the well-known harmonic oscillator formula

$$\hbar\omega = (\hbar^2/m)(n + 3/2)/ < r^2 >, \tag{1}$$

where $< r^2 >^{1/2}$ is the average size of the proton or neutron shell of interest derived by the packing of the polyhedral shells, when their vertices stand for the proton or neutron most probable positions and the nucleons possess finite sizes.

The present approach has proved to be very successful, but in addition it is very interesting according to the following reasoning (from here till the end of the introduction). The polyhedra employed come as a direct consequence of the fermionic nature of the nucleons, but they are not permanent like in crystals. The vertices of each polyhedron are just the most probable positions of the nucleons accommodated by this polyhedron. Moreover, the nucleons accommodated by a specific polyhedron are simultaneously assumed at their most probable positions (polyhedra vertices), a situation which reappears periodically. Thus, in an instant of time the polyhedra are formed and for this instant we have the Leech equilibrium. Since the nuclear properties remain the same in all instants of time, we take advantage of this Leech instant of time to determine the nuclear properties.

As known, the two basic models in nuclear structure are the independent particle model and the collective model. However, the first model assumes that the nucleons in a nucleus are free of force and thus freely move in the independent orbitals of a potential, while the second model assumes that the nucleons strongly interact with each other favoring a collective (rotational or vibrational) behavior. That is, these two important models have, at first glance, contradicting assumptions and still are very successful, each describing different nuclear properties. In the framework of the present model this long-standing contradiction is easily removed. In the present work we always have an independent particle motion of the nucleons, however, their most probable positions of the independently moving particles are instantly interacting, leading to equilibrium, polyhedral forms. Since there we have equilibrium of forces, the situation is equivalent to having no forces at all. The force among nucleons is in true action, if the particles are found out of equilibrium, forcing them to go to equilibrium. The polyhedra are like snap shots of the nuclear shells periodically reappearing. Thus, these polyhedral forms permit collectivity without assuming permanent, strong interacting forces among the nucleons. While nucleons have an independent particle motion, they can simultaneously participate in a collective motion.

2. The Model

In Fig. 4 the polyhedra employed in the model to present the average forms of all nuclear shells up to A = 310 are shown. Specifically, at the top of each block of the figure the name of the polyhedron shown in this block (left) and the quantum states of the nucleons accommodated by this polyhedron (right) are given. At the bottom of the block, four other characteristics of this polyhedron are listed. Specifically, in a

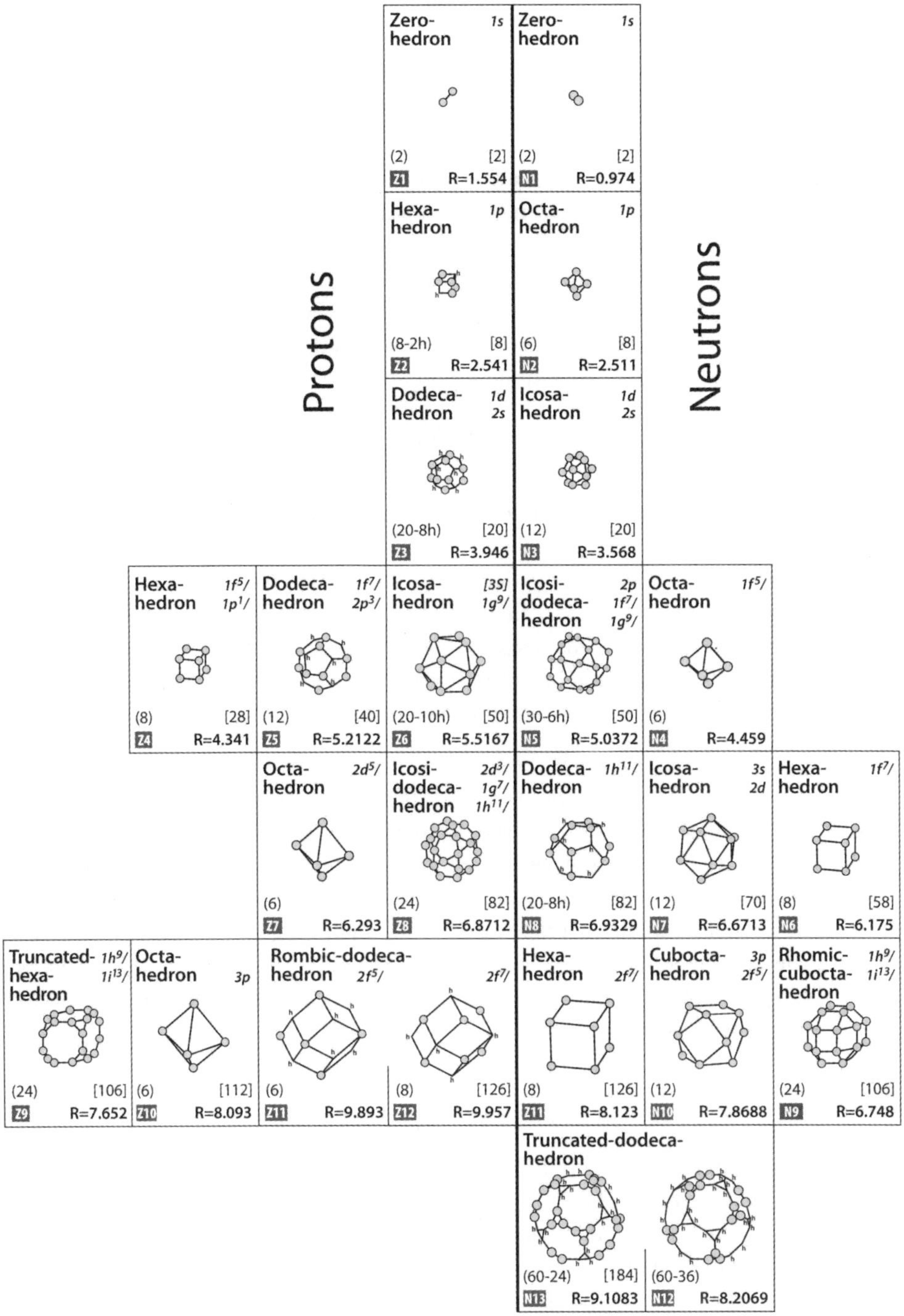

Fig. 4. Pictorial average nuclear structure of all closed shells up to A=310. The numbers in brackets given close to the bottom of the blocks of the figure coincide with the experimentally known magic and semi-magic numbers.

black square at the left of each block the numbering of this polyhedron in successive order proceeded with the letter Z for protons and with the letter N for neutrons is given. Over this square the number of the polyhedral vertices, and the number of its possible unoccupied vertices characterized as wholes, (h), are also given in parentheses. At the right of each block the radius of this polyhedron in units Fermi is listed. Over this radius the cumulative number of vertices of all previous polyhedra and this polyhedron is also given in brackets. This radius results from the packing of the polyhedral shells for the specific sizes of proton and neutron bags (assumed at the polyhedra vertices) discussed shortly.

We assume a harmonic oscillator potential for the nucleons of each proton or neutron shell (and not a harmonic oscillator potential common for all nucleons in a nucleus as usual). That is, we assume a potential of the form given in Eq. (2)

$$V(r) = -v_0 + \frac{1}{2}m\omega^2 r^2, \tag{2}$$

where v_0 and ω are different for each proton or neutron shell. In other words, the potential form is the same for all nucleons, but its parameters v_0 and ω are different for each shell.

According to the above brief discussion, the initial total number of potential parameters involved in the treatment of a nucleus is twice the number of shells involved in the structure of this nucleus. However, due to two physical assumptions discussed below, the final number of potential parameters involved in the study of any nucleus is very small. These assumptions are:

- The ω, or $\hbar\omega$, for each shell is determined according to Eq. (1) and thus depends strictly from the size of this shell. The sizes of all shells are common for all nuclei and are derived form the packing of the polyhedral shells themselves discussed earlier. Due to the high symmetries of the polyhedral shells and to their high symmetric packing, the only parameters necessary in determining the average sizes of all shells are the average distances of the proton-proton, proton-neutron, and neutron-neutron pairs in contact. If $r_p = 0.860\ fm$ and $r_n = 0.974\ fm$ are the average sizes of proton and neutron bags, then

$$d_{nn} = 2r_n = 1.948\ fm \quad \text{and}$$

$$d_{pn} = r_p + r_n = 1.834\ fm$$

  However, d_{pp} could be different from $2r_p$ due to the fact that protons are charged particles. Thus, d_{pp} is taken as a new parameter, i.e., $d_{pp} = 1.800\ fm$. This quantity is consistent with the average charge radius of a proton, i.e., $r_{p,ch} = 0.9\ fm$, known from the literature.
- The parameter of the depth of the potential for each shell is determined according to the assumption expressed by Eq. (3).

$$E_j = v_j - \hbar\omega_j(n_j + 3/2) = v_i - \hbar\omega_i(n_i + 3/2) = E_i$$

or

$$v_j = v_i - \hbar\omega_i(n_i + 3/2) + \hbar\omega_j(n_j + 3/2) = V_0 + \hbar\omega_j(n_j + 3/2). \qquad (3)$$

This assumption implies that despite the different potentials for the different shells, all nucleons in a nucleus are equally bound in their own potentials (beyond of course Coulomb force, spin-orbit interaction, and odd-even effect.) From Eq. (3) it is apparent that, since all $\hbar\omega$ are determined according to the three parameters d discussed above, the depth of the potential for each shell can be defined with the knowledge of only one parameter V_0.

Solving the Schroedinger's equation for the potential of Eq. (2), we obtain[10] the general expression for the wave functions given by Eq. (4).

$$L_k^{\ell+1/2}(z) = [(1/2 + 1/2\ell + 1)]2/k!(\ell + 3/2)_1F_1(-k; \ell + 3/2; z) \qquad (4)$$

where

$$_1F_1(a; c; z) = 1 + a/cz + a(a+1)z^2/[c(c+1)2!] + a(a+1)(a+2)z^3/c(c+1)(c+2)3! + \ldots$$

and $c = \ell + 3/2$, $a = -k$, $z + 2ar^2$ and $a = m\omega/2\hbar$.

The series terminates with the term

$$(-1)^k[(c)/(c+k)]z^k$$

and the various quantum numbers involved obey the conditions

$$n = 0, 1, 2, \ldots; \quad k = 1, 1, 2, \ldots; \quad k \leq n/2,$$
$$\ell = 0, 1, 2, \ldots; \quad \ell \leq n; \quad \ell = n - 2k$$

The explicit forms for the individual wave functions could be found in different books of Quantum Mechanics and Nuclear Physics. However, usually only the first few of them are listed there. Thus, we think (for the interest of the general reader) that it is instructive to give the full list of the wave functions involved in the study of nuclei up to A = 210 in Table 1. Due to the fact that in the model $\hbar\omega$ is different for the different shells, the wave functions with the same ℓ are not orthogonal . For these wave functions the Gram-Smidth's technique has been applied. Furthermore, in order for the orthogonalization procedure to preserve the initial values of polyhedral radii, the parameters of orthogonalization are determined under the condition that the maxima of probability for the orthogonalized wave functions are at the vertices of the same equilibrium polyhedra with the same size as before the orthogonalization. As applications of the model presented here, we provide the verification of magic and semi-magic numbers, and samples of nuclear radii and binding energies. These are among the main properties of a nucleus, but they also provide a very good test for our wave functions, which can be applied to any other nuclear property.[11] General comparative comments on the model can be found in Ref. 12.

$$R1s = (128/)^{1/4}a^{3/4}e^{-ar^2}$$
$$R2s = (128/)^{1/4}a^{3/4}(3/2)^{1/2}(1 - 4ar^2/3)e^{-ar^2}$$
$$R3s = (128/)^{1/4}a^{3/4}(15/8)^{1/2}(1 - 8ar^2/3 + 16a^2r^4/15)e^{-ar^2}$$
$$R4s = (128/)^{1/4}a^{3/4}(35/16)^{1/2}(1 - 4ar^2 + 16a^2r^4/5 - 64a^3r^6/105)e^{-ar^2}$$
$$R1p = (128/)^{1/4}a^{5/4}(2/3)^{1/2}2^{1/2}re^{-ar^2}$$
$$R2p = (128/)^{1/4}a^{5/4}(5/3)^{1/2}2^{1/2}r(1 - 4ar^2/5)e^{-ar^2}$$
$$R3p = (128/)^{1/4}a^{5/4}(35/12)^{1/2}2^{1/2}r(1 - 8ar^2/5 + 16a^2r^4/35)e^{-ar^2}$$
$$R1d = (128/)^{1/4}a^{7/4}(4/15)^{1/2}2r^2e^{-ar^2}$$
$$R2d = (128/)^{1/4}a^{7/4}(14/15)^{1/2}2r^2(1 - 4ar^2/7)e^{-ar^2}$$
$$R3d = (128/)^{1/4}a^{7/4}(63/30)^{1/2}2r^2(1 - 8ar^2/7 + 16a^2r^4/63)e^{-ar^2}$$
$$R1f = (128/)^{1/4}a^{9/4}(8/105)^{1/2}2^{3/2}r^3e^{-ar^2}$$
$$R2f = (128/)^{1/4}a^{9/4}(36/105)^{1/2}2^{3/2}r^3(1 - 4ar^2/9)e^{-ar^2}$$
$$R1g = (128/)^{1/4}a^{11/4}(16/945)^{1/2}2^2r^4e^{-ar^2}$$
$$R2g = (128/)^{1/4}a^{11/4}(88/945)^{1/2}2^2r^4(1 - 4ar^2/11)e^{-ar^2}$$
$$R1h = (128/)^{1/4}a^{13/4}(32/10395)^{1/2}2^{5/2}r^5e^{-ar^2}$$
$$R1i = (128/)^{1/4}a^{15/4}(64/135135)^{1/2}2^3r^6e^{-ar^2}$$
$$R1j = (128/)^{1/4}a^{17/4}(128/2027025)^{1/2}2^{7/2}r^7e^{-ar^2}$$

3. Verification of Magic and Semi-Magic Numbers

By observing the numbers in brackets in all blocks of Fig. 4, we observe that the numbers which are close and in both sides of the vertical, heavy, central line between proton and neutron polyhedra coincide with the magic numbers 2, 8, 20, 50, 82, 126 for both protons and neutrons and 184 for neutrons only. These numbers for the remaining polyhedra lying in horizontal lines coincide with semi-magic numbers like 28, 40, 106, 112 for protons and 58, 70 and 106 for neutrons. The filling of polyhedra Z7, Z11 and N4, N10 do not lead to semi-magic numbers since their filling simultaneously takes place with their neighboring polyhedra Z8, Z12 and N5, N11, respectively. Thus, the model predicts the new magic numbers 126 for protons and 184 for neutrons, as the conventional shell model does. New semi-magic numbers for neutrons beyond ^{208}Pb are not yet definite since the choice of the relevant polyhedra requires additional research.

4. Nuclear Radii

Since the wave functions are known in the framework of the model, the nuclear radii can be calculated straightforwardly by using these wave functions. However, due to the way these wave functions have been correlated with the size of nuclear shells through Eq. (1), the charge radii can equivalently be calculated by a simple application of Eq. (5).

$$< r^2 >_{ch} = \sum_i r_i^2/Z + r^2_{ch.proton} - r^2_{ch.neutron} * N/Z, \qquad (5)$$

where $r_{ch.proton} = 0.9\ fm$ and $r_{ch.neutron} = 0.34\ fm$ from the literature. By applying Eq. (5) we construct Table 2, where samples of radii before and after ^{208}Pb are presented together with the corresponding experimental values for comparison. The very good agreements of the model predictions with the experimental values are apparent for both regions before and after ^{208}Pb.

Nucleus	Model	Exper.	Nucleus	Model	Exper.
^{40}Ca	3.49	3.48	^{209}Bi	5.55	5.53
^{58}Ni	3.78	3.79	^{232}Th	5.78	5.72
^{90}Zr	4.27	4.27	^{235}U	5.83	5.83
^{108}Sn	4.57	4.56	^{244}Pu	5.87	5.86
^{208}Pb	5.51	5.51	^{243}Am	5.89	5.90

5. Binding Energies

In the model considered here dealing only with even-even nuclei the binding energy of a nucleus is given by Eq. (6)

$$E_B = 1/2\sum_{ij}(v_{Zi} * Z + v_{Nj} * N) - 3/4\sum_{ij}\hbar\omega_i(n_j + 3/2)$$
$$+\sum_i V_{LiSi} - \sum_{ij} E_{Cij}. \tag{6}$$

All v_{Zi} and v_{Nj} come from Eq. (3) with respect to the common parameter V_0, all $\hbar\omega_i$ come from Eq. (1) where the necessary radius of a polyhedral shell is given at the bottom of the relevant block of Fig. 4, n_j is the principal quantum number, the spin-orbit term is given by Eq. (7)

$$\sum_i V_{LiSi} = \sum_i (\hbar\omega_i)^2/(\hbar_2/m)l_i s_i, \tag{7}$$

with $\lambda = 0.6$, and the Coulomb term $\sum_{ij} E_{Cij}$ is given by Eq. (8)

$$\sum_{ij} E_{Cij} = \sum_{ij} e^2/d_{ij}. \tag{8}$$

In Eq. (8) d_{ij} is the distance between any two protons calculated from the coordinates of the proton polyhedral vertices for the radii of proton polyhedra given at the bottom of each relevant block of Fig. 4.

The coefficients 1/2 and 3/4 in Eq. (6) serve to avoid the double counting, since the real forces are two-body forces.

For wave functions which need orthogonalization the term with $\hbar\omega_i$ in both Eqs. (3) and (6) has to be readjusted accordingly.

Calculations of binding energies by using Eq. (6) for a sample of closed shell nuclei (^{40}Ca, ^{90}Zr, and ^{208}Pb) are given in Table 3 together with experimental values for comparison. In this table only binding energies for nuclei before ^{208}Pb are examined, since the neutron polyhedra involved in (super heavy) nuclei beyond this nucleus are still under investigation.

Table 3. Single particle state and total energies in MeV.

State	Energy	Occupation	Total energy per state	Binding energy		
				^{40}Ca	^{90}Zr	^{208}Pb
1d5/	13.096	6	78.58	78.58	78.58	78.58
	11.910		71.46	71.46	71.46	71.46
1d3/	9.951	4	39.80	39.80	39.80	39.80
	7.211		28.84	28.84	28.84	28.84
2s1/	12.393	2	24.79	24.79	24.79	24.79
	10.479		20.96	20.96	20.96	20.96
1f7/	13.292	8	106.34		106.34	106.34
	12.904		103.23		103.23	103.23
1f5/	10.900	6	65.40		65.40	65.40
	6.900		41.40		41.40	41.40
2p3/	16.453	4	65.81		65.81	65.81
	15.860		63.44		63.44	63.44
2p1/	8.677	2	17,35		17,35	17,35
	14.687		29.37		29.37	29.37
1g9/	11.315	10	113.15		99.84	113.15
	99.84		99.84			99.84
1g7/	7.656	8	61.25			61.25
	4.726		37.81			37.81
2d5/	14.700	6	88.20			88.20
	17.517		105.10			105.10
2d3/	14.299	4	57.20			57.20
	6.592		66.37			66.37
3s1/	11.80	2	23.60			23.60
	19.582		19.16			19.16
1h11/	11.896	12	142.75			142.75
	12.030		144.36			144.36
1h9/	8.866	10	88.66			88.66
2f7/	16.521	8	132.17			132.17
2f5/	15.677	6	94.06			94.06
3p3/	18.067	4	72.27			72.27
3p1/	17.203	2	34.41			34.41
1i13/	9.225	14	129.15			129.15
Core				140.19	140.19	140.19
Coul.				-64.15	-222.33	-757.02
$E_B mod.$				340.47	774.47	1649.39
$E_B exp.$				342.06	783.91	1636.50

In Table 3 the single particle energies (including spin-orbit) for all proton and neutron single-particle states up to ^{208}Pb are given [derived from Eqs. (1)–(4) and (6)–(8)]. Specifically, for each state (first column) the numerical value of energy for protons and underneath for neutrons are listed (second column). For all nuclei of Table 3 ^{16}O is taken as a core. This means that the experimental binding energy (127.62 MeV) and the Coulomb energy (12.57 MeV from the model) are assumed known. ^{16}O has a particle-hole structure and, in general, nuclei with N, Z < 8 do not contain enough nucleons in order for a central potential to be a good approximation. Also, an additional reason is that mainly in this region of nuclei the halo phenomenon is present,[13,14] which requires special attention.

In column 4 of Table 3 the energy of all nucleons in a state (see occupation number in column 3) is given again in two rows for each state as above. In the remaining columns 5–7 the energies from column 2 relevant to ^{40}Ca, ^{90}Zr, and ^{208}Pb are repeated. At the bottom of the table the core energy of ^{16}O (140,19 MeV) is listed for each for the three nuclei. In the next row the Coulomb energy for each of these nuclei, calculated in the model, is given. Finally, the binding energies predicted by the model and the corresponding experimental energies are listed. For each nucleus the good agreement between these two values of energy is apparent.

As obvious from Table 4 the binding energy of any nucleus examined is obtained by summing up the energies of single particle states involved in this nucleus. This means that the model employed here considers exclusively an independent particle motion for all nucleons.

Besides the binding energies of nuclei included in Table 4, **in principle**, the binding energy of any nucleus up to ^{208}Pb with known quantum states could be obtained from the single particle energies given in this table. In other words, the model considers that for each shell, we have a complete saturation of nuclear forces after its completion. However, since particle-hole structure (like in ^{16}O), deformation and other phenomena appear in nuclei, the above calculation of binding energies for any nucleus could lead to a first estimation of this energy. For a better estimation of the binding energy of a nucleus one should consider any known observable of this nucleus.

The above estimation of binding energies assumes that the nuclear states are associated with the polyhedra vertices in the way given in Fig. 4, which corresponds to a situation of complete shells. Of course, we could have particle-hole structure in a nucleus, mixing of states etc. In general, these behaviors of a nucleus can be nicely dealt with in the framework of the model, but they do not constitute main part of the present work. However, we would like to comment briefly below on the cases of ^{108}Sn and isotopes of iron and nikel.

Following the experimental spins for nucleons as we approach ^{108}Sn we notice that, while the protons are at the state 1g9/2 (as expected), the neutrons are rather in the 2d state instead of the 1g7/2 state. Thus, in our calculations for this nucleus (with 8 neutrons in the fifth shell) we consider 6 neutrons in the 2d5/2 state and

2 neutrons in the 2d3/2 state whose binding energies can be taken from Table 3. Repeating the same procedure as for the above closed shell nuclei we obtain 917.31 MeV as binding energy for ^{108}Sn, which is very close to its experimental binding energy 914.66 MeV. In these calculations the Coulomb energy 330.92 MeV coming from the model was employed.

For isotopes of Fe and Ni (which are open shell nuclei) we found that the protons after the magic number 20 do not have most probable positions on the polyhedron Z5, like here the case where the 1f2d1g9/2 shell is complete as shown in Fig. 4, but on the polyhedron Z4. Indeed, the situation presented in Fig. 4 maximizes the energy of the whole shell and energetically is more favored the protons in the state 1f7/2 to be on Z5 and the protons in the states 2p1/2 and 1f5/2 on Z4, than the opposite. For these nuclei the binding energies calculated are of similar accuracy as those presented here. In general, for situations where the assignment of quantum states to polyhedral vertices is different to the situation presented in Fig. 4 for closed shells, always the nucleons of a sub shell are assigned to the same polyhedron.

For radii and binding energies better values are expected to be obtained in the future, if the basic parameters of the model r_p and r_n are re-evaluated. The given values of these parameters were determined a long time ago [8], when the model was much less advanced than today.

6. Parameters of the Model

They are only five parameters in the model:

- The potential parameter $V_0 = 40$ Mev
- The spin-orbit coefficient $\lambda = 0, 6$ and
- The three size parameters d_{pp} =1.800 fm, d_{pn} =1.834 fm and $d_{nn} = 1.948$ fm, which define the central distances of two nucleon bags in contact, i.e., of two proton bags, of a proton bag and of a neutron bag , and of two neutron bags in contact.

Someone might think that the three size parameters could have been known from Particle Physics, an expectation, which could reduce the parameters from 5 to 2! The odd-even effect is not discussed here since we restrict the present study in even-even nuclei. Obviously, if this effect is considered an additional parameter is expected. Calculations of binding energies by using Eq. (6) for a sample of closed shell nuclei (40 Ca, 90 Zr and 208 Pb) are given in Table 3 together with experimental values for comparison.

7. Conclusion

- From the present work it is apparent that the Multi-harmonic Shell Model employed here offers a new look at the nuclear structure throughout the periodic table of the elements including super heavy nuclei.

- The consideration of nucleons with finite sizes and their fermionic nature are indispensable in the model.
- Besides the good results, the model provides a pictorial description of nuclear structure.
- The magic numbers are well verified without dependence on strong spin-orbit coupling, as in the conventional shell model.
- Strong support of the model is offered by the fact that the quantization of orbital angular momentum is precisely inherent to the symmetry of polyhedra employed by the model to present the average forms of nuclear shells.[3–6]

References

1. J. Leech, *Mathematical Gazette* **41**, 81 (1957).
2. G.S. Anagnostatos, *Lett. Nuovo Cimento* **22**, 507 (1978).
3. G.S. Anagnostatos, *Lett. Nuovo Cimento* **28**, 573 (1980).
4. G.S. Anagnostatos, *Lett. Nuovo Cimento* **29**,188 (1980).
5. G.S. Anagnostatos, J, Yapitzakis, and A. Kyritsis, *Lett. Nuovo Cimento* **32**, 332 (1981).
6. G.S. Anagnostatos and C.N. Panos, *Phys. Rev. C* 26, 260 (1982).
7. G.S. Anagnostatos, *Int. J. Theor. Phys.* **24**, 579 (1985).
8. G.S. Anagnostatos, *Can. J. Phys.* **70**, 361 (1992).
9. W.F. Hornyak, *Nuclear Structure* (Academic, New York, 1975).
10. J. Dabrowski, J. Rozynek and G.S. Anagnostatos, *Eur. Phys. J. A* **14**, 125 (2002).
11. P.E. Hodgson, *Contemp. Phys.* **35**, 329 (1994).
12. G.S. Anagnostatos, A.N. Antonov, P. Ginis, J. Giapitzakis, M.K. Gaidarov and A. Vassiliou, *Phys. Rev. C* **58**, 2115 (1998).
13. G.S. Anagnostatos, A.N. Antonov, P. Ginis, J. Giapitzakis and M.K. Gaidarov, *J. Phys. G* **25**, 69 (1999).
14. C.W. Sherwin, *Introduction to Quantum Mechanics* (Holt Rinehart and Winston, New York, 1959).

ASYMMETRIC NUCLEAR MATTER: A VARIATIONAL APPROACH

S. SARANGI

ICFAI Institute of Science & Technology, Bhubaneswar-751010, India

P. K. PANDA

Indian Association for the Cultivation of Sciences, Jadavpur, Kolkata-700 032, India

S. K. SAHU

Physics Department, Banki College, Banki-754008, Cuttack, India

L. MAHARANA

Physics Department, Utkal University, Bhubaneswar-751004, India

Received 31 July 2008

We discuss here a self-consistent method to calculate the properties of the cold asymmetric nuclear matter. In this model, the nuclear matter is dressed with s-wave pion pairs and the nucleon-nucleon (N-N) interaction is mediated by these pion pairs, ω and ρ mesons. The parameters of these interactions are calculated self-consistently to obtain the saturation properties like equilibrium binding energy, pressure, compressibility and symmetry energy. The computed equation of state is then used in the Tolman-Oppenheimer-Volkoff (TOV) equation to study the mass and radius of a neutron star in the pure neutron matter limit.

1. Introduction

The search for an appropriate nuclear equation of state has been an area of considerable research interest because of its wide and far reaching relevance in heavy ion collision experiments and nuclear astrophysics. In particular, the studies in two obvious limits, namely, the symmetric nuclear matter (SNM) and the pure neutron matter (PNM) have helped constrain several properties of nuclear matter such as binding energy per nucleon, compressibility modulus, symmetry energy and its density dependence at nuclear saturation density ρ_0[1–3] to varying degrees of success. Of late, the avaliability of flow data from heavy ion collision experiments and phenomenological data from observation of compact stars have renewed the efforts to further constrain these properties and to explore their density and isospin content (asymmetry) variation behaviours.[4–7]

One of the fundamental concerns in the construction of nuclear equation of state is the parametrization of the nucleon-nucleon (N-N) interaction. Different approaches have been developed to address this problem. These methods can be broadly classified into three general types,[8] namely, the *ab initio* methods, the effective field theory approaches and calculations based on phenomenological density functionals. The *ab initio* methods include the Brueckner-Hartree-Fock (BHF)[9–11] approach, the (relativistic) Dirac-Brueckner-Hartree-Fock (DBHF)[12–16] calculations, the Green Function Monte-Carlo (GFMC)[17–19] method using the basic N-N interactions given by boson exchange potentials. The other approach of this type, also known as the variational approach, is pioneered by the Argonne Group.[20,21] This method is also based on basic two-body (N-N) interactions in a non-relativistic formalism with relativistic effects introduced successively at later stages. The effective field theory (EFT) approaches are based on density functional theories[22,23] like chiral perturbation theory.[24,25] These calculations involve a few density dependent model parameters evaluated iteratively. The third type of approach, namely, the calculations based on phenomenological density functionals include models with effective density dependent interactions such as Gogny or Skyrme forces[26] and the relativistic mean field (RMF) models.[27–30] The parameters of these models are evaluated by carefully fitting the bulk properties of nuclear matter and properties of closed shell nuclei to experimental values. Our work presented here belongs to this class of approaches in the non-relativistic approximation.

The RMF models represent the N-N interactions through the coupling of nucleons with isoscalar scalar σ mesons, isoscalar vector ω mesons, isovector vector ρ mesons and the photon quanta besides the self- and cross-interactions among these mesons.[29] Nuclear equations of state have also been constructed using the quark meson coupling model (QMC)[31] where baryons are described as systems of non-overlapping MIT bags which interact through effective scalar and vector mean fields, very much in the same way as in the RMF model. The QMC model has also been applied to study the asymmetric nuclear matter at finite temperature.[32]

It has been shown earlier,[33,34] that the medium and long range attraction effect simulted by the σ mesons in RMF theory can also be produced by the s-wave pion pairs. This "dressing" of nucleons by pion pairs has also been applied to study the properties of deuteron[35] and ^{4}He.[36] On this basis, we start with a nonrelativistic Hamiltonian density with πN interaction. The $\omega-$repulsion and the isospin asymmetry part of the NN interaction are parametrized by two additional terms representing the coupling of nucleons with the ω and the ρ mesons respectively. The parameters of these interactions are then evaluated self-consistently by using the saturation properties like binding energy per nucleon, pressure, compressibility and the symmetry energy. The equation of state (EOS) of asymmetric nuclear matter is subsequently calculated and compared with the results of other independent approaches available in current literature. The EOS of pure neutron matter is also used to calculate the mass and radius of a neutron star. We organize the paper as follows: In Sec. 2, we present the theoretical formalism of the asymmetric nuclear matter as outlined above. The results are presented and discussed in Sec. 3. Finally, in the last section the concluding remarks are drawn indicating the future outlook of the model.

2. Formalism

We start with the effective pion nucleon Hamiltonian

$$\mathcal{H}(\mathbf{x}) = \mathcal{H}_N(\mathbf{x}) + \mathcal{H}_{int}(\mathbf{x}) + \mathcal{H}_M(\mathbf{x}), \tag{1}$$

where the free nucleon part $\mathcal{H}_N(\mathbf{x})$ is given by

$$\mathcal{H}_N(\mathbf{x}) = \psi^\dagger(\mathbf{x})\ \varepsilon_x\ \psi(\mathbf{x}), \tag{2}$$

the free meson part $\mathcal{H}_M(\mathbf{x})$ is defined as

$$\mathcal{H}_M(\mathbf{x}) = \frac{1}{2}\left[\dot{\varphi}_i^2 + (\boldsymbol{\nabla}\varphi_i)\cdot(\boldsymbol{\nabla}\varphi_i) + m^2\varphi_i^2\right], \tag{3}$$

and the πN interaction[33] is provided by

$$\mathcal{H}_{int}(\mathbf{x}) = \psi^\dagger(\mathbf{x})\left[-\frac{iG}{2\epsilon_x}\boldsymbol{\sigma}\cdot\mathbf{p}\ \varphi + \frac{G^2}{2\epsilon_x}\varphi^2\right]\psi(\mathbf{x}). \tag{4}$$

In Eqs. (2) and (4), ψ represents the non-relativistic two component spin-isospin quartet nucleon field. The single particle nucleon energy operator ϵ_x is given by $\epsilon_x = (M^2 - \boldsymbol{\nabla}_x^2)^{1/2}$ with nucleon mass M and the pion-nucleon coupling constant G. The isospin triplet pion fields of mass m are represented by φ.

We expand the pion field operator $\varphi_i(\mathbf{x})$ in terms of the creation and annihilation operators of off-mass shell pions satisfying equal time algebra as

$$\varphi_i(\mathbf{x}) = \frac{1}{\sqrt{2\omega_x}}(a_i(\mathbf{x})^\dagger + a_i(\mathbf{x})), \qquad \dot{\varphi}_i(\mathbf{x}) = i\sqrt{\frac{\omega_x}{2}}(a_i(\mathbf{x})^\dagger - a_i(\mathbf{x})), \tag{5}$$

with energy $\omega_x = (m^2 - \boldsymbol{\nabla}_x^2)^{1/2}$ in the perturbative basis. We continue to use the perturbative basis, but note that since we take an arbitrary number of pions in the unitary transformation U in equation (7) as given later, the results would be non-perturbative. The expectation value of the first term of $\mathcal{H}_{int}(\mathbf{x})$ in Eq. (4) vanishes and the pion pair of the second term provides the isoscalar scalar interaction of nucleons thereby simulating the effects of σ-mesons. A pion-pair creation operator given as

$$B^\dagger = \frac{1}{2}\int f(\mathbf{k})\ a_i(\mathbf{k})^\dagger\ a_i(-\mathbf{k})^\dagger d\mathbf{k}, \tag{6}$$

is then constructed in momentum space with the ansatz function $f(\mathbf{k})$ to be determined later.

We then define the unitary transformation U as

$$U = e^{(B^\dagger - B)}, \tag{7}$$

and note that U, operating on vacuum, creates an arbitrarily large number of scalar isospin singlet pairs of pions. The "pion dressing" of nuclear matter is then introduced through the state

$$|f\rangle = U|vac\rangle = e^{(B^\dagger - B)}|vac\rangle, \tag{8}$$

where U constitutes a Bogoliubov transformtion given by

$$U^\dagger\ a_i(\mathbf{k})U = (\cosh\ f(\mathbf{k}))\ a_i(\mathbf{k}) + (\sinh\ f(\mathbf{k}))\ a_i(-\mathbf{k})^\dagger. \tag{9}$$

We then proceed to calculate the energy expectation values. We consider N nucleons occupying a spherical volume of radius R such that the density $\rho = N/(\frac{4}{3}\pi R^3)$ remains constant as $(N,\ R) \to \infty$ and we ignore the surface effects. We describe the system with a density operator $\hat{\rho}_N$ such that its matrix elements are given by[33]

$$\rho_{\alpha\beta}(\mathbf{x},\mathbf{y}) = Tr[\hat{\rho}_N\ \psi_\beta(\mathbf{y})^\dagger \psi_\alpha(\mathbf{x})], \tag{10}$$

and

$$Tr[\hat{\rho}_N \hat{N}] = \int \rho_{\alpha\alpha}(\mathbf{x},\mathbf{x})d\mathbf{x} = N = \rho V. \tag{11}$$

We obtain the free nucleon energy density

$$h_f = \langle f|Tr[\hat{\rho}_N \mathcal{H}_N(\mathbf{x})]|f\rangle = \sum_{\tau=n,p} \frac{\gamma k_f^{\tau 3}}{6\pi^2}\left(M + \frac{3}{10}\frac{k_f^{\tau 2}}{M}\right). \tag{12}$$

In the above equation, the spin degeneracy factor $\gamma = 2$, the index τ runs over the isospin degrees of freedom n and p and k_f^τ represents the Fermi momenta of the nucleons. For asymmetric nuclear matter, we define the neutron and proton densities ρ_n and ρ_p respectively over the same spherical volume such that the nucleon density $\rho = \rho_n + \rho_p$. The Fermi momenta k_f^τ are related to neutron and proton densities by the relation $k_f^\tau = (6\pi^2\rho_\tau/\gamma)^{\frac{1}{3}}$. We also define the asymmetry parameter $y = (\rho_n - \rho_p)/\rho$. It can be easily seen that $\rho_\tau = \frac{\rho}{2}(1 \pm y)$ for $\tau = n, p$ respectively.

Using the operator expansion of Eq. (5), the free pion part of the Hamiltonian as given in Eq. (3) can be written as

$$\mathcal{H}_M(\mathbf{x}) = a_i(\mathbf{x})^\dagger\ \omega_x\ a_i(\mathbf{x}). \tag{13}$$

The free pion kinetic energy density is given by

$$h_k = \langle f|\mathcal{H}_M(\mathbf{x})|f\rangle = \frac{3}{(2\pi)^3}\int d\mathbf{k}\ \omega(\mathbf{k})\ \sinh^2\ f(\mathbf{k}), \tag{14}$$

where $\omega(\mathbf{k}) = \sqrt{\mathbf{k}^2 + m^2}$. Using $\epsilon_x \simeq M$ in the nonrelativistic limit, the interaction energy density h_{int} can be written from Eq. (4) as

$$h_{int} = \langle f|Tr[\hat{\rho}_N\ \mathcal{H}_{int}(\mathbf{x})]|f\rangle \simeq \frac{G^2\rho}{2M}\langle f| : \varphi_i(\mathbf{x})\varphi_i(\mathbf{x}) : |f\rangle. \tag{15}$$

Using the Eqs. (7), (8) and (9), we have from Eq. (15)

$$h_{int} = \frac{G^2\rho}{2M}\left(\frac{3}{(2\pi)^3}\int \frac{d\mathbf{k}}{\omega(\mathbf{k})}\left(\frac{\sinh 2f(\mathbf{k})}{2} + \sinh^2\ f(\mathbf{k})\right)\right). \tag{16}$$

The pion field dependent energy density terms add up to give $h_m(= h_k + h_{int})$ which is to be optimized with respect to the ansatz function $f(\mathbf{k})$ for its evaluation. However, this ansatz function yields a divergent value for h_m. This happens because we have taken the pions to be point like and have assumed that they can approach as near each other as they like, which is physically inaccurate. Therefore, we introduce a phenomenological repulsion energy between the pions of a pair given by

$$h_m^R = \frac{3a}{(2\pi)^3} \int (\sinh^2 f(\mathbf{k}))\, e^{R_\pi^2 k^2} d\mathbf{k}, \tag{17}$$

where the two parameters a and R_π correspond to the strength and length scale, repectively, of the repulsion and are to be determined self-consistently later. Thus the pion field dependent term of the total energy density becomes $h_m = h_k + h_{int} + h_m^R$. Then the optimization of h_m with respect to $f(\mathbf{k})$ yields

$$\tanh 2f(\mathbf{k}) = -\frac{G^2\rho}{2M} \cdot \frac{1}{\omega^2(\mathbf{k}) + \frac{G^2\rho}{2M} + a\omega(\mathbf{k})e^{R_\pi^2 k^2}}. \tag{18}$$

The expectation value of the pion field dependent parts of the total Hamiltonian density of Eq. (1) alongwith the modification introduced by the phenomenological term h_m^R becomes

$$h_m = -\frac{3}{2}\frac{1}{(2\pi)^3}\left(\frac{G^2}{2M}\right)^2 \rho\Big[\rho_n I_n + \rho_p I_p\Big] \tag{19}$$

with the integrals I_τ $(\tau = n, p)$ given by

$$I_\tau = \int_0^{k_f^\tau} \frac{4\pi k^2 dk}{\omega^2}\left[\frac{1}{(\omega + ae^{R_\pi^2 k^2})^{1/2}(\omega + ae^{R_\pi^2 k^2} + \frac{G^2\rho}{M\omega})^{1/2} + (\omega + ae^{R_\pi^2 k^2}) + \frac{G^2\rho}{2M\omega}}\right] \tag{20}$$

and $\omega = \omega(\mathbf{k})$.

We now introduce the energy of ω repulsion by the simple form

$$h_\omega = \lambda_\omega \rho^2, \tag{21}$$

where the parameter λ_ω corresponds to the strength of the interaction at constant density and is to be evaluated later. We note that Eq. (21) can arise from a Hamiltonian density given in terms of a local potential $v_R(\mathbf{x})$ as

$$\mathcal{H}_R(\mathbf{x}) = \psi(\mathbf{x})^\dagger \psi(\mathbf{x}) \int v_R(\mathbf{x}-\mathbf{y})\psi(\mathbf{y})^\dagger \psi(\mathbf{y})d\mathbf{y}, \tag{22}$$

where, when density is constant, we in fact have

$$\lambda_\omega = \int v_R(\mathbf{x})d\mathbf{x} \; .$$

The isospin dependent interaction is mediated by the isovector vector ρ mesons. We represent the contribution due to this interaction, in a manner similar to the ω-meson energy, by the term

$$h_\rho = \lambda_\rho \rho_3^2 \tag{23}$$

where $\rho_3 = (\rho_n - \rho_p)$ and the strength parameter $\lambda_\rho$is to be determined as described later. Thus we finally write down the binding energy per nucleon E_B of the cold asymmetric nuclear matter:

$$E_B = \frac{\varepsilon}{\rho} - M, \tag{24}$$

where $\varepsilon = (h_f + h_m + h_\omega + h_\rho)$ is the energy density. The expression for ε contains the four model parameters a, R_π, λ_ω and λ_ρ as introduced above. These parameters are then determined self-consistently through the saturation properties of nuclear matter. The pressure P, compressibility modulus K and the symmetry energy E_{sym} are given by the standard relations:

$$P = \rho^2 \frac{\partial(\varepsilon/\rho)}{\partial\rho} \tag{25}$$

$$K = 9\rho^2 \frac{\partial^2(\varepsilon/\rho)}{\partial\rho^2} \tag{26}$$

$$E_{sym} = \left(\frac{1}{2}\frac{\partial^2(\varepsilon/\rho)}{\partial y^2}\right)_{y=0}. \tag{27}$$

The effective mass M^* is given by $M^* = M + V_s$ with $V_s = (h_{int} + h_m^R)/\rho$.

3. Results and Discussion

We now discuss the results obtained in our calculations and compare with those available in literature. The four parameters of the model are fixed by self-consistently solving Eqs. (24) through (27) for the respective properties of nuclear matter at saturation density $\rho_0 = 0.15$ fm^{-3}. While pressure P vanishes at saturation density for symmetric nuclear matter (SNM), the values of binding energy per nucleon and symmetry energy are chosen to be -16 MeV and 31 MeV respectively. In the numerical calculations, we have used the nucleon mass $M = 940$ MeV, the meson masses $m = 140$ MeV, $m_\omega = 783$ MeV and $m_\rho = 770$ MeV and the $\pi - N$ coupling constant $G^2/4\pi = 14.6$. In order to ascertain the dependence of compressibility modulus on the parameter values, we vary the K value over a range 210 MeV to 280 MeV for the symmetric nuclear matter ($y = 0$) and evaluate the parameters.

It may be noted that this is the range of the compressibility value which is under discussion in the current literature. For K values in the range 210 MeV to 250 MeV, the program does not converge. The solutions begin to converge for compressibility

Table 1. Parameters of the model obtained by solving the Eqs. (24)–(27) self consistently at saturation density.

a (MeV)	R_π (fm)	λ_ω (fm^2)	λ_ρ (fm^2)
16.98	1.42	3.10	0.65

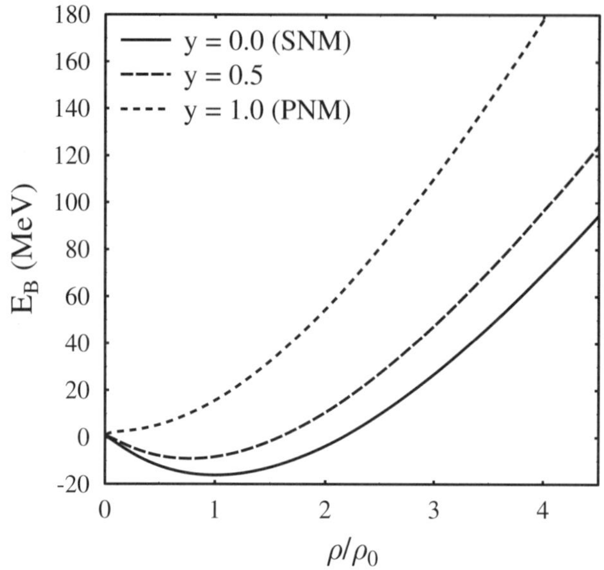

Fig. 1. The binding energy per nucleon E_B as a function of relative nucleon density ρ/ρ_0 calculated for different values of the asymmetry parameter y. The values $y = 0.0$ and 1.0 correspond to symmetric nuclear matter (SNM) and pure neutron matter (PNM) respectively.

modulus K around 258 MeV. We choose the value $K = 260$ MeV for our calculations. In Table 1 we present the four free parameters of the model for ready reference.

For this set of parameter values the effective mass of nucleons at saturation density is found to be $M^*/M = 0.81$. In the Fig. 1, we present the binding energy per nucleon E_B calculated for different values of the asymmetry parameter y as a function of the relative nuclear density ρ/ρ_0. The values $y = 0.0$ and 1.0 correspond to SNM and PNM respectively. As expected, the binding energy per nucleon E_B of SNM initially decreases with increase in density, reaches a minimum at $\rho = \rho_0$ and then increases. In case of PNM, the binding energy increases monotonically with increasing density in consistence with its well known behaviour. In Fig. 2(a), we compare the E_B of SNM as a function of the nucleon density with a few representative results in the literature, namely, the Walecka model[27] (long-short dashed curve), the DBHF calculations of Li *et al.* with Bonn A potential (short-dashed curve) (data for both the models are taken from Ref. 13) and the variational A18 + δv + UIX* (corrected) model of Akmal *at al.* (APR)[21] (long-dashed curve). While the Walecka and Bonn A models are relativistic, the variational model is nonrelativistic with relativistic effects and three body correlations introduced successively. Our model produces an EOS softer than that of Walecka and Bonn A, but stiffer than the variational calculation results of the Argonne group. It is well-known that the Walecka model yields a very high compressibility K. However, its improvised versions developed later with self- and cross-couplings of the meson fields have been able to bring down the compressibility modulus in the ball park of 230 ± 10 MeV.[7] Our model yields nuclear matter saturation properties correctly along with the compressibility of $K = 260$ MeV which is resonably close to the empirical data. In Fig. 2(b), we plot E_B as a function of the relative nucleon density for PNM. Similar to the SNM case, our EOS is softer than that of Walecka and Bonn A models, but stiffer than the variational model. We use this EOS to calculate the mass and radius of a neutron star of PNM as discussed later.

The density dependence of pressure of SNM and PNM are calculated using the Eq. (25). These results are plotted (solid blue curves) in Figs. 3(a) and (b). Recently, Danielewicz *et al.*[4] have deduced the empirical bounds on the EOS in the density range of $2 < \rho/\rho_0 < 4.6$ by analysing the flow data of matter from the fireball of Au+Au heavy ion collision experiments both for SNM and PNM. These bounds are represented by the color-filled and shaded regions of the two figures. These bounds rule out both the "very stiff" and the "very soft" classes of EOSs produced, for example, by some variants of RMF calculations and Fermi motion of a pure neutron gas.[4] As shown in these figures, the EOS of SNM and PNM generated by our model are consistent with both the bounds.

The potentials per nucleon in our model can be defined from the meson dependent energy terms of Eqs. (19), (21) and (23). Contribution to potential from the scalar part of the meson interaction is due to the pion condensates and is given by $V_s = (h_{int} + h_m^R)/\rho$ as defined earlier. The contribution by vector mesons has two components, namely, due to the ω and the ρ mesons and is given by $V_v = V_\omega + V_\rho = (h_\omega + h_\rho)/\rho$. In the Figs. 4 (a) and (b), we plot V_s and V_v as functions of relative density ρ/ρ_0 calculated for PNM (Fig. 4(a)) and for SNM (Fig. 4(b)) respectively. The magnitudes of the potentials calculated by our model are weaker compared to those produced by DBHF calculations with Bonn A interaction[13] as shown in both the panels of Fig. 4. In Fig. 4(a), we show the contributions to the repulsive vector potential due to ω mesons (short-dashed curve), ρ mesons (long-dashed curve) and their combined contribution (long-short-dashed curve). The contribution due to ρ mesons rises linearly at a slow rate and has a low contribution at saturation density. This indicates that major contribution to the short-range repulsion part of nuclear force is from ω meson interaction.

Knowledge of density dependence of symmetry energy is expected to play a key role in understanding the structure and properties of neutron-rich nuclei and neutron stars at densities above and below the saturation density. Therefore this problem has been receiving considerable attention of late. Several theoretical and experimental investigations addressing this problem have been reported (Refs. 3, 8, 39 and references therein). While the results of independent studies show reasonable consistency at sub-saturation densities $\rho \leq \rho_0$, they are at wide variance with each other at supra-saturation densities $\rho > \rho_0$. This wide variation has given rise to the so-called classification of "soft" and "stiff" dependence of symmetry energy on density.[38,39]

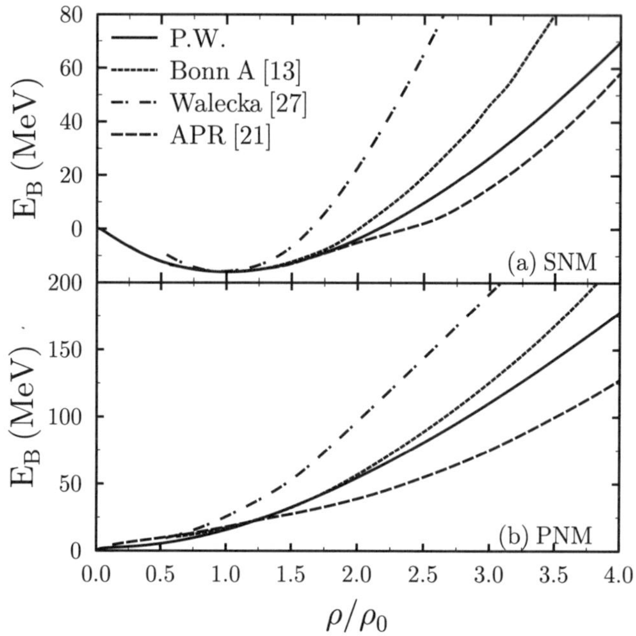

Fig. 2. (a)The binding energy per nucleon E_B as a function of relative nucleon density ρ/ρ_0 for SNM. The results of present work (P.W.) are compared with the results of DBHF calculations with Bonn A potential,[13] the variational calculations of the Argonne group[21] and the Walecka model.[27] The data for the Bonn A and Walecka model curves are taken from.[13] (b) Same as Fig. (2a), but for PNM.

Figure 5 shows a representation of the spectrum of such results along with the results of the present work (solid blue curve). While the Gogny and Skyrme forces (dark rib-dotted and dotted curves respectively with data taken from Refs. 8, 39) produce "soft" dependence on one end, the NL3 force (dot-dashed curve with data taken from Ref. 8) produces a very "stiff" dependence on the other end. The analysis of experimental and simulation studies of intermediate energy heavy-ion reactions as reported by Shetty *et al.*[39] (red triangles and long-short-dashed red curve repectively), results of DBHF calculations of Li *et al.* and Huber *et al.*[13,29,40] (rib-dashed and magenta ribbed curve), variational model[3,21] (short-dashed curve), RMF calculations with nonlinear Walecka model including ρ mesons by Liu *et al.*[30] (long-dashed green curve) as shown in Fig. 5 suggest "stiff" dependence with various degrees of stiffness. The experimental results (represented by the red triangles with data taken from Shetty *et al.*[39]) are derived from the isoscaling parameter α which, in turn, is obtained from relative isotopic yields due to multifragmentation of excited nuclei produced by bombarding beams of ^{58}Fe and ^{58}Ni on ^{58}Fe and ^{58}Ni

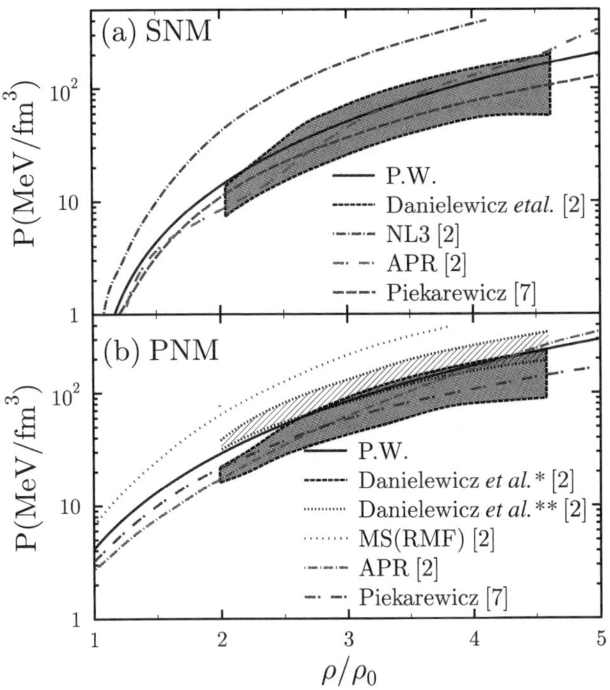

Fig. 3. (Color online) (a) The pressure as a function of relative nucleon density for SNM as generated by the present work (P.W.) (solid blue curve). The color-filled region in green corresponds to the bounds deduced from experimental flow data and simulations studies by Danielewicz *et al.*[4] The data for the curves corresponding to RMF(NL3) calculations and the variational calculations of Akmal *et al.* (APR) are taken from Ref. 4. (b) Pressure as a function of relative nucleon density for PNM. The shaded region and the color-filled region in green correspond to the bounds deduced by Danielewicz *et al.* using the "stiff" and "soft" parametrizations of Prakash *et al.*.[37] Our EOS is consistent with these bounds in the cases of both SNM and PNM.

targets. Shetty *et al.* have shown that the results of multifragmentation simulation studies carried out with Antisymmetrized Molecular Dynamics (AMD) model using Gogny-AS interaction and Statistical Multifragmentation Model (SMM) are consistent with the above-mentioned experimental results and suggest (as shown by the red long-short-dashed curve) a moderately stiff dependence of the symmetry energy on density. Our results (represented by the solid blue curve) calculated using Eq. (27) are consistent with these results at subsaturation densities but are stiffer at supra-saturation densities. More observational or experimental information is required to be built into our model to further constrain the symmetry energy at higher densities. In Fig. 5, the curve due to Huber *et al.*[40] (with data taken from Ref. 29) correspond to their DBHF 'HD' model calculations which involves only the σ, ω and ρ mesons. Similarly the long-dashed green curve due to Liu *et al.*[30] is from the basic non-linear Walecka model with σ, ω and ρ mesons. Our formalism is

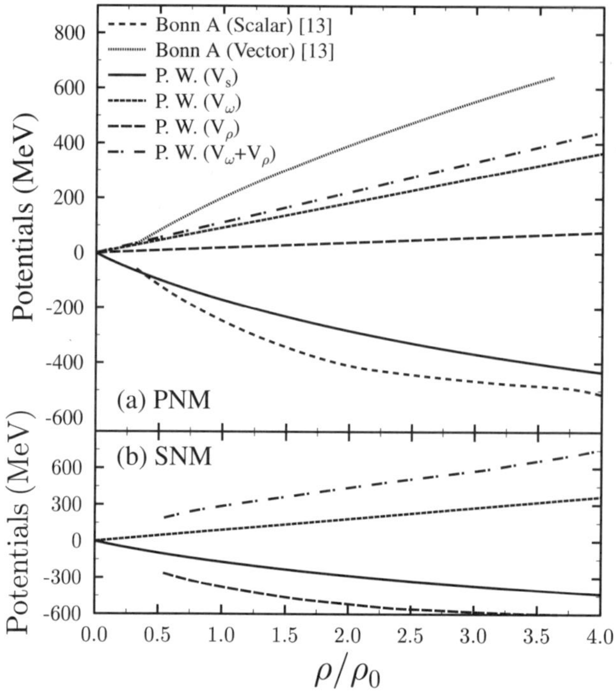

Fig. 4. (a) The potentials V_s, V_ω and V_ρ (as defined in the text) in PNM as calculated by our model are compared with the Bonn A results of Li *et al.*[13] The contributions made by the ω-meson (short-dashed curve) and ρ-meson (long-dashed curve) mediated interactions are distinctly shown for comparison. (b) The potentials in SNM. Because of isospin symmetry, V_ρ (see text for definition) vanishes. Both the scalar (solid curve) and vector (short-dashed curve) potentials produced by our calculations are weaker in magnitude compared to those of Bonn A calculations.

the closest to these two models with the exception that in our model the effect of σ mesons is simulated by the π meson condensates. It is also noteworthy that our results are consistent with these results for densities upto $2\rho_0$.

The wide variation of density dependence of symmetry energy at supra-saturation densities has given rise to the need of constraining it. As discussed by Shetty *et al*,[39] a general functional form $E_{sym} = E^0_{sym}(\rho/\rho_0)^\gamma$ has emerged. Studies by various groups have produced the fits with $E^0_{sym} \sim 31 - 33$ MeV and $\gamma \sim 0.55 - 1.05$. A similar parametrization of the E_{sym} produced by our EOS with $E^0_{sym} = 31$ MeV yields the exponent parameter $\gamma = 0.85$.

We next use the equation of state for PNM derived by our model in the Tolman-Oppenheimer-Volkoff (TOV) equation to calculate the mass and radius of a PNM neutron star. The mass and radius of the star are found to be $2.25M_\odot$ and 11.7 km respectively.

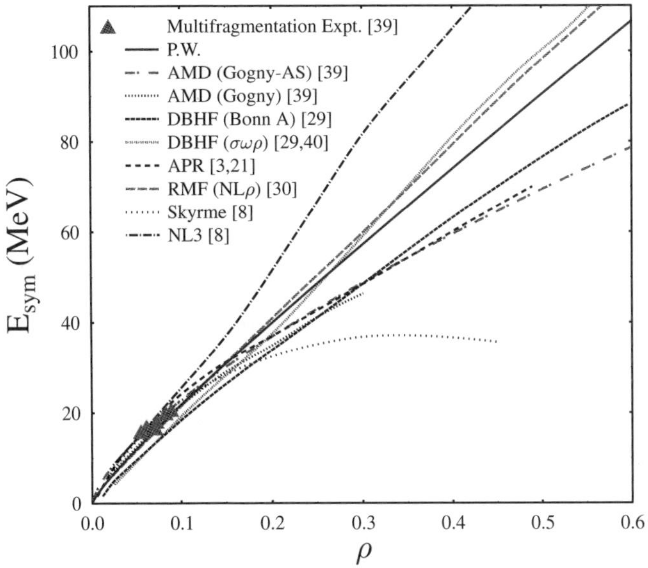

Fig. 5. (Color online) Symmetry energy E_{sym} calculated from the EOS (as in Eq. (27)) (P.W.) (solid blue line) is plotted as a function of density along with results of other groups. The data for experimental points and the results of the antisymmetrized molecular dynamics (AMD) simulations with Gogny-AS and Gogny interactions are taken from Shetty *et al*,[39] DBHF (Bonn A) results are taken from Ref. 29, RMF (NLρ) data are from Ref. 30, the variational model of Akmal *et al.* (APR)[21] results are from Ref. 3, DBHF ($\sigma\omega\rho$) model of Huber *et al.*[40] data are from Ref. 29, the Skyrme amd NL3 results are from Ref. 8. Our result shows consistency with those of other groups and corroborates the moderately "stiff" dependence of E_{sym} as advocated by Shetty *et al.*[39]

4. Conclusion

In this work we have presented a quantum mechanical nonperturbative formalism to study cold asymmetric nuclear matter using a variational method. The system is assumed to be a collection of nucleons interacting via exchange of π pairs, ω and ρ mesons. The equation of state (EOS) for different values of asymmetry parameter is derived from the dynamics of the interacting system in a self-consistent manner. This formalism yields results similar to those of the *ab initio* DBHF models, variational models and the RMF models without invoking the σ mesons. The compressibility modulus and effective mass are found to be $K = 260$ MeV and $M^*/M = 0.81$ respectively. The symmetry energy calculated from the EOS suggests a moderately "stiff" dependence at supra-saturation densities and corroborates the recent arguments of Shetty *et al.*[39] A parametrization of the density dependence of symmetry energy of the form $E_{sym} = E^0_{sym}(\rho/\rho_0)^\gamma$ with the symmetry energy E^0_{sym} at saturation density being 31 MeV produces $\gamma = 0.85$. The EOS of pure neutron matter (PNM) derived by the formalism yields the mass and radius of a PNM neutron star to be $2.25M_\odot$ and 11.7 km respectively.

Acknowledgments

P.K.P would like to acknowledge Julian Schwinger foundation for financial support. P.K.P wishes to thank Professor F.B. Malik and Professor Virulh Sa-yakanit for inviting the CMT31 workshop. The authors are also thankful to Professor S.P. Misra for many useful discussions.

References

1. M. Prakash, I. Bombaci, M Prakash, P.J. Ellis and J.M. Lattimer, *Phys. Rep.* **280**, 1 (1997).
2. J.M. Lattimer and M. Prakash, *Phys. Rep.* **333**, 121 (2000); *Astrophys. J.* **550**, 426 (2001); *Science* **304**, 536 (2004).
3. A. W. Steiner, M. Prakash, J.M. Lattimer and P.J. Ellis, *Phys. Rep.* **441**, 325 (2005).
4. P.Danielewicz, R. Lacey and W.G. Lynch, *Science* **298**, 1592 (2002).
5. D.J. Nice *et al.*, *Astrophys. J.* **634**, 1242 (2005).
6. T. Klähn *et al.*, *Phys. Rev. C* **74**, 035802 (2006).
7. J. Piekarewicz, *Phys. Rev. C* **76**, 064310 (2007).
8. C. Fuchs and H.H. Wolter, *Eur. Phys. J. A* **30**, 5 (2006).
9. M. Jaminon and C. Mahaux, *Phys. Rev. C* **40**, 354 (1989).
10. X.R. Zhou, G.F. Burgio, U. Lombardo, H.-J. Schulze and W. Zuo, *Phys. Rev. C* **69**, 018801 (2004).
11. M.Baldo and C. Maieron, *J. Phys. G* **34**, R243 (2007).
12. R. Brockmann and R. Machleidt, *Phys. Rev. C* **42**, 1965 (1990).
13. G.Q. Li, R. Machleidt and R. Brockmann, *Phys. Rev. C* **45**, 2782 (1992).
14. F. de Jong and H. Lenske, *Phys. Rev. C* **58**, 890 (1998).
15. T. Gross-Boelting, C. Fuchs and A. Faessler, *Nucl. Phys. A* **648**, 890 (1999).
16. E.N.E van Dalen, C. Fuchs and A. Faessler, *Nucl. Phys. A* **744**, 227 (2004); *Eur. Phys. J. A* **31**, 29 (2007).
17. J.Carlson, J. Morales, V.R. Pandharipande and D.G. Ravenhall, *Phys. Rev. C* **68**, 025802 (2003).
18. W.H. Dickhoff and C. Barbieri, *Prog. Part. Nucl. Phys.* **52**, 377 (2004).
19. A. Fabrocini, S. Fantoni, A.Y. Illarionov and K.E. Schmidt, *Phys. Rev. Lett.* **95**, 192501 (2005).
20. A. Akmal and V.R. Pandharipande, *Phys. Rev.* **C 56**, 2261 (1997).
21. A. Akmal, V.R. Pandharipande and D.G. Ravenhall *Phys. Rev. C* **58**, 1804 (1998).
22. B.D. Serot and J.D. Walecka, *Int. J. Mod. Phys. E* **E 6**, 515 (1997).
23. R.J Furnstahl, *Lect. Notes Phys.* **641**, 1 (2004)
24. M. Lutz, B. Friman and Ch. Appel, *Phys. Lett B* **474**, 7 (2000).
25. P. Finelli, N. Kaiser, D. Vretenar and W. Weise, *Eur. Phys. J A* **17**, 573, (2003); *Nucl. Phys. A* **735**, 449 (2004).
26. M. Bender, P.-H.Heenen and P.-G. Reinhard, *Rev. Mod. Phys.* **75**, 2003.
27. J.D. Walecka, *Ann. Phys.* (N.Y.) **83**, 491 (1974); B.D. Serot and J.D. Walecka, *Adv. Nucl. Phys.* **16**, 1 (1986).
28. P. Ring, *Prog. Part. Nucl. Phys.* **73**, 193 (1996); Y.K. Gambhir and P. Ring, *Phys. Lett. B* **202**, 2 (1988).
29. J.K. Bunta and Š. Gmuca, *Phys. Rev. C* **68**, 054318 (2003); S. Gmuca, *J. Phys. G* **17**, 1115 (1991).
30. B. Liu, M. D Toro, V. Greco, C. W. Shen, E. G. Zhao and B. X. Sun, e-print Arxiv No. nucl-th/0702064.

31. K. Saito and A.W. Thomas, *Phys. Lett. B* **327**, 9 (1994); **335**, 17 (1994); **363**, 157 (1995); *Phys. Rev. C* **C 52**, 2789 (1995); P.A.M. Guichon, K. Saito, E. Rodionov and A.W. Thomas, *Nucl. Phys. A* **601** 349 (1996); P.K. Panda, A. Mishra, J.M. Eisenberg and W. Greiner, *Phys. Rev. C* **56**, 3134 (1997).
32. P.K. Panda, G. Krein, D.P. Menezes and C. Providência, *Phys. Rev. C* **68**, 015201 (2003).
33. A. Mishra, H. Mishra and S.P. Misra, *Int. J. Mod. Phys. A* **7**, 3391 (1990).
34. H. Mishra, S.P. Misra, P.K. Panda and B. K. Parida, *Int. J. Mod. Phys. E* **2**, 405 (1992).
35. P.K. Panda, S.P. Misra and R. Sahu, *Phys. Rev. C* **45**, 2079 (1992).
36. P.K. Panda, S.K. Patra, S.P. Misra and R. Sahu, *Int. J. Mod. Phys. E* **5**, 575 (1996).
37. M. Prakash, T.L. Ainsworth and J.M. Lattimer, *Phys. Rev. Lett.* **61**, 2518 (1988).
38. J. R. Stone, J. C. Miller, R. Koncewicz, P.D. Stevenson and M. R. Strayer, *Phys. Rev. C* **68**, 034324 (2003).
39. D.V. Shetty, S.J. Yennello and G.A. Souliotis, *Phys. Rev. C* **76**, 024606 (2007)
40. H. Huber, F. Weber and M. K. Weigel, *Phys. Lett. B* **317**, 485 (1993); H. Huber, F. Weber and M. K. Weigel, *Phys. Rev. C* **51**, 1790 (1995).

PROGRESS IN THE DESCRIPTION OF NUCLEAR PHYSICS FROM LATTICE QCD

A. S. B. TARIQ*
Department of Physics, Rajshahi University,
Rajshahi 6205, Bangladesh
asbtariq@ru.ac.bd

Received 31 July 2008

It has been a natural desire for a long time to be able to describe nuclear physics in terms of the fundamental strong interaction. Recently some significant progress has been made in this area in terms of lattice QCD calculations of simple nuclear physics processes such as nucleon nucleon scattering. An attempt is made to introduce the progress made in this area, to an audience composed mainly of many-body theorists (non-lattice QCD and even non-particle/nuclear physics) interested in inter-disciplinary approaches.

Keywords: Nuclear physics; lattice QCD; nucleon-nucleon potential; hyperon-nucleon potential.

1. Introduction

In the subatomic world we have the three domains of (a) the force between quarks described in quantum chromodynamics (QCD), (b) the force between hadrons made of quarks, often termed hadron physics and (c) the force between nuclei made of nucleons studied in nuclear physcis. Hadron physics in a sense sits in the middle overlapping both nuclear physics and QCD.

As the van der Waals force is the residual of the electrostatic force within molecules, the nuclear force is the residual of the force between quarks, above the length scales of quark-gluon interactions, nuclear physics is simply an effective theory of QCD. Therefore, it should, in principle, be possible to give a description of nuclear physics starting from QCD.

However, the issue is that, this connection has not been easy to establish. Rather, the elusiveness of this connection is evident from textbook quotes such as,[1]

> ... a quantitative connection between nuclear force and quark-quark interaction is still lacking. The root of the problem is, as usual, the difficulty of carrying out QCD calculations at the low energies where nuclear physics

*No originality is claimed in this work. Rather an attempt is made to introduce, for a non-lattice QCD, rather non-particle/nuclear physics audience, the progress made in this area. Most work discussed hear is done by two groups in the USA (M. Savage, Washington and others) and Japan (led by S. Aoki, Tsukuba).

> operates ... in linking the quark-quark interaction with the interaction between nucleons other methods of attack, such as lattice gauge calculations and soliton models, are yet to be fully explored.

As we know, the strong coupling blows up at large distances or low momenta, making the problem non-perturbative. It is not possible to use familiar perturbative techniques and as mentioned above, it is here that lattice QCD comes in. In lattice QCD, the field theory is formulated on a discrete hypercubic space-time lattice — providing a non-perturbative discretization of the path integral. This is then solved numerically using large computers. In fact, lattice QCD can be seen as an effective theory of continuum QCD with the lattice spacing acting like a UV cutoff, but with gauge invariance conserved. There are excellent books and notes for an introduction to lattice QCD to which a reader can refer to for further details.

However, lattice QCD simulations also have practical limitations. Foremost of these is that, even present day simulations can be done only at unphysically large light (u,d) quark masses. When one refers to simulations being done at *a pion mass of, say, 400 MeV*, this has to be compared to the physical pion mass for an idea of how heavy the simulated light quarks are. Simulations are also done on a limited volume of 3–4 fm lengths along spatial dimensions. Intriguingly the present volumes are good for nuclear simulations at present quark masses; at realistic masses larger volumes are needed.

Coming to nuclear physics prospects of lattice QCD, on the one hand, lattice calculations on common nuclear quantities have not been easily forthcoming. However, on the other hand, there is hope that lattice QCD will provide information on nuclear physics quantities with little or no experimental data, e.g. dense nuclear matter. This is relevant to supernova remnants — determining whether it ends up as a black hole or a neutron star. Again compressibility of nuclear matter depends on the composition and here experimental information is lacking.

When it comes to neutron star nuclear matter, it is well known how electron capture becomes energetically favorable to turn protons into neutrons. However, usually other possible hadronic components (kaons, sigmas etc.) are neglected. Kaplan and Nelson[2,3] showed that K^-n interaction can reduce the mass of the kaon, and Brown[4] showed that if the K^- mass gets lower than the electron chemical potential, which again is raised at higher nuclear densities, there will be condensation of kaons. Neutron star nuclear matter then becomes: $p + n$+kaon condensate, which has a softer equation of state. To study these, the density dependence of kaon-nucleon interaction is required, obtainable in principle from lattice QCD. The case for $n\Sigma^-$ is also quite similar.

2. Effective Theory Approach

To match lattice calculations to nuclear physics, one approach has been of matching with effective theories. This has been a long series of work starting from the pioneering works of Kaplan, Savage and Wise[5–7] followed over many years by collaborators

Beane, Bedaque, Detmold, Orginos, Parreno, van Kolck, Luu, Torok, Walker-Loud and others.[8–13] A recent review is available.[14] The aim has been to establish a rigorous pathway from QCD to nuclei by performing lattice calculations of some nuclear quantities and from these calculations, determining counter-terms in appropriate low-energy effective field theories (EFTs). Then the EFT is used to compute quantities of interest in relatively simple systems and subsequently matched to nuclear physics.

This programme may be interesting, for example, in relation with the description of the shell model as an effective theory.[15] Here the starting point is the large model space of harmonic oscillator levels. One can match chiral perturbation theory and nucleon-nucleon EFTs in this space and systematically reduce model space integrating out higher harmonic oscillator levels *a la* renormalization group, but for bound states. Renormalization of the operators and the Hamiltonian leaves one with the small low energy space (shell model) and a large excluded space. Evolution is such that levels and matrix elements in shell model reproduce those in the full space.

In a sense this is like, rather than waiting for lattice calculations to come down to physical masses, to try take theory predictions up to simulated masses. The progress attained in this approach, coupled with the continuous improvement of lattice simulations can be observed from the two sets of figures given in Fig. 1. In the figures presented in 2005, the lattice data was still far away from theory. But for the same figures presented in 2007 one can see that the lattice data is now near the border of applicability of EFT predictions.

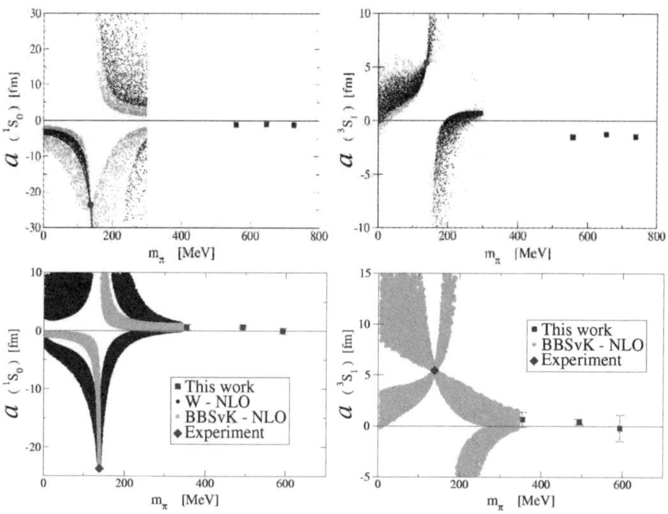

Fig. 1. Allowed regions for the NN scattering lengths in the s-wave channels as a function of the pion mass at NLO in the NN EFT compared to simulated points as presented in 2005[13] (top row) and 2007[14] (bottom row). Figures reproduced from these sources.

3. NN Potentials from Lattice QCD

Phenomenologically, nucleon-nucleon potentials are pretty well-known. The question is: Can these be reproduced from lattice QCD? In attempting to address this question, lattice QCD is *speaking the language* of nuclear physicists.

NN potentials have three regions. The tail and intermediate region are understood from one- and two- pion exchange (OPE, 2PE) etc. However, the repulsive core is not theoretically well-founded. This is in fact a region where quark and gluon degrees of freedom should be important.

The core region is important not only for the academic interest of understanding it from first principle theoretical calculations. The hard core is related to issues like nuclear saturation / compressibility, the maximum mass of neutron stars, the ignition of type II supernovae, and by extension, in fact, to the origin of life itself! (Since the ignition of type II supernovae is responsible for dispersion of the elements throughout the universe.

To calculate this potential in lattice QCD, there are two alternative approaches:

3.1. *Method I: With heavy quark*

This method is similar to calculation of the static quark- antiquark potential. Static (infinitely heavy) quarks also define distance between hadrons. If one calculates the energy of a two hadron system for different separations, then the potential can be defined as the energy difference $V_{BB} - 2V_B$. Several studies[8,16,17] have been undertaken in this approach. However, the outcome does not seem to be conclusive yet.

3.2. *Method II: From wavefunction*

The Salpeter-Bethe equation[18] describes the bound states of a two-body quantum mechanical system. Examples of systems desribed by this equation are the positronium, bound state of an electron-positron pair; and, in condensed matter physics, the exciton, bound state of an electron-hole pair. Aoki *et al.*[19] used an idea based on the Salpeter-Bethe wavefunction/amplitude for pion scattering lengths.

First a wave function is defined by

$$\varphi(\mathbf{r}, E) = \langle 0|N(\mathbf{x}, 0)N(\mathbf{y}, 0)|2N; E\rangle, \qquad \mathbf{r} = \mathbf{x} - \mathbf{y}, \tag{1}$$

where $N(\mathbf{x}, t)$ is an interpolating field of the nucleon, $|2N; E\rangle$ is a $2N$ state with energy E below the inelastic threshold.

To calculate this wavefunction/amplitude in lattice QCD, one uses four point correlation function

$$F_{\rm NN}(\mathbf{x}, \mathbf{y}, t; t_0) \equiv \langle 0|N^i_\alpha(\mathbf{x}, t)N^j_\beta(\mathbf{y}, t)\mathcal{J}_{\rm NN}|0\rangle \tag{2}$$

where $\mathcal{J}_{\rm NN}(t_0) = P^I_{ij}P^S_{\alpha\beta}\mathcal{N}^i_\alpha\mathcal{N}^j_\beta$, Ps being (iso)spin projection operators. Then the potential is defined through the Schrödinger equation

$$[H_0 + V(\mathbf{r})]\varphi(\mathbf{r}, E) = E\varphi(\mathbf{r}, E), \tag{3}$$

leading to

$$V(\mathbf{r}) = \frac{(E - H_0)\varphi(\mathbf{r}, E)}{\varphi(\mathbf{r}, E)}. \tag{4}$$

A recent review of the developments along this direction can be found in Ref. 20

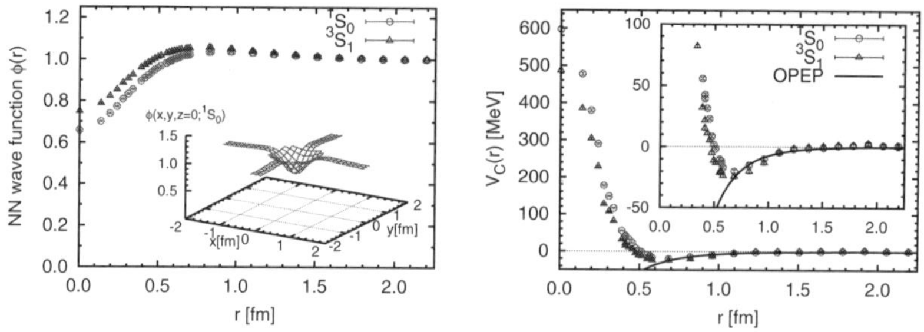

Fig. 2. Singlet and triplet nucleon-nucleon wavefunctions (left) and potentials (right) reproduced from Ishii *et al.*[21]

Recently, this was applied for the first time to extract a nucleon-nucleon potential by Ishii *et al.*[21] Their obtained wavefunctions (singlet and triplet) and potentials are reproduced in Fig. 2. The potential obtained from wavefunction reproduces the long-range and attractive parts and and the repulsive core, latter two being rather qualitative. However, it might be noted that, while the long range part matches OPEP predictions, the attractive and core parts are of smaller than expected magnitude. This simulation was done for a pion mass of around 500 MeV. To explore

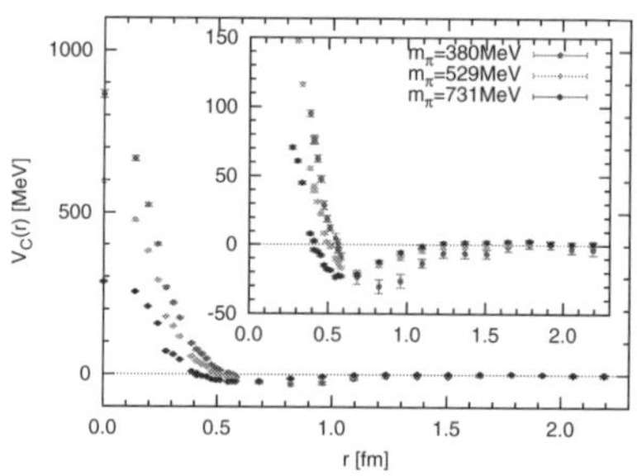

Fig. 3. Singlet potential at three light quark masses corresponding to pion masses of around 700, 500 and 400 MeV.[21]

this further, they[21] tried three values of quark mass corresponding to pion masses of around 700, 500 and 400 MeV. The obtained potentials are plotted in Fig. 3. With lighter quark mass, well gets slightly deeper, core gets more repulsive. So, at least things are on the right track!

Why is this interesting? After a long time, lattice QCD results are *speaking the language of nuclear physicists.* Though the results are preliminary and at unphysically large quark masses, the trend seems to suggest an approach to the correct result in the limit of physically light quark masses. For the NN case we may have more information from experiment than from lattice QCD, making it an academic exercise. However, there are areas where experimental data are not so readily available, but lattice QCD can give meaningful predictions.

3.3. *Y-N potentials*

Hypernuclei are on the frontiers of present-day nuclear physics. In contrast to the NN case, there is very limited experimental information on the hyperon-nucleon (YN) and YY potentials, e.g. we really do not know much about the potential between and nucleon and a Ξ or a Σ. As mentioned earlier, these interactions are important to ascertain the composition of neutron star nuclear matter. Nemura *et al.*[22] have studied this with the same formulation as in the NN case. Figure 4 shows the wavefunctions for singlet and triplet channel. Quark masses corresponding to pion masses of around 370 MeV. These are only first results, but it looks like there is more exciting time ahead!

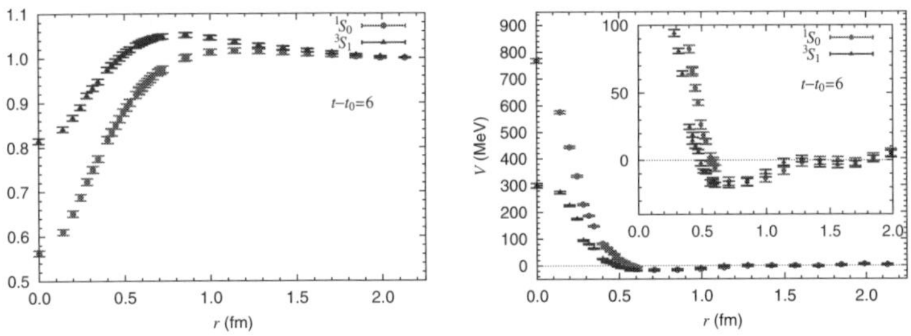

Fig. 4. Singlet and triplet hyperon-nucleon wavefunctions (left) and potentials (right) reproduced from Nemura *et al.*[22]

4. Summary

There are first glimpses of nuclear physics as understood by nuclear physicists emerging from lattice QCD. Exciting results are expected soon, particularly in the areas of hyperon-nucleon interaction. These results will be qualitative, at least, initially. Nevertheless these should provide important input to many-body calculations relevant to neutron star nuclear matter etc.

Acknowledgments

Support of CMT31 organizers and sponsors, the Schwinger Foundation and Professor F.B. Malik for funding participation at CMT31 is thankfully acknowledged.

References

1. S.S.M. Wong, *Introductory Nuclear Physics* (Prentice-Hall India, 1990).
2. D. B. Kaplan and A. E. Nelson, *Phys. Lett. B* **175**, 57 (1986).
3. D. B. Kaplan and A. E. Nelson, *Nucl. Phys. A* **479**, 273 (1988).
4. G. E. Brown, *Nucl. Phys. A* **574**, 217 (1994).
5. D. B. Kaplan, M. J. Savage and M. B. Wise, *Nucl. Phys. B* **478**, 629 (1996) [arXiv:nucl-th/9605002].
6. D. B. Kaplan, M. J. Savage and M. B. Wise, *Phys. Lett. B* **424**, 390 (1998) [arXiv:nucl-th/9801034].
7. D. B. Kaplan, M. J. Savage and M. B. Wise, *Nucl. Phys. B* **534**, 329 (1998) [arXiv:nucl-th/9802075].
8. W. Detmold, K. Orginos and M. J. Savage, *Phys. Rev. D* **76**, 114503 (2007) [arXiv:hep-lat/0703009].
9. S. R. Beane, T. C. Luu, K. Orginos, A. Parreno, M. J. Savage, A. Torok and A. Walker-Loud [NPLQCD Collaboration], arXiv:0709.1169 [hep-lat].
10. S. R. Beane, P. F. Bedaque, K. Orginos and M. J. Savage, *Phys. Rev. D* **75**, 094501 (2007) [arXiv:hep-lat/0606023].
11. S. R. Beane, P. F. Bedaque, T. C. Luu, K. Orginos, E. Pallante, A. Parreno and M. J. Savage, *Phys. Rev. D* **74**, 114503 (2006) [arXiv:hep-lat/0607036].
12. S. R. Beane, P. F. Bedaque, M. J. Savage and U. van Kolck, *Nucl. Phys. A* **700**, 377 (2002) [arXiv:nucl-th/0104030].
13. M. J. Savage, PoS **LAT2005**, 020 (2006) [arXiv:hep-lat/0509048].
14. K. Orginos, *Eur. Phys. J. A* **31**, 799 (2007).
15. W. C. Haxton and C. L. Song, *Phys. Rev. Lett.* **84**, 5484 (2000) [arXiv:nucl-th/9907097].
16. T. Doi, T. T. Takahashi and H. Suganuma, *AIP Conf. Proc.* **842**, 246 (2006) [arXiv:hep-lat/0601008].
17. T. T. Takahashi, T. Doi and H. Suganuma, *AIP Conf. Proc.* **842**, 249 (2006) [arXiv:hep-lat/0601006].
18. E. E. Salpeter and H. A. Bethe, *Phys. Rev.* **84**, 1232 (1951).
19. S. Aoki *et al.* [CP-PACS Collaboration], Phys. Rev. D **71**, 094504 (2005) [arXiv:hep-lat/0503025].
20. S. Aoki, PoS **LAT2007**, 002 (2007) [arXiv:0711.2151 [hep-lat]].
21. N. Ishii, S. Aoki and T. Hatsuda, PoS **LATTICE2007**, 146 (2006) [arXiv:0710.4422 [hep-lat]].
22. H. Nemura, N. Ishii, S. Aoki and T. Hatsuda, PoS **LAT2007**, 156 (2007) [arXiv:0710.3622 [hep-lat]].

DILUTE ALPHA-PARTICLE CONDENSATION IN ^{12}C AND ^{16}O

T. YAMADA

Laboratory of Physics, Kanto Gakuin University, Yokohama 236-8501, Japan
yamada@kanto-gakuin.ac.jp

Y. FUNAKI

Nishina Center for Accelerator-Based Science, The Institute of Physical and Chemical Research (RIKEN), Wako 351-0098, Japan

H. HORIUCHI and A. TOHSAKI

Research Center for Nuclear Physics (RCNP), Osaka University, Osaka 567-0047, Japan

G. RÖPKE

Institut für Physik, Universität Rostock, D-18051 Rostock, Germany

P. SCHUCK

Institut de Physique Nucléaire, CNRS, UMR 8608, Orsay, F-91406, France, and Université Paris-Sud, Orsay, F-91505, France

Received 31 July 2008

Alpha particle condensation is studied in ^{12}C and ^{16}O with the orthogonality condition model (OCM). The OCM equation is an approximation of the equation of motion of α bosons based on microscopic theories. We demonstrate that 1) the Hoyle state (the 0_2^+ state at 7.65 MeV in ^{12}C), located just above the 3α disintegration threshold has a 3α-particle condensate character, in which 3α particles occupy an identical $0S$ orbit with 70 % occupancy, forming a dilute gas-like configuration, and 2) the 0_6^+ state at $E_x = 15.1$ MeV in ^{16}O, appearing above the 4α breakup threshold, is a strong candidate with dilute 4α structure of the condensate-type.

Keywords: α particle condensation; nuclear system; microscopic cluster model.

1. Introduction

Four-nucleon correlations of the α-cluster-type play an important role in the structure of light nuclei.[1,2] Recently, a particular interest has been paid to the occurrence of α-particle condensation in finite nuclei.[3] This is the counterpart of a condensation in infinite α matter expected at low density.[4] The α condensation in nuclei is most likely to appear in excited states with low density in the vicinity of the $N\alpha$ threshold in self conjugate $4N$ nuclei.[3,5] An example of such an excited state with low density is the Hoyle state (the 0_2^+ state at 7.65 MeV in ^{12}C), which has a loosely bound 3α cluster structure found by very successful microscopic α-cluster-model calculations.[6–9] Recently the state has been reinvestigated from the viewpoint of the α-particle condensation, and now it is well established by many theoretical works, as having a 3α-particle condensate character, in which 3α particles occupy an identical $0S$ orbit forming a dilute gas-like configuration.[3,10,12,13]

The establishment of this novel aspect of the Hoyle state naturally leads us to the speculation about 4α-particle condensation in ^{16}O. The spectrum in ^{16}O is very well reproduced up to about 13 MeV excitation energy including the ground state with a semi-microscopic cluster model, i.e., the $\alpha+^{12}$C OCM (Orthogonality Condition Model).[14] In particular, this model calculation, as well as that of $\alpha+^{12}$C GCM (Generator Coordinate Method),[15] demonstrates that the 0_2^+ state at 6.05 MeV and the 0_3^+ state at 12.05 MeV have $\alpha+^{12}$C structures[16] where the α-particle orbits around the ^{12}C ground state-core in an S-wave and in a D-wave, respectively. Consistent results were later obtained by the 4α OCM calculation within the harmonic oscillator basis.[17] However, the model space adopted in Refs. 14, 15, 17 is not sufficient to explore the 4α gas-like configuration. On the other hand, the 4α-particle condensate state was first investigated in Ref. 3 and its existence was predicted around the 4α threshold with a new type of microscopic wave function of α-particle condensate character. While that wave function can well describe the dilute α cluster states as well as shell-model-like states, other structures such as $\alpha+^{12}$C clustering are smeared out and only incorporated in an average way. Since there exists no calculation, so far, which reproduces both the 4α gas and $\alpha+^{12}$C cluster structures simultaneously, it is crucial to perform a full calculation, which will give a decisive benchmark for the existence of the 4α-particle condensate state from a theoretical point of view. The 4α OCM with the Gaussian expansion method is much appropriate to explore such interesting structures in ^{16}O.

The purpose of the present paper is to demonstrate the present results of the study of α condensation in ^{12}C and ^{16}O with the $N\alpha$ OCM ($N = 3, 4$). The OCM equation is an approximation of the equation of motion for $N\alpha$ bosons based on the microscopic theory. We formulate the $N\alpha$ OCM in §2, and the results and discussion on ^{12}C and ^{16}O are given in §3. Finally, a summary is presented in §4.

2. Formulation

2.1. *Mapping of the fermionic $N\alpha$ wave function onto a $N\alpha$ boson wave function*

In the microscopic $N\alpha$ cluster model, the total wave function, $\Psi_J^{(\mathrm{F})}$, with the total angular momentum J is given as

$$\Psi_J^{(\mathrm{F})} = \mathcal{A}\left\{\prod_{i=1}^{N}\phi_{\alpha_i}^{(\mathrm{int})}\chi_J(\boldsymbol{r})\right\} = \int d\boldsymbol{a}\Psi_J^{(\mathrm{F})}(\boldsymbol{a})\chi(\boldsymbol{a}), \tag{1}$$

where $\phi_\alpha^{(\mathrm{int})}$ denotes the intrinsic wave function of the α particle with the simple $(0s)^4$ shell-model configuration, and χ_J represents the relative wave function with a set of the relative Jacobi coordinates, $\boldsymbol{r} = \{\boldsymbol{r}_1, \ldots, \boldsymbol{r}_{N-1}\}$, with respect to the c.m. of α clusters. The antisymmetrization among $4N$ nucleons is properly taken into account in terms of the operator $\mathcal{A}$. The function $\Psi_J^{(\mathrm{F})}(\boldsymbol{a})$ in Eq. (1) is defined as

$$\Psi_J^{(\mathrm{F})}(\boldsymbol{a}) = \mathcal{A}\left\{\prod_{i=1}^{N} \phi_{\alpha_i}^{(\mathrm{int})} \prod_{j=1}^{N-1} \delta(\boldsymbol{r}_j - \boldsymbol{a}_j)\right\}, \tag{2}$$

which describes the α-cluster state located at the relative positions specified by the Jacobi parameter coordinate $\boldsymbol{a} = \{\boldsymbol{a}_1, \boldsymbol{a}_2, \cdots, \boldsymbol{a}_{N-1}\}$.

The total Hamiltonian for $4N$ fermions is given as

$$H = \sum_{i=1}^{4N} t_i - T_{\mathrm{cm}} + \sum_{i<j=1}^{4N} \left(v_{ij} + v_{ij}^{\mathrm{Coul}}\right), \tag{3}$$

where t_i and T_{cm} denote, respectively, the kinetic energy operator of the ith nucleon and of the center-of-mass of the total system. The nucleon-nucleon interaction (Coulomb interaction) between the ith and jth nucleons is expressed as v_{ij} (v_{ij}^{Coul}).

The Schrödinger equation for the fermionic $N\alpha$ system is

$$H\Psi_J^{(\mathrm{F})} = E\Psi_J^{(\mathrm{F})}. \tag{4}$$

Substituting the total wave function of Eq. (1) into Eq. (4), we obtain the equation of motion for the relative wave function χ_J,

$$\int d\boldsymbol{a}' \left\{H(\boldsymbol{a}, \boldsymbol{a}') - EN(\boldsymbol{a}, \boldsymbol{a}')\right\} \chi_J(\boldsymbol{a}') = 0, \qquad \text{or} \qquad (\mathcal{H} - \mathcal{N})\,\chi_J = 0, \tag{5}$$

which is called the RGM (resonating group method) equation.[18] The Hamiltonian and norm kernels, $H(\boldsymbol{a}, \boldsymbol{a}')$ and $N(\boldsymbol{a}, \boldsymbol{a}')$, are defined as

$$\left\{\begin{matrix} H(\boldsymbol{a}, \boldsymbol{a}') \\ N(\boldsymbol{a}, \boldsymbol{a}') \end{matrix}\right\} = \langle \Psi_J^{(\mathrm{F})}(\boldsymbol{a}) \mid \left\{\begin{matrix} H \\ 1 \end{matrix}\right\} \mid \Psi_J^{(\mathrm{F})}(\boldsymbol{a}')\rangle. \tag{6}$$

Recalling the normalization condition $\langle \Psi_J^{(\mathrm{F})} \mid \Psi_J^{(\mathrm{F})}\rangle = 1$ for the total wave function in Eq. (1), the normalization of χ_J in Eq. (5) is given by $\langle \chi_J(\boldsymbol{a}')|N(\boldsymbol{a}, \boldsymbol{a}')|\chi_J(\boldsymbol{a}')\rangle = 1$. This suggests that an $N\alpha$ boson wave function $\Phi_J^{(\mathrm{B})}$ corresponding to the fermionic wave function $\Psi_J^{(\mathrm{F})}$ should be taken to be[12,13]

$$\Phi_J^{(\mathrm{B})}(\boldsymbol{a}) = \mathcal{N}^{1/2}\chi_J = \int d\boldsymbol{a}' N^{1/2}(\boldsymbol{a}, \boldsymbol{a}')\chi_J(\boldsymbol{a}'), \qquad \int d\boldsymbol{a} \left|\Phi_J^{(\mathrm{B})}(\boldsymbol{a})\right|^2 = 1. \tag{7}$$

It is noted that the boson wave function $\Phi_J^{(\mathrm{B})}(\boldsymbol{a})$ has only the Jacobi coordinates $\boldsymbol{a} = \{\boldsymbol{a}_1, \boldsymbol{a}_2, \cdots, \boldsymbol{a}_{N-1}\}$ of the $N\alpha$ system and the internal coordinates in the α cluster are integrated out completely. In addition, $\Phi_J^{(\mathrm{B})}(\boldsymbol{a})$ is totally symmetric for any two-α-cluster exchange. Thus, it has bosonic property. From the RGM equation (5), $\Phi_J^{(\mathrm{B})}$ should satisfy the following equation

$$\left(\mathcal{N}^{-1/2}\mathcal{H}\mathcal{N}^{-1/2} - E\right)\Phi_J^{(\mathrm{B})} = 0. \tag{8}$$

Thus, we can interpret $\mathcal{N}^{-1/2}\mathcal{H}\mathcal{N}^{-1/2}$ as the $N\alpha$ boson Hamiltonian, and Equation (8) is the equation of motion for the $N\alpha$ boson wave function.

If the eigenvalue problem for the norm kernel

$$\mathcal{N}u_\lambda = \int d\boldsymbol{a}' N(\boldsymbol{a}, \boldsymbol{a}')u_\lambda(\boldsymbol{a}') = \lambda u_\lambda(\boldsymbol{a}) \tag{9}$$

is solved, the boson wave function $\Phi_J^{(\mathrm{B})}(\boldsymbol{a})$ in Eq. (7), then, is obtained as

$$\Phi_J^{(\mathrm{B})}(\boldsymbol{a}) = \sum_\lambda \sqrt{\lambda} u_\lambda(\boldsymbol{a})\langle u_\lambda \mid \chi_J\rangle. \tag{10}$$

The eigenvalue of the norm kernel, λ, is non-negative, and the eigenfunction u_λ with $\lambda = 0$ is called the Pauli-forbidden state, which satisfies the condition $\mathcal{A}\{\prod_{i=1}^N \phi_{\alpha_i}^{(\mathrm{int})} u_\lambda\} = 0$. The boson wave function $\Phi_J^{(\mathrm{B})}$, thus, has no component of the Pauli-forbidden state,

$$\langle u_\lambda \mid \Phi_J^{(\mathrm{B})}\rangle = 0 \quad \text{for} \quad u_\lambda \quad \text{with} \quad \lambda = 0. \tag{11}$$

2.2. *$N\alpha$ OCM with bosonic properties*

In order to obtain the boson wave function $\Phi_J^{(\mathrm{B})}$ in Eq. (7), we need to solve the RGM equation (5) and the eigenvalue equation of the norm kernel (9) or to solve directly the equation of motion (8). Solving the eigenvalue equation of the norm kernel, however, is difficult in general even for the 3α case. In addition, computational problems are encountered for solving the $N\alpha$ RGM equation (5) for $N \geq 4$. Thus, it is requested to use more feasible frameworks for the study of the bosonic properties and the amount of α condensation for the $N\alpha$ system. In the present study, we take the orthogonality condition model (OCM)[19] as one of the more feasible frameworks. The OCM scheme, which is an approximation to the RGM, is known to describe nicely the structure of low-lying states in light nuclei.[2,19] The essential properties of the $N\alpha$ boson wave function $\Phi_J^{(\mathrm{B})}$, as mentioned in §2.1, can be taken into account in OCM in a simple manner. We will demonstrate this briefly.

In OCM, the α cluster is treated as a point-like particle. We approximate the $N\alpha$ boson Hamiltonian (non-local) in Eq. (8) by an effective (local) one $H^{(\mathrm{OCM})}$,

$$\mathcal{N}^{-1/2}\mathcal{H}\mathcal{N}^{-1/2} \sim H^{(\mathrm{OCM})} \tag{12}$$

$$H^{(\mathrm{OCM})} \equiv \sum_{i=1}^N T_i - T_{\mathrm{cm}} + \sum_{i<j=1}^N V_{2\alpha}^{\mathrm{eff}}(i,j) + \sum_{i<j<k=1}^N V_{3\alpha}^{\mathrm{eff}}(i,j,k), \tag{13}$$

where T_i denotes the kinetic energy of the ith α cluster, and the center-of-mass kinetic energy T_{cm} is subtracted from the Hamiltonian. The effective 2α and 3α potentials are presented as $V_{2\alpha}^{\mathrm{eff}}$ and $V_{3\alpha}^{\mathrm{eff}}$, respectively. Referring to the RGM framework in Eqs. (8)–(11), the equation of the relative motions for the $N\alpha$ particles with $H^{(\mathrm{OCM})}$, called the OCM equation, is expressed as

$$\left\{H^{(\mathrm{OCM})} - E\right\}\Phi_J = 0 \quad \text{with} \quad \langle u_F \mid \Phi_J\rangle = 0, \tag{14}$$

where u_F denotes the Pauli-forbidden state, satisfying the following condition, $\mathcal{A}\{\prod_{i=1}^{N}\phi_{\alpha_i}^{(\text{int})}u_F\} = 0$. The orthogonality condition $\langle u_F \mid \Phi_J\rangle = 0$ in Eq. (14) corresponds to Eq. (11). The bosonic property of the wave function Φ_J can be taken into account by symmetrizing the wave function with respect to any 2α-particle exchange, $\Phi_J = \mathcal{S}\Phi_J(1, 2, \cdots, N)$, where $\mathcal{S}$ denotes the symmetrization operator. It is noted that the complete overlapped state of the $N\alpha$ particles is forbidden within the present framework because of the Pauli-blocking effect in Eq. (14), although we take into account the bosonic statistics for the constituent α particles.

In order to discuss the bosonic properties such as the degree of α condensation in a nucleus, it is needed to extract information on the single α-particle orbits and its occupation probabilities in the nucleus from the total wave function Φ_J. The one-particle density matrix for the $N\alpha$ system is very useful for this.[12,13,20] The single α-particle density matrix is given as

$$\rho(\boldsymbol{r}, \boldsymbol{r}') = \langle\Phi_J \mid \sum_{i=1}^{N} \mid \delta(\boldsymbol{r}_i^{(\text{cm})} - \boldsymbol{r}')\rangle\langle\delta(\boldsymbol{r}_i^{(\text{cm})} - \boldsymbol{r}) \mid \mid \Phi_J\rangle, \tag{15}$$

where $\boldsymbol{r}_i^{(\text{cm})}$ is the relative coordinate of the ith α particle with respect to the center-of-mass coordinate of the total system. It is noted that the diagonal matrix element reduces to the single α-particle density $\rho(\boldsymbol{r}, \boldsymbol{r}' = \boldsymbol{r}) = \rho(\boldsymbol{r})$. The single α-particle orbit and its occupation number in the nucleus can be evaluated by solving the eigenvalue equation of the one-particle density matrix

$$\int d\boldsymbol{r}'\rho(\boldsymbol{r}, \boldsymbol{r}')\varphi_\mu(\boldsymbol{r}') = \mu\varphi_\mu(\boldsymbol{r}), \tag{16}$$

where the eigenvalue μ presents the occupation number. The eigenfunction φ_μ denotes the single-α orbital wave function in a nucleus with the argument of the intrinsic coordinate ($\boldsymbol{r}_\alpha^{(\text{cm})}$) of an arbitrary α particle in a nucleus measured from the center-of-mass coordinate. The ratio μ/N represents the occupation probability of an α particle in the orbit φ_μ. The spectrum of the occupation probabilities offers important information on the occupancy of the single α-particle orbit in a nucleus.

2.3. *Effective 2α potentials with OCM*

In the present study, we use an effective 2α potential which reproduces the observed $\alpha\alpha$ scattering phase shifts (S-, D- and G-waves) and the resonant ground-state energy of ^{8}Be within the 2α OCM framework.[13] The effective 2α potential and Coulomb potential, $V_{2\alpha}$ and $V_{2\alpha}^{\text{Coul}}$, are constructed with the folding procedure, where we fold the modified Hasegawa-Nagata effective NN interaction (MHN)[21] [Schmid-Wildermuth one (SW)[22]] and the pp Coulomb potential with the α-cluster density.

Only using the effective 2α potential leads to an overbinding energy for the ground state of ^{12}C within the frame of the 3α OCM.[13] Thus, we introduce an effective, repulsive, 3α potential, $V_{3\alpha}$, of the one-range Gaussian type, in addition

Table 1. Calculated energies ($E_{3\alpha}$) and nuclear radii (R) for the 0_1^+ and 0_2^+ states of ^{12}C together with the α-α and α-2α rms distances ($R_{\alpha\alpha}$ and $R_{\alpha-2\alpha}$). The energy $E_{3\alpha}$ is one measured from the 3α threshold.

J^π	$E_{3\alpha}^{cal}$ [MeV]	$E_{3\alpha}^{exp}$ [MeV]	R [fm]	$R_{\alpha\alpha}$ [fm]	$R_{\alpha-2\alpha}$ [fm]
0_1^+	−7.27	−7.27	2.44	3.02	2.61
0_2^+	0.85	0.38	4.31	6.84	5.93

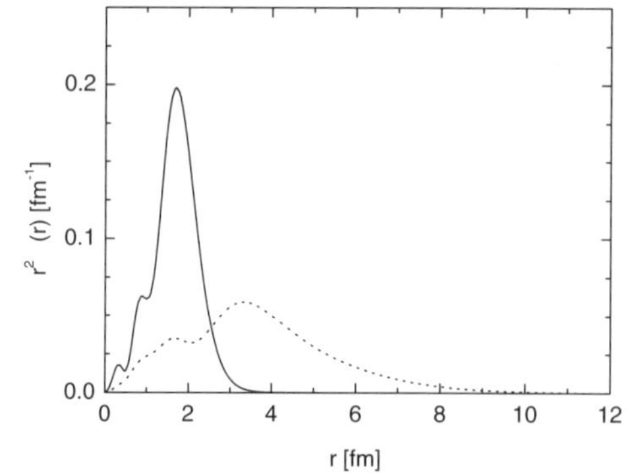

Fig. 1. Density distribution of the α particle for the 0_1^+ (solid line) and 0_2^+ (dotted) states.

to the 2α potential. Including the 3α potential, the energy of the ground state of ^{12}C is reproduced with respect to the 3α threshold, together with the nuclear radius. In addition, we introduce a phenomenological 4α potential of the one-range Gaussian type so as to adjust the ground state energy of ^{16}O with the 4α OCM.[23] This four-body potential is short-range, and, hence, this only acts in compact configurations. It is noted that the three-body and four-body potentials introduced here are regarded as substitutes representing the density dependence of the effective 2α interactions.

3. Results and Discussion

The 3α (4α) OCM equation[13,23] in Eq. (14) for ^{12}C (^{16}O) is solved with the Gaussian expansion method,[24] where a large 3α (4α) model space is spanned by Gaussian basis functions. The orthogonality condition $\langle u_F \mid \Phi_J \rangle = 0$ in Eq. (14) is taken into account with the Kukulin's projection method.[25]

3.1. 0^+ *states in* ^{12}C

The results for the energy, measured from the 3α threshold, and nuclear radii for the ground (0_1^+) and excited states (0_2^+) of ^{12}C are presented in Table 1. The energy for the ground state is reproduced well, and the corresponding nuclear radius, 2.44 fm, is in good agreement with the experimental charge radius (2.4829 ± 0.019 fm) with an error of about 2%. Figure 1 shows the density distribution of the α particle

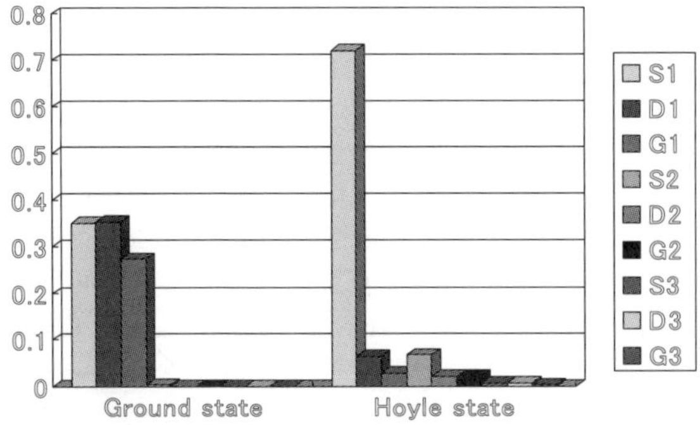

Fig. 2. Occupation of the single-α orbitals of the ground state of ^{12}C compared with the Hoyle state. For explanation, see the text.

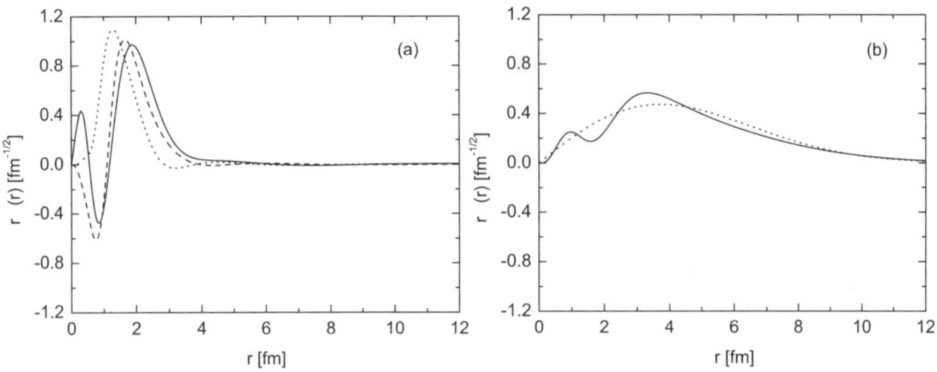

Fig. 3. Radial parts of the single α orbits, (a) $S1$ (solid line), $D1$ (dashed) and $G1$ (dotted), in the 0_1^+ state, and (b) the $S1$ (solid) orbit in the 0_2^+ state and S-wave Gaussian function (dotted), $r\varphi_{0s}$, with the size parameter $a = 0.038$ fm^{-2} (see text).

for the 0_1^+ state of ^{12}C. We see a prominent peak at $r \sim 2$ fm, which demonstrates clearly the shell-model-like compact structure of the ground state of ^{12}C.

As for the 0_2^+ state, the energy measured from the 3α threshold is $E_{3\alpha}$=0.86 MeV (E_x=8.13 MeV), which agrees well with the experimental data $E_{3\alpha}^{exp}$=0.38 MeV (E_x^{exp}=7.65 MeV). The calculated nuclear radius for the 0_2^+ state is as large as 4.31 fm (see Table 1). This means that the state has a dilute 3α structure, although our nuclear radius is a little larger than that in Ref. 3. The density distribution of the α particle for the 0_2^+ state is illustrated in Fig. 2. In comparison with that for the ground state, we can easily recognize the dilute structure of the 0_2^+ state, which is in contrast with the compact structure of the ground state.

It is instructive to study the single α-particle orbits (eigenfunctions) and occupation probabilities (eigenvalues) of the one-body density matrix in Eq. (15). The results of the diagonalization of the latter are shown in Fig. 2. As for the ground state, the occupation probabilities spread out over S, D and G waves, but they

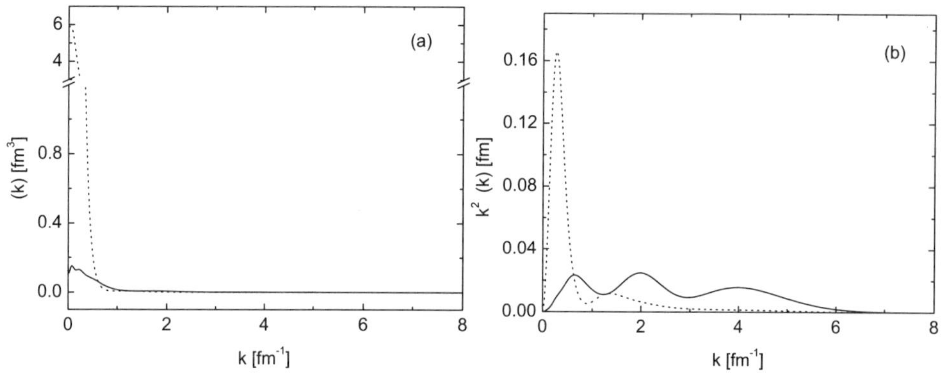

Fig. 4. Momentum distribution of the α particle, (a) $\rho(k)$ and (b) $k^2\rho(k)$, for the 0_1^+ (solid line) and 0_2^+ (dotted) states.

are concentrated to the first orbits, $S1$, $D1$ and $G1$ orbits, respectively, where Lk denotes the kth orbit for the L wave. The occupation probabilities are about 30 % for all orbits. This result is expected from the fact that the ground-state wave function is of nuclear SU(3)-like character, $\mathrm{SU(3)}[f](\lambda\mu)_{J^\pi} = [444](04)_{0^+}$ with quanta Q=8, where the SU(3) bases with $Q < 8$ correspond to the Pauli-forbidden states. Since the $\mathrm{SU(3)}[444](04)_{0^+}$ state is the eigenfunction of the 3α RGM norm kernel in Eq. (9), it can be regarded as the 3α boson wave function with $Q = 8$, see Eq. (10). The state is described as

$$|[444](04)\rangle_{0^+} = \sqrt{64/225}|2s2S\rangle - \sqrt{80/225}|1d1D)\rangle + \sqrt{81/225}|0g0G\rangle, \quad (17)$$

where $|(n\ell)(NL)\rangle$ presents the basis function $|u_{n\ell}(\boldsymbol{r}_{2\alpha})u_{NL}(\boldsymbol{r}_{\alpha-2\alpha})\rangle$ with $2n+\ell+2N+L=8$, and $u_{n\ell}$ (u_{NL}) denotes the harmonic oscillator wave function with the number of nodes n (N) and orbital angular momentum ℓ (L) referring to the coordinate vector $\boldsymbol{r}_{2\alpha}$ ($\boldsymbol{r}_{\alpha-2\alpha}$) between 2α clusters (between the center-of-mass for the 2α clusters and the third α cluster). From the definition of the one-body density matrix in Eq. (15), the single-α orbits and occupation probabilities for the SU(3) state in Eq. (17) are given as follows: $64/255 \sim 28\%$ for S-orbit, $80/225 \sim 36\%$ for D-orbit, and $81/225 \sim 36\%$ for G-orbit. Thus, we can understand the reason why the $S1$, $D1$ and $G1$ orbits in Fig. 2 have about 30% occupation probabilities each.

Figure 3(a) demonstrates the radial parts for the $S1$-, $D1$- and G-orbits, the number of nodes of which are two, one and zero, respectively. Reflecting the SU(3) character, the radial behaviors of the three orbits are similar to those of the harmonic oscillator wave functions (u_{NL}) with $Q = 4$, u_{02}, u_{21} and u_{40}, respectively. We see that the radial parts of the single α-particle orbits oscillate widely in the inside region ($r < 4$ fm). This is due to the strong Pauli blocking effect for the ground state with the compact shell-model-like structure.

Concerning the 0_2^+ state, the occupation probabilities are shown in Fig. 2. We see a strong concentration on a single orbit: the occupation probability of the $S1$ orbit is largest, amounting to about 70%, and those for other orbits are very small.

This means that each of the three α particles in the 0_2^+ state is in the $S1$ orbit with occupation probability as large as about 70%. The radial behavior of the $S1$ orbit is illustrated with the solid line in Fig. 3(b). We see no nodal behavior but small oscillations in the inner region ($r < 4$ fm) and a long tail up to $r \sim 10$ fm. For reference, the radial behavior of the S-wave Gaussian function, $\varphi_{0s}(r) = N_{0s}(a)\exp(-ar^2)$, is drawn with the dashed line in Fig. 3(b), where the size parameter a is chosen to be 0.038 fm^{-2}, and $N_{0s}(a)$ denotes the normalization factor. The radial behavior of the S_1 orbit is similar to that of the S-wave Gaussian function, in particular, in the outer region ($r > 4$ fm), whereas a slight oscillation of the former around the latter can be seen in the inner region ($r < 4$ fm).

The momentum distributions of the α particles, $\rho(k)$ and $k^2 \times \rho(k)$, are shown for the 0_1^+ and 0_2^+ states in Fig. 4. Reflecting the dilute structure of the 0_2^+ state, we see a strong concentration of the momentum distribution in the $k < 1$ fm^{-1} region, and the behavior of $\rho(k)$ is of the δ-function type, similar to the momentum distribution of the dilute neutral atomic condensate states at very low temperature trapped by the external magnetic field.[26] On the other hand, the ground state has higher momentum component up to $k \sim 6$ fm^{-1} as seen from the behavior of $k^2 \times \rho(k)$ reflecting the compact structure. The above results for the radial behavior of the S_1 orbit, occupation probability and momentum distribution for the 0_2^+ state leads us to conclude that this state is similar to an ideal dilute 3α condensate with as much as about 70% occupation probability.

3.2. 0^+ *states in* ^{16}O

Figure 5 shows the energy spectrum with $J^\pi = 0^+$, obtained by solving the 4α OCM equation.[23] We can reproduce the full spectrum of 0^+ states, and tentatively make a one-to-one correspondence of those states with the six lowest 0^+ states of the experimental spectrum. In view of the complexity of the situation, the agreement is considered to be very satisfactory.

We show in Table 2 the calculated rms radii and monopole matrix elements to the ground state, together with the experimental values. The M(E0) values for the $0_2^+, 0_3^+$, and 0_5^+ states are consistent with the corresponding experimental values. As mentioned above, the structures of the 0_2^+ and 0_3^+ states are well established as having the $\alpha+^{12}$C(0_1^+) and $\alpha+^{12}$C(2_1^+) cluster structures,[14,15,17] respectively. These structures of the 0_2^+ and 0_3^+ states are confirmed in the present calculation. We also mention that the ground state has a shell-model configuration within the present framework, the calculated radius agreeing with the observed one (2.71 fm).

On the contrary, the structures of the observed 0_4^+, 0_5^+ and 0_6^+ states in Fig. 5 have, in the past, not clearly been understood, since they have never been discussed with the previous cluster model calculations.[14,15,17] Although Ref. 3 predicts the 4α condensate state around the 4α threshold, it is not clear which of those states corresponds to the condensate state. For example, the 0_4^+ state has been considered as one of the candidates,[27] and also the 0_5^+ state with large $M(E0)$ value,[28] since the strong monopole transition implies a developed cluster structure.[29] This ambiguous

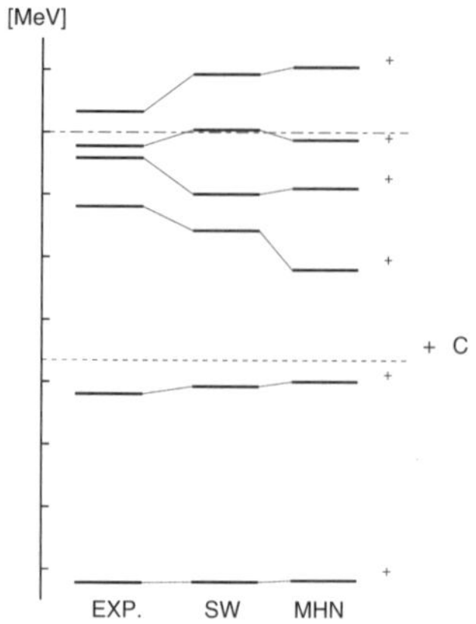

Fig. 5. Comparison of energy spectra between experiment and the present calculation. Two kinds of effective two-body nucleon-nucleon forces MHN and SW are adopted (see text). Dotted and dash-dotted lines denote the α+^{12}C and 4α thresholds, respectively. The assignments with experiment are tentative, see, however, detailed discussion in the text.

Table 2. The rms radii R and monopole transition matrix elements to the ground state M(E0), respectively, together with the experimental data.

	R [fm]		M(E0) [fm^2]		$R_{\text{exp.}}$ [fm]	$M(\text{E0})_{\text{exp.}}$ [fm^2]
	SW	MHN	SW	MHN		
0_1^+	2.7	2.7			2.71 ± 0.02	
0_2^+	3.0	3.0	4.1	3.9		3.55 ± 0.21
0_3^+	2.9	3.1	2.6	2.4		4.03 ± 0.09
0_4^+	4.0	4.0	3.0	2.4		no data
0_5^+	3.1	3.1	3.0	2.6		3.3 ± 0.7
0_6^+	5.0	5.6	0.5	1.0		no data

situation stems from the fact that the α-condensate-type wave function in Ref. 3 can only reproduce the ground state and the 4α condensate state, but not the α+^{12}C configurations responsible for a large part of the ^{16}O spectrum up to the 0_6^+ state.

As shown in Fig. 5, the present calculation succeeded, for the first time, to reproduce the 0_4^+, 0_5^+ and 0_6^+ states, together with the 0_1^+, 0_2^+ and 0_3^+ states. This puts us in a favorable position to discuss the 4α condensate state, expected to exist around the 4α threshold. In Table 2, the largest rms value of about 5 fm is found for the 0_6^+ state. Compared with the relatively smaller rms radii of the 0_4^+ and 0_5^+ states, this large size suggests that the 0_6^+ state may be composed of a weakly interacting gas of α particles of the condensate type.

While a large size is generally necessary for forming an α condensate, the best way for its identification is to investigate the single-α orbit and its occupation probability, which can be obtained by diagonalizing the one-body (α) density matrix[12,13] in Eq. (15). As a result of the calculation, a large occupation probability of 61% of the lowest $0S$-orbit is found for the 0_6^+ state, whereas the other five 0^+ states all

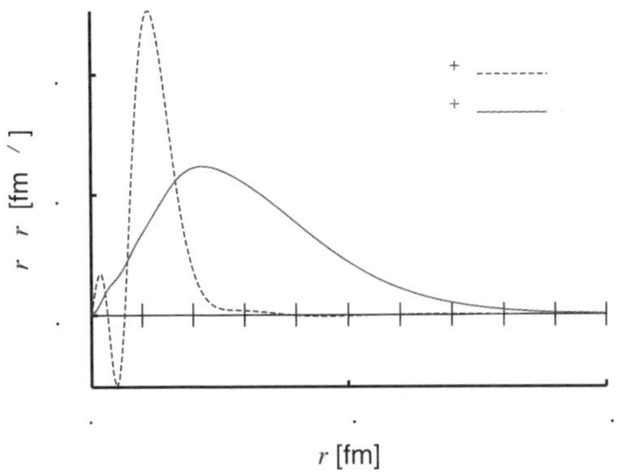

Fig. 6. The radial parts of S-wave single-α orbits belonging to the largest occupation number, for the ground and 0_6^+ states with MHN force.

have appreciably smaller values, at most 25% (0_2^+). The corresponding single-α S orbit is shown in Fig. 6. It has a strong spatially extended behavior without any node ($0S$). This indicates that α particles are condensed into the very dilute $0S$ single-α orbit. Thus, the 0_6^+ state clearly has 4α condensate character. We should note that the orbit is very similar to the single-α orbit of the Hoyle state[12,13] in Fig. 3(b). We also show in Fig. 6 the single-α orbit for the ground state. It has maximum amplitude at around 3 fm and oscillations in the interior with two nodal ($2S$) behavior, due to the Pauli principle and reflecting the shell-model configuration.

The α decay width is a very important information to identify the 0_6^+ state from the experimental point of view. It can be estimated, based on the R-matrix theory, with use of the 4α OCM wave function. We find that the total α decay width of the 0_6^+ state is as small as 50 keV (experimental value: 166 keV). This means that the state can be observed as a quasi-stable state. Thus, the width, as well as the excitation energy, are consistent with the observed data. All the characteristics found from our OCM calculation, therefore, indicate that the 6th 0^+ state shall be identified with the experimental 0_6^+ state at 15.1 MeV and that it is the α-condensate state of ^{16}O.

Finally we discuss the structures of the 0_4^+ and 0_5^+ states. Our present calculations show that the 0_4^+ and 0_5^+ states mainly have α+^{12}C(0_1^+) structure with higher nodal behavior and α+^{12}C(1^-) structure, respectively. Further details will be given in a forthcoming extended paper. The calculated width of the 0_4^+ is $\sim$ 150 keV, which is much larger than that found for the 0_5^+ state $\sim$ 50 keV. Both are qualitatively consistent with the corresponding experimental data, 600 keV and 185 keV, respectively. The reason why the width of the 0_4^+ state is larger than that of the 0_5^+ state, though the 0_4^+ state has lower excitation energy, is due to the fact that the former has a much larger component of the α+^{12}C(0_1^+) decay channel, reflecting the characteristic structure of the 0_4^+ state. The 4α condensate state, thus, should not be assigned to the 0_4^+ or 0_5^+ state but very likely to the 0_6^+ state.

4. Summary

Alpha particle condensation was discussed in ^{12}C and ^{16}O with OCM. We found that 1) the Hoyle state (the 0_2^+ state at 7.65 MeV in ^{12}C), located just above the 3α disintegration threshold has a 3α-particle condensate character, in which 3α particles occupy an identical $0S$ orbit with 70 % occupancy, forming a dilute gas-like configuration, and 2) the 0_6^+ state at $E_x = 15.1$ MeV in ^{16}O, appearing above the 4α breakup threshold, is a strong candidate with dilute 4α structure of the condensate-type. Further experimental information is very much needed to confirm the existence of the novel 4α condensate state.

References

1. K. Wildermuth and Y. C. Tang, *A Unified Theory of the Nucleus* (Vieweg, Braunschweig, 1977).
2. K. Ikeda *et al.*, *Prog. Theor. Phys. Suppl.* **68**, 1 (1980).
3. A. Tohsaki *et al.*, *Phys. Rev. Let.* **87**, 192501 (2001).
4. G. Röpke *et al.*, *Phys. Rev. Lett.* **80**, 3177 (1998).
5. T. Yamada *et al.*, *Phys. Rev. C* **69**, 024309 (2004).
6. H. Horiuchi, *Prog. Theor. Phys.* **51**, 1266 (1974); **53**, 447 (1975).
7. Y. Fukushima *et al.*, *Suppl. of J. Phys. Soc. Japan* **44**, 225 (1978); M. Kamimura, *Nucl. Phys. A* **351**, 456 (1981).
8. E. Uegaki *et al.*, *Prog. Theor. Phys.* **57**, 1262 (1977); E. Uegaki *et al.*, *Prog. Theor. Phys.* **59**, 1031 (1978); **62**, 1621 (1979).
9. P. Descouvemont *et al.*, *Phys. Rev. C* **36**, 54 (1987).
10. Y. Funaki *et al.*, *Phys. Rev. C* **67**, 051306(R) (2003).
11. M. Chernykh *et al.*, *Phys. Rev. Lett.* **98**, 032501(2007).
12. H. Matsumura *et al.*, *Nucl. Phys. A* **739**, 238 (2004).
13. T. Yamada *et al.*, *Euro. Phys. J. A* **26**, 185 (2005).
14. Y. Suzuki, *Prog. Theor. Phys.* **55**, 1751 (1976); **56**, 111 (1976).
15. M. Libert-Heinemann *et al.*, *Nucl. Phys. A* **339**, 429 (1980).
16. H. Horiuchi *et al.*, *Prog. Theor. Phys.* **40**, 277 (1968).
17. K. Fukatsu *et al.*, *Prog. Theor. Phys.* **87**, 151 (1992).
18. H. Horiuchi, *Prog. Theor. Phys. Phys. Suppl.* **62**, 90 (1977).
19. S. Saito, *Prog. Theor. Phys.* **40**, 893 (1968); **41**, 705 (1969); *Prog. Theor. Phys. Suppl.* **62**, 11 (1977).
20. T. Yamada *et al.*, arXiv:cond/mat0804.1672.
21. A. Hasegawa *et al.*, *Prog. Theor. Phys.* **45**, 1786 (1971); F. Tanabe *et al.*, *ibid.* **53**, 677 (1975).
22. E. W. Schmid *et al.*, *Nucl. Phys.* **26**, 463 (1961).
23. Y. Funaki *et al.*, arXiv:nucl/th0802.3246.
24. M. Kamimura, *Phys. Rev. A* **38**, 621 (1988); E. Hiyama *et al.*, *Prog. Part. Nucl. Phys.* **51**, 223 (2003).
25. V. I. Kukulin *et al.*, *Nucl. Phys. A* **417**, 128 (1984).
26. F. Dalfovo *et al.*, *Rev. Mod. Phys.* **71**, 463 (1999).
27. T. Wakasa *et al.*, *Phys. Lett. B* **653**, 173 (2007).
28. M. Stroetzel *et al.*, *Phys. Lett.* **29B**, 306 (1969).
29. T. Yamada *et al.*, arXiv:nucl/th0703045.

JAYNES-CUMMINGS MODEL AS A CASE STUDY FOR THE DERIVATION OF TIME-DEPENDENT SCHRÖDINGER EQUATION

SUPITCH KHEMMANI*

Department of Physics, Faculty of Science, Srinakharinwirot University, 114 Sukhumvit 23, Bangkok 10110, Thailand

supitch@swu.ac.th

VIRULH SA-YAKANIT

Center of Excellence in Forum for Theoretical Science, Faculty of Science, Chulalongkorn University, Bangkok 10330, Thailand

virulh.s@chula.ac.th

Received 31 July 2008

The derivation of a time-dependent Schrödinger equation (TDSE) from a time-independent Schrödinger equation (TISE) in the coherent state representation is considered for the special case of a simple coupled atom-field system described by the soluble Jaynes-Cummings model. The derivation shows why, from the outset, a linear combination of energy eigenstates, instead of a single state, must be used in order to obtain a TDSE for general states. Moreover, this study leads to a method of solving a TDSE by simply solving a TISE.

Keywords: Jaynes-Cummings model; semiclassical limit; time-independent and time-dependent Schrödinger equations; Floquet theory.

1. Introduction

In 1926, Schrödinger[1] first derived the time-independent Schrödinger equation (TISE) from a variational principle and showed that quantization of energy can be formulated as an eigenvalue problem. Since his TISE applies to time-independent Hamiltonians only, to be applied to time-dependent Hamiltonians he had to derive the time-dependent Schrödinger equation (TDSE). After two attempts,[2,3] he arrived at the TDSE we know today, involving only a first time derivative. The TDSE was first applied to the problem of interaction of an atom with the external classical electromagnetic field.[3] From this first application, it was recognized that the time-dependence arises from classical equations (the Maxwell equations) and in this

*Corresponding author.

sense the TDSE is a mixed classical-quantum equation. In 1926, Born[4] quantized a particle beam and then described quantum transitions by using the TISE. In 1931, Mott[5] showed that when the beam energy far exceeds the atomic exciation energies, Born's TISE for quantized beam-atom system can be reduced to a TDSE for the atom alone, in which the beam move on a classical straight-line path. This aspect of the approximate nature of the TDSE appears to have been neglected and in Refs. 6 and 7, it was sought to re-emphasize this by generalizing Mott's analysis. It was shown how, beginning with a closed quantum composite at fixed energy described by a TISE, the TDSE for a part of the composite (the quantum system) can be derived in the limit that the rest of the system (the environment) becomes classical. Although this derivation showed very well how time enters the quantum system, it failed at turning points of the (classical) environment as the WKB approximation was used and, therefore, it failed to provide a global time in the TDSE. For the special case that the environment is the single-model boson field, it was shown in Ref. 8 how this turning point problem can be overcome by using a coherent state representation of the field state. Note that, in this special case, it is quite natural to obtain a global time for the TDSE from the coherent state representation since the coherent state is the quasi-classical state whose classical limit is exact. In contrast, it may be impossible to eliminate the turning point problem for a general environment since quasi-classical states in general cases are, up to now, not available.

In this paper, we will investigate again the derivation using a coherent state representation as in Ref. 8 in detail when the quantum system has only two levels and the environment is the single-mode quantized field within the framework of the soluble Jaynes-Cummings model.[9] Since the solution of the Jaynes-Cummings model is completely known, we can show that the derivation in Ref. 8, where only one fixed energy eigenvector is used at the beginning, leads to the TDSE for very specific states rather than general states. In order to obtain the TDSE for general states, one should begin the derivation with a linear combination (with respect to the dimension of the Hilbert space of the quantum system) of energy eigenvectors.[10] Moreover, as a consequence of this study, we can also show how to use the classical limiting processes in order to obtain the state at any time t without directly using the TDSE.

2. Semiclassical Model Versus Fully Quantum Mechanical Model and a Condition on the Semiclassical Limit

Without loss of generality, let us consider a classical monochromatic linear polarized electromagnetic field propagating in z-direction, interacting with a two-level system having the center of mass $\mathbf{r}_{\rm cm}$ in a cavity of volume V. The electric field at the two-level system in this volume is (in the dipole approximation)[9]

$$\mathbf{E}_{\rm cl}(z,t) = \hat{\mathbf{x}}q(t)(2\Omega^2/\epsilon_0 V)^{1/2}\sin(kz_{\rm cm})\,, \tag{1}$$

which has the corresponding quantized form

$$\mathbf{E}(z,t) = \hat{\mathbf{x}}\varepsilon_\Omega[a(t) + a^\dagger(t)]\sin(kz_{\text{cm}})\,. \tag{2}$$

Here, $q(t) = q_0\cos(\Omega t - \phi)$ is the measure of the field amplitude, $\hat{\mathbf{x}}$ is a polarization, Ω is the single-mode field oscillation frequency, ϕ is the initial phase, $z_{\text{cm}} = \mathbf{r}_{\text{cm}} \cdot \hat{\mathbf{z}}$, $k = \Omega/c$ is the wave vector, $\varepsilon_\Omega = (\hbar\Omega/\epsilon_0 V)^{1/2}$ is the electric field per photon. In rotating wave approximation, the Hamiltonian of a two-level system interacting with the electromagnetic field given in (1) is[9]

$$H_{\text{sc}}(t) = \frac{\hbar\omega}{2}\sigma_z - \frac{\wp E_0}{2}(\sigma_+ \mathrm{e}^{-i(\Omega t-\phi)} + \sigma_- \mathrm{e}^{i(\Omega t-\phi)})\,, \tag{3}$$

where $\hbar\omega$ is the energy difference of the two levels, $\wp$ is the component of a matrix element of dipole along the field polarization which can be chosen, without loss of generality, to be real by choosing the appropriate atomic quantization axis. Here, $E_0 = q_0\sqrt{2\Omega^2/\epsilon_0 V}\sin(kz_{\text{cm}})$ and $\sigma_z, \sigma_+, \sigma_-$ are standard Pauli spin matrices. We call $H_{\text{sc}}(t)$ in (3) the semiclassical Hamiltonian for the "semiclassical model" as the two-level system is quantized while the field is classical. This Hamiltonian generates dynamics of the system such that the state of a two-level system evolves according to[9]

$$|\psi(t)\rangle = C_a(t)\mathrm{e}^{i\delta t/2}\mathrm{e}^{-i\omega t/2}\,|a\rangle + C_b(t)\mathrm{e}^{-i\delta t/2}\mathrm{e}^{i\omega t/2}\,|b\rangle\,,$$

where

$$\begin{bmatrix} C_a(t) \\ C_b(t) \end{bmatrix} = \begin{bmatrix} \cos(Rt/2) - i\delta R^{-1}\sin(Rt/2) & iR_0R^{-1}\mathrm{e}^{i\phi}\sin(Rt/2) \\ iR_0R^{-1}\mathrm{e}^{-i\phi}\sin(Rt/2) & \cos(Rt/2) + i\delta R^{-1}\sin(Rt/2) \end{bmatrix} \begin{bmatrix} C_a \\ C_b \end{bmatrix}. \tag{4}$$

Here, $\delta = \omega - \Omega$ is the detuning, $R_0 = \wp E_0/\hbar$ is the Rabi flopping frequency, $R = \sqrt{\delta^2 + R_0^2}$ is the generalized Rabi flopping frequency, and $|a\rangle$ and $|b\rangle$ denotes respectively the upper and lower states of a two-level system. Similar to (3), the equivalent full quantum time-independent Hamiltonian of a two-level system interacting with the single-mode quantized field given in (2) is (in the rotating wave approximation)[9]

$$H = \frac{\hbar\omega}{2}\sigma_z + \hbar\Omega a^\dagger a + \hbar g(a\sigma_+ + a^\dagger\sigma_-)\,, \tag{5}$$

where $g = -(\wp\varepsilon_\Omega/\hbar)\sin(kz_{\text{cm}})$ is the atom-field coupling strength. Equation (5) is known as the Jaynes-Cummings model which is the simple fully quantum mechanical model used to described a coupled atom-field system. This Jaynes-Cummings Hamiltonian generates dynamics of the system, similar to (4), such that the state of a system evolves according to[9]

$$\begin{aligned} |\Psi(t)\rangle = {} & \left(\sum_{n=0}^{\infty} C_{an}(t)\mathrm{e}^{i\delta t/2}\mathrm{e}^{-iE_{an}t/\hbar}\,|an\rangle + C_{bn+1}(t)\mathrm{e}^{-i\delta t/2}\mathrm{e}^{-iE_{bn+1}t/\hbar}\,|bn+1\rangle\right) \\ & + C_{b0}(t)\mathrm{e}^{-i\delta t/2}\mathrm{e}^{i\omega t/2}\,|b0\rangle\,, \end{aligned}$$

where

$$\begin{bmatrix} C_{an}(t) \\ C_{bn+1}(t) \end{bmatrix} = \begin{bmatrix} \cos(R_n t/2) - i\delta R_n^{-1}\sin(R_n t/2) & -2igR_n^{-1}\sqrt{n+1}\sin(R_n t/2) \\ -2igR_n^{-1}\sqrt{n+1}\sin(R_n t/2) & \cos(R_n t/2) + i\delta R_n^{-1}\sin(R_n t/2) \end{bmatrix} \begin{bmatrix} C_{an}(0) \\ C_{bn+1}(0) \end{bmatrix}. \tag{6}$$

Here $R_n = \sqrt{\delta^2 + 4g^2(n+1)}$ is the quantized generalized Rabi flopping frequency, $E_{an} = n\hbar\Omega + \hbar\omega/2$ and $E_{bn+1} = (n+1)\hbar\Omega - \hbar\omega/2$ are the unperturbed energy eigenvalues. Comparing (4) and (6), one can see that the flopping between upper and lower states of a two-level system predicted by the semiclassical model and fully quantum mechanical model look the same if the semiclassical Rabi frequency R_0 is replaced by $-2g\sqrt{n+1}$. The situation where this replacement holds is given by the following condition.

Condition: Let $\bar{n} \geq 0$ be the average photon number. Then, we have

$$-2g\sqrt{\bar{n}} = R_0 \quad \text{if and only if} \quad \bar{n}\hbar\Omega = \mathcal{E}_{\text{cl}}\,, \tag{7}$$

where $\mathcal{E}_{\text{cl}}$ denotes the classical energy. This requirement can be simply proved by using the relations $\mathcal{E}_{\text{cl}} = (\Omega q_0)^2/2$ and $E_0 = q_0\sqrt{2\Omega^2/\epsilon_0 V}\sin(kz_{\text{cm}})$. Physically, this condition states that R_0 and $-2g\sqrt{\bar{n}}$ can be used interchangeably when the average photon number is large enough so that a field energy reaches certain classical values (a classical limit of field). We call this condition the "condition on the semiclassical limit" since only the classical limit of the field part of a coupled atom-field system is taken.

3. The Derivation of TDSE from TISE: An Illustration by the Jaynes-Cummings Model

Let us first consider the system composing of a quantum system and an environment. When the environment becomes sufficiently large, the wave function of the whole system can be written in an adiabatic-like form[6]

$$\Psi(x, R) = \chi(R)\psi(x; R), \tag{8}$$

where $\chi(R)$ is the wave function of the environment and $\psi(x; R)$ is the wave function of the quantum system depending parametrically on environment coordinate R. Similar to (8), the adiabatic-like form in the coherent state representation for the coupled atom-field system is[8]

$$\langle\alpha|\Psi\rangle = \mathrm{e}^{-|\alpha|^2/2}\chi(\alpha^*)|\psi(\alpha^*)\rangle. \tag{9}$$

Here $|\Psi\rangle$ is the state of the whole system and $|\psi(\alpha^*)\rangle$ is the state of an atom (two-level system) depending parametrically on α^* of the field environment. Now, if we let $|\Psi\rangle = |in\rangle$ $(i = 1, 2)$ be an energy eigenvector of the Jaynes-Cummings Hamiltonian H in (5) with the corresponding energy eigenvalues E_{in} i.e., we have the TISE $H|in\rangle = E_{in}|in\rangle$ $(i = 1, 2)$, then (similar to Eqs. (36) and (39) in Ref. 8)

$\langle\alpha|H - E_{in}|in\rangle = 0$ leads to

$$\left[\frac{\hbar\omega}{2}\sigma_z + \hbar g(\alpha\sigma_+ + \alpha^*\sigma_-) + (n\hbar\Omega - E_{in})\right]|\psi^{(i)}(\alpha^*)\rangle$$
$$= \left(-\hbar g\sigma_+\frac{\partial}{\partial\alpha^*} - \hbar\Omega\alpha^*\frac{\partial}{\partial\alpha^*}\right)|\psi^{(i)}(\alpha^*)\rangle\,;\; i = 1, 2\,. \qquad (10)$$

Here

$$|1n\rangle = \sin\theta_n|an\rangle + \cos\theta_n|bn+1\rangle \quad \text{with} \quad E_{1n} = (n+1/2)\hbar\Omega + \hbar R_n/2\,,$$
$$|2n\rangle = \cos\theta_n|an\rangle - \sin\theta_n|bn+1\rangle \quad \text{with} \quad E_{2n} = (n+1/2)\hbar\Omega - \hbar R_n/2\,,$$

where

$$\sin\theta_n = \frac{2g\sqrt{n+1}}{\sqrt{(R_n-\delta)^2 + 4g^2(n+1)}} \quad \text{and} \quad \cos\theta_n = \frac{(R_n-\delta)}{\sqrt{(R_n-\delta)^2 + 4g^2(n+1)}}\,.$$

In (10), $|\psi^{(i)}(\alpha^*)\rangle$ can be obtained via (9), i.e., $\langle\alpha|in\rangle = e^{-|\alpha|^2/2}\chi(\alpha^*)|\psi^{(i)}(\alpha^*)\rangle$ where, from the definition of $|in\rangle$ above, $\chi(\alpha^*) = (\alpha^*)^n/\sqrt{n!}$ and

$$|\psi^{(1)}(\alpha^*)\rangle = \sin\theta_n|a\rangle + \frac{\alpha^*}{\sqrt{n+1}}\cos\theta_n|b\rangle,$$
$$|\psi^{(2)}(\alpha^*)\rangle = \cos\theta_n|a\rangle - \frac{\alpha^*}{\sqrt{n+1}}\sin\theta_n|b\rangle. \qquad (11)$$

In the classical limit of field, the time parameter t of the field environment enters the two-level system through a complex trajectory $\alpha^*(t)$,[8] where $\alpha(t) = \alpha_0 e^{-i(\Omega t-\phi)}$; $\alpha_0 = \sqrt{\bar{n}}$ with $\bar{n}\hbar\Omega = \mathcal{E}_{\rm cl}$. Now, replacing α and α^* in (10) by $\alpha(t)$ and $\alpha^*(t)$ and also using the condition (7), we have

$$\left[\frac{\hbar\omega}{2}\sigma_z - \frac{\wp E_0}{2}(\sigma_+ e^{-i(\Omega t-\phi)} + \sigma_- e^{i(\Omega t-\phi)}) - \frac{\hbar}{2}(\Omega \pm R)\right]|\psi^{(1,2)}(t)\rangle = i\hbar\frac{\partial}{\partial t}|\psi^{(1,2)}(t)\rangle, \qquad (12)$$

where superscripts 1 and 2 on $|\psi(t)\rangle$ correspond to + and − in the term $-\hbar(\Omega \pm R)/2$ respectively. Here $|\psi^{(1,2)}(t)\rangle$ is defined through a complex trajectory $\alpha^*(t)$ i.e., $|\psi^{(1,2)}(t)\rangle := |\psi^{(1,2)}(\alpha^*(t))\rangle$ and from (11) and condition (7), one can obtain $|\psi^{(1,2)}(t)\rangle$ as

$$|\psi^{(1)}(t)\rangle = \frac{-R_0}{\sqrt{(R-\delta)^2 + R_0^2}}|a\rangle + \frac{(R-\delta)e^{i(\Omega t-\phi)}}{\sqrt{(R-\delta)^2 + R_0^2}}|b\rangle,$$
$$|\psi^{(2)}(t)\rangle = \frac{(R-\delta)}{\sqrt{(R-\delta)^2 + R_0^2}}|a\rangle + \frac{R_0 e^{i(\Omega t-\phi)}}{\sqrt{(R-\delta)^2 + R_0^2}}|b\rangle. \qquad (13)$$

Note that the tiny term $n\hbar\Omega - E_{in}$ in (10), which has the corresponding classical form $-\hbar(\Omega \pm R)/2$ in (12), comes from the term $\hbar\Omega\alpha^*\partial\chi(\alpha^*)/\partial\alpha^* - E_{in}$ which has earlier been neglected.[8] Although we keep this tiny term, it can still be removed by the simple phase transformation. However, we will show next that this phase transformation, although it does not change any physical state, leads to the quasienergy terms in the general Floquet's solution. Now, let us define the phase transformation by $|\widetilde{\psi}^{(1,2)}(t)\rangle = \exp[-i(\Omega \pm R)t/2]|\psi^{(1,2)}(t)\rangle$ so that the term $-\hbar(\Omega \pm R)/2$ in (12) is removed and thus leading to the TDSE of the semiclassical model

$$H_{\rm sc}(t)|\widetilde{\psi}^{(1,2)}(t)\rangle = i\hbar\frac{\partial}{\partial t}|\widetilde{\psi}^{(1,2)}(t)\rangle\,, \tag{14}$$

where $H_{\rm sc}$ is the semiclassical Hamiltonian defined in (3). Similar to TDSE Eq. (42) in Ref. 8, we have derived TDSE (14) by the same analysis but, with the help of the soluble Jaynes-Cummings model, we now know the exact form of the state satisfying this TDSE, i.e., $|\widetilde{\psi}^{(1,2)}(t)\rangle$ where $|\psi^{(1,2)}(t)\rangle$ is defined in (13). From (13), it is clear that $|\psi^{(1,2)}(t)\rangle$, and thus $|\widetilde{\psi}^{(1,2)}(t)\rangle$, is the very "specific state" (it is the state satisfying TDSE with the specific initial condition) as there is no Rabi's oscillation, i.e., $|\langle a|\widetilde{\psi}^{(1,2)}(t)\rangle|^2 = |\langle b|\widetilde{\psi}^{(1,2)}(t)\rangle|^2 = 0\ \forall t \geq 0$. Now, the problem of our derivation of TDSE arises because the TDSE we have here is the TDSE for specific states rather than general states as it should be. To solve this problem, notice that $|\widetilde{\psi}^{(1)}(t)\rangle$ and $|\widetilde{\psi}^{(2)}(t)\rangle$ both satisfy the TDSE (14) and $\langle\widetilde{\psi}^{(i)}(t)|\widetilde{\psi}^{(j)}(t)\rangle = \delta_{ij}\ \forall t \geq 0$ which implies that they can form a basis for the two-dimensional Hilbert space of a two-level system and, therefore, any state in this Hilbert space can be written as a unique linear combination of them. Hence, in order to get the TDSE for general states, we now begin the derivation with the linear combination (with respect to the dimension of the Hilbert space of a two-level system) $c_1|1n\rangle + c_2|2n\rangle$ (instead of only $|1n\rangle$ or $|2n\rangle$) so that, by following the previous analysis, $\langle\alpha|[c_1(H - E_{1n})|1n\rangle + c_2(H - E_{2n})|2n\rangle] = 0$ leads to the TDSE for general state $|\psi(t)\rangle$ i.e.,

$$H_{\rm sc}(t)|\psi(t)\rangle = i\hbar\frac{\partial}{\partial t}|\psi(t)\rangle, \tag{15}$$

where $|\psi(t)\rangle = c_1|\widetilde{\psi}^{(1)}(t)\rangle + c_2|\widetilde{\psi}^{(2)}(t)\rangle$. This general state $|\psi(t)\rangle$ can also be written by using the phase transformation defined above as

$$|\psi(t)\rangle = c_1 \mathrm{e}^{-i[\hbar(\Omega+R)/2]t/\hbar}|\psi^{(1)}(t)\rangle + c_2 \mathrm{e}^{-i[\hbar(\Omega-R)/2]t/\hbar}|\psi^{(2)}(t)\rangle. \tag{16}$$

Notice that $|\psi^{(1,2)}(t+T)\rangle = |\psi^{(1,2)}(t)\rangle$; $T = 2\pi/\Omega$ which implies that the state $|\psi^{(1,2)}(t)\rangle$ is periodic in time with period T. The solution in the form of (16) is known as the "Floquet solution" which comes originally from the Floquet's theorem[11] describing the general form of the solution of a differential equation with periodic coefficient (in our case, $H_{\rm sc}(t)$ is periodic in time). In (16), the periodic state $|\psi^{(1,2)}(t)\rangle$ is called the "steady state"[12] or "Floquet state" and the corresponding term $\hbar(\Omega \pm R)/2$ in the exponent is called the "quasienergy".[12]

4. Solution of TDSE from TISE

From the Floquet's solution (16), it is worth to note here that $|\psi^{(i)}(t)\rangle$ $(i = 1, 2)$ comes from the classical limit of the state $|\psi^{(i)}(\alpha^*)\rangle$ which is obtained from the extraction of $|in\rangle$, i.e., $\langle\alpha|in\rangle = \mathrm{e}^{-|\alpha|^2/2}[(\alpha^*)^n/\sqrt{n!}]|\psi^{(i)}(\alpha^*)\rangle$. Moreover, the quasienergies $\hbar(\Omega + R)/2$ and $\hbar(\Omega - R)/2$ come from the classical limit of the tiny terms $n\hbar\Omega - E_{1n}$ and $n\hbar\Omega - E_{2n}$ respectively. In this way, without directly solving the TDSE, one can obtain the solution of TDSE by starting with the Floquet form of the solution and then obtaining the Floquet states and quasienergies from the eigenvectors and eigenvalues of TISE via the classical limiting processes. The discussion above can be summarized in the figure below.

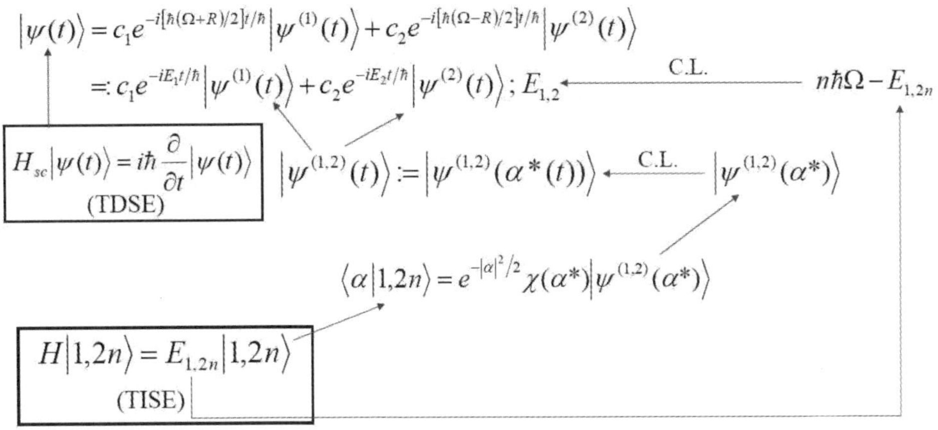

Fig. 1. The chart describing how to obtain the solution of TDSE from TISE via the classical limiting (C.L.) processes.

5. Conclusion

We studied the semiclassical and fully quantum Jaynes-Cummings models and stated the condition on the semiclassical limit. We then used this condition to investigate the derivation of time-dependent Schrödinger equation (TDSE) from time-independent Schrödinger equation (TISE) in the coherent state representation within the framework of the soluble Jaynes-Cummings model. It was shown why, from the outset, the linear combination of energy eigenstates is needed in order to derive the TDSE for general states. Moreover, this investigation also leads to the indirect method of solving TDSE from TISE via the classical limiting processes. It is interesting to generalize this method to any system with a Hamiltonian periodic in time because the solution of TDSE, in this case, can always be written in the form of Floquet's solution.

Acknowledgments

We would like to thank Professor John S. Briggs for continuing interest in this work. We would also like to thank Professor Walter T. Strunz for helpful discussions and Mr. Suttipong Locharoenrat for Latex typing. We acknowledge the financial support of the Research Grant for New Scholar (MRG5080009) of the Thailand Research Fund and the Commission on Higher Education. S. Khemmani also wishes to thank the Faculty of Science, Srinakharinwirot University for partial support.

References

1. E. Schrödinger, *Ann. der Phys.* **79**, 361 (1926).
2. E. Schrödinger, *Ann. der Phys.* **79**, 489 (1926).
3. E. Schrödinger, *Ann. der Phys.* **81**, 109 (1926).
4. M. Born, *Zeit. Phys.* **38**, 803 (1926).
5. N. F. Mott, *Proc. Camb. Phil. Soc.* **27**, 553 (1931).
6. J. S. Briggs and J. M. Rost, *Eur. Phys. Jour. D* **10**, 311 (2000).
7. J. S. Briggs, S. Boonchui and S. Khemmani, *J. Phys. A: Math. Theor.* **40**, 1289 (2007).
8. L. Braun, W. T. Strunz and J. S. Briggs, *Phys. Rev. A* **70**, 033814 (2004).
9. See, for example, Pierre Meystre and Murray Sargent III, *Elements of Quantum Optics* (Springer-Verlag, Berlin, 3rd ed. 1999); Stephan M. Barnett and Paul M. Radmore, *Methods in Theoritical Quantum Optics* (Oxford, New York, 1997), or other works on quantum optics.
10. This idea is, in fact, a motivation for the derivation made in the appendix of Ref. 7.
11. J. H. Shirley, *Phys. Rev.* **138**, 979 (1965) and some references therein.
12. H. Sambe, *Phys. Rev. A* **7**, 2203 (1973).

TUNNELING AND HOPPING BETWEEN DOMAINS IN THE METAL-INSULATOR TRANSITION IN TWO-DIMENSIONS

DAVID NEILSON

Dipartimento di Fisica, Università di Camerino, I-62032 Camerino, Italy
NEST CNR-INFM, I-56126 Pisa, Italy
School of Physics, University of New South Wales, Sydney 2052, Australia
david.neilson@unicam.it

ALEX HAMILTON

School of Physics, University of New South Wales, Sydney 2052, Australia
alex.hamilton@unsw.edu.au

Received 31 July 2008

Motivated by recent surprising experimental results for temperature dependent resistivities in 2D mesoscopic electron systems, we investigate transport in these systems by percolation connected through a network of metallic domains. The size of the domains is determined by the level of disorder in the system and by the strength of the electron correlations. In the insulating phase the metallic domains are connected for transport by two competing mechanisms, thermally activated hopping and quantum tunneling. We calculate the transmission across the potential barriers that separate the metallic domains. Using recent data from transport measurements in mesoscopic 2D systems, we obtain the observed saturation of the temperature dependent resistivity at $T \sim 1$ K and consistent values for the size of the domains and for magnitude of the average variation in the random disorder potential.

Keywords: Metal-insulator transition; electron transport; mesoscopic; domains; tunneling.

In two-dimensional (2D) electron systems, the random potential fluctuations associated with disorder will always localize the electrons if the interactions between the electrons are neglected. It is the competition between the disorder and the electron correlations arising from the electron-electron interactions that is crucial for a 2D metal-insulator transition.[1]

Random potential fluctuations may create a percolative second-order transition in the transport coefficients. It has been proposed near the metal-insulator transition that these potential fluctuations create domains (regions of slightly lower and slightly higher density).[2] When the electron density is decreased this can lead to a classical metal-insulator transition.[3] Experimental evidence for localized domains has been seen in gated GaAs heterostructures using near-field spectroscopy with sub-wavelength resolution.[4] Negatively charged exciton luminescence is used to image the spatial distribution of the electrons in a 2D layer. In the range of gate voltages where the conductivity drops, the electrons have been shown to be localized inside the potential fluctuations of the remote ionized donors. The spectral signature of the regions filled with localized electrons was very different from the regions containing conducting electrons.

Within the domain model it is assumed that the random potential associated with the impurities $V_{imp}(\mathbf{r})$ varies smoothly. When the linear dimensions of the domains ℓ_V exceed the correlation lengths, the local energy density is given by $E(\mu - V_{imp}(\mathbf{r}))$, for chemical potential μ. Such a model was employed by Shimshoni *et al.*[5] to study transport properties near the phase transition for quantum Hall insulators and superconductors. Neilson *et al.*[6] used the domains model to study transport near the 2D metal-insulator transition. A $V_{imp}(\mathbf{r})$ which varies smoothly on a scale long compared with correlation lengths leads to spatial inhomogeneities in the local electron density. As the insulator-metal transition is density driven, these variations in the electron density can lead to the formation of domains of coexisting metal and insulating phases. The metallic domains form in regions of higher electron density $V_{imp}(\mathbf{r}) < \mu$, with the insulating domains occupying the remaining regions.

For strong disorder, insulating variable-range hopping transport is observed, with interaction effects leading only to a modification of the single-particle density of states.[7] If the disorder is mainly long-range, then at low electron densities the system is expected to become increasingly inhomogeneous. In this case transport behaves in accordance with classical percolation law, and this masks the interactions between electrons.[8]

Recently Baenninger *et al.*[9] used a new experimental approach to investigate interaction effects in the presence of disorder by minimizing the effects of the long-range disorder and focusing only on the short-range fluctuations. They used mesoscopic 2D electron systems of width 8 μm with lengths from 0.5 to 2 μm. Such short lengths reduce the impact on transport of long-range disorder, which for modulation doped GaAs/AlGaAs heterojunctions varies on a length scale greater than 0.5 μm.[10,11] For a device size smaller than the length scale of long-range disorder, the short-range fluctuations of order of the width of the spacer layer are dominant and the effect of charge inhomogeneity is reduced. This permitted Baenninger *et al.* to focus on the effect of short-range fluctuations.

The electron density was tuned from 0.9 to 1.7×10^{10} cm^{-2} using a metal top gate. For GaAs this corresponds to $4 \leq r_s \leq 5.6$ and a Fermi temperature of $T_F = 3.7$ to 7 K. The temperature range was from 60 mK to 4.4 K so the electrons are degenerate. The spacer layer thickness was $40 < \delta < 80$ nm.

At relatively high temperatures exceeding 1 K, the resistivity was observed to follow the expected exponential activation $\rho(T) \sim \exp[E_0/k_B T]$ characteristic of insulating behavior. However for temperatures lower than 1 K, Ref. 9 found, surprisingly, that when T was further decreased, the $\rho(T)$ saturated or in some cases was even observed to decrease. This behavior is difficult to reconcile with thermal activated hopping.

To understand this phenomenon we apply the model in Ref. 5 to determine the transport properties on this length scale. For 2D systems, the percolation threshold occurs when the total areas of the two types of domains are equal, that is the critical area fraction $p = 1/2$, where p is the total area fraction of the insulating phase. For this unique value of p, both the metal and the insulating domains contain at least one connected path across the whole sample. Precisely at the transition all neighboring domains will touch just at one point. Since the impurity potential increases away from the contact point when we move into either of the two adjacent metal domains and decreases when we move into either of the insulating domains, it follows that $V_{imp}(\mathbf{r})$ forms a saddle point centered on the contact point.

As the electron density decreases on the insulating side of the transition, the metallic phase will retreat from the center of the junction and will be replaced by the insulating phase. On this side of the transition $p > 1/2$. The area of the insulating phase exceeds that of the metallic phase by an amount equal to the area defined by the metallic boundary and the lines where the domain boundaries had been located when p was equal to 1/2. Denoting the excess area by δA and assuming a hyperbolic form for metallic boundary near the junction, one obtains $\delta A = (1/2)\log(\ell_V/d)\ d^2$, where d is the average distance separating the closest metallic phase domains. The mean free path $\ell_V >> d$. If there are N_{sp} junctions in a total area A then the total excess area fraction of the insulating phase will be

$$\begin{aligned} p &= 1/2 + \gamma d^2 \\ \gamma &= (N_{sp}/2A)\log(\ell_v/d)\ . \end{aligned} \tag{1}$$

Writing the densities within the metal and insulating domains as $n_M(p)$ and $n_I(p)$, the total density in the insulator with $p > 1/2$ is

$$\begin{aligned} n(p) &= pn_I(p) + (1-p)n_M(p) \\ &= n^c + (p-1/2)\Delta n^c + (p-1/2)^2\Delta(n^c)'\ . \end{aligned} \tag{2}$$

n^c is the density at the transition, $\Delta n^c = n_I^c - n_M^c$ is the density discontinuity between the insulating and metal phases at the transition, and $\Delta(n^c)' = d\Delta(n)/dp|_{n^c}$. Using Eqs. 1 and 2, we can expand the density n in even powers of the junction width,

$$\begin{aligned} n(d) &= n^c + \alpha d^2 + \beta d^4 + ... \\ \alpha &= \gamma\Delta n^c; \quad \beta = \gamma^2\Delta(n^c)'\ . \end{aligned} \tag{3}$$

We can determine the resistivity from the transmission across the quantum junctions as a function of d. The primary contribution to the resistivity comes from the saddle points of the potential near $V_{imp}(r) \sim \mu_{crit}$. We assume that all quantum interference effects take place on the length scale of the tunnel barrier so that there is no coherence between tunneling events. For a finite width distribution of junction resistances in a 2D array, the total resistance is given by the resistance of the typical junction.[12] The resistivity depends on the transmission rate $\mathcal{T}$ through the junction: $\rho(T) = (1 - \mathcal{T})/\mathcal{T}$, in units of $\hbar/e^2$.

At relatively high temperatures, transport across the junction is by thermally activated hopping. The transmission $\mathcal{T} \sim \int_{V_{barr}}^{\infty} \exp(-E/k_B T)\mathrm{d}E$ is then

$$\mathcal{T} = \mathcal{T}_0 \exp\left(-\frac{V_{barr}}{k_B T} d^2\right) \tag{4}$$

where $\mathcal{T}_0$ is the transmission for $d = 0$, that is, at the transition. V_{barr} is the average barrier height across the junction and is related to the curvature of the impurity potential at the junction V'', by $V_{barr} = (1/2)V''d^2$. On the other hand, at zero temperature the tunneling transmission through the barrier is given by

$$\mathcal{T} = \mathcal{T}_0 \exp(-2S(d)) \sim \mathcal{T}_0 \exp\left(-S''d^2\right) , \tag{5}$$

where the action across barrier is $S(d) \simeq S(0) + \frac{1}{2}S''d^2$. The S'' is the second derivative of the action which can be expressed as $S'' = \sqrt{mV''}$. Thus Eq. 5 can be written

$$\begin{aligned} \mathcal{T} &= \mathcal{T}_0 \exp\left(-(\pi/2\hbar)\sqrt{mV''}d^2\right) \\ &= \mathcal{T}_0 \exp\left[-\frac{\pi\sqrt{2mV_{barr}}}{\hbar} d\right] . \end{aligned} \tag{6}$$

Table 1. Best fit values for tunneling distance d and resulting average barrier height V_{barr} from $\rho(T)$ data in Ref. 9. Column 1 lists the Figure number of the $\rho(T)$ data from Ref. 9.

Fig.	n (10^{10}cm^{-2})	d (nm)	V_{barr} (meV)
1d		24	0.15
1e		26	0.19
2a	0.94	47	0.26
2b	1.15	28	0.21
2c	1.39	21	0.12
2d	1.65	9.9	0.03

Combining these results, the resistivity can be written

$$\rho(T, n) = \mathcal{T}_0^{-1} \exp\left[\frac{\pi\sqrt{2mV_{barr}}}{\hbar} d\right] - 1 \; ; \qquad \text{for } T = 0 \tag{7a}$$

$$= \mathcal{T}_0^{-1} \exp\left[\frac{V_{barr}}{k_B T}\right] - 1 \; ; \qquad \text{for } \textit{large } T \tag{7b}$$

We fitted the measured resistivity at the lowest T in Ref. 9 to Eq. (7a), and the large T temperature-dependent resistivity data from the exponential section of each resistivity curve in Ref. 9 to Eq. (7b). We fixed the fitting parameters V'' and $\mathcal{T}_0$ to be independent of electron density for one sample. From the combined data in Figs. 1d-e, and 2a-d of Ref. 9 we obtained best fit values $V'' = 0.35\ \mu\text{eVnm}^{-2}$ and $\mathcal{T}_0 = 0.17$. Table 1 gives the best fit values for the fitting parameter d and the resulting V_{barr} for the data from each of the figures of Ref. 9.

We estimate the crossover temperature T_θ between these two effects by identifying T_θ with the temperature at which quantum tunneling takes over as preferred transport mechanism. This occurs when the tunneling transmission rate between domains equals the rate from transmission by thermally activated hopping. From Eqs. 4 and 6 we obtain the condition on T_θ,

$$\pi\sqrt{mV''}/\hbar = V''/(2k_B T_\theta)\ . \tag{8}$$

With our fitted value $V'' = 0.35\ \mu\text{eVnm}^{-2}$, Eq. 8 gives $T_\theta = 1.2$ K. This value is in good agreement with the crossover temperatures reported in Ref. 9.

We use a power law to interpolate between the low and high T limits of Eq. 7. Figure 1 shows typical results in which we compare Fig. 1(d) from Ref. 9 (points) to our best fit with Eq. 7.

The values of V_{barr} in Table 1 are much smaller than the variations $\Delta V_{disorder}$ measured in macroscopic samples by Finkelstein *et al.*[10] and by Chakraborty *et al.*[11] of $\Delta V_{disorder} \sim 2 - 5$meV. The values of the length scale of the disorder variation, $\ell_{disorder} \sim 0.5 - 1.5\ \mu$m, reported in Refs. 10 and 11 are also much greater than the d in Table 1. This is consistent with the proposal that in the mesoscopic samples of Ref. 9 of sizes smaller than the length scale of long-range disorder, it is short-range fluctuations of $\mathcal{O}(\delta)$ that dominate.

Our physical picture for the contribution to the resistivity is thermally activated hopping at relatively large temperatures, given by Eq. (7b), and quantum tunneling, given by the zero-temperature expression Eq. (7a), when the temperature is sufficiently low.

If we substitute in Eq. 8 values for $V_{barr} = \Delta V_{disorder} \sim 2 - 5$ meV and $d \sim 0.5 - 1.5\ \mu$m for the macroscopic samples from Refs. 10 and 9, we obtain much smaller curvatures of the disorder potential, $V'' \sim 0.002 - 0.04\ \mu\text{eVnm}^{-2}$. When substituted in Eq. 8 these values give lower crossover temperatures, $T_\theta \leq 100$ mK, suggesting that the saturation of the resistivity observed in Ref. 9 at $T_\theta \sim 1$ K, could also be present in macroscopic samples but at a significantly lower T_θ.

We note that the analysis does not require extreme proximity to the percolation transition. The crucial assumption is that the insulating component is sufficiently insulating that the transport is dominated by paths that avoid the insulating region as much as possible. This analysis therefore holds well beyond the critical dynamical scaling regime.

In conclusion, a model of formation of domains of metallic and insulating regions near the phase boundary for the metal-insulator transition due to the randomly

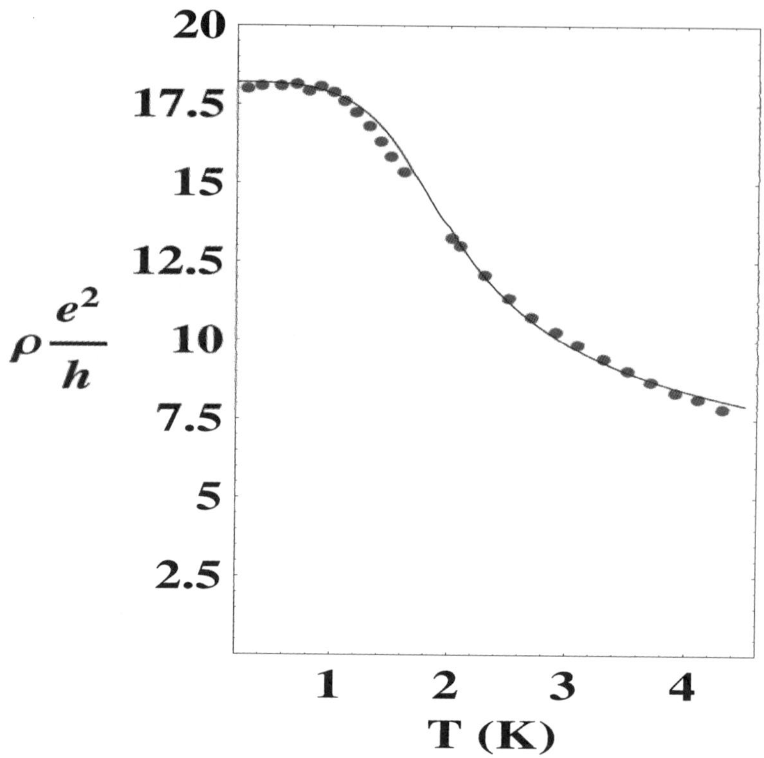

Fig. 1. Fit of temperature dependence of resistivity using Eq. 7. See text.

fluctuating impurity potential leads to calculated resistivities with properties that are in agreement with the new and surprising results for the temperature dependent resistivity in mesoscopic samples. The crossover to saturation of the resistivity for temperatures below $T_\theta \sim 1$ is understood as a switching on of quantum tunneling for temperatures $T < T_\theta$. Because of the shorter-ranged variations of the random impurity potential for the mesoscopic systems, quantum tunneling switches on at temperatures higher than those for macroscopic samples.

Acknowledgments

We thank Arindam Ghosh for useful discussions.

References

1. D. Simonian, S.V. Kravchenko, M.P. Sarachik and V.M. Pudalov, *Phys. Rev. Lett.* **79**, 2304 (1997)
2. A.L. Efros, *Solid State Commun.* **70**, 253 (1989); J.H. Davies, J.A. Nixon and H.U.

Baranger (unpublished); B.I. Shklovskii and A.L. Efros, *Electronic Properties of Doped Semiconductors* (Springer-Verlag, Berlin, 1984).
3. J.A. Nixon and J.H. Davies, *Phys. Rev. B* **41**, 7929 (1990).
4. G. Eytan, Y. Yayan, M. Rappaport, H. Shtrikman and I. Bar-Joseph, *Phys. Rev. Lett.* **81**, 1666 (1998).
5. Efrat Shimshoni, Assa Auerbach and Aharon Kapitulnik, *Phys. Rev. Lett.* **80**, 3352 (1998).
6. D. Neilson, J.S. Thakur, and E. Tosatti, *Aust. J. Phys.* **53**, 531 (2000).
7. A.L. Efros and B. I. Shklovskii, *J. Phys. C* **8**, L49 (1975).
8. S. Das Sarma, M. P. Lilly, E.H. Hwang, L. N. Pfeiffer, K.W. West and J. L. Reno, *Phys. Rev. Lett.* **94**, 136401 (2005).
9. Matthias Baenninger, Arindam Ghosh, Michael Pepper, Harvey E. Beere, Ian Farrer and David A. Ritchie, *Phys. Rev. Lett.* **100**, 016805 (2008).
10. G. Finkelstein, P.I. Glicofridis, R.C. Ashoori and M. Shayegan, *Science* **289**, 90 (2000).
11. S. Chakraborty, I.J. Maasilta, S.H. Tessmer and M.R. Melloch, *Phys. Rev. B* **69**, 073308 (2004).
12. E. Shimshoni and A. Auerbach, *Phys. Rev. B* **58**, 9817 (1997).

BIRKHOFF THEOREM AND ERGOMETER: MEETING OF TWO CULTURES

M. HOWARD LEE

Department of Physics and Astronomy, University of Georgia, Athens, Georgia 30602, USA and Korea Institute for Advanced Study, Seoul 130-012, Korea*

MHLee@uga.edu

Received 31 July 2008

There are two approaches to understanding Boltzmann's ergodic hypothesis in statistical mechanics. The first one, purely mathematical, goes by way of theorems while the second one relies on physical measurements. By its own nature the former is universal whereas the latter is specific to a system. By all account they seem orthogonal to each other. But should not they meet at the end? If, for example, both conclude that the hypothesis is not valid in a given system, should not their conclusions be compatible? We illustrate in this work how the two cultures meet in the physics of ergodicity.

Keywords: Birkhoff's theorem; ergometer; ergodicity.

1. Introduction

One may perhaps say that the rationale behind the ergodic hypothesis (EH) that Boltzmann proposed in 1871 is comprised of the following five disparate parts: (a) Observation is a time average. (b) Time averages are generally difficult to calculate. (c) Ensemble averages are much easier to calculate than time averages. (d) Time averages may be equal to ensemble averages. (e) Observation is thus an ensemble average. Since Boltzmann's time, EH has become, as we all know, one of the foundations of statistical mechanics. We have been routinely calculating ensemble averages to obtain thermodynamic functions.

Is EH (part-d) really valid? Is it valid as widely as we would like to believe? How could the two unrelated averages be equal? What might be the root or roots of this equalitity? There are math theorems which are said to prove EH.[1] But as we know, math theorems are couched in abstract conditions and the ergodic theorems are no different. They do not readily lend themselves to any thermodynamic levers by which EH could be demonstrated. In addition these theorems rely on concepts of classical physics. No one I believe knows whether they can be made to go beyond. Yet we usually accept EH to be valid even in non-classical systems too. Under these circumstances, it is curious that we should believe in EH. This belief seems to be born less of reason and more of faith.

---

*Permanent address.

Recently a physical theory of EH has been developed,[2] which employs an approach orthogonal to the math approach. It is system specific, hence not universal, but is capable of providing physical insight into what makes the two averages to be equal. By combining the results of the two approaches, we hope to shed light on the validity of EH and, equally important, on the domain of validity of EH.

2. Math Approach

A modern definition for EH may be stated as follows. Let A be a dynamical variable of a *macroscopic* system denoted by a Hamiltonian H and let the system be in *thermal equilibrium*. Then, EH asserts that

$$\lim_{T\to\infty} \frac{1}{T} \int_0^T \langle A\,(t)\rangle \, dt = \langle A\rangle \,, \tag{1}$$

where the brackets denote an ensemble average. The averages on $A(t)$ on the lhs of (1) mean that all initial values of $A(t)$ are considered. Thus, both sides are being compared at the same temperature. At this stage it ought to be stressed that the original intent of EH was for a macroscopic system. In statistical mechanics it means that the thermodynamic limit is being taken. Applying EH to a finite system as has been done would seem to be at best a misuse or at worst a corruption of the original concept.

Many mathematicians have studied EH.[1] Perhaps the most relevant for us may be the work of Birkhoff,[3] which we will briefly state in simple terms. It is assumed that $f(t) = {}_{-}\langle A\,(t)\rangle$ is a finite integrable function in phase space Γ, a measure preserving space, due to Liouville's theorem. Then it is proved that the time average of $f(t)$ exists not everywhere but "almost" everywhere. It means that the time average does not exist over any set of phase points of measure zero. A good example which yields a set of measure zero is the Cantor set. The time average that exists almost everywhere is equal to the phase space average of f if and only if the phase space is *metrically transitive*, i.e., not decomposable. This is said to prove EH. This kind of proof must be satisfying to the mathematical mind. But the proof gives no obvious means — something the physical mind needs — with which to establish ergodicity in a specific physical system.

3. Physical Approach

In a physical approach to EH, we regard (1) to be an assertion that is testable by measurement. Imagine that a macroscopic system in thermal equilibrium is scattered by a weak external probe (e or n beams, X-rays etc.). A scattering process is described by two quantities — energy and momentum transfers $\hbar\omega$ and $\hbar k$, respectively.

Suppose a probe field couples to a dynamical variable A of the system. Then lhs can represent an inelastic scattering process at a fixed k in which $\omega \to 0$ while the rhs an elastic scattering process at the same value of k. Under what condition would they be the same? Since the comparison must be system dependent, the condition for EH would also be system dependent.

Let $h_A(t)$ be a time dependent probe field, which couples to A. The total energy at time t may be written as

$$H'_A(t) = H(A, B, C, \ldots) + h_A(t)\, A, \tag{2}$$

where $A, B, C, \ldots$ are possible dynamical variables of H, but only A is excited by the probe. Observe that EH refers to A, not to the other variables. The existence of other variables raises the possibility that even if A is ergodic, B or the others may not be so. That is, suppose another probe say $h_B(t)$ is turned on. Our measurement may say that B is not ergodic. In other words EH is not a general statement about a system but a specific statement about a variable of a system, which can be excited by an external probe.

In this physical picture, a system is perturbed through a dynamical variable by a time dependent external field. The perturbation energy imparted to the system will become delocalized throughout if the variable is not in a stationary state of the system (i.e., not a constant of motion.) If not in a stationary state, it will evolve in time. We can measure the average of this time evolution $\langle A(t)\rangle_{H'_{A(t)}}$. The time average of this quantity is precisely one side of EH, the lhs of (1). If the same system is perturbed through the same dynamical variable now by a time independent external field, we can also measure $\langle A\rangle_{H'_A}$, where by H'_A we mean (2) in which $h_A(t)$ is replaced by h_A. This static measurement is the other side of EH, the rhs of (1). Will they be equal to each other as EH asserts? We can perform these measurements and compare for different variables A, B, or C, etc. of the same system. We can repeat the same measurement for a different set of variables of another system. This is how one might proceed to verify the validity of EH, clearly a physical approach to testing the validity of EH.

The simplest may be to take the external field to be weak enough so that we can use the formalism of linear response theory,[4] which is now well established and widely applied in dynamic studies of many-body systems. Van Hove, Kubo and others have shown that $I_K(\omega)$ the intensity of absorption in inelastic scattering processes is given by $\frac{Im}{\chi}{}_k(\omega)$ the imaginary part of the dynamic response function. If a system is homogeneous and the scattering is causal, the dynamic response function is just the Fourier transform of $\chi_k(t)$, where suppressing k now and in units where $\hbar = 1$,

$$\begin{aligned}\chi_A(t) &= i\,\langle[A(t), A]\rangle_H \quad if \quad t > 0\\ &= 0 \quad \text{if otherwise},\end{aligned} \tag{3}$$

where

$$A(t) = \exp iHtA\exp -iHt. \tag{4}$$

Thus in the lineary response view, EH (1) is transformed to:

$$\lim_{T\to\infty}\frac{1}{T}\int_0^T\int_0^t \chi_A(t-t')\,dt'dt = \chi_A, \tag{5}$$

where χ_A means the static response function, commonly denoted as $\chi_A = (A, A)$, where the inner product is the Kubo scalar product.

It has already been shown[2] that if (5) is to hold, the following condition must be satisfied: $0 < W < \infty$, where

$$W(A) = \int_0^\infty r_A(t)\,dt, \tag{6}$$

where $r_A(t) = (A(t), A)\,/\,(A, A)$, the normalized autocorrelation function of A.[5–7] It is evident that if $W(A)$ is to be finite, $r_A(t)$ must vanish sufficiently fast as $t \to \infty$. This behavior of the autocorrelation function is termed *irreversible*.[8] The ergodicity depends on two necessary properties, the first being irreversibility and the second being delocalizability. The second property is a consequence of the perturbation on the system in the scattering processes. The two necessary properties together constitute a sufficient condition for ergodicity. It also implies that ergodicity is contained in irreversibility.[8]

Also note that W is self diffusivity.[9] If it is to be finite, it must be of normal diffusion.[10] In physical theory, W has been dubbed an *ergometer* since it can tell or detect ergodicity in a system.[11] Being a device albeit gedanken it measures the existence of ergodicity system by system. Its approach to establishing ergodicity is evidently very different from the math approach as we shall further discuss below.

4. Meeting of Two Cultures

The math approach is universal by the conditions of its theorems, not referring to any one system. The physical approach is specific since it establishes ergodicity system by system. Clearly the two approaches are orthogonal. But should not they meet at the end? If the ergometer were to say that a variable of a system is not ergodic, should not one find at least one violation in the conditions of an ergodic theorem? If so, it would represent a meeting of the two cultures, arriving from different directions.[11]

Let us be more specific. Suppose one calculates $W(A)$ for dynamical variable A of a system and finds that it is ergodic. That is to say, the autocorrelation function of A is irreversible and the delocalization of the perturbation energy is complete. The ergometer's reading must then imply that there is just one invariant in it. It means that the phase space is not decomposable, thus metrically transitive. It must correspond to a complete delocalization of the perturbation energy.

Conversely if $W(B)$ of the *same* system is calculated and found *not* ergodic, the ergometer would say that the delocalization is only partial. It must mean that there are at least two invariants or the phase space is decomposable hence not transitive.

Suppose we take another possibility. Let A' be a variable of another system. By calculating $W(A')$, we find it to be not ergodic because the delocalization does not appear robust but the rate of delocalization seems to diminish with the distance from the point of perturbation. Seemingly there is only one invariant, but it could be running over a set of measure zero. In the following section we shall illustrate the correspondence by means of a many-body model.

5. Ergometry on a Many-Body Model

To be specific, we shall consider a well known model — the spin-1/2 XY model in 1d,

$$H = -J \sum_{i=1}^{N} \left(s_i^x s_{i+1}^x + s_i^y s_{i+1}^y \right), \tag{7}$$

where s_i^x and s_i^y are the x and y component of the spin-1/2 operator at site i in a one-dimensional periodic lattice, such that $[s_i^x, s_i^y] = i\hbar s_i^z$, s_i^z being the z component, J the coupling constant between the nearest neighbors. We impose periodic boundary conditions, e.g., $s_{N+1}^x = s_1^x$.

For this spin-chain model, there are two possible independent dynamical variables: $A = s_i^x$ or equivalently s_i^y and $B = s_i^z$, where site i may be anyone of N sites. Observe that the variable B as defined above is not explicitly contained in H. See (7). Thus classically it would be a null operator but quantum mechanically it is implicitly present in the Hilbert space of the spin-1/2 operator.

At infinite temperature the autocorrelations of A and B are both known. We begin with $r_A(t) = J_0(t)$, in the units of $J = 1/\sqrt{2}$, where J_0 is the Bessel function.[12] We find that $W(A) = \sqrt{\pi/2}$, hence $A = s_i^x$ for this model is ergodic. We now turned to $r_B(t) = [J_0(t)]^2$ in the units of $J = 1$.[13] It is not difficult to show that $W(B) = \infty$, so that $B = s_i^z$ of the same model is not ergodic. Why this contrasting results on ergodicity in the same model? Since both A and B are irreversible, the difference must arise from delocalizability for A being complete and for B not being complete. In terms of Birkhoff's conditions, there is one invariant for A but at least two for B. Can we identify them separately?

If the external field $h_A(t)$ is turned on which couples to a spin s_1^x say, the perturbation energy is delocalized from site 1 to 2, site 2 to 3 etc. because s_1^x and others are not in a stationary state of H. In time the delocalization spreads with $N \to \infty$, attaining ergodicity. One may say after Birkhoff that there can be just one invariant in it.

If an external field $h_B(t)$ is now turned on, the physics is very different since the field does not directly couple to anyone of spins of the model. The effect of the field is to create a vortex-like motion along the z direction in spin space as may be seen by

$$d/dt s_i^z = i \left(s_i^y s_j^x - s_i^x s_j^y \right), \tag{8}$$

where $j = i+$ or -1. The vortex-like motion is perpendicular to the transverse direction of the spin interaction. Thus, the effect of the perturbation by $h_B(t)$ is to create motions in the orthogonal direction to the interaction energy plane, leaving it unaffected by the perturbation. One may say after Birkhoff that this is not the realization of one invariant, but at least two. This situation recalls another rather similar: If a Brownian particle is perturbed, the effect of the perturbation is to create a ballistic movement of the Brownian particle without affecting its internal degrees of freedom.[14]

A fundamental difference between A and B may also be seen by the following: If $B' = \sum_i s_i^z$, $B'(t) = B'$, a constant of motion. There is no time evolution in this variable. But if $A' = \sum_i s_i^x$, A' is not a constant of motion. Thus there is a time evolution in it.

6. Concluding Remarks

We showed that the math and physics approaches to proving ergodicity are totally different in spirit and body. It may thus be said that they are orthogonal. Nonetheless we have been able to show that in the end they are arriving at the same end. The meeting of the two cultures has been explicitly demonstrated through a many body model. Although the math approach due to Birkhoff[3] is meant for a classical many-body system, we find that the concepts from the math approach are applicable to a quantum analog

Physicists may say that a theorem without devices to test its conditions is an impotent knowledge. If we had no means to measure angles, the Pyathagorean theorem would be a statement in the math realm. In the same vein, it is the ergometer which takes Birkhoff theorem out into the physical arena.

Birkhoff says that EH is valid provided that certain conditions are met. He says nothing about what class of physical systems can qualify. At a first glance it is disappointing that the theorem does not say anything about this information critical to physics. A reflection would convince us that no math theorems would do that or are expected to do that. A physical problem has to be solved by the physics mind molded and tempered by the heat of the physical reality.

Acknowledgments

The author wishes to thank the US Army Research Office for providing partial travel support to present this work at CMT31 in Bangkok, Thailand in Decemeber 2007. The author also wishes to thank Ms. Teresa Lopes for her kind assistance in the preparation of this work.

References

1. I. E. Farquhar, *Ergodic Theory in Statistical Mechanics* (Wiley, NY, 1964). See sec. 3.4 and App. 2.

2. M.H. Lee, *Phys. Rev. Lett.* **87**, 250601 (2001).
3. G. D. Birkhoff, *Proc. Nat. Acad. Sci. (USA)* **17**, 656 (1931).
4. R. Kubo, *Rep. Prog. Phys.* **29**, 255 (1966), H. Nakano, *Int. J. Mod. Phys. B* **7**, 239 (1993). For a recent review, see A.L. Kuzemsky, *Int. J. Mod. Phys. B* **19**, 1029 (2005).
5. A. L. Kuzemsky, *Int. J. Mod. Phys. B* **21**, 2821 (2007).
6. R.M. Yulmatyev, A.V. Mokshin and P. Hănggi, *Phys. Rev. E* **68**, 051201 (2003). A.V. Mokshin, R.M. Yulmatyev and P. Hănggi, *Phys. Rev. Lett.* **95**, 200601 (2005).
7. U. Balucani, M.H. Lee and V. Tognetti, *Phys. Rep.* **373**, 409 (2003).
8. M.H. Lee, *Phys. Rev. Lett.* **98**, 190601 (2007). Also see T. Prosen, *J, Phys. A* **40**, 7881 (2007).
9. M.H. Lee, *Phys. Rev. Lett.* **85**, 2422 (2000).
10. R. Morgado, F. A. Oliveira, G.G. Bartolouni and A. Hansen, *Phys. Rev. Lett.* **89**, 100601 (2002). M.H. Vainstein, I.V.L. Costa, R. Morgado and F.A. Oliveira, *Europhys. Lett.* **73**, 726 (2006). J.D. Bao,Y.Z. Zhuo, F.A. Oliveira and P. Hănggi, *Phys. Rev. E* **74**, 061111 (2006).

TOWARDS BALLISTIC TRANSPORT IN GRAPHENE

XU DU, IVAN SKACHKO and EVA Y. ANDREI
Department of Physics and Astronomy, Rutgers University,
Piscataway, New Jersey 08854 USA

Received 31 July 2008

Graphene is a fascinating material for exploring fundamental science questions as well as a potential building block for novel electronic applications. In order to realize the full potential of this material the fabrication techniques of graphene devices, still in their infancy, need to be refined to better isolate the graphene layer from the environment. We present results from a study on the influence of extrinsic factors on the quality of graphene devices including material defects, lithography, doping by metallic leads and the substrate. The main finding is that trapped Coulomb scatterers associated with the substrate are the primary factor reducing the quality of graphene devices. A fabrication scheme is proposed to produce high quality graphene devices dependably and reproducibly. In these devices, the transport properties approach theoretical predictions of ballistic transport.

Keywords: Graphene; transport.

1. Introduction

The discovery of techniques to isolate and study graphene, a one-atom thick layer of crystalline carbon,[1–3] has stimulated a massive effort to understand its electronic properties.[4,5] As a Dirac fermion system with linear energy dispersion, electron-hole symmetry and internal degree of freedom (pseudo spin), graphene promises intriguing physical properties such as electronic negative index of refraction,[6] specular Andreev reflections in graphene-superconductor junctions,[7,8] evanescent transport,[9] anomalous phonon softening,[10] etc. As an electronic material, graphene exhibits many desirable properties such as high mobility, thin body, low carrier density (tunable by electric field gating), and compatibility with the top-down fabrication scheme.

Despite the high expectations for ideal graphene devices, the commonly used fabrication techniques yield samples with large uncontrollable variations and middling quality as characterized by mobility, gate control, and minimum conductivity. It is widely believed that this lower than expected performance is not intrinsic, but rather due to extrinsic factors such as material quality and ambient environment.[11,12] But it is not clear how the different factors contribute to the deterioration of the transport properties in graphene and how to asses their relative importance. Nor is there a scheme in place to produce high quality graphene devices reliably and reproducibly. Here we report results of a study that gauged the impact of various extrinsic factors on the quality of graphene devices. The basic strategy was to isolate the influence of each of the extrinsic factor on the transport properties of a freshly prepared graphene device by deliberately magnifying the corresponding factor to dominate all others. This method was used to study material defects, lithographic residues, invasiveness of metallic contacts, substrate. We find that, while all factors contribute to the degradation of the graphene device, trapped charges in the substrates are the main contributor. Based on our observations, a scheme is proposed for future fabrication of ultrahigh quality graphene devices.

2. Experiments and Results

Graphene is deposited using the method of mechanical exfoliation.[1] Prior to graphene deposition, the Si/SiO_2 substrates are baked in forming gas (Ar/H_2) at 200 C for one hour to remove water and organic residue. A thin foil of highly oriented pyrolytic graphite (neutron detector quality) is peeled from the bulk material using scotch tape and transferred onto the Si/SiO_2 substrate. Pressure is then applied onto the graphite foil using compressed high purity nitrogen gas through a stainless steel needle, for ~5 seconds. The foil is then removed from the substrate and the substrate is carefully checked under an optical microscope for candidates of single layer graphene. This process is repeated until a few graphene flakes are identified. AFM inspection is then used to confirm the single graphene layer followed immediately by coating with PMMA resist. The electrical contacts and leads are then fabricated with standard e-beam lithography techniques. To remove the organic and water residues, the samples are baked in forming gas (Ar/H_2) at 200 C for 1 hour.

To test the impact of the material quality on transport properties of the graphene devices, we created defects in the graphene films with DC hydrogen glow discharge. For this purpose, pre-characterized graphene devices were placed ~5 cm away from a circular electrode biased at 500 V DC potential generating a glow discharge at a hydrogen pressure of 150 mTorr. The damage produced by 10 seconds exposure to the glow discharge is then characterized by comparing the gate voltage dependence of the sample resistance before and after irradiation. No noticeable change in the shape of the flake could be detected by optical microscopy or through SEM. We therefore attribute any changes in the transport properties to microscopic (atomic) defects from the irradiation. Figure 1(a) shows a typical result of the glow discharge tests. We note that the irradiation causes a significant shift of the Dirac point indicative of hole-doping, presumably due to the trapping of the positively charged hydrogen ions. The large increase of sample resistance throughout the whole range of gate voltage suggests a large increase in the density of scattering centers. Since the samples are in the diffusive regime, we were able to separate the Coulomb scattering contribution to the resistance (which yields gate-independent mobility) from the short-range scattering contribution (gate-independent resistivity), as shown in

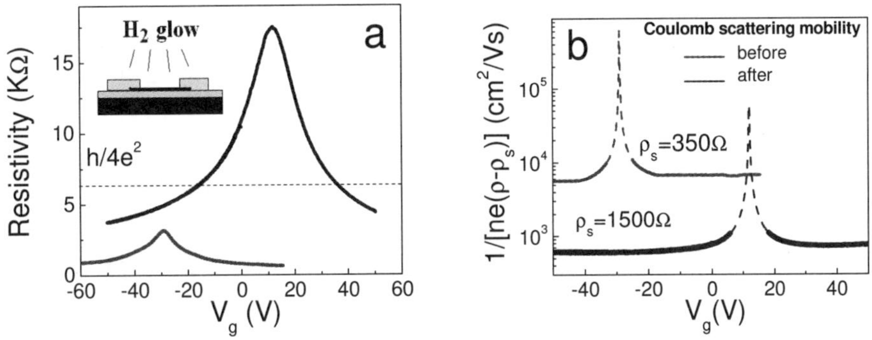

Fig. 1. (Color online) (a) Gate voltage dependence of resisitivity before (red) and after (blue) the H2 glow discharge test. Inset. Scheme of the glow discharge treatment setup. (b) Gate dependence of the estimated Coulomb scattering mobility before and after the glow discharge test. The dashed lines indicate e-h puddle regime in which the mobility cannot be estimated.

Fig. 1(b). Here the Coulomb scattering mobility is estimated as $\mu_C = 1/(\rho - \rho_S)ne$ where $n \sim 7.4 \times 10^{10}|V_g - V_D|\,\mathrm{cm}^{-2}$ is the carrier density, V_D the gate voltage at maximum resistivity. and ρ_s is the short range scattering resistivity which is chosen so that the Coulomb scattering mobility μ_C is roughly independent of V_g outside the e-h puddle regime.[2,13] It is clear that irradiation with the glow discharge introduces both Coulomb and short-range scatterers: for the sample in Fig. 1, the short-range resistivity increases from 350 C to 1500 C, and the Coulomb scattering mobility decreased from 6000 cm^2/Vs to 600 cm^2/Vs.

For graphene deposited on SiO_2 substrates, the typical observed maximum resistivity is $\rho_{\max} \leq h/4e^2$. This is also the case for the sample shown here before the glow discharge treatment. After the irradiation, however the maximum resistivity became significantly larger than $h/4e^2$. Although the glow discharge caused damage to the material to an extent that is unlikely to appear in the commercially available high quality graphite, the test here suggests that the maximum resistivity in the graphene is not universal, but depends on material quality. In the extreme case of graphene with high defect density, the resistivity will consistently exceed that of high quality graphene throughout the entire doping range including the Dirac point.

Next, we discuss the impact of the lithographic process. Again, we start with pre-measured freshly made graphene devices. Subsequently the graphene is subjected to two lithographical processes: e-beam resist (PMMA) coating and Acetone stripping. Following every step of the treatment, the gate dependence of the resistivity is measured and compared to the pre-treatment measurement as illustrated in Fig. 2(a). We note that the PMMA coat shifts the Dirac point to negative gate voltage, indicating electron doping. Further analysis of the $\rho(V_g)$ curve suggests that the decrease in mobility associated with the PMMA coating is rather small ($\sim$20%) and can be attributed to Coulomb scattering. After the PMMA is removed with Acetone, the position of the Dirac point almost recovers to zero gate voltage and

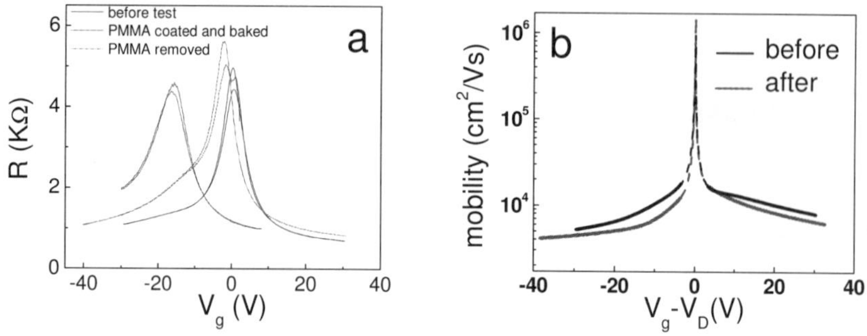

Fig. 2. (a) Gate voltage dependence of the resistivity at different stages of the lithography process, indicated by the legends. (b) Gate voltage dependence of mobility before the PMMA coating test and after the PMMA removal.

the entire $\rho(V_g)$ curve regains its pre-treatment shape with a slightly reduced mobility. These tests suggest that standard lithographic procedures do not significantly reduce device quality.

We next discuss the effect of contact leads. We focused on short aluminum-graphene-aluminum junctions with voltage leads that run across the sample width — 2-lead geometry, measuring the $\rho(V_g)$ curves as a function of lead separation. For lead separation shorter than 1 μm, we observe clear particle-hole asymmetry in the $\rho(V_g)$ curves. For the shortest junctions, $L = 300$ nm, the asymmetry is most pronounced we observe a hump forming on the hole branch. The asymmetry is not observed in samples with Hall bar lead geometry where the voltage leads do not run across the sample, or in samples with long channel length ($L \gg 1$ μm). This suggests that the asymmetry is not intrinsic but rather that it is associated with the invasiveness of the contact leads. A possible explanation is that the asymmetry arises from doping by the metallic leads resulting in a locally altered Fermi energy relative to the Dirac point. This scenario can be modeled by considering the graphene channel as consisting of 3 sections: 2 doped sections near the contacts flanking one undoped graphene section. For long junctions where the center section is much longer than the doped sections, the undoped part of the channel dominates the total resistance, and the measured 2-terminal $\rho(V_g)$ dependence is roughly symmetric. As the junction becomes shorter, the contributions from the contact-doped sections become more important. The different positions of the Dirac point in the lead segments is different and in the undoped segment will thus cause an asymmetry in the $\rho(V_g)$ curve arising from the 2 Dirac points. For the central segment the Dirac point is rather sharp and sits at $V_g \sim 0$, while close to the leads it is broader (due to the variation of the Dirac point energy as a function of the distance from the leads) and its position is shifted towards large negative values of V_g (electron doped).

In order to understand the effect of aluminum contacts on the Fermi energy in graphene we coated a freshly made graphene device with a partially oxidized aluminum oxide film. We used e-beam evaporation of aluminum in oxygen envi-

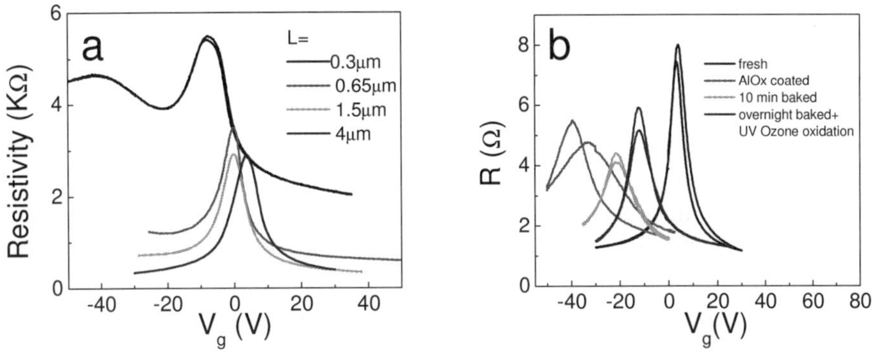

Fig. 3. (a) Gate voltage dependence of resistivity for various samples with different channel length as indicated in the legends. (b) Resistance as a function of gate voltage for a graphene device. The device was measured before AlOx coating (black). Then it was coated with incompletely oxidized AlOx, baked in oxygen for 10 min (100 C), and finally further oxidized by overnight oxygen baking and ozone oxidation. At the end of each step, the sample was measured to characterize its quality.

ronment, keeping the graphene sample in a local oxygen pressure of 1e-4 Torr. The aluminum was evaporated at a rate of ~ 0.3 Å/sec, which was mostly not yet fully oxidized when coated onto the graphene device. Aluminum oxide films grown by this method were characterized and controlled to be transparent and insulating below its breakdown field. Yet incompletely oxidized aluminum particles, which are still present in these films, act as dopants. Figure 3(b) shows the $\rho(V_g)$ dependence of the graphene device before aluminum oxide coating, after the aluminum oxide coating, and after 2 steps of further oxidation of the aluminum oxide film. After the initial aluminum oxide coating we observe a large shift of the maximum in the $\rho(V_g)$ curve to negative gate voltages indicating electron doping. The hysteresis is a result of un-settled charges in the aluminum oxide film. This doping effect can be attributed to incompletely oxidized particles, as confirmed by reduction of the effect upon further oxidation of the aluminum oxide film. This test indicates that contact with the aluminum dopes introduces electron doping in graphene. However, in order to understand this doping effect quantitatively further experimental and theoretical studies are needed.

Next we discuss how the SiO_2 substrates affect the quality of graphene devices. The deterioration of the transport properties of graphene in the presence of the Si/SiO_2 substrate is primarily due to trapped charges either within the SiO_2 substrate or at the interface between graphene and SiO_2. The resulting charge inhomogeneity leads to enhanced long range Coulomb scattering[14] which becomes especially important near the neutrality point where screening is poor. In addition, the atomic roughness of the substrate introduces short range scattering centers and may contribute to quench-condensation of ripples within the graphene layer.[15] To study the effect of the substrate on the quality of graphene, we developed techniques for fabricating suspended graphene (SG) devices with multiple contacts, which allowed transport measurements for sample characterization. The SG devices were fabricated from conventional non-suspended graphene (NSG) devices with Au/Ti leads deposited on Si/SiO_2 (300 nm) substrates. The freshly prepared NSG device is coated with PMMA followed by an additional e-beam lithography step to open two small windows (typically $0.2 \sim 0.5$ μm squares) in the PMMA on the two sides of the graphene channel (illustrated in Fig. 1). The samples were then immersed in 7:1 (NH_4F: HF) buffered oxide etch (BOE). Etching was done at 25°C for 6.5 min. Due to weak coupling between graphene and the substrate, capillary action draws the etchant underneath the whole graphene film. Therefore, etching actually starts in the entire graphene channel shortly after the sample is immersed. Because the etching is isotropic it causes the whole device (graphene together with the leads attached to it) to become suspended. When the etching is complete, the etchant is replaced by DI water, followed by hot acetone (to remove the PMMA) and finally hot isopropanol, with the sample remaining in the liquid at all times. Finally, the sample is taken out of the isopropanol and left to dry. At this point one would expect that the very fragile suspended device would be destroyed by wicking of the liquid as it evaporates. However, due to the small surface tension of hot isopropanol, devices with channel length smaller than 1 μm were found to survive the process with a high success rate. Figure 4 shows the SEM image of an actual SG device. The SG samples were baked in forming gas (Ar/H_2) at 200°C for 1 hour to remove any remaining organic residue and water molecules. In order to avoid contamination this step was promptly followed by transfer to the low pressure measurement cell.

To characterize the relation between the gate voltage and the induced carrier density, low temperature magneto-transport measurements were carried out. The carrier density at applied gate voltages were related to the Shubnikov-deHaas (ShdH) oscillations by $n = (4e/h)B_F(V_g)$, where B_F is the frequency of the ShdH oscillation. Here we found from the $n(V_g)$ relation that the dielectric constant of

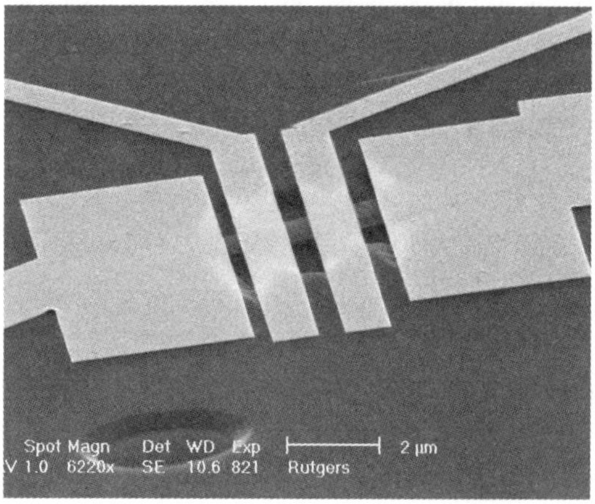

Fig. 4. SEM image of a suspended graphene device.

the gate dielectric is close to 1 (as is for vacuum), indicating complete removal of the SiO_2 underneath the graphene channel.

To characterize the quality of the SG devices, we focused on the carrier density dependence of the resistivity, $\rho(n)$, in zero magnetic field in the temperature range 4.2 K to 250 K [Fig. 5(a)]. Compared to NSG devices with similar geometry, the SG samples show much sharper carrier density dependence of the resistivity around the DP at low temperatures. At the lowest temperature, 4.2 K, the hole branch half width at half maximum (HWHM) is $\delta V_g \sim 0.15$ V, $\delta n \sim 3.2\,10^9 \text{cm}^{-2}$ for the sample with channel length $L = 0.5$ μm, is almost one order of magnitude narrower than that of the best NSG samples published so far.[4,16] This is directly

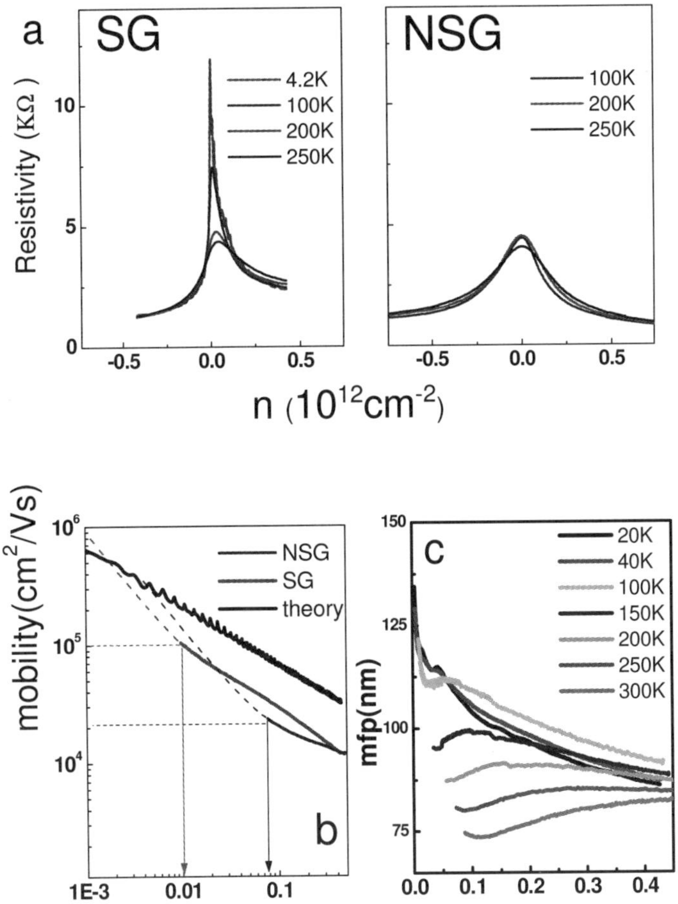

Fig. 5. (a) Side-bt-side comparison between carrier density dependence of the resistivity at various temperatures for NSG and SG devices with 500 nm channel length. (b) Carrier density dependence fo mobility for 500 nm NSG and SG at $T = 100$ K compared with calculated values for a ballistic graphene device with the same geometry. (c) Carrier density dependence of mean free path for SG device at various temperatures.

seen [Fig. 5(b)] in the side-by-side comparison of the $\rho(n)$ curves for SG and NSG samples of the same size and taken from the same graphite crystal. The HWHM of the NSG sample shown here is ?$\delta V_g \sim 3$ V ($\delta n \sim 2.2\,10^{11}$cm^{-2}). The sharp carrier density dependence of the resistivity is a direct consequence of the greatly reduced potential fluctuations at the DP. These potential fluctuations induce electron-hole puddles and broaden the gate (carrier density) dependence of the resistivity at the charge neutrality point clearly seen in the NSG device. The reduction of the potential fluctuation and electron-hole puddles also results in the strong temperature dependence of the maximum resistivity at the DP in the SG samples. This is in stark contrast to NSG samples, where the maximum resistivity saturates below ~20 0K,[17] because the residual carrier population in the electron-hole puddles induced by the potential fluctuations is much larger than the thermally activated carriers at these temperatures

A direct consequence of the low level of charge inhomogeneity in the SG samples is that one can follow the intrinsic transport properties of Dirac fermions much closer to the DP than is possible with any NSG samples fabricated to date. In Fig. 5(c) we compare the carrier density dependence of mobility $\mu = \sigma/ne$ for SG and NSG samples to that of the calculated mobility in a ballistic device. For $T < 100$ K, at low carrier densities (just outside the puddle regime) the maximum mobility of the SG samples exceeds 100,000 cm^2/Vs. compared to ~2,000–20,000 cm^2/Vs in the best NSG samples. We note that the SG mobility is approaching the calculated value of ideal ballistic devices. Since at low densities the mobility is mostly determined by Coulomb scattering[18] (short range scattering is very weak near the DP due to the small density of states[19]), the difference in mobility between the SG and NSG samples is naturally attributed to substrate-induced charge inhomogeneity. The removal of the substrate in the SG samples eliminates the primary source of Coulomb scattering, the trapped charges. At high carrier densities ($n > 4 \times 10^{11}$ cm^{-2}), the mobility in the two types of samples becomes comparable (~10000 cm^2/Vs) indicating that short range scattering becomes dominant. The short range scattering can be attributed to imperfections in the graphene layer reflecting defects in the parent graphite crystal or could be introduced during the fabrication process. Both sources of defects can in principle be reduced to produce SG samples with even better quality.

Figure 5(c) illustrates the density dependence of the mean free path $mfp = \sigma h/2e^2 k_F$ for the SG sample at the indicated temperatures. The negative slope and absence of T dependence for $T < 100$ K, suggest that scattering is predominantly by short range scatterers. For $T > 100$ K, the slopes become increasingly positive, suggesting thermally induced long range scattering. Such long range scattering cannot be attributed to charged impurities because such mechanism is expected to be independent of T.[17] Possible explanations include scattering by thermally excited ripples[15] and ripple induced charge inhomogeneity.[20] However, more work is needed to understand the scattering mechanism in this regime.

3. Discussion

We studied the effect of extrinsic factors on the quality of the graphene devices including material quality, lithography process, metallic leads and substrate. This work suggests that material defects in graphite can reduce the mobility and possibly the minimum conductivity. Higher mobility can be achieved by using better quality graphite. The lithographic process leaves a polymer residuewhich may induce Coulomb scattering. However, our tests suggested that the effect of the residue is minor and that it can in principle be eliminated by removing the residue with an appropriate cleaning method, such as baking in forming gas or UHV baking. The contact leads can play an important role and can modify the transport properties of short graphene junctions. Further study needs to be carried out to fully understand this effect and to find the most favorable lead geometry and material.

Our work suggests that the most important limitation on the quality of graphene devices is imposed by trapped charges associated with the substrate. These could be charged scatterers trapped inside the SiO_2 dielectric or at the graphene-substrate interface. The former may be reduced by pre-treatment of the substrate (annealing, for example) before the graphene deposition. The latter may be the most important reason for the observed large variations in device quality, because graphene deposition, hence the trapped charged scatterers (from water molecules, organic contamination, etc.) between the deposited graphene and the substrate, vary greatly from device to device. To reduce these effects it is important that the deposition be done in a clean and well controlled manner. It is also important to notice that even though the SG devices show excellent quality at low temperatures, their room temperature characteristics appear to be similar to that of the NSG devices, possibly due to the thermally induced corrugation (ripples) scattering. Therefore, in order to fabricate high quality room temperature graphene devices, it is still necessary to have a substrate in order to suppress the ripples.

In summary, the work described here indicates a number of necessary conditions that are needed to achieve high quality graphene devices reproducibly. (a) High quality single crystal graphite should be used for extracting graphene layers. (b) Trapped charges in the gate dielectric have to be eliminated — this could in principle be accomplished by a proper choice a dielectric substrate, but more work is needed to identify such substrtae. (c) Trapped impurities at the graphene-substrate interface, such water and organic residue, should be avoided by carrying out the deposition in a controlled and clean environment. We expect that devices fabricated under these guidelines will exhibit many novel physical properties that are expected to emerge when the motion of Dirac fermion is ballistic.

Acknowledgments

We thank G. Li, Z. Chen for discussions; S. W. Cheong and M. Gershenson for use of AFM and e-beam; V. Kiryukhin for HOPG crystals; A. H. Castr-Neto and F.

Guinea for useful discussions. Work supported by DOE, Office of the Army and ICAM.

References

1. K. S. Novoselov, A. K. Geim, S. V. Morozov *et al.*, *Science* **306**, 666 (2004).
2. K. S. Novoselov, A. K. Geim, S. V. Morozov *et al.*, *Nature* **438**, 197 (2005).
3. C. Berger, Z. Song, T. Li *et al.*, *J. Phys. Chem. B* **108**, 19912 (2004).
4. A. K. Geim and K. S. Novoselov, *Nature Mater.* **6**, 183 (2007).
5. A. H. Castro Neto, F. Guinea, N. M. R. Peres *et al.*, *Rev. Mod. Phys.* (to appear) (2007).
6. V. V. Cheianov, V. Fal'ko and B. L. Altshuter, *Science* **315**, 1252 (2007).
7. J. C. Cuevas and A. L. Yeyati, *Phys. Rev. B* **74**, 180501 (2006).
8. C. W. J. Beenakker, *Phys. Rev. Lett.* **97**, 067007 (2006).
9. J. Tworzydlo, B. Trauzettel, M. Titov *et al.*, *Phys. Rev. Lett.* **96**, 246802 (2006).
10. J. Yan, E. A. Henriksen, P. Kim *et al.*, arXiv:0712.3879v1 (2007).
11. X. Du, I. Skachko and E. Y. Andrei, eprint arXiv: 0710.4984 (2007).
12. L. DiCarlo, J. R. Williams, Y. Zhang *et al.*, eprint arXiv: 0711.3206 (2007).
13. J. Martin, N. Akerman, G. Ulbricht *et al.*, *Nature Phys.* **4**, 144 (2008).
14. S. Cho and M. S. Fuhrer, arXiv:0705.3239v2 (2007).
15. M. I. Katsnelson and A. K. Geim, *Phil. Trans. Roy. Soc. A* **366**, 195 (2008).
16. Y. W. Tan, Y. Zhang, K. Bolotin *et al.*, *Phys. Rev. Lett.* **99**, 246803 (2007).
17. S. V. Morozov, K. S. Novoselov, M. I. Katsnelson *et al.*, *Phys. Rev. Lett.* **100**, 016602 (2008).
18. E. H. Hwang, S. Adam and S. D. Sarma, *Phys. Rev. Lett.* **98**, 186806 (2007).
19. T. Stauber, N. M. R. Peres and F. Guinea, eprint arXiv: 0707.3004 (2007).
20. L. Brey and J. J. Palacios, *Phys. Rev. B* **77**, 041403 (2008).

EMPIRICAL ASPECTS OF STATISTICAL MECHANICS' AXIOMATICS

A. PLASTINO and E. M. F. CURADO

It is here shown how to use pieces of macroscopic thermodynamics to generate microscopic probability distributions for generalized ensembles, thereby directly connecting macro-state-axiomatics with microscopic results.

Keywords: Axiomatics for statistical mechanics; maximum entropy; thermodynamics.

1. Introduction

We "rationally understand" some physical problem when we are able to place it within the scope and context of a specific theory.[1] In turn, we have a theory when we can derive all the interesting results from a small set of axioms.[1] Examples: von Neumann's axioms for Quantum Mechanics, Maxwell's equations for Electromagnetism, Euclid axioms for classical geometry, etc.

Boltzmann's main goal in inventing Statistical Mechanics (SM) was to explain thermodynamics.[1,2] The first SM theory was that of Gibbs, on the basis of four ensemble related postulates.[1–5] The other great SM theory is that of Jaynes' based upon the axiom (derived from information theory[6]) that ignorance is to be extremized (with suitable constraints).[6]

Thermodynamics itself has also been axiomatized using four macroscopic postulates.[7] The axioms of these three theories (i.e., two for SM plus Thermodynamics) belong to different worlds altogether. This is quite right from an epistemological viewpoint, but a question lingers: would it be possible to have a SM derived from axioms that speak the same language as that of thermodynamics? To what extent is this possible? We intend to provide answers below. We start by providing a brief review of the concomitant three theories.

2. Brief Thermodynamics' Sketch

Thermodynamics can be thought of as a formal logical structure whose *axioms* are empirical facts.[7] The ones below are equivalent to the celebrated three laws.[7]

- Associate unique internal energy E value to each system's state. Work is the energy difference $E_1 - E_2$ surmounted in taking the system from state 1 to state 2 while adiabatically enclosed.
- Equilibrium states are uniquely determined by E and a set of extensive parameters A_i, $i = 1, \ldots, n$.

- For every system there exists a state function S, a function of E and the A_i that always grows if internal constraints are removed. S is a monotonous (growing) function of E.
- Both S and $[\frac{\partial E}{\partial S}]_{A_1,\ldots,A_n}$ vanish for the state of minimum energy and are ≥ 0 for all other states.

If we know $S(E, A_1, \ldots, A_n)$ (or, equivalently because of monotonicity, $E(S, A_1, \ldots, A_n)$) we have a *complete* thermodynamic description of a system. It is often experimentally more convenient to work with intensive variables.

Let define $S \equiv A_0$. The intensive variable associated to the extensive A_i is:

$$P_j = \left[\frac{\partial E}{\partial A_j}\right]_{S,A_1,\ldots,A_{j-1},A_{j+1},\ldots,A_n}$$

$$T \equiv P_0 = \left[\frac{\partial E}{\partial S}\right]_{A_1,\ldots,A_n}.$$

Any one of the Legendre transforms that replaces any s extensive variables by their associated intensive ones

$$L_{r_1,\ldots,r_s} = E - \sum_j P_j\, A_j, \quad (j = r_1, \ldots, r_s)$$

contains the same information as either S or E. The transform $L_{r_1,\ldots,r_s}$ is a function of $n-s$ extensive and s intensive variables. This is called the Legendre invariant structure of thermodynamics.

Let us have a closer look to the Second Postulate stating that equilibrium states are uniquely determined by E and a set of extensive parameters A_i, $i = 1, \ldots, n$. It follows that:

$$dE = \sum_j \frac{\partial E}{\partial A_j}\, dA_j. \tag{1}$$

The third Postulate, in turn, can be decomposed into three statements: 3a) For every system there exists a state function S, a function of E and the $A_i \Rightarrow S = S(E, A_1, \ldots, A_M)$. 3b) S always grows if internal constraints are removed. 3c) S is a monotonous (growing) function of $E \Rightarrow E = E(S, A_1, \ldots, A_M) \Rightarrow$ [cf. Eq. (1)]

$$dE = \sum_j \frac{\partial E}{\partial S}\, dS + \frac{\partial E}{\partial A_j}\, dA_j \tag{2}$$

$$dE = TdS + \sum_j \frac{\partial E}{\partial A_j}\, dA_j = TdS + \sum_j P_j dA_j. \tag{3}$$

Equation (3) is the combination second axiom plus the 3a) and 3c)-subaxioms, that are *existence* subaxioms, while 3b) is a physical one (telling what happens if you do something).

3. Gibbs' Statistical Mechanics

SM constructs microscopic realizations of Thermodynamics, that some people call mechanical analogues. The first complete formal such theory is due to Gibbs (1902),[5] based on the notion of *ensemble.* In the 1970s, Jaynes reformulated statistical mechanics on the basis of Information Theory.[6] Gibbs' 1902-ensemble postulates can be stated as follows[5]:

1) The physical probability p of finding our classical system at a small neighborhood of the phase-space point G at time t equals the mathematical probability P of finding that system at that same phase-space neighborhood, *within the ensemble*, at that time.

2) A priory one has equal probabilities for phase-space points.

3) P depends only on E.

4) This dependence is of exponential form.

4. Information Theory

On the basis of four postulates, Information Theory (IT) associates a degree of knowledge (or ignorance) to any normalized probability distribution $p(i)$, $(i = 1,\ldots,N)$, determined by a functional of the $\{p_i\}$ called the information measure I.[6,8] IT's first axiom states that I is a function only of the $p(i)$. The second axioms tells us that I is an absolute maximum for the uniform probability distribution. The third IT-postulate states that I is not modified if an $N+1$ event of probability zero is added. Finally, the last IT-axiom establishes a composition rule,[8] that refers to two given independent sub-systems $[\Sigma^1,\ \{p^1(i)\}]$ and $[\Sigma^2,\ \{p^2(j)\}]$ of $[\Sigma,\ \{p(i,j)\}]$ such that

$$p(i,j) = p^1(i)\,p^2(j).$$

Consider the conditional probability distribution (PD) $Q(j|i)$ of the event j in system 2 for fixed i in system 1. To this PD one associates the information measure $I[Q]$. Clearly, $p(i,j) = p^1(i)\,Q(j|i)$ and the appropriate composition axiom writes $I(p) = I(p^1) + \sum_i p^1(i)\,I\big(Q(j|i)\big)$. Out of the 4 axioms one deduces the celebrated Shannon's information measure, the only one that satisfies them[8]:

$$I_S\{p(i)\} = -\sum_{i=1}^{N} p(i)\ln\left[p(i)\right].$$

5. Jaynes Reformulation of Statistical Mechanics

Statistical Mechanics and thereby Thermodynamics can be formulated on the basis of Information Theory if the statistical operator $\hat{\rho}$ is obtained by recourse to

the MAXIMUM ENTROPY PRINCIPLE (MaxEnt).[6] It can be formulated in the following fashion:
Assume your prior knowledge about the system is given by the values of M expectation values $< R_1 >, \ldots, < R_M >$ of M quantum operators $\mathcal{A}_j$. Then $\hat{\rho}$ is uniquely fixed by extremizing $I(\hat{\rho})$ subject to the constraints given by the M conditions

$$< R_j >= Tr[\hat{\rho}\,\hat{A}_j], \tag{4}$$

(entailing the introduction of M associated Lagrange multipliers λ_i). The extremizing process yields a unique measure $I_{maximal}$. In the process one discovers that $I \equiv S$, the equilibrium entropy, if our prior knowledge $< R_1 >, \ldots, < R_M >$ refers to extensive quantities. $I_{maximal}$, once determined, *yields complete thermodynamical information with respect to the system of interest.*[6]

6. Our New "Thermal-Flavored" Axiomatics for Statistical Mechanics

We first establish that the extensive thermal quantities A_j of Sec. 2 have as quantum counterparts the M operators $\mathcal{A}_j$ of the preceding section and advance afterwards the following two new axioms for statistical mechanics:
First axiom:

$$dE = TdS + \sum_j P_j dR_j. \tag{5}$$

Second axiom: If there are W microscopic accessible states labeled by i, of microscopic probability p_i, then

$$S_{E;\forall R_j} = S_{E;\forall R_j}(p_1, p_2, \ldots, p_W). \tag{6}$$

(Note: this is just Boltzmann's conjecture, pure and simple![9])

We will show below that these two axioms are equivalent to MaxEnt. How do we prove it? Consider a generic change $p_i \to p_i + dp_i$ constrained by (5). Obviously, S, R_j, and E will change accordingly.
We need the following ingredients: an *arbitrary smooth function $f(p_i)$, such that $p_i f(p_i)$ is a concave function, and, say, n quantities R_ν that represent mean values of extensive physical quantities $\langle \mathcal{A}_\nu \rangle$, that take, for the micro-state i, the value a_i^ν with probability p_i*. Also another arbitrary smooth, monotonous function $g(p_i)$ ($g(0) = 0$; $g(1) = 1$). The process $p_i \to p_i + dp_i$ generates corresponding changes dS, dR_ν, and dE in, respectively, S, the R_ν, and E, for

$$S = \sum_i p_i f(p_i);\; R_\nu \equiv \langle \mathcal{A}_\nu \rangle,$$

$$E = \sum_i \epsilon_i\, g(p_i);\; R_\nu = \sum_i a_i^\nu\, g(p_i).$$

$$\sum_i dp_i = 0.$$

Notice that the essential point that we are here introducing is that of enforcing *obedience to* the relation

$$dE - TdS + \sum_{\nu=1}^{n} dA_\nu \lambda_\nu = 0, \tag{7}$$

with T the temperature. The above process leads, up to first order in the dp_i, straightforwardly to three relations[10,11]

$$\begin{aligned} \textstyle\sum_i [C_i^{(1)} + C_i^{(2)}] dp_i &= \textstyle\sum K_i dp_i = 0 \\ C_i^{(1)} &= [\textstyle\sum_{\nu=1}^{M} \lambda_\nu \, a_i^\nu + \epsilon_i] \, g'(p_i) \\ C_i^{(2)} &= -kT[f(p_i) + p_i \, f'(p_i)], \end{aligned} \tag{8}$$

where primes denote p_i-derivatives. From Eqs. (8) one deduces one and just one probability distribution p_i.[10,11]

Alternatively, extremize S with the constraints in E, A_ν and normalization. The ensuing variational equation can be easily seen to yield[10,11] the three Eqs. (8) as well! The equivalence between MaxEnt and our two new axioms above is thereby proved.

7. Conclusion

The set of equations (8), derived from our two axioms, yields the same p_i-expression that Jaynes' MaxEnt's axiomatics for statistical mechanics. MaxEnt is thus seen to be equivalent to:

(1) Boltzmann's conjecture of an underlying microstate PD plus

(2): $dE = TdS + \sum_j P_j dA_j$.

These last two postulates are intuitively intelligible, which is NOT the case neither of Gibbs' ensemble ones nor of Jaynes' extremizing of the Observer's ignorance.

References

1. R. B. Lindsay and H. Margenau, *Foundations of Physics* (Dover, NY, 1957).
2. R.K. Pathria, *Statistical Mechanics* (Pergamon Press, Exeter, 1993); H. B. Callen, *Thermodynamics* (J. Wiley, NY, 1960).
3. F. Reif, *Statistical and Thermal Physics* (McGraw-Hill, NY, 1965).
4. J. J. Sakurai, *Modern Quantum Mechanics* (Benjamin, Menlo Park, Ca., 1985).
5. J. Willard Gibbs, *Elementary Principles in Statistical Mechanics* (New Haven, Yale University Press, 1902).
6. E. T. Jaynes, *Phys. Rev.* **106**, 620 (1957); **108Statistics and Statistical Physics**, ed. R. D. Rosenkrantz (Reidel, Dordrecht, Boston, 1987); A. Katz, *Principles of Statistical Mechanics, The Information Theory Approach* (Freeman and Co., San Francisco, 1967); A. Plastino and A. R. Plastino, *Braz. J. of Phys.* **29**, 50 (1999).
7. E. A. Desloge, *Thermal Physics* (NY, Holt, Rhinehart and Winston, 1968).

8. T. M. Cover and J. A. Thomas, *Elements of Information Theory* (J. Wiley, NY, 1991).
9. D. Lindley, *Boltzmann's Atom* (The Free Press, NY, 2001).
10. E. Curado and A. Plastino, *Phys. Rev. E* **72**, 047103–107 (2005).
11. E. Curado and A. Plastino, *Physica A* **365**, 24–32 (2006).

FERROMAGNETISM IN DOPED SEMICONDUCTORS WITHOUT MAGNETIC IONS

R. N. BHATT

Department of Electrical Engineering, Princeton University, and Princeton Center for Theoretical Science, Jadwin Hall, Princeton, NJ 08544 USA

ERIK NIELSEN

Department of Electrical Engineering, Princeton University, Princeton, NJ 08544 USA

Received 12 May 2008

While ferromagnetism has been obtained above 100 K in doped semiconductors with magnetic ions such as $Ga_{1-x}Mn_xAs$, bulk doped semiconductors in the absence of magnetic ions have shown no tendency towards ferromagnetism. We re-examine the non-magnetic doped semiconductor system at low carrier densities in terms of a generalized Hubbard model. Using exact diagonalization of the many-body Hamiltonian for finite clusters, we find that the system exhibits significant ferromagnetic tendencies at nanoscales, in a region of parameter space not accessible to bulk systems, but achievable in quantum dots and heterostructures. Implications for studying these effects in experimentally realizable systems, as well as the possibility of true (macroscopic) ferromagnetism in these systems is discussed.

Keywords: Disordered Hubbard model; Nagaoka ferromagnetism; shallow impurities; quantum dots.

1. Introduction

Doped semiconductors offer a versatile system for the study of the effects of electron correlation and disorder in a controlled manner. The fact that shallow dopants give rise to a new length, the Bohr radius of the bound carrier, which is unrelated to the interatomic distances in the host lattice, and can indeed be very different, adds to this versatility.

For the physics arising from the dopants, the host semiconductor, with its band gap, and lack of low energy excitations, just plays the role of a vacuum. In the dilute limit, at low temperatures, the dopant carriers can be described in terms of generalized hydrogenic wavefunctions, centered at the (random) sites of the dopant atoms.[1] A hydrogenic center in a semiconductor, like the hydrogen atom, is known to bind up to two electrons. The one electron case is described by the exactly soluble hydrogen atom problem, with a 1s ground state, decaying exponentially with distance, on the scale of a renormalized Bohr radius given by $a_B = \epsilon\hbar^2/m^*e^2$, and an effective Rydberg $\mathrm{Ry}^* = m^*e^4/2\epsilon^2\hbar^2$, where ϵ is the static dielectric constant of the host semiconductor and m^* the effective mass of the appropriate band (conduction band for n-type dopants, and valence band for p-type). With typical masses $m^* \sim 0.05$ to 0.5 times the free electron mass, and $\epsilon \sim 10-20$, the Bohr radius a_B ranges from $10-500$ Å, and Ry* is typically in the meV to tens of meV range, much below the band gap of the host semiconductor. Realistic effects like mass anisotropy, band degeneracy etc. need to be included for quantitative calculations, but do not make essential differences to the phenomena we wish to describe.[2,3] Thus the physics of the impurity electrons is on a much lower energy scale, with the host semiconductor essentially acting as an inert medium (vacuum), as stated earlier.

The hydrogenic center with two electrons corresponds to the H^- ion. It has a ground state which is a spin singlet, with a binding energy of $\approx 0.0555\,\mathrm{Ry}^*$.[4] Thus the simples model to describe the center, must include an electron-electron repulsion term (an on-site "Hubbard U"), of an amount $U = 0.945\,\mathrm{Ry}^*$, in addition to the $1\,\mathrm{Ry}^*$ one-electron binding energy. This picture, which neglects all orbitally excited states of the hydrogenic center, is often applied to an ensemble of dopants as well, giving rise to a variant of the famous Hubbard model,[5,6] so often used in many-body physics to describe correlated electron systems, including high temperature superconductors,[7–9] vanadium oxides,[10,11] nickel sulphide-selenide alloys[12–14] and other Mott insulators, with or without doping.

With its putative wide applicability, the Hubbard model has been explored extensively since it was formulated forty-five years back.[5] While there are many aspects of the model, which serves as a paradigm for correlated electronic materials, which are of interest, including the nature of the excitations, density of states, spectral weight, transport, optical and magnetic behavior,[15–18] here we will concentrate on the ground state (i.e., $T = 0$) phase diagram, in particular the nature of magnetic correlations in the ground state. Further, we will concentrate on the system at low dopant densities, which in Hubbard parlance corresponds to large Hubbard U. (Even though the Hubbard U in our model is practically density independent, its effect depends on the ratio between U and the hopping integral between neighboring sites, denoted by t. t depends strongly, indeed exponentially, on density, because of the exponential, tight-binding like nature of the hydrogenic impurity wavefunctions). At these densities corresponding to large U/t, the doped semiconductor system is still insulating, i.e., we are at densities well below the Mott density n_c, which in three dimensions is given by the famous Mott criterion[19] $n_c^{1/3} a_B \approx 0.25$.

Most of the research on the Hubbard model is for a regular lattice, i.e., without disorder. Such a situation may be achieved in semiconductors by placing dopants on a superlattice, a capability that has only very recently been demonstrated.[20] Clearly, the details of the zero temperature magnetic phase diagram will depend on the lattice. The conceptually easiest case is that of bipartite lattices, for which the phenomenon of magnetic frustration is absent, at least for the canonical starting point of one electron per site (the so-called half-filled case), as exists for simple hypercubic lattices. (We will consider 2 dimensional square and 3 dimensional simple cubic lattices, for concreteness). In these cases, the ground state is that of a simple Neel antiferromagnet, with spins on the two sublattices pointing up and down respectively, so the spin arrangement alternates as you go along the principal crystallographic axes (x, y, and z — in the case of 3D). In the large U limit, this is easily understood as arising from the antiferromagnetic exchange $J \sim t^2/U$ between nearest neighbor sites, due to virtual hopping of electrons between the sites in question for antiparallel spins, but which is prohibited by Pauli principle for parallel spins (ferromagnetic configuration).

However, when one deviates from half-filling (one electron per site), thereby creating, for fillings slightly less than half, empty sites (holes), it becomes advantageous to have a ferromagnetic configuration of occupied sites, because the hole can hop from site to site without scrambling the (ferromagnetic) spin configuration. For an antiferromagnetic configuration, however, such a motion is always accompanied by an alteration in spin configuration of the lattice, leading to an increase in energy, which is disfavored. In the large U limit, the gain in hole kinetic energy due to delocalization is $\sim t$, whereas the loss of magnetic energy is $\sim J$, i.e., $\sim t^2/U$, so for large enough hole concentrations δ, so $t\delta > J$ (i.e., $\delta > t/U$), one would expect a magnetic transition from an antiferromagnetic to a ferromagnetic phase. The effect is especially pronounced at large U, since the hole density needed can be rather small. Indeed, one of the early results, due to Nagaoka,[21] was that for $U = \infty$, a single hole suffices to produce a ferromagnetic ground state, a result that goes by the name of Nagaoka Theorem.

Unfortunately, despite the existence of the Nagaoka Theorem, a Nagaoka ferromagnet has not been seen to-date in any experimental material. Most oxide and chacogenide systems, including the cuprate superconductors, do not have a large enough U/t, and finding a naturally occurring material with a large enough U/t seems not likely in the near future. This is where the versatility of doped semiconductors can be brought to bear. Ideally, one would like to perform measurements on artificially created dopant superlattices of the type recently reported, with several well thought out lattice structures most conducive to the Nagaoka phenomenon. However, until engineering such structures becomes routine, and indeed for the past four decades before such structures were realized, one has to make do with naturally forming doped semiconductors, as well as structures that are within the realm of realizable materials. In all such cases, the dopants are distributed randomly, adding what turns out to be an important ingredient — disorder — to the Hubbard-like description of the system.

There are issues with the Nagaoka ferromagnet on the theoretical front as well. Lieb and coworkers[22] have shown that the ground state is a singlet (i.e., zero spin, and hence clearly not a ferromagnet) in the one-dimensional Hubbard model with boundary conditions requiring the wavefunction or its derivative to vanish at the boundary, while recently Haerter and Shastry[23] showed that hopping of the opposite sign, an itinerant hole gives rise to antiferromagnetism, rather than ferromagnetism. Finally, several people have pointed out to the possibility of phase separation into carrier rich and carrier poor regions instead of a uniformly doped state, in the case of the Hubbard model *without* disorder. In such a case, the carrier rich state is seen as ferromagnetic (or ferrimagnetic), while the other phase is usually antiferromagnetic with zero or low carrier density. Arguments for such a separation on mesoscopic scales have also been advanced.[24]

It is not clear whether any of these results for uniform system are relevant for a Hubbard model with randomly positioned sites. First, to the extent that the itinerancy of the hole is dependent on the local lattice geometry, a random structure will necessarily show spatial inhomogeneity in the magnetic structure. A locally fluctuating carrier density is also expected, which may in fact wash out any distinction between phase separation at macroscopic and mesoscopic length scales for such systems. The question that then remains is whether the ground state has macroscopic spin degeneracy or not. In our work, we restrict this to an even more basic question – does there exist, even on the nanoscale, large spin degeneracy of clusters of hydrogenic centers, using an appropriate Hubbard-like description for the system in question?

Before we embark on details of our calculations, it is appropriate to review what is known in the case of randomly positioned centers. Indeed, following the result of Nagaoka, there were several investigations of the magnetic behavior of n-doped semiconductors at low doping (insulating phase), both uncompensated[3,25–28] (donors only, corresponding to one electron per site, or the half-filled case in Hubbard parlance), as well as compensated[29] (donors plus a smaller density of acceptors, which captured some of the donor electrons, and thus yielded a system with less than one electron per donor site, or hole-doped away from half-filling in Hubbard language), with the aim of seeing if the system became ferromagnetic at low temperatures. Around the same time, alternative theories[30,31] of ferromagnetism of Anderson localized electrons, were also proposed. Both theoretical scenarios notwithstanding, the experimental systems showed no sign of ferromagnetism down to millikelvin temperatures, though the nearest neighbor coupling though distributed broadly, was characterized by a median value of 1–10 K. Indeed, both the uncompensated and compensated systems exhibited a significantly lowered magnetic susceptibility compared to the high temperature paramagnetic Curie result, by factors of 10–50, indicating that the system was predominantly characterized by *antiferromagnetic* correlations, both at half-filling and away from it.

The understanding of what really happens in the half-filled case was provided by Bhatt and Lee,[32,33] who showed, using a perturbative renormalization group method appropriate for the large amount of disorder present in the actual system, that the system is best viewed as a valence-bond glass in which pairs of spins form singlets (see also Ref. 34) in a hierarchical fashion, resulting in what has been dubbed a valence-bond glass,[35,36] random singlet[37] or the Bhatt-Lee (BL) phase.[38,39] There is no long-range antiferromagnetic order, and the susceptibility remains strongly temperature dependent down to the lowest accessible temperatures ~10 mK, below which coupling to nuclear spins sets in, which is not included in our model description.

The change from a long-range ordered magnetic phase like the Neel antiferromagnet to the BL valence-bond glass with only short range magnetic order helps explain the absence of ferromagnetism with hole doping — the hole gets localized on one (or a few) valence bonds, and is unable to move long enough distances to disrupt the (now-local) magnetic arrangements, unlike the hole in the uniform antiferromagnet. In short, BL destroys the possibility of the Nagaoka ferromagnet for compensated semiconductors.

It may appear that disorder such as due to random positions of the centers "destroys" ferromagnetism. However, in the past decade, ferromagnetism has been realized in a family of diluted magnetic semiconductors, such as $Ga_{1-x}Mn_xAs$,[40,41] in which the transition metal atom (Mn in this case) acts as both a dopant (p-type) and a local moment (coming from the unfilled d-shell of the atom). These systems also have substantial disorder, both in position of the dopant atom and due to anti-site defects in the semiconductor itself (e.g. As on Ga sites). Thus, disorder by itself does not always destroy ferromagnetism, and may in some cases even enhance the ferromagnetic transition temperature.[42,43]

The big difference between the conventional "non-magnetic" doped semiconductor and the diluted magnetic semiconductor is in the presence of two different length scales in the latter – the Bohr radius of the Mn hole wavefunction is ~10 Å, whereas the spin on the Mn has an extent ~1–2 Å, so the hole wavefunction extends over several Mn spins, a phenomenon which only gets more accentuated as the hole delocalizes at higher Mn density x.

Using the standard Hubbard model representation (with its implicit electron-hole symmetry around half-filling) for hydrogenic centers in semiconductors would suggest that ferromagnetism is precluded for all parts of the phase diagram for randomly positioned centers. However, for hydrogenic centers, this argument fails to take into account the rather weak binding energy of the H^- ion, because of which the two-electron center has a much greater radius (by a factor of 3–4) than the neutral atom. As a consequence, an ensemble of hydrogenic centers, when represented by a Hubbard model, must properly include this asymmetry, at the very least, by requiring the hopping term between centers be occupation dependent, thereby destroying electron-hole symmetry of the standard Hubbard model around half-filling (one electron per site). The fact that the ratio of the two radii is large, and the hopping integral in the low density regime is exponentially dependent on the radius, gives rise to the hope that a doped semiconductor with filling of *more than one electron* per donor site, should be in a quite different regime of parameters than the conventional compensated semiconductor with *less than one electron* per site. Such a regime, while not obtainable in bulk doped semiconductors, should be realizable in semiconductor heterostructures, as well as quantum dots.

Another way to view this asymmetry is that for the more than half-filled case, the electronic wavefunctions are more extended; consequently, there is *less randomness* in the model, and greater hope for the Nagaoka phenomenon to take place. Here we describe our efforts to study a generalized Hubbard model, which takes into account this effect, and discuss the results obtained to-date, and its consequences for the doped semiconductor system, doped with standard, non-transition-metal, shallow dopants.

2. Model

Originally proposed in the early 1960s, the Hubbard model[5] combines tight binding hopping between nearest neighbors on a lattice with an on-site Coulomb repulsion between electrons in the same orbital. Though it is one of the simplest interacting models, its on-site intra-orbital correlations are believed to be the most important source of correlations in solids. the Hubbard model on a lattice of N sites is given by:

$$\mathcal{H} = -t \sum_{\langle i,j \rangle \sigma} \left(c_{i\sigma}^{\dagger} c_{j\sigma} + \text{h.c.} \right) + U \sum_{i=1}^{N_s} n_{i\uparrow} n_{i\downarrow} \tag{1}$$

where i and j range from 1 to N, and the first sum is over all distinct nearest neighbor pairs. The operators $c_{i\sigma}^{\dagger}$ and $c_{i\sigma}$ create and annihilate respectively an electron of spin $\sigma \in \{\uparrow, \downarrow\}$ on site i. The hopping amplitude t includes a minus sign, so that for the familiar example of the tight-binding model with hydrogenic wavefunctions,

$$t(r) = 2(1 + r/a_{\text{B}}) \exp(-r/a_{\text{B}}) \tag{2}$$

is positive.[44]

Mean field theory can be used to obtain a first approximation of the model's magnetic phase diagram. We use a self-consistent method to solve for the ground state energy and wavefunction of systems initialized in either a paramagnetic, antiferromagnetic, or ferromagnetic configurations. By comparing the resulting energies and spin-spin correlation functions, we obtain the system's zero temperature magnetic phase diagram. Our analysis does not include the possibility of phase separation,[24] which could substantially alter the simple phase diagram given here if it exists. However, based on the results of 10- and 16-site square lattices, Dagotto *et al.*,[45] argue that phase separation is generally absent in the Hubbard model, at least at short length scales.

As pointed out in the introduction, to correctly model hydrogenic centers we must account for the relatively weak binding of the H^- ion. Equation (1) does not account for this fundamental property, and the Hamiltonian must be modified. It is much easier for an electron on a doubly-occupied hydrogenic center to hop away to a neighboring site than it is for the electron on a singly-occupied site to make a similar hop. This implies that the hopping amplitude seen by an itinerant electron, hopping around in a background of singly-occupied (i.e., neutral hydrogen) sites, is larger than that seen by a hole in a similar background. In Hubbard parlance, near half-filling the hopping amplitude for an electron is much larger than for a hole. At the very least, the different radii of the doubly vs. singly occupied sites suggests that we modify Eq. (1) by adding occupation-dependence to the kinetic term:

$$\mathcal{H} = -\sum_{\langle i,j\rangle\sigma}\left(t(n_i,n_j)c^{\dagger}_{i\sigma}c_{j\sigma} + \text{h.c.}\right) + U\sum_i n_{i\uparrow}n_{i\downarrow} \tag{3}$$

where n_i is the total occupation of site i, and the hopping now has occupation dependence given by the piecewise function (and shown pictorially to the right):

$$t(n_i,n_j) = \begin{cases} \tilde{t} & (n_j = 2, n_i = 1) \\ t & \text{otherwise} \end{cases} \tag{4}$$

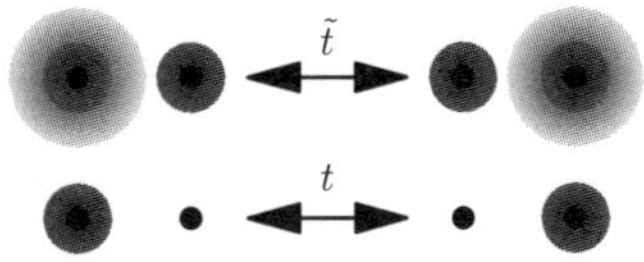

where $\tilde{t}$ is larger (and because of the exponential dependence of t on radius a_B [see Eq. 2], can be much larger) than t. In a random system, the kinetic hopping amplitude must also be site-dependent, specifically a function of the site separation: $t_{ij} = t\left(|\vec{r}_i - \vec{r}_j|\right)$. The resulting model, which we shall use in the remainder of the paper, is given by the Hamiltonian:

$$\mathcal{H} = -\sum_{i,j,\sigma}\left(t_{ij}(n_i,n_j)c^{\dagger}_{i\sigma}c_{j\sigma} + \text{h.c.}\right) + U\sum_i n_{i\uparrow}n_{i\downarrow} \tag{5}$$

where n_i is the total occupation of site i, and t_{ij} is given by:

$$t_{ij}(n_i,n_j) = \begin{cases} \tilde{t}_{ij} & (n_j = 2, n_i = 1) \\ t_{ij} & \text{otherwise} \end{cases} \tag{6}$$

where again $c^{\dagger}_{i\sigma}$ ($c_{i\sigma}$) is the electron creation (annihilation) operator on site i with spin σ, the t_{ij} and $\tilde{t}_{ij}$ are hopping amplitudes, and U is the strength of the on-site Coulomb repulsion.

3. Results

The original Hubbard model, Eq. (1), has been solved using a mean-field approximation for the simple square and cubic lattices to obtain a first approximation of the Hubbard model's zero temperature magnetic phase diagram. Figure 1 show these results, which agree qualitatively with the extensive work in two dimensions by Hirsch.[46] We are concerned with the ferromagnetic-antiferromagnetic transition at low doping and large U. At zero doping (half-filling), the system is an antiferromagnet for all values of U/t due to the effective exchange interaction (from an electron hopping to a neighboring site and back) and an absence of mobile carriers. As the

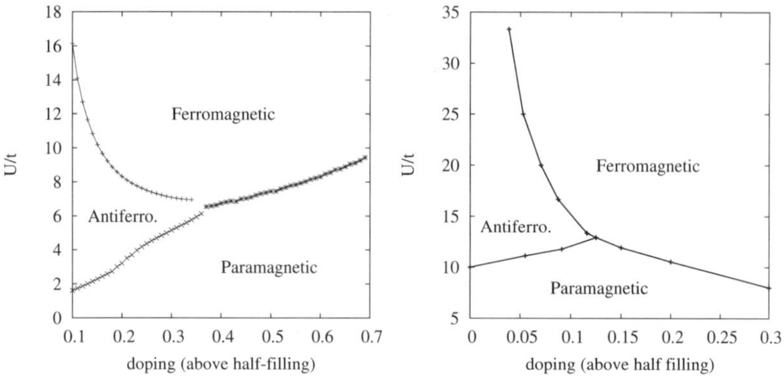

Fig. 1. Mean-field theory approximations of magnetic phase diagram for the Hubbard model on a 10×10 square lattice (left) and a $4 \times 4 \times 4$ simple cubic lattice (right).

doping is increased, we find that for large enough U/t ferromagnetism exists. We therefore expect in this regime a transition to a ground state which is ferromagnetic on some mesoscopic or macroscopic length scale (even though its precise location in phase space depends on the lattice structure and requires more careful work).

Moving beyond a mean-field analysis, the generalized Hamiltonian (5) is solved by exact diagonalization on finite lattices (with periodic boundary conditions) and small clusters. We use a symmetry-optimized Lanczos method, allowing the solution of systems with 10-16 sites. On finite square and honeycomb lattices, where there is a single $(t,\tilde{t})$ pair, we find that with one electron above half-filling and fixed $\tilde{t}/t$ a critical value of $t/U = (t/U)_c$ above which the ground state is ferromagnetic. As $\tilde{t}/t$ is increased we find that $(t/U)_c$ decreases and the region of ferromagnetism enlarges. This qualitative behavior is found for all sizes of square and honeycomb lattices (8, 10, and 16-site square lattices;[47] 6 and 10-site honeycomb lattices). We also find this behavior in these lattices for certain dopings of greater than one electron above half-filling, though no general trend is observed.

Next we consider finite clusters, defined as those systems possessing less symmetry than a finite lattice. This results in their having more hopping parameters, and makes their solution more computationally demanding. We have solved many small clusters which have two independent pairs of hopping amplitudes $(t_1, \tilde{t}_1)$ and $(t_2, \tilde{t}_2)$. Under the constraint $\tilde{t}_1/t_1 = \tilde{t}_2/t_2$ we determine the ground state spin of each cluster for $0.01 < t/U < 0.1$, $0 < t_2/t_1 < 1$, and $1 < \tilde{t}_i/t_i < 10$. Results for a selected group of clusters is shown in Fig. 2, which lists the maximal ground state spin for each cluster geometry when doped with up to two electrons or holes. Note that high-spin ground states appear for clusters doped with a single electron above half-filling.We will take a detailed look at two clusters which show behavior typical of the high-spin clusters observed.

The first is cluster 5 of Fig. 2: a 6-site cluster formed by linking together two triangles (also shown on the right in Fig. 3). The model corresponding to this clusters

#	Geometry	S^{2h}_{max}	S^{1h}_{max}	S^{1e}_{max}	S^{2e}_{max}
1	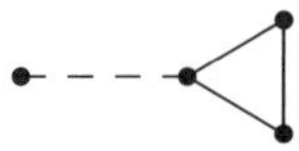	0	$\frac{1}{2}$	$\frac{1}{2}$	1
2		$\frac{1}{2}$	1	2	$\frac{1}{2}$
3		$\frac{1}{2}$	1	1	$\frac{1}{2}$
4	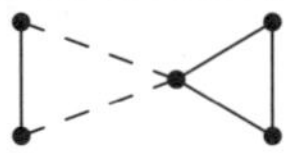	$\frac{1}{2}$	1	1	$\frac{1}{2}$
5	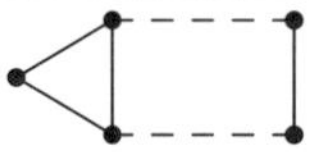	$\frac{1}{2}$	1	2	$\frac{1}{2}$
6		1	$\frac{1}{2}$	$\frac{1}{2}$	1
7	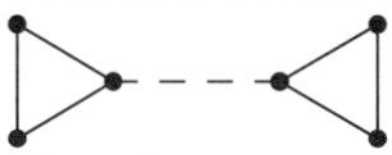	1	$\frac{1}{2}$	$\frac{3}{2}$	0
8	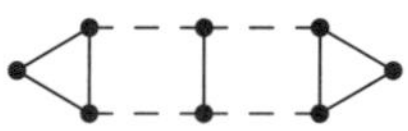	1	$\frac{3}{2}$	$\frac{7}{2}$	3

Fig. 2. Geometries and maximal ground state spin for a selected group of clusters. The maximal spin is computed over the parameter range $0.01 < t/U < 0.1$, $0 < t_2/t_1 < 1$, and $1 < \tilde{t}_i/t_i < 10$.

has two different hopping amplitude pairs, $t_1, \tilde{t}_1$ (solid lines) and $t_2, \tilde{t}_2$ (dashed lines). We fix U/t_1 and vary $\tilde{t}_1/t_1 = \tilde{t}_2/t_2$ and t_2/t_1 to produce the plot in Fig. 3 (where different lines indicate different values of U/t_1). For any U/t, the extent of the $S = \frac{5}{2}$ region increases as $\tilde{t}_i/t_i$ increases, reducing the size of the lower spin regions. Spin 5/2 is the largest spin possible in the Hubbard model with 6 sites and 7 electrons, so the $S = \frac{5}{2}$ region marks a fully polarized, completely ferromagnetic ground state. In the $U = 50, 100$ cases, the only lower-spin region is $S = \frac{3}{2}$ whereas when $U = 20$ at t_2/t_1 near 1 there exists a $S = \frac{1}{2}$ region. It is typical that stronger electron correlation (i.e., larger values of U/t) result in greater ground state spin polarization. In our study of many clusters we find some correlation between the number of triangles in a two-dimensional cluster and its ground state spin: larger numbers of triangles tend to yield greater ground state spin.

The second cluster can be thought of as a geometrical distortion of cluster 8 in Fig. 2. It is shown on the right in Fig. 4, and has three distinct t-parameter pairs: $t_1, \tilde{t}_1$ (solid lines) $t_2, \tilde{t}_2$ (dashed lines), and $t_3, \tilde{t}_3$ (dotted lines). We fix $U/t_1 = 20, 50, 100$ and $t_2/t_1 = 0.3$ and vary $\tilde{t}_i/t_i$ and t_3/t_1 to create the ground state spin diagram of Fig. 4. In this case there is a sharp transition from a ground state with

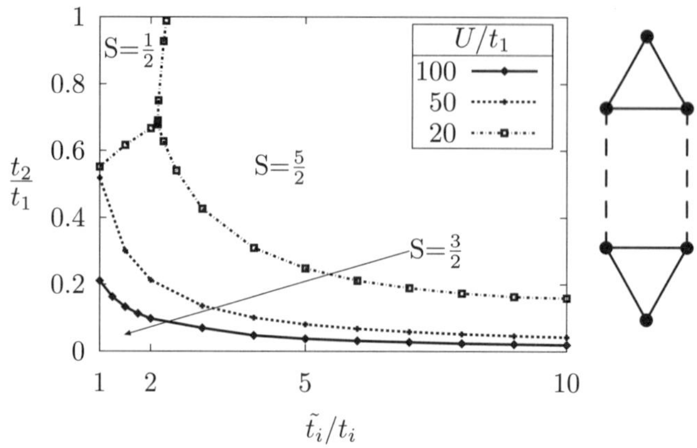

Fig. 3. Ground state spin diagram for two-triangle cluster shown to the right with 7 electrons. In the diagram on the right, solid lines denote t_1 hopping and dashed lines denote t_2 hopping.

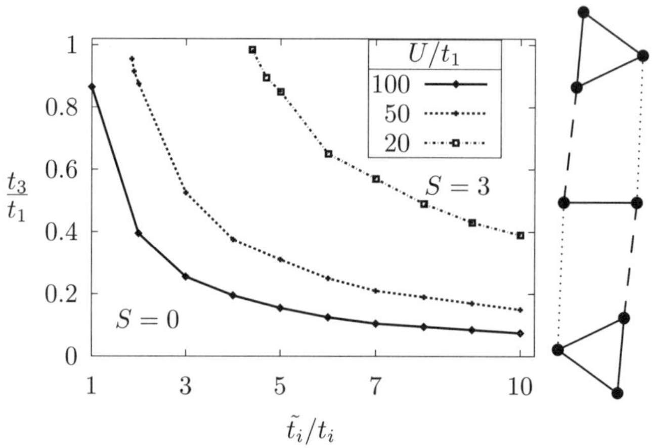

Fig. 4. Ground state spin diagram of cluster shown to the right with 10 electrons. $t_2/t_1 = 0.3$ In the diagram on the right, solid lines denote t_1 hopping, dashed lines denote t_2 hopping, and dotted lines denote t_3 hopping.

minimal spin $S = 0$ to one with maximal spin $S = 3$. As in the previous cluster, we see the ferromagnetic range of t_3/t_1 increase as either $\tilde{t}_i/t_i$ or U/t_1 increase.

4. Conclusion

We have motivated and analyzed a generalized Hubbard model which characterizes the inherent electron-hole asymmetry present in systems of hydrogenic centers (particularly semiconductors with shallow n-type dopants). We find in both finite lattices and clusters with large U/t, that as the strength of electron-hole asymmetry $\tilde{t}/t$ is increased there is a significant increase in the likelihood of a spin-polarized (Nagaoka) ground state if the system is doped *above half-filling*. Thus there is an important difference between doping a semiconductor above or below half-filling, and it follows to observe this Nagaoka ferromagnetism, experiments should be done by injecting additional electrons into the system (beyond those supplied by the donors). This is possible in doped quantum dots or in modulated structures with dopants in both quantum wells and barrier regions, but not in true bulk doped semiconductors. That ferromagnetism seen on small length scales in simulation at small electron doping and large U/t raises the question of its occurrence on mesoscopic or even macroscopic scales. As a first step towards answering this question, we have analyzed ensembles of clusters generated in a random poisson distributed system in two dimensions, as may be expected in heterostructures.[48] A full answer, however, requires going beyond the small sizes possible with exact diagonalization methods, using more approximate methods such as density matrix or perturbative renormalization group methods in combination with other numerical techniques. Nonetheless, we have shown that high spin states do occur frequently for "electron" doping (i.e., above half-filling) for these Hubbard models. Even if ferromagnetism exists only on small length scales, there should be a substantial asymmetry between the magnetic response of systems doped above half-filling and those doped below half-filling (traditional compensated). The former should have a larger susceptibility in the paramagnetic phase at low temperatures, a prediction which can be experimentally tested using a 2D heterostructure with n-type dopants in the wells, or n-doped quantum dots, using gates to tune the electron density.

Acknowledgments

This research was supported by NSF DMR-0213706.

References

1. W. Kohn, *Solid State Physics V* (Academic Press, Inc., New York, 1957), 257.
2. G. A. Thomas, M. Capizzi, F. DeRosa, R. N. Bhatt and T. M. Rice, *Phys. Rev. B* **23** 5472 (1981).
3. K. Andres, R. N. Bhatt, P. Goalwin, T. M. Rice and R. E. Walstedt, *Phys. Rev. B* **24**, 244 (1981).
4. H. A. Bethe and E. E. Salpeter, *Quantum Mechanics of 1 and 2 Electron Atoms* (Springer, 1977).
5. J. Hubbard, *Proc. Roy. Soc. A* **276**, 238 (1963).
6. M. C. Gutzwiller, *Phys. Rev. Lett.* **10**, 159 (1963).
7. P. W. Anderson, *Science* **235**, 1196 (1987).
8. F. C. Zhang and T. M. Rice, *Phys. Rev. B* **37**, 3759 (1988).
9. P. A. Lee, N. Nagaosa and X.-G. Wen, *Rev. Mod. Phys.* **78**, 17 (2006).
10. D. B. McWhan, A. Menth, J. P. Remeika, W. F. Brinkman and T. M. Rice, *Phys. Rev. B* **7**, 1920 (1973).
11. S. A. Carter, J. Yang, T. F. Rosenbaum, J. Spalek and J. M. Honig, *Phys. Rev. B* **43**, 607 (1991).
12. S. Ogawa, *J. Appl. Phys.* **50**, 2308 (1979).
13. T. Thio and J. W. Bennett, *Phys. Rev. B* **50**, 10574 (1994).

14. T. Thio and J. W. Bennett, *Phys. Rev. B* **52**, 3555 (1995).
15. A. Georges, G. Kotliar, W. Krauth and M. J. Rozenberg, *Rev. Mod. Phys.* **68**, 13 (1996).
16. M. Ulmke, V. Janiš and D. Vollhardt, *Phys. Rev. B* **51**, 10411 (1995).
17. N. Chandra, M. Kollar and D. Vollhardt, *Phys. Rev. B* **59**, 10541 (1999).
18. M. Eckstein, M. Kollar, M. Potthoff and D. Vollhardt, *Phys. Rev. B* **75**, 125103 (2007).
19. N. F. Mott, *Metal-Insulator Transitions*, 2nd edn. (Taylor and Francis London, 1990), and references therein.
20. S. R. Schofied, N. J. Curson, M. Y. Simmons, F. J. Ruess, T. Hallam, L. Oberbeck and R. G. Clark, *Phys. Rev. Lett.* **91**, 136104 (2003).
21. Y. Nagaoka, *Phys. Rev.* **147**, 392 (1966).
22. E. Lieb and D. Mattis, *Phys. Rev.* **125**, 164 (1962).
23. J. O. Haerter and B. S. Shastry, *Phys. Rev. Lett.* **95**, 087202 (2005).
24. E. Eisenberg, R. Berkovits, D. A. Huse and B. L. Altshuler, *Phys. Rev. B* **65**, 134437 (2002).
25. W. Sasaki and J. Kinoshita, *Jour. Phys. Soc. of Japan* **25**, 1622 (1968).
26. W. Sasaki, *J. Phys. (Paris) Colloq.* **37**, C4-307 (1976).
27. H. Ue and S. Maekawa, *Phys. Rev. B* **3**, 4232 (1971).
28. J. P. Quirt and J. R. Marko, *Phys. Rev. B* **7**, 3842 (1973).
29. M. J. Hirsch, D. F. Holcomb, R. N. Bhatt and M. A. Paalanen, *Phys. Rev. Lett.* **68**, 1418 (1992).
30. H. Kamimura, *Proc. 19th Scottish Universities Summer School in Physics*, eds. L. R. Friedman and D. P. Tunstall (SUSSP, 1978), pp. 327-68.
31. H. Kamimura, in *Electron-Electron Interactions in Disordered Systems*, eds. M. Pollack and A. L. Efros (North-Holland 1985), pp. 555-617.
32. R. N. Bhatt and P. A. Lee, *J. Appl. Phys.* **52**, 1703 (1981).
33. R. N. Bhatt and P. A. Lee, *Phys. Rev. Lett.* **48**, 344 (1982).
34. R. N. Bhatt, *Physica Scripta* **T14**, 7 (1986).
35. R. N. Bhatt, *Proc. 20th Int. Conf. on Physics of Semiconductors*, eds. E. M. Anastassakis and J. D. Joannopoulos (World Scientific, Singapore, 1990), p. 2633.
36. R. N. Bhatt, M. A. Paalanen and S. Sachdev, *Jour. de Physique* **49**, C8-1179 (1988).
37. D. S. Fisher, *Phys. Rev. B* **50** 3799 (1994).
38. D. Holcomb in Localization and interaction in disordered metals and doped semiconductors, *Proc. 31st Scottish Universities Summer School in Physics*, ed. D. M. Finlayson (SUSSP 1986), p. 313.
39. M. A. Paalanen, J. E. Graebner, R. N. Bhatt and S. Sachdev, *Phys. Rev. Lett.* **61**, 597 (1988).
40. H. Ohno, *Science* **281**, 951 (1998).
41. D. Chiba, K. Takamura, F. Matsukura and H. Ohno, *Appl. Phys. Lett.* **82**, 3020 (2003).
42. M. Berciu and R. N. Bhatt, *Phys. Rev. Lett.* **87**, 107203 (2001).
43. M. P. Kennett, M. Berciu and R. N. Bhatt, *Phys. Rev. B* **66**, 045207 (2002).
44. See, *e.g.*, R. N. Bhatt, *Phys. Rev. B* **24**, 3630 (1981).
45. E. Dagotto, A. Moreo, F. Ortolani, B. Poilblanc and J. Riera, *Phys. Rev. B* **45**, 10741 (1992).
46. J. E. Hirsch, *Phys. Rev. B* **31**, 4403 (1985).
47. E. Nielsen and R. N. Bhatt, *Phys. Rev. B* **76**, 161202 (2007).
48. E. Nielsen and R. N. Bhatt, in preparation.

KIDNEY-BOOJUM-LIKE SOLUTIONS AND EXACT SHAPE EQUATION OF SOLID-LIKE DOMAINS IN LIPID MONOLAYER

HUAN TONG
Center for Advanced Study, Tsinghua University
Beijing, 100084, China

FEI LIU
Center for Advanced Study, Tsinghua University
Beijing, 100084, China

MITSUMASA IWAMOTO
Department of Physical Electronics, Tokyo Institute of Technology
Tokyo, 2-12-1 O-okayama, Meguro-ku, 152-8552, Japan

ZHONG-CAN OU-YANG*
Institute of Theoretical Physics, The Chinese Academy of Sciences
Beijing, P.O.Box 2735, 100080, China
oy@itp.ac.cn

Received 31 July 2008

The shape of solid lipid monolayer domain surrounded by a fluid phase is of considerable interest from physical and mathematical points of view. Here we report two new results about this topic. First, we obtain an exact analytical solution to an approximated shape equation that was derived by us recently [*Phys. Rev. Lett.* **93**, *206101 (2004)*]. This solution can well describe the kidney- and boojum-like domains that abound in lipid monolayer. Second, we derive an exact domain shape equation by a direct variation of domain energy without any artificial cutoff. We find that no continuous solutions satisfies this shape equation due to the divergence of its coefficients, which is rooted in the continuous description of electrostatic dipoles.

Keywords: Solid lipid monolayer domain; energy variation; shape equation.

1. Introduction

In the past two decades, due to the development of many sophisticated microscopic techniques, such as florescence microscopy (FM)[1,2] and Brewster angle microscopy

*Institute of Theoretical Physics, The Chinese Academy of Sciences, P.O. Box 2735, 100080, Beijing, China

(BAM),[3] the experimental study of shape transition in lipid monolayer domains at the air-water interface has revealed a diversity of domain shapes from round structures to complex labyrinthine patterns.[4–10] Nowadays this study becomes one of the main projects of shape science. Theoretical understanding of lipid monolayer domain formation is based on a competition between electrostatic repulsion and line tension proposed by Andelman *et al.*,[11,12] and is investigated extensively by McConnel group[13–15] and many others.[16–20] The line tension γ (> 0) acts to minimize domain boundary $\vec{r}(s)$ and s being the arc-length element. In contrast, the repulsive dipole-dipole interaction tends to separate the dipoles in maximal. Therefore, considerable task of the mentioned works is the calculation of the free energy

$$F = \gamma \oint dl + F_\perp + F_\parallel \tag{1}$$

for comparison of the energies of various domain shapes or for evaluation of the stability of a given shape. Here $F_\perp$ and $F_\parallel$ are the components of the dipole-dipole interaction energy normal and parallel to the water surface, respectively. They can be written as double line integrals following:

$$F_\perp = -\frac{\mu_\perp^2}{2} \oint \oint \frac{\mathbf{t}(l) \cdot \mathbf{t}(s)}{|\mathbf{r}(l) - \mathbf{r}(s)|} dl ds \tag{2}$$

and

$$F_\parallel = \frac{\mu_\parallel^2}{2} \oint \oint \frac{(\mathbf{t}(l) \cdot \mathbf{y}_0)(\mathbf{t}(s) \cdot \mathbf{y}_0)}{|\mathbf{r}(l) - \mathbf{r}(s)|}, \tag{3}$$

where $\mathbf{t} = d\mathbf{r}/ds = (\cos\phi(s), \sin\phi(s))$ is the unit tangential vector, $\mathbf{y}_0$ is the unit vector of the y axis, and $\mu_\perp$, and $\mu_\parallel$ are respectively the dipolar density components normal and parallel to the monolayer. Here all dipoles are thought to uniformly tilt along x axis with constant angle θ, i.e., $\mu = (\mu_\perp, \mathbf{0}, \mu_\parallel) = \mu_0(\sin\theta, \mathbf{0}, \cos\theta)$; see Fig. 1.

Recently, our works[21,22] showed that, by employing the Taylor expansion, the double-line integrals in Eq. (1) could be well approximated by a sum of an additionally negative line tension and a curvature-elastic energy of the domain boundary. An approximated shape equation was derived by a variation of the approximated energy. Many important physical properties and experimental phenomena about the two-dimension (2D) monolayer domain were understood analytically and quantitatively. Although this significant progress was made, we still face the core problems in the domain shape studies, i.e., are there analytical solutions to the approximated shape equation? We have partially investigated this issue by substituting a given shape, e.g., a circle into the equation and searching the existence conditions. From rigorous mathematics viewpoint, such a kind of efforts is not yet relevant to the real proof. In this work we find that the approximated shape equation indeed has analytical solutions under some certain physical conditions.

On the other hand, one may ask why we do not variate the Eqs. (2) and (3) directly without any approximation. The main reason may be that these two energies are in fact ill-defined in mathematics. We can readily see from the term of

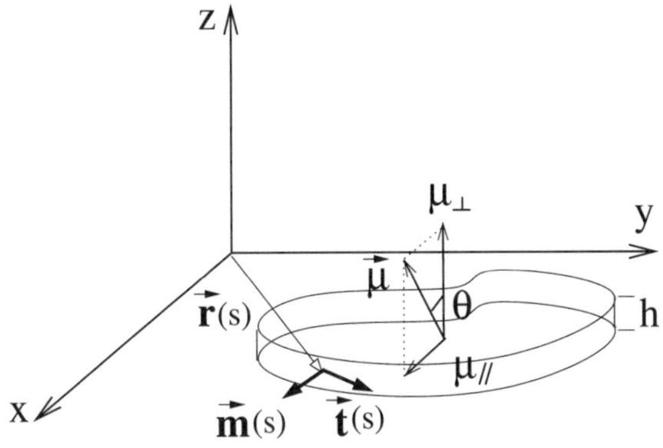

Fig. 1. Schematic illustration of the shape of a monolayer domain, whose dipoles $\vec{\mu}$ uniformly tilt along x-direction with the same tilted angle θ. The geometric quantities describing its boundary curve are also showed. h represents the thickness of the monolayer.

$1/|\mathbf{r}(l) - \mathbf{r}(s)|$ in these integrals. This nonphysical result arises from the use of a continuous description of dipoles distribution. Interestingly, this circumstance is not fresh in electromagnetics in that $F_\perp$ has the same formula with the Neumann's formula for the coefficient of self-inductance of a single coil except for the prefactors.[23,24] To prevent the divergence of Eq. (1), in all existing literature including us, different kinds of cutoff in the distance were employed.[13–17,25] For instance, McConnel group modified the free energy formulas by changing $|\mathbf{r}(l) - \mathbf{r}(s)|$ to $\sqrt{[\mathbf{r}(l) - \mathbf{r}(s)]^2 + \Delta^2}$.[14,15] However, a divergent energy formula does not mean the absence of finite shape equation. In this work, we derive a general shape equation by a direct variation of Eq. (1), $\delta F = 0$. We think this derivation should be meaningful. First of all, to the best of our knowledge, no exact shape equation of the solid domain has been formally reported in literature until now. In all previous works, the discussed equilibrium domain shapes were either predetermined according to observation, or were derived by approximated free energies as we did previously. Second, an exact shape equation would useful in practice, e.g., to investigate whether a well-defined shape equation exists, to evaluate the quality of various approximation schemes. Finally, we cannot exclude the possibility that the general shape equation can be solved in future.

2. The Approximated Shape Equation and its Exact Solution

We first briefly review the approximation of Eq. (1) and its corresponding shape equation. The details could be found in our previous works.[21,22]

In order to take into account the difference in Gibbs free energy density of g_0 between outer (fluid) and inner (solid) phases and the domain shape transition induced by surface pressure of compression, Π, we introduced an additional term $\Delta P \oint dA$ in Eq. (1); here $\Delta P = \Pi - g_0$ and $g_0 > 0$ for the solid phase is more stable than the fluid one.

The double-line integrals in Eqs. (2) and (3) could be rewritten as

$$-\frac{\mu^2}{2}\oint\left[\oint\frac{\mathbf{t}(s)\cdot\mathbf{t}(s+x)}{|\mathbf{r}(s+x)-\mathbf{r}(s)|}dx\right]ds, \tag{4}$$

and

$$\frac{\mu_{\|}^2}{2}\oint\left[\oint\frac{\mathbf{t}(s+x)\cdot\mathbf{y}_0\mathbf{t}(s)\cdot\mathbf{y}_0}{|\mathbf{r}(s+x)-\mathbf{r}(s)|}dx\right]ds, \tag{5}$$

where the range of arc-variable $x \equiv l - s$ is $[h, L]$, L and h represent the boundary length and the nonzero monolayer thickness of the domain as a cutoff,[26] respectively; see Fig. 1. Employing the Frenet formulas[27] in the plane and the Taylor's expansion of $\mathbf{r}$ with respect to the variable x, we had

$$\mathbf{r}(s+x) = \mathbf{r}(s) + \mathbf{t}(s)x + \frac{1}{2}\kappa(s)\mathbf{m}(s)x^2 + \cdots, \tag{6}$$

and

$$\mathbf{t}(s+x) = \mathbf{t}(s) + \kappa(s)\mathbf{m}(s)x + \frac{1}{2}\left[\kappa_s(s)\mathbf{m}(s) - \kappa(s)^2\mathbf{t}(s)\right]x^2 + \cdots, \tag{7}$$

where $\kappa(s)$ is the curvature of the boundary curve at s, $\kappa_s = d\kappa(s)/ds$. Substituting Eqs. (6) and (7) into Eqs. (4) and (5), the inner integrals (the brackets) are a sum of $c_i(s)x^i, i = -1, 0, 1, ...,$ and $c_i(s)$ are the functions of the curvature $\kappa(s)$ and/or its derivatives with different orders. If we consider a somewhat smooth curve and only take into account the terms of x till the first order, we obtained approximated expressions of the double-line integrals given by

$$F_{\perp} \approx -\tfrac{1}{2}\mu_{\perp}^2 \ln\tfrac{L}{h}\oint ds + \tfrac{11}{96}\mu_{\perp}^2 L^2 \oint \kappa(s)^2 ds, \tag{8}$$

and

$$F_{\|} \approx \tfrac{1}{2}\mu_{\|}^2 \ln\tfrac{L}{h}\oint \sin^2\phi(s)ds - \tfrac{1}{192}\mu_{\|}^2 L^2 \oint\left[11 + 13\cos 2\phi(s)\right]\kappa(s)^2 ds. \tag{9}$$

To obtain the shape equation, we derived the first variation of the approximated domain energy (1) and had

$$\Delta P - \Lambda\kappa + \alpha\kappa^3 + \beta\kappa_{ss} + \sigma\kappa\kappa_s = 0, \tag{10}$$

where $\kappa_{ss} = d^2\kappa(s)/ds^2$, and the definitions of the coefficients are

$$\begin{aligned}
\Lambda &= \gamma - \frac{\mu_{\perp}^2}{2}\ln\frac{Le}{h} + \frac{\mu_{\|}^2(1+3\cos 2\phi)}{4}\ln\frac{L}{h} + \frac{11}{48}\mu_{\perp}^2 L\oint\kappa(s)^2 ds + \\
&\quad \frac{\mu_{\|}^2}{2L}\oint\sin^2\phi(s)ds - \frac{1}{96}\mu_{\|}^2 L\oint(11 + 13\cos 2\phi)\kappa(s)^2 ds, \\
\alpha &= \frac{11}{96}\mu_{\perp}^2 L^2 + \mu_{\|}^2 L^2\frac{39\cos 2\phi - 11}{192}, \\
\beta &= \frac{11}{48}\mu_{\perp}^2 L^2 - \mu_{\|}^2 L^2\frac{11 + 13\cos 2\phi}{96}, \\
\sigma &= -\frac{13}{24}\mu_{\|}^2 L^2 \sin 2\phi,
\end{aligned} \tag{11}$$

respectively. Equation (10) and its coefficients can be further simplified if we only concern about the case of $\mu_{\parallel} = 0$:

$$\Delta P - \Lambda_0 \kappa + \alpha_0 \kappa^3 + 2\alpha_0 \kappa_{ss} = 0, \tag{12}$$

where

$$\alpha_0 = \frac{11}{96}\mu_{\perp}^2 L^2,$$
$$\Lambda_0 = \gamma - \frac{\mu_{\perp}^2}{2}\ln\frac{eL}{h} + \frac{11}{48}L\mu_{\perp}^2 \oint \kappa(s)^2 ds.$$

At first sight, both Eqs. (10) and (12) are highly nonlinear and are hardly solved in analytical way. We have studied them by only substituting a circular domain and investigating the stable and instable conditions between the radius of a circle and the pressure ΔP and line tension γ.[21,22] Here we will show shortly that Eq. (12) has an exact solution if $\Delta P = 0$. In physics the vanishing ΔP means that the solid and fluid phases of lipid monolayer coexist and the π_c-plateau presents in the $\Pi - A$ isotherm experiment.[6,7]

After simple transformations, we can rewrite the first and second derivations of κ with respect to s as follows:

$$\kappa_s = -(1/2)d\kappa^2/d\phi,$$
$$\kappa_{ss} = (1/2)\kappa d^2\kappa^2/d\phi^2.$$

Substituting these geometrical relations into Eq. (12), and multiplying both sides of the equation with $d\kappa^2/d\phi$ we obtain its first integral

$$(\frac{d\kappa^2}{d\phi})^2 = (\frac{2\Lambda_0}{\alpha_0})\kappa^2 - \kappa^4 + C, \tag{13}$$

where C is constant of integration. For a domain $\mathbf{r}(s)$ is a finite close curve and κ may take extreme values at some points at which $d\kappa^2/d\phi \propto \kappa_s = 0$. By choosing $s = 0$ at a certain extreme point, we then rewrite the above equation as

$$\left[\frac{d}{d\phi}\left(\frac{\kappa^2}{a^2} - 1\right)\right]^2 = \left(\frac{\kappa_0^2}{a^2} - 1\right)^2 - \left(\frac{\kappa^2}{a^2} - 1\right)^2, \tag{14}$$

where $\kappa_0 = \kappa(0)$ and $a = \sqrt{\Lambda_0/\alpha_0}$. If we further choose appropriate coordinates and let $\phi(0) = \pi/2$ there, we finally have

$$\kappa(s) = -\frac{d\phi}{ds} = -a\sqrt{1 + b\sin\phi(s)} \tag{15}$$

where $b = (\kappa_0/a)^2 - 1$. Obviously, if $b = 0$ the above equation gives a circle with radius a. Moreover, Eq. (15) can describe both round shapes with convex or concave cusp. As illustrations, Fig. 2 shows numerical results of the equation with $a = 1$, $b = \pm 0.7$, and $\pm\phi(0)$. Here we have chosen $s = 0$ at the cusp of the boundary and $\vec{m}(0)$ along y. We see that $\phi(0)$ with opposite signs can yield very distinct domain shapes (the upper and low curves in the figure). These shapes were indeed

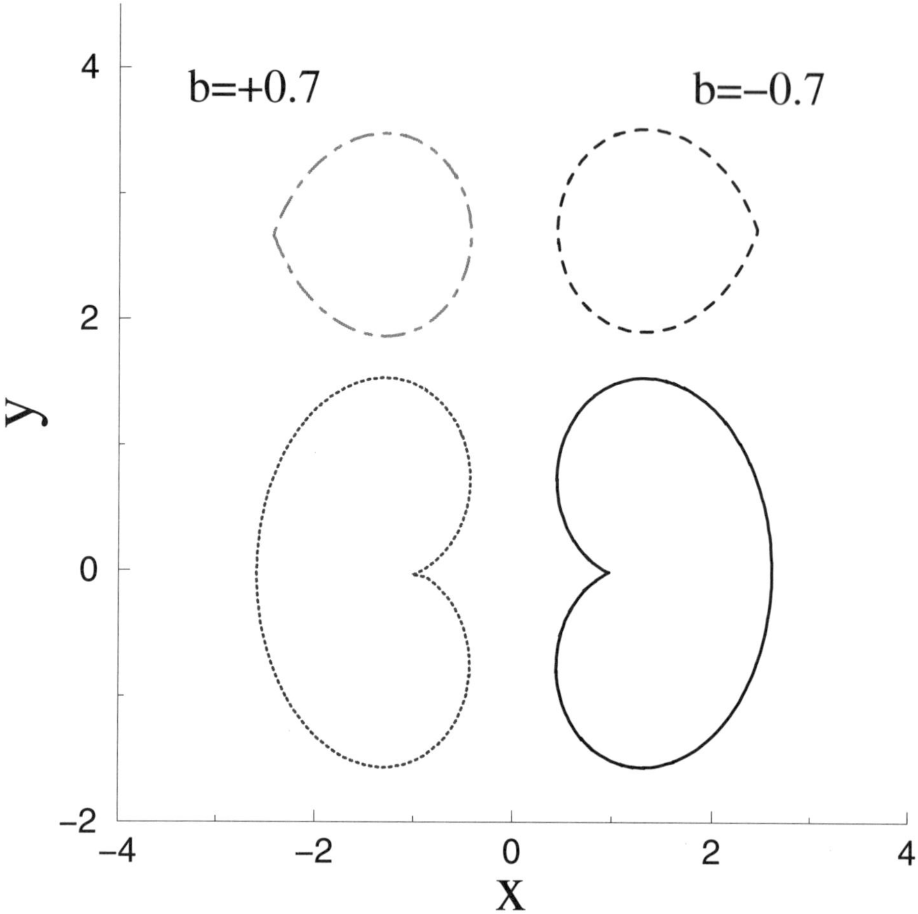

Fig. 2. Illustration of boojum- (upper) and kidney-like (low) shapes calculation by Eq. (15) with $a = 1$, $b = \pm 0.7$ and $\pm\phi(0)$. The different signs of b gives the left- and right- "kidneys" and "boojums", respectively.

observed at the Π_c plateau in the dipamitoylphosphatidylcholine (DPPC) monolayer experiments,[6] and the upper shapes in Fig. 2 were called boojum-like therein. Hence, the solution (15) is not only exact but also satisfactorily predicts the interesting shapes observed in the experiment.

3. Exact Shape Equation

In this section, we study the exact shape equation of the free energy (1) by variation without any approximation. To derive the equilibrium shape equation of a domain with boundary curve $\mathbf{r}(s)$, we consider the slightly distorted domains

$$\mathbf{r}'(s) = \mathbf{r}(s) + \psi(s)\mathbf{m}(s), \tag{16}$$

where $\mathbf{m} = (\sin\phi(s), -\cos\phi(s))$ is the unit outward normal vector of $\mathbf{r}(s)$, and $\psi(s)$ is a sufficiently small smooth variation function. The first-order variation of Eq. (1) includes four parts:

$$\delta F^{(1)} = \Delta P \delta^{(1)} \int dA + \gamma \delta^{(1)} \oint ds + \delta^{(1)} F_{\perp} + \delta^{(1)} F_{\parallel}. \tag{17}$$

The first two variations of the domain area and the boundary length have been given in our previous work.[22] We list them below for the self-consistence:

$$\oint \delta^{(1)} ds = \oint -\kappa(s)\psi(s) ds \tag{18}$$

and

$$\int \delta^{(1)} dA = \int \psi(s) ds. \tag{19}$$

The variations of other two terms are relatively complex. The main steps are shown in the Appendix. According to Eqs. (A.6) and (A.10), the first variations of $F_{\perp}$ and $F_{\parallel}$ are

$$\delta^{(1)} F_{\perp} = -\mu_{\perp}^2 \oint \oint \frac{[\mathbf{r}(l) - \mathbf{r}(s)] \cdot \mathbf{m}(l)}{|\mathbf{r}(l) - \mathbf{r}(s)|^3} \psi(s) dl ds \tag{20}$$

and

$$\delta^{(1)} F_{\parallel} = \mu_{\parallel}^2 \oint \oint \frac{[\mathbf{r}(l) - \mathbf{r}(s)] \cdot \mathbf{x}_0 \mathbf{m}(l) \cdot \mathbf{x}_0}{|\mathbf{r}(l) - \mathbf{r}(s)|^3} \psi(s) dl ds, \tag{21}$$

respectively, where $\mathbf{x}_0$ is the unit vector of the x-axis. Hence, if a curve $\mathbf{r}(s)$ describes a boundary of an equilibrium domain, the first variation of its corresponding free energy must vanish, namely, $\delta^{(1)} F = 0$ for any infinitesimal function $\psi(s)$. Combining these variation equations (18)–(21), we obtain the exact shape equation of the energy equation (1) as follows,

$$\begin{aligned} \Delta P - \gamma\kappa(s) - \mu_{\perp}^2 \oint \frac{[\mathbf{r}(l) - \mathbf{r}(s)] \cdot \mathbf{m}(l)}{|\mathbf{r}(l) - \mathbf{r}(s)|^3} dl \\ + \mu_{\parallel}^2 \oint \frac{[\mathbf{r}(l) - \mathbf{r}(s)] \cdot \mathbf{x}_0 \mathbf{m}(l) \cdot \mathbf{x}_0}{|\mathbf{r}(l) - \mathbf{r}(s)|^3} dl = 0. \end{aligned} \tag{22}$$

This equation is a differential and integral equation and seems to be impossible to solve analytically. Importantly, the presence of $1/|\mathbf{r}(l) - \mathbf{r}(s)|^3$ in Eq. (22) implies that the exact shape equation is ill-defined for any continuous boundary curves. As an illustration, if we substitute a test circular solution of $\mathbf{r}(s) = \rho_0(\sin\phi(s), -\cos\phi(s))$ with radius ρ_0, Eq. (22) results in the following relation:

$$\Delta P + \gamma\rho_o^{-1} - \mu_{\perp}^2 2^{-3/2} \rho_0^{-1} \alpha + \mu_{\parallel}^2 2^{-3/2} \rho_0^{-1} \beta \sin\phi(s) = 0, \tag{23}$$

with two infinite integrals $\alpha = \int_0^{2\pi}(1/\sqrt{1-\cos\theta})d\theta$ and $\beta = \int_0^{2\pi}(\cos\theta/\sqrt{1-\cos\theta})d\theta$. Hence, Eq. (23) reveals that circle is not rigorous solution to Eq. (22). Even if the exact shape equation (22) has some flaws in definition, it is yet exact if we introduce a certain cutoff in the integrations, which is always plausible in physics. Of course, the potential of the equation in application would largely rely on whether we can solve it numerically or analytically.

4. Conclusion

In this work, we report two new results in the study of the shapes of solid lipid monolayer domains at the air-water interface. We find an exact analytical solution to an approximated shape equation. This solution can well describe the kidney- and boojum-like domains that abound in lipid monolayer. In addition, we first derive an exact domain shape equation by a direct variation of the domain energy without any approximation. Our results would be useful in the real experimental studies of solid domain.

Appendix A. The first-order variations of $F_\perp$ and $F_\parallel$

The first-order variation of the free energy $F_\perp$ can be written as

$$\delta F_\perp^{(1)} = -\frac{\mu_\perp^2}{2}\oint\left\{\delta^{(1)}ds\oint\frac{\mathbf{t}(l)\cdot\mathbf{t}(s)}{|\mathbf{r}(l)-\mathbf{r}(s)|}dl + ds\oint\delta^{(1)}\left[\frac{\mathbf{t}(l)\cdot\mathbf{t}(s)}{|\mathbf{r}(l)-\mathbf{r}(s)|}dl\right]\right\}. \quad \text{(A.1)}$$

We have given the expressions of ds' and $\mathbf{t}'$ for a slightly distorted boundary Eq. (16)[22] and list them below for self-consistence:

$$ds' = \left[(1-\kappa\psi)^2 + \psi_s^2\right]^{\frac{1}{2}} ds, \quad \text{(A.2)}$$

and

$$\mathbf{t}' = \frac{ds}{ds'}\left[(1-\kappa\psi)\mathbf{t} + \psi_s\mathbf{m}\right], \quad \text{(A.3)}$$

where $\psi_s = d\psi/ds$. According to the definition of the variation, $\delta ds = ds' - ds$. We readily obtain the first-order variation of the distance element $\delta^{(1)}ds = -\kappa(s)\psi(s)ds$. To obtain the first-order variation of the integral in Eq. A.1, we need calculate

$$\delta\left[\frac{\mathbf{t}(l)\cdot\mathbf{t}(s)}{|\mathbf{r}(l)-\mathbf{r}(s)|}dl\right] = \frac{\mathbf{t}'(l)\cdot\mathbf{t}'(s)}{|\mathbf{r}'(l)-\mathbf{r}'(s)|}dl' - \frac{\mathbf{t}(l)\cdot\mathbf{t}(s)}{|\mathbf{r}(l)-\mathbf{r}(s)|}dl. \quad \text{(A.4)}$$

Substituting Eqs. (A.2) and (A.3) into above equation and applying the Taylor's expansion with respect to small $\psi(s)$, Eq. (A.4) can be approximated by

$$-\frac{\mathbf{t}(l)\cdot\mathbf{t}(s)}{|\mathbf{r}(l)-\mathbf{r}(s)|}dl + \frac{\mathbf{t}(s)\cdot\mathbf{m}(l)}{|\mathbf{r}(l)-\mathbf{r}(s)|}\psi_l(l)dl + \frac{\mathbf{t}(l)\cdot\mathbf{m}(s)}{|\mathbf{r}(l)-\mathbf{r}(s)|}\psi_s(s)dl \quad \text{(A.5)}$$
$$-\frac{(\mathbf{r}(l)-\mathbf{r}(s))\cdot\mathbf{m}(l)\mathbf{t}(l)\cdot\mathbf{t}(s)}{|\mathbf{r}(l)-\mathbf{r}(s)|^3}\psi(l)dl - \frac{(\mathbf{r}(s)-\mathbf{r}(l))\cdot\mathbf{m}(s)\mathbf{t}(l)\cdot\mathbf{t}(s)}{|\mathbf{r}(l)-\mathbf{r}(s)|^3}\psi(s)dl.$$

Substituting Eq. (A.5) and $\delta^{(1)}ds$ into Eq. (A.1), we finally obtain the first-order variation of the free energy $F_\perp$ as follows:

$$\delta^{(1)}F_\perp = -\mu_\perp^2 \oint\oint \frac{[\mathbf{r}(l)-\mathbf{r}(s)]\cdot\mathbf{m}(l)}{|\mathbf{r}(l)-\mathbf{r}(s)|^3}\psi(s)dlds. \tag{A.6}$$

In deriving the above equation, we have used several mathematical manipulations, including the exchange of the variables s and l, an identity $\oint f(s)\psi_s ds = -\oint f_s(s)\psi ds$ for any smooth function $f(s)$, and $\mathbf{m}(l) = -\mathbf{t}(l)\cdot\mathbf{m}(s)\mathbf{t}(s) + \mathbf{t}(l)\cdot\mathbf{t}(s)\mathbf{m}(s)$.

Similarly, the first-order variation of the energy $F_\parallel$ can be written as

$$\delta^{(1)}F_\parallel = \frac{\mu_\parallel^2}{2}\oint\left\{\delta^{(1)}ds\oint\frac{\mathbf{t}(l)\cdot\mathbf{y}_0\mathbf{t}(s)\cdot\mathbf{y}_0}{|\mathbf{r}(l)-\mathbf{r}(s)|}dl + ds\delta^{(1)}\left[\frac{\mathbf{t}(l)\cdot\mathbf{y}_0\mathbf{t}(s)\cdot\mathbf{y}_0}{|\mathbf{r}(l)-\mathbf{r}(s)|}dl\right]\right\}. \tag{A.7}$$

Hence, we need calculate a new variation of

$$\delta\left[\frac{\mathbf{t}(l)\cdot\mathbf{y}_0\mathbf{t}(s)\cdot\mathbf{y}_0}{|\mathbf{r}(l)-\mathbf{r}(s)|}dl\right] = \frac{\mathbf{t}'(l)\cdot\mathbf{y}_0\mathbf{t}'(s)\cdot\mathbf{y}_0}{|\mathbf{r}'(l)-\mathbf{r}'(s)|}dl' - \frac{\mathbf{t}(l)\cdot\mathbf{y}_0\mathbf{t}(s)\cdot\mathbf{y}_0}{|\mathbf{r}(l)-\mathbf{r}(s)|}dl. \tag{A.8}$$

We still substitute Eqs. (A.2) and (A.3) into above equation and apply the Taylor's expansion with respect to small $\psi(s)$ to obtain an approximation of Eq. (A.8). After a long calculation, we have the following approximation:

$$\begin{aligned}
&-\frac{\mathbf{t}(s)\cdot\mathbf{y}_0\mathbf{t}(l)\cdot\mathbf{y}_0}{|\mathbf{r}(l)-\mathbf{r}(s)|}\kappa(l)\psi(l)dl + \frac{\mathbf{t}(s)\cdot\mathbf{y}_0\mathbf{m}(l)\cdot\mathbf{y}_0}{|\mathbf{r}(l)-\mathbf{r}(s)|}\psi_l(l)dl\\
&+\frac{\mathbf{t}(l)\cdot\mathbf{y}_0\mathbf{m}(s)\cdot\mathbf{y}_0}{|\mathbf{r}(l)-\mathbf{r}(s)|}\psi_s(s)dl - \frac{[\mathbf{r}(l)-\mathbf{r}(s)]\cdot\mathbf{m}(l)\mathbf{t}(s)\cdot\mathbf{y}_0\mathbf{t}(l)\cdot\mathbf{y}_0}{|\mathbf{r}(l)-\mathbf{r}(s)|^3}\psi(l)dl\\
&-\frac{[\mathbf{r}(s)-\mathbf{r}(l)]\cdot\mathbf{m}(s)\mathbf{t}(s)\cdot\mathbf{y}_0\mathbf{t}(l)\cdot\mathbf{y}_0}{|\mathbf{r}(l)-\mathbf{r}(s)|^3}\psi(s)dl.
\end{aligned} \tag{A.9}$$

Substituting Eq. (A.9) and $\delta^{(1)}ds$ into Eq. (A.7), we finally obtain the first-order variation of the free energy $F_\parallel$ as follows:

$$\delta^{(1)}F_\parallel = \mu_\parallel^2 \oint\oint \frac{[\mathbf{r}(l)-\mathbf{r}(s)]\cdot\mathbf{x}_0\mathbf{m}(l)\cdot\mathbf{x}_0}{|\mathbf{r}(l)-\mathbf{r}(s)|^3}\psi(s)dlds. \tag{A.10}$$

In addition to the exchange of variables and the identity mentioned before, we use other two identities in the deriving, which are $\mathbf{x}_0 = \mathbf{m}(s)\cdot\mathbf{y}_0\mathbf{t}(s) - \mathbf{t}(s)\cdot\mathbf{y}_0\mathbf{m}(s)$ and $\mathbf{t}(l)\cdot\mathbf{y}_0 = -\mathbf{m}(l)\cdot\mathbf{x}_0$, respectively.

References

1. V. Tscharher and H. M. McConnel, *Biophys. J.* **36**, 409 (1981).
2. M. Losche, E. Sackmann and H. Mohwald, Ber. Bunsen-Ges. *Phys. Chem.* **87**, 848 (1983).
3. S. Henon and J. Meunier, *Rev. Sci. Instrum.* **62**, 936 (1991).
4. C. W. McConlogue and T. K. Vanderlick, *Langmuir* **13**, 7158 (1991).
5. C. W. McConlogue and T. K. Vanderlick, *Langmuir* **15**, 234 (1999).
6. K. Kjaer and J. A. Niesen, *Phys. Rev. Lett.* **58**, 2224 (1987).

7. K. J. Stine and D. T. Stratmann, *Langmuir* **8**, 2509 (1992).
8. D. K. Schwartz, M. W. Tsao, C. M. Knobler and J. C. Knobler, *J. Chem. Phys.* **101**, 8258 (1994).
9. M. Weis and H. M. McConnel, *Nature (London)* **310**, 47 (1984).
10. M. Weis and H. M. McConnel, *J. Phys. Chem.* **89**, 4453 (1985).
11. D. Andelman, F. Brochard, P. G. De Gennes and J. F. Joanny, *Acad. Sci. Ser.* **C301**, 675 (1985).
12. D. Andelman, F. Brochard and J. F. Joanny, *J. Chem. Phys.* **86**, 3673 (1987).
13. D. J. Keller, J. P. Korb and H, M. McConnel, *J. Phys. Chem.* **91**, 6417 (1987).
14. H. M. McConnell and V. T. Moy, *J. Phys. Chem.* **92**, 4520 (1988).
15. H. M. McConnel, *J. Phys. Chem.* **94**, 4728 (1990).
16. T. K. Vanderlick and H. Mohwald, *J. Phys. Chem.* **94**, 886 (1990).
17. M. A. Mayer and T. K. Vanderlick, *Langmuir* **8**, 13131 (1992).
18. J. M. Deutch and F. E. Low, *J. Phys. Chem.* **96**, 7097 (1990).
19. J. M. Miranda, *J. Phys. Chem. B* **103**, 1303 (1999).
20. J. M. Miranda and M. Widom, *Phys. Rev. E* **55**, 3758 (1993).
21. M. Iwamoto and Z.-C. Ou-Yang, *Phys. Rev. Lett.* **93**, 206101 (2004).
22. M. Iwamoto, F. Liu and Z.-C. Ou-Yang, *J. Chem. Phys.* **125**, 224701 (2006).
23. *Die mathmatischen Gesetze der inducirten eekrischen strome*, Belrin Acad. Wissen., Abh. 1845; Ostwald's Klassiker d. exakten Wissen, W. Engeman, Leipzig, 1892, No. 36, P. Moon and D. E. Spencer, Foundations of Electrodynamics (D.Van Nostrand Co.New Jersey, 1960).
24. W. Pauli, *Pauli Lecture on Physics, Vol.1: Electrodynamics*, ed. by C. P. Enz; Trans. by S. Margulies and H. R. Lewis (MIT Press, Cambridge, MA, 1979), pp. 93.
25. P. D. Duncan and P. J. Camp, *J. Chem. Phys.* **121**, 11322 (2004).
26. S.A. Langer, R.E. Goldstein, and D.P. Jackson, *Phys. Rev. A* **46**, 4894 (1992).
27. M.P. do Carmo, *Differential Geometry of Curves and Surfaces* (Prentice-Hall, 1976).

EXCITED STATE PROCESSES IN PHOTOSYNTHESIS MOLECULES

HENRIK BOHR, PER GREISEN and BARRY MALIC*

QuP, Phys. Dept., Tech. Univ. of Denmark, DK-2800, Lyngby, Denmark
**Dept. of Physics, Washington University, St. Louis, Mo 63130 USA*

Received 31 July 2008

A study of electronic processes in the chlorophyll and carothenoid molecules of the photo-reaction center II is presented with the focus on the electronic excitations and charge transfer in the photosynthetic process. Several novel ideas are mentioned especially concerning the electron replenishment and nuclear vibrational excitations. The study is build mainly on numerical quantum calculations of electronic structures of molecules in the photo-reaction center.

1. Introduction

In the present article we shall discuss certain features in the photosynthetic process seen from the viewpoint of condensed matter physics. First we review (non-professionally) the basics of photo-systems and their molecular biology.[2] For further information there are excellent reviews on the matter both in proceedings[3,4] and in books or periodicals.[2,5,6]

The sun is the primary source of energy for almost all life on earth and it supplies this energy in the full electromagnetic spectrum besides other radiation. The energy from the sunlight is supplied to the biosphere through the photosynthetic process, either directly (by incident photons) or indirectly (by climate changes, evaporation etc.) Needless to say, all organisms require energy for the chemical reactions of their life processes where chemical reactions are involved with reproduction, growth and maintenance.

Photosynthesis is the essential source of energy for plants that use solar energy to produce carbohydrates (glucose) to be stored in bonds for later needs of energy to the cell.[4] For animals the energy is supplied mostly by the reverse process of photosynthesis, respiration, although photo-radiation is important also as a secondary means. Below we give the fundamental chemical formulas for 1) photosynthesis and 2) respiration where water and carbon dioxide (and light) in the former case is supplied while oxygen and glucose is the product opposite to the latter case where oxygen and glucose is supplied and where the products are water and carbon dioxide. In a simple way, we can write the overall chemical reaction as:

$$h\nu + 6CO_2 + 6H_2O ==> C_6H_{12}O_6 + 6O_2$$

$$6O_2 + C_6H_{12}O_6 ==> 6CO_2 + 6H_2O + Calories$$

Later we shall specify the reaction in terms of electrons.

Plants receive light at daytime when the photosynthetic process is active but at night-time the metabolic processes basically become active, as an overall tendency.[4]

The photosynthetic process is one of the most effective and sophisticated energy harvesting processes known in nature with a quantum efficiency of around 95%[8,9] in getting photo-energy converted into electrical energy. Besides, there is the remarkable property of the molecules in photosynthesis that they are able to collect and transfer more than 99% of the collected solar energy to their reaction center.[10,11]

2. Bio-Organisms Using Photosynthesis

There are three types of organisms that use photosynthesis: Archaea, Bacteria, and Eucarya of which the last is considered higher level organisms. They all convert light into chemical free energy. The more primitive Archaea includes the halobacteria that also convert light into chemical energy but without oxidation/reduction chemistry and hence no use of CO_2 as a carbon source.[4]

The higher photosynthetic organisms can be divided into oxygenic, photosynthetic organisms such as plants and algae and an-oxygenic photosynthetic organisms being certain bacteria. The former class has organisms that reduce CO_2 into carbohydrate by extraction of electrons from H_2O giving O_2 and H^+ while the latter class, believed to be more ancient, involves extraction of electrons from molecules other than water and without oxygen.[5]

The process of photosynthesis in plants and algae is located in certain organelles, the chloroplasts, which consists of a "light" reaction part involving transfer of electrons and protons and a "dark" reaction part that involves the biosynthesis of carbohydrates from CO_2[4] also called carbon fixation.

We shall in this article mostly be concerned with the "light" reaction which occurs in a membrane system comprising proteins, electron carriers embedded in lipid molecules building up a membrane that divides the space into inner and outer domains and through which molecules or ions can pass. The protein complex, the light harvesting complex, gives a scafold for the organic compounds, the chlorophylls and peridinins. The electron carriers are aromatic groups and metallic ion complexes. The protein complex controls the electron pathway entering the complex via an antenna molecule, peridinin or carothenoid, that absorps the photons and thereby causes the excitation of electrons. The electrons, or actually the charge displacements caused by the excitations, are then being transfered from one carrier to another.

3. The Basic Molecules of Photosynthesis

There are today crystallographic X-ray diffraction structures of the photosynthetic reaction center in the protein databases, e.g. from the organism *Rhodospirillum molischianum*.[13] In Fig. 1 the molecular system is illustrated. The center of the protein with ligands is anchored in the membrane and which, in the case of plants, are the chloroplasts anchored in the thylakoid membrane (the membrane containing light harvesting complex, electron transport chain and the ATP synthase). The small plackets in Fig. 1 are the chlorophyll molecules that are the focus of the present study of calculations.

In Fig. 1(b) the chlorophyll molecules from the light harvesting complex are arranged in a ring around[25] where they receive incoming photons from peridinin or carotenoid.[12] We shall propose an overall picture of the electronics of the process of photosynthesis concerning the chlorophyll and try to give arguments for that.

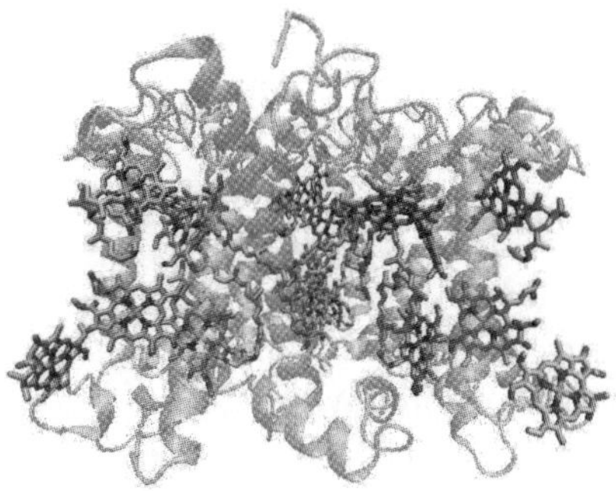

(a) Light-harvesting complex

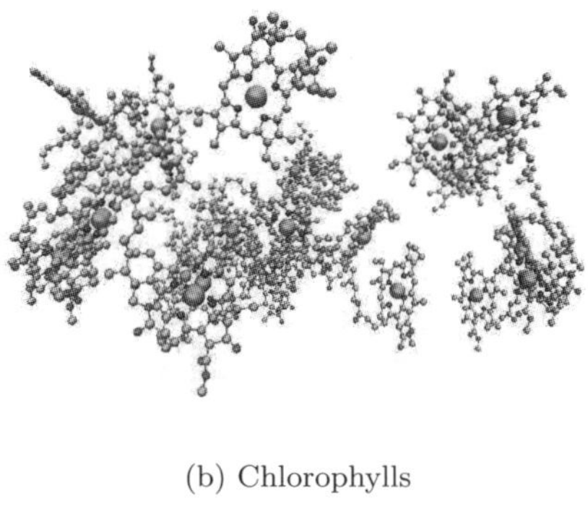

(b) Chlorophylls

Fig. 1. (a) Shows the pea light-harvesting complex at 2.5 Å resolution(PDB ID : 2BHW)[24] The protein is represented as NewCartoon and the chlorophyll a/b using licorice. (b) Shows the chlorophyll molecules from the crystal structure represented in licorice. The chlorophylls are organised in two ring structures, one upper and one lower. The figures were generated using VMD.[23]

Basically, an incoming photon will excite the antenna molecule and create an exciton moving to the chlorophyll ring which acts as a storage ring consisting of 16 chlorophyll molecules arranged in asymmetric pairs of almost parallel plackets. Such a storage ring of smaller rings can accumulating up to 8 electrons and reject bunches of 4 electrons that are carried away by certain proteins to another similar reaction center. This storage ring, or rather sink, is very dependent on the nature of the excited states of the chlorophyll and carotenoid molecules and that, in turn, is very dependent on the excitation gab derived from HOMO-LUMO gab calculations. The ionization energy of the relevant quantum states can also be derived as discussed later where quantum calculations especially are carried out to answer these questions.

3.1. *The electronic transfer mechanisms in the peridinin and chlorophyll molecules*

A much studied process is that of the photo-absorption of the antenna molecules, peridinin or carotenoid.[14,15] They can, upon the absorption of a photon, shuffel excitons between conjugated π-bonds, alternating double and single carbon bonds thereby creating excitons that can propagate down to the attached chlorophyll placket molecule. The direction of the propagation of the exciton is dependent on the dipole orientation of the peridinin or carothenoid molecule that points towards the chlorophyll molecule it is attached to.

The transfer along the antenna and from the antenna to a chlorophyll placket and from one placket to another in the chlorophyll ring system is usually considered to involve a Coulomb interaction and which is a multipole-multipole interaction

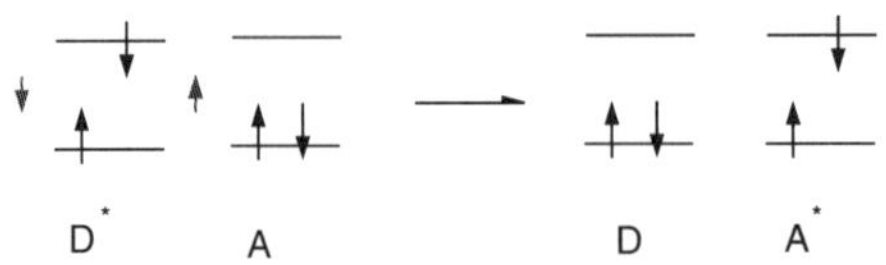

(a) Förster

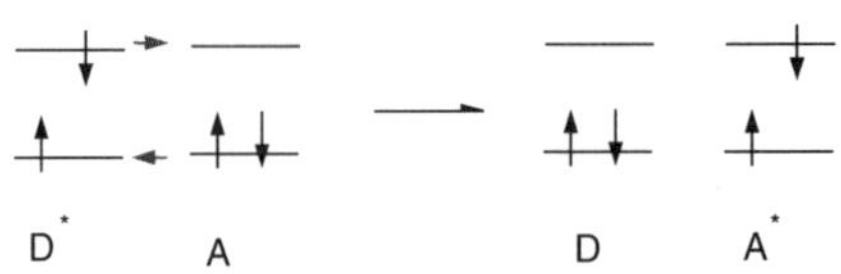

(b) Dexter

Fig. 2. (a) Illustrates the Förster mechanism where the excited electron is transfered due to dipole coupling. (b) Illustrates the Dexter mechanism where an excited electron is transferred.

that de-excites an initially excited electron on a donor molecule and directly excites an electron on the acceptor molecule and is usually called Förster mechanism.[16] In the case of a transfer between two molecules such a transfer between the donor and acceptor is usually considered to be according to the Dexter mechanism[17] when an excited electron is exchanged for a ground state electron or a hole going oppositely. (See Fig. 2 for schematic illustration). In the latter case of transfer between different molecules we find it improbable that electrons are running between molecules like fixed entities. We shall instead propose that energy and excitation is transfered through electromagnetic dipole radiation which we shall argue for in the next chapter.

We shall first briefly discuss the physics behind the Förster mechanism. As mentioned above the mechanism is responsible for a donor chromophore with electrons in excited states to be able to transfer energy by a long-range, non-radiative dipole-dipole interaction to another acceptor chromophore within a distance of about 10 *nm*. This remarkable mechanism is typically being used for resonance fluorescence phenomena[18] (or FRET, Fluorescence Resonance Energy Transfer) but in our case the focus is on chlorophyll molecules. The efficiency, E, of the mechanism is often expressed in terms of the quantum yield of the energy transfer as

$$E = \frac{k_{ET}}{k_f + k_{ET} + \sum_i k_i} \tag{1}$$

where k_{ET} is the rate of energy transfer, k_f is the radiative decay rate and k_i is the rate constant for other de-excitation processes. In terms of the donor-acceptor separation distance a one can also write the efficiency E depending on the inverse 6th power law of a:

$$E = \frac{1}{1 + (a/a_o)^6} \tag{2}$$

where a_o is a normalization factor (separation of the donor-acceptor at which the resonance transfer is 50%) and is called the Förster distance. Even more interesting is that this distance, a_o is related to the overlap integral between the donor emission spectrum and the acceptor absorption spectrum along with the mutual molecular orientation. The overlap integral is expressed as

$$L = \int \epsilon_d(\nu)\alpha_a(\nu)\nu d\nu \tag{3}$$

where ϵ_d and the α_a are respectively the emission spectrum of the donor and the absoption spectrum (molar extinction coefficient) of the acceptor. This leads to a relation between the Förster distance and the overlap integral:

$$a_o^6 = 8.810^{-28}\kappa^2 n^{-4} F_l L \tag{4}$$

κ is the dipole orientation factor often set to $\kappa^2 = 2/3$.[16] F_l is the fluorescence quantum yield of the donor in the absence of the acceptor and n is the refractive index of the medium.

Similar to the Förster mechanism of electron transfer (displacement) is the Dexter mechanism, which as described above involves both transfer of electron and hole and hence works predominantly on shorter distances. The formula for the efficiency of the Dexter mechanism is similar to the Förster that, however, has an $1/a^6$ behavior rather than an exponential e^{-a} as in Dexter mechanism.[17]

One can compare the radial behavior of the two mechanisms and at what distances they dominate. Basically Dexter dominates at distances $a < 10\AA$ and Förster mechanism at distances $20 < a < 100\AA$.[1] The Förster and the Dexter mechanisms help adding to the strong efficiency of the photosynthesis process with a total quantum efficiency of around 99%. The efficiency of the two mechanisms can also be expressed in terms of fluorescence lifetime of the donor molecule:

$$E = (1 - \tau_D^{ua}/\tau_D^{a}) \tag{5}$$

where τ_D^a and τ_D^{ua} are the fluorescence lifetimes of the donor state with and without the presence of an acceptor state. Such number is expected to be very low since the usual quantum yield or the fluorescence efficiency is given by the number n_{pe} of photons emitted divided by the number n_{pa} of photons absorbed, i.e.,

$$Q = \frac{n_{pe}}{n_{pa}} \tag{6}$$

which is a number close to 1. In the case of chlorophyll we expect the Förster and Dexter mechanisms to contribute much less than 100%[7] to the total quantum efficiency and thus, in order to explain this 99% quantum efficiency mentioned above we need another component which we shall explain in the next chapter.

4. The Energy Transfer by Radiation in Photosynthesis

We shall now try to formulate the entire photosynthesis process from a more abstract physical side. In such formulation the chloroplast is considered a whole molecular entity with chemical bonds and metal ions. In short our story is as follows:

Basically, charges are being transferred around in the chloroplast system by the same mechanism of electric dipoles and multi-poles as in the Förster mechanism.[16,21] However, we believe that there also appears to be electromagnetic radiation as a means of charge transfer and, most importantly, as a result of electron excitation that gives rise to nuclear vibration, which is in the far infrared region and to some extend verified in references.[10,11]

As we argue, light being incident on a leaf with chlorophyll exhibits typical resonant absorption peaks, particularly around 400 and 680 nm, see Refs. 10 so

$$E = h\nu = 1.237 \cdot 10^3 eV/\lambda \tag{7}$$

which correspond respectively to 3.093 and 1.767 eV.

These energies could induce electronic excitations in many ligands chl(A,B,...,Z) that are constituents of chlorophylls. Thus, symbolically one has

$$h\nu + Chl(A, B, C, .., Z) \rightarrow Chl(A^*, B, .., Z) \tag{8}$$

where $Chl(A, B, ...)$ stands for the electrons $A, B, ...$ in the chlorophyll molecule, Chl, and A^* stands for excitation of electron A.

The electronic excitation could be Frank-Condon type or excimer type since electronic clouds of neighboring atoms overlap. In either case, the electronically excited state of one of the molecules can pass the electron to another one in excited state forming a kind of dipole. Before doing so, it could de-excite along the vibronic states emitting infrared radiation. Thus,

$$Chl(A^*, B, C, ..., Z) \rightarrow Chl(A^+, B^*, C^-, ..., Z). \tag{9}$$

The electronically excited negative ion $(C^*)^-$ could de-excite to lower states leading ultimately to an isomeric state i, $(C_i^*)^-$. At this stage one has a dipole $(A^+, ..., C_i^{*-})$. The distance between two nuclei at this stage is affected and oscillates about their previous mean position, effectively creating **oscillating – dipoles**. This is supported by the experiments of Refs. 10, 11. In effect one has now a series of oscillating dipoles of dipole length l about 2 to 3 or (0.2 to 0.3) femto seconds(fm). The components of electric field emitted by such an oscillating dipole in classical electrodynamics are[19] (z-axis is the axis of the dipole, θ is with respect to that), for $\lambda \gg l$:

$$E_r = \frac{I_o l}{4\pi} 2\eta cos\theta [\frac{1}{r^2} - \frac{i}{kr^3}] e^{-ikr} \tag{10}$$

$$E_\theta = \frac{I_o l}{4\pi} 2\eta cos\theta [\frac{ik}{r} + \frac{1}{kr^2} - \frac{i}{kr^3}] e^{-ikr} \tag{11}$$

where $\eta = \sqrt{\mu/\epsilon}$ (μ is permeability in Henry/m and ϵ is permittivity in Farad/m) which is the intrinsic impedance of the medium, k, the wave number $= 2\pi/\lambda, \lambda$ being the wavelength radiated and I_o is the current [redacted] in this case the rate of production of dipole.

Assuming that the current is uniform and [redacted]q is the charge at the endpoints, in the frequency domain ω, one can write $I_o = i\omega q$ and use $\eta/k = (\omega\epsilon)^{-1}$.

Then for the emitted radiation, with $\lambda/2\pi \gg r$ we have:

$$E \cong \frac{ql}{4\pi\epsilon r^3} [2cos(\theta)\hat{e_r} + sin(\theta)\hat{e_\theta}] \tag{12}$$

and in the same approximation:

$$H = \frac{I_o l}{4\pi r^2} sin(\theta)\hat{e_\theta}. \tag{13}$$

The average Poynting vector, $< S >$, is defined as $\frac{1}{2}$ times the real part of the Poynting vector $\mathbf{S} = (\mathbf{E} \times \mathbf{H})^*$, so therefore, in terms of w/m^2:

$$< S >= \frac{1}{8} \eta sin^2\theta \left(\frac{I_o l}{\lambda r}\right) \hat{a_r} \tag{14}$$

and the total radiated power (in Watts), P, is thus:

$$P = 40\pi^2 I_o^2 \left(\frac{l}{\lambda}\right)^2. \tag{15}$$

In the case $l \cong \frac{1}{1000}\lambda$; $P \cong (40\pi\pi^2)I_o^2 10^{-6}(I_o$ in Amps) then $I_o \approx q\nu$ where ν is the frequency of vibration of 2 nuclei which can be estimated from vibrational spectra of one of the ligands. One can get this number from the vibrational spectra of aromatic molecules. The electric charge, q, is about $1.602 \cdot 10^{-19}C$ for a single dipole. In the case of 10^{20} dipoles per cm^3, this is a substantial energy.

As mentioned above there are also radiation caused by nucleic vibration in infrared frequencies. They are playing an important role contributing to the breaking up of water in the process:

$$energy + 6CO_2 + 8H_2O ==> C_6H_{12}O_6 + 7O_2 + 4H^+ + 4e^-$$

where the manganese atoms in the center of the chloroplast play an important role of oxidation. The electromagnetic radiation is directed towards the center causing electron transports and oxidation as described above. The chlorophyll molecules are fixed in pairs (in a so-called $C2$ pseudo symmetry where each partner being rotated 180 degrees compared to the other). These pairs are then arranged almost in parallel around a ring (altogether 8 pairs) and will have electrons excited and de-excited arranged as a series of positive and negative charges working as a series of dipoles which effectively acts as capacitors adding up to the picture of chlorophyll molecules arranged as a storage ring of electrons. The issue here, from a theoretical solid-state point of view, is really how localized these kind of transfer mechanisms are in this "electron storage ring". We expect that, after the electrons have been transfered to the chlorophyll ring they become de-localized and eventually be "spiled out" in portions of $4e^-$ to the docking quinone-cytochrome molecule-complex to be carried away.

5. The Structure and Organization of Chlorophyll

The chlorophyll molecule is composed of a ring system with four nitrogens coordinating a magnesium ion and is part of the light harvesting complex. A single chlorophyll molecule, which is pictured in Fig. 3, has a ring structure with 5 pyrine

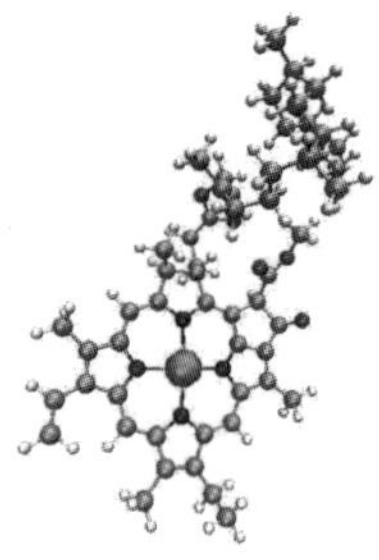

Fig. 3. (Color online) It shows the chlorophyll molecule where the green sphere is the Mg^{2+} coordinated by four nitrogen atoms(blue spheres). Carbon atoms are colored cyan, hydrogen white, and oxygens are colored red. Figure was generated using VMD.

rings arranged inside the ring around a magnesium atom that coordinates to the neighboring nitrogens.[13] The tail of the molecule ankers to the protein complex. The chlorophylls are arranged in two layers within the thylakoid membrane where each layer is located near either side of the membrane pointing toward the stromal or lumenal surface. In each layer neighboring chlorophylls are located 11.26 Å away from each other measured from the Mg^{2+} ion. These chlorophyll molecules are slightly asymmetric and are positioned in pairs of mirror images and arranged around a ring of 8 mirror image pairs. The molecules are located with a pseudo C2-symmetry between each pair of chlorophylls.[25]

5.1. *Computation of chlorophyll analog*

We start by performing calculation of a single chlorophyll analog molecule where the tail of molecule has been removed resulting in 82 atoms. The main function of the tail is anchoring of the head group so for transfer properties it can be neglected. Tentatively we could calculate a HOMO-LUMO gap to be 3–4 eV. Such a HOMO (Highest Occupied Molecular Orbital) - LUMO (Lowest Un-occupied Molecular Orbital) is somewhat artificial in standard DFT formalism unless time-dependent methods are being used.

The chlorophyll analog was optimized using restricted Hartree–Fock with the 3-21G** basis set implemented in the DFT based Gamess-US[26] containing time-dependent and excited state techniques.

6. ESR Experiments on the Chloroplast System

In order to see if we could verify experimentally any of the assumptions made in the presented model we arranged a preliminary experiments on chloroplast systems from plants in an ESR (Electron Spin Resonance) apparatus and preferably operated at low temperature. Thus we should, for example, be able to see if there was a difference in the signal from the unpaired electrons from day time with incident light and night time of no incident light. The most clear signal of electronic spin is expected to come from the manganese atoms. We have made recordings from periods (5 min. each) with incident full-spectral light and with no light at the sample cell situated in the magnet. One can observe a clear difference in the recording of the two types of periods.

7. Conclusion

A model has been proposed for the electronics of the photosynthetic process involving carothenoid and chlorophyll molecules. In the model the chlorophyll receive electrons from carotenoid after being photo-excited. However part of the energy is mediated in charge transfer and part of the energy is mediated in electromagnetic radiation. The chlorophyll molecules, arranged in a "storage" ring around a manganese atom, will transfer excitations of electrons alternating between π and π^*

states exchanging quanta by a Förster-like mechanism that can account for 80–90% of the energy output. The rest is explained by EM radiation originating from nuclear vibrations.

References

1. P Andrew and L. Barnes, *Science* 306 (2004).
2. J. Amesz, *Photosynthesis* (Elsevier, Amsterdam, 1987).
3. J. Barber (ed.), *The Photosystems: Structure, Function and Molecular Biology* (Elsevier, Amsterdam, 1992) and J. Barber (ed.), *Photosynthesis and the Environment* (Kluver Academics, 1996).
4. J. Whitmarsh and Govindjee, *Concepts in Photobiology: Photosynthesis and Photomorphogenesis*, eds G.S. Singhal *et al.* (Narosa Publishers, New Delhi, 2006), pp. 11–51.
5. P. H. Raven and G. B. Johnson, Photosynthesis, in *Biology*, 4th edn. (WCB, Wm. C. Brown Publishers, Dubuque, USA, 1996), pp. 209–232.
6. P.R. Chitnis, *Annu. Rev. Plant Physiol. Plant Mol. Biol.* **52**, 593–626 (2001).
7. V. Sundström, T. Pullerits and R. van Grondelle, *J. Phys. Chem. B* **103**, 2327–2346 (1999).
8. R. J. Sension, *Nature* **446**, 740 (2007).
9. G. S. Engel, T. R. Calhoun, E. l. Read, T. K. Ahn, T. Mancal, Y. C. Cheng, R. E. Blankenship and G. R. Fleming, *Nature* **446**, 782 (2007).
10. T. Brixner *et al.*, *Nature* **434**, 625 (2005).
11. C. Day, *Physics Today July* 23 (2005).
12. A. Damjanovic, T. Ritz and K. Schulten, *Phys. Rev. E* **59**, 3293 (1999).
13. J. Koepke, X. Hu, C. Muenke, K. Schulten and H. Michel, *Structure* **4**, 581 (1996).
14. D. Zigmantas, T. Polivka, R. G. Hiller, A. Yartsev and V. Sundström, *J. Phys. Chem. A* **105**, 10296 (2001).
15. H.M. Waswani, N. E. Holt and G. R. Fleming, *Pure Appl. Chem.* **77**, 925 (2005).
16. T. Förster, *Ann. Phys. (Leipzig)* **2**, 55 (1948).
17. D. L. Dexter, *J. Chem. Phys.* **21**, 836 (1963).
18. J. R. Lakowicz, *Principle of Fluorescence Spectroscopy* , 2nd edn. (Plenum Publisher, 1999).
19. K. R. Demarest, *Engineering Electromagnetics* (Prentice Hall, 1998), p. 576.
20. H. G. Bohr and F. B. Malik, ıtPhys. Lett. A (2007).
21. G. Cinque, R. Croce, A. Holzwarth, R. Bassi, *Biophysical Journal* (2000).
22. T. Hino, E. Kanamori, J.R. Shen and T. Kouyama, *Acta Crystallogr. D Biol. Crystallogr.* (2004).
23. W. Humphrey, A. Dalke and K. Schulten, *Journal of Molecular Graphics* (1996).
24. J. Standfuss, A.C. Terwisscha van Scheltinga, M. Lamborghini and W. Kühlbrandt, *EMBO J.* (2005).
25. Z. Liu, H. Yan, K. Wang, T. Kuang, L. Gui, X. An and W. Chang, *Nature* (2004).
26. M. F. Schmidt, K. K. Baldridge, J. A. Boatz, S. T. Elbert, M. S. Gordon, J. H. Jensen, S. Koseki, N. Matsunaga, K. A. Nguyen, S. J. Su, T. L. Windus, M. Dupuis and J. A. Montgomery, *J. Comput. Chem.* (1993).

OPTICAL AND TRANSPORT PROPERTIES IN DENSE PLASMAS COLLISION FREQUENCY FROM BULK TO CLUSTER

H. REINHOLZ, T. RAITZA and G. RÖPKE
Universität Rostock, Institut für Physik, D-18051 Rostock, Germany
heidi.reinholz@uni-rostock.de

I. V. MOROZOV
Joint Institute for High Temperatures of RAS, Izhorskaya 13/2, Moscow 125412, Russia
bogous@orc.ru

Received 31 July 2008

The dielectric function of dense plasmas is treated within a many-particle linear response theory beyond the RPA. In the long-wavelength limit, the dynamical collision frequency can be introduced which is expressed in terms of momentum and force auto-correlation functions (ACF). Analytical expressions for the collision frequency are considered for bulk plasmas, and reasonable agreement with MD simulations is found. Different applications such as Thomson scattering, reflectivity, electric and magnetic transport properties are discussed. In particular, experimental results for the static conductivity of inert gas plasmas are now well described.

The transition from bulk properties to finite cluster properties is of particular interest. Within semiclassical MD simulations, single-time characteristics as well as two-time correlation functions are evaluated and analyzed. In particular, the Laplace transform of current and force ACFs show typical structures which are interpreted as collective modes of the microplasma. The damping rates of these modes are size dependent. They increase for the transition from small clusters to bulk plasmas.

Keywords: Laser excited clusters; dielectric function; molecular dynamics simulations; absorption; dynamical conductivity.

1. Introduction

The plasma state is the most abundant state of matter in the universe, and the properties of plasma are subject of many investigations. The approach to the plasma state considered in the present paper is based on a quantum statistical treatment of a charged particle system, where interactions are governed by the Coulomb law. In contrast to high-temperature, low-density ideal plasmas where correlations can be neglected, we are interested in moderately to strongly coupled plasmas where the interaction energy $e^2/(4\pi\epsilon_0 d)$ of particles with charge e, at average distance d given by the electron density n_e, is comparable or larger than the thermal energy

$k_B T = \beta^{-1}$. The plasma parameter

$$\Gamma = \frac{e^2 \beta}{4\pi\varepsilon_0} \left(\frac{4\pi n_e}{3}\right)^{1/3} \tag{1}$$

characterizes the nonideality of the plasma and is in the order of 1 or larger for non-ideal plasmas.

Strongly coupled plasmas occur in astrophysical objects such as compact stars and planets. In laboratory experiments with schock compressed or laser excited condensed matter, plasmas within the warm dense matter regime are produced. They are characterized by a density of $10^{20} - 10^{26}$ particles/cm^3 and temperatures in the region of $10^3 - 10^8$ K. Such plasmas are of relevance for inertial fusion experiments which is one of the emerging fields in recent research. Another application is the production of nanoplasmas by irradiation of matter with high intense, short pulse laser in the visible or VUV/X-ray region. Currently, the FLASH facility at DESY, Hamburg, is available with wavelengths ranging from 7–50 nm in the VUV region.[1] A XFEL at Hamburg is proposed[2] as well as a free electron laser at the Stanford Linear Acceleration Center (SLAC).[3] In particular, we will consider laser excited small clusters. A fascinating issue is the transition from condensed matter as described in solid state physics to the so-called warm dense matter.

For the sake of simplicity, we consider here only singly charged ions and restrict ourselves to the nonrelativistic case. We are interested in the dielectric properties ruling the interaction of the strongly coupled Coulomb system with the radiation field.[4] Optical properties are of interest for plasma diagnostics but also for excitation and absorption processes. Thus the approach given here allows to consider the static and dynamic conductivity, scattering processes like Thomson scattering,[5–7] reflectivity,[8,9] bremsstrahlung radiation[10,11] and optical transitions.[12,13] Further effects which arise in magnetic fields such as Hall effect or spectral line shapes in strong magnetic fields can also be considered,[14] but will not be presented here.

A more delicate question is the relation between classical and quantum description. Quantum effects have two origins. On the one hand, we have the statistical treatment where the thermal wavelength occurs, describing the degeneracy of the constituents of the plasma, in particular electrons. Here we can introduce the degeneracy parameter

$$\Theta = \frac{2m_e}{\beta\hbar^2} \left(3\pi^2 n_e\right)^{-2/3}, \tag{2}$$

describing the ratio of thermal energy to the Fermi energy. On the other hand, the quantum effect is related to the interaction where we have bound states and scattering phase shifts to be calculated from a quantum approach. A corresponding parameter is the ratio of the thermal energy to the binding energy.

We can include all quantum effects in a systematic way using a quantum statistical approach as shown in Sec. 2. Within linear response theory, transport coefficients are expressed in terms of equilibrium correlation functions which can be evaluated using the Green function formalism. However, this is a perturbative approach with respect to the interaction, and can be applied only for $\Gamma \leq 1$. Alternatively, MD simulations are applicable for the evaluation of equilibrium correlation functions at any coupling strengths, but are based on a classical treatment. To understand the optical properties such as absorption, excitation, reflection, scattering and emission of light we investigate different ACFs such as dipole, current and force, see Sec. 3.

Of particular interest are microplasmas produced in small clusters if they are irradiated by intense short-pulse laser beams. Recently, MD simulations codes have been developed to describe the properties and the dynamical evolution of such excited clusters.[15,16] Vlasov and VUU simulations[17] were also applied to study the interaction of short laser pulses with clusters. Calculations for Na icosahedrons are shown in Sec. 4. Compared with bulk plasmas, the ACFs show a more complex structure which is related to different, single-particle properties, Sec. 5, as well as collective, excitation modes, Sec. 6. The position and the damping rates of the excitations are discussed in Sec. 7. Varying the number of atoms bound in a cluster, we will investigate size effects for the cluster properties and compare with bulk properties.

2. Dielectric and Optical Response

The response of homogeneous, two-component plasma to external fields is given by the dielectric function

$$\epsilon(\mathbf{k},\omega) = 1 + \frac{i}{\epsilon_0\omega}\sigma(\mathbf{k},\omega) = 1 - \frac{\omega_{\rm pl}^2}{\omega(\omega - i\nu(\mathbf{k},\omega))}\,, \tag{3}$$

where the plasma frequency is given by $\omega_{\rm pl} = \sqrt{n_{\rm e}e^2/\epsilon_0 m_e}$. Assuming local thermal equilibrium, and according to the fluctuation dissipation theorem,[4,8,18] the calculation of dynamical conductivity $\sigma(\mathbf{k},\omega)$ and dynamical collision frequency $\nu(\mathbf{k},\omega)$ can be performed within linear response theory evaluating the current-current correlation function[19] or, more generally, force-force and dipole-dipole ACFs.

In particular, we will consider the long-wavelength limit $k \to 0$ where only the frequency dependence remains. The frequency-dependent conductivity is given by equilibrium ACFs of the current $\vec{j}$ or the velocity $\dot{\vec{R}}$ of the charge carriers (because of the large ion to electron mass fraction, we can restrict ourselves to electrons only when considering electronic properties)

$$\sigma(\omega) = \beta\Omega\langle\vec{j};\vec{j}\rangle_{\omega+i\eta} = \frac{e^2\beta}{\Omega}\langle\dot{\vec{R}}\;;\dot{\vec{R}}\rangle_{\omega+i\eta}\,,$$

where the Laplace transform of the correlation functions are defined by equilibrium averages according to

$$\langle A;B\rangle_z = \int_{-\infty}^{0} dt\, e^{izt}\int_0^\beta d\tau\langle A(t-i\hbar\tau)B\rangle\,. \tag{4}$$

The equilibrium correlation functions can be evaluated using quantum statistical methods such as thermodynamic Green functions. Within perturbation theory, a diagram representation can be given. Performing appropriate partial summations, different effects such as dynamical screening, strong collisions or higher correlations are included. This approach treats quantum effects in a rigorous way. Analytical expressions in the weak-coupling limit can be found, but are questionable for coupling parameters $\Gamma \geq 1$. On the other hand, MD simulations can be performed based on a classical description, calculating the trajectory of the system in the phase space of N particles and replacing the ensemble average by the time average, details are given below in Sec. 4. Using an appropriate pseudopotential, quantum effects can be included, and good agreement of results between analytical quantum statistical calculations and MD simulations has been found.[20]

Numerically as well as analytically, it is of advantage to use the force ACFs instead of the current ACFs since they are directly related to the dynamical collision frequency[22] for the different relavant scattering mechanisms between electrons and electrons (ee), ions (ei) and atoms (ea), respectively,

$$\nu(\omega) = \frac{\beta}{n_\mathrm{e}\Omega}\left\langle \dot{\vec{P}}; \dot{\vec{P}} \right\rangle_{\omega+i\eta}, \qquad \dot{\vec{P}} = \vec{F} = \vec{F}_\mathrm{ei} + \vec{F}_\mathrm{ee} + \vec{F}_\mathrm{ea}\,. \tag{5}$$

This formalism has been applied to different phenomena such as the conductivity in fully ionized plasmas, or to the evaluation of optical properties such as refraction index $n(\omega)$ and absorption coefficient $\alpha(\omega)$ for bulk according to

$$\lim_{k\to 0} \epsilon(\mathbf{k},\omega) = \left(n(\omega) + \frac{ic}{2\omega}\alpha(\omega)\right)^2 \tag{6}$$

which allows to calculate bremsstrahlung[10] and spectral line profiles.[12]

The dynamical structure factor has been evaluated according to

$$S(\mathbf{k},\omega) = \frac{\epsilon_0 \hbar k^2}{\pi\, n_\mathrm{e} e^2}\,\frac{1}{\mathrm{e}^{-\beta\hbar\omega} - 1}\,\mathrm{Im}\,\epsilon^{-1}(\mathbf{k},\omega)\,, \tag{7}$$

and comparison with MD simulations has been performed.[20–22] An important issue is the account of correlations in determining Thomson scattering cross section

$$\frac{d^2\sigma}{d\Omega\, d\omega} = \frac{8\pi\, r_\mathrm{e}^2}{3}\frac{k_1}{k_0} S(k,\omega) \tag{8}$$

where the transfer wave number $\vec{k} = \vec{k}_0 - \vec{k}_1$ is given by the difference between incident and scattered wave number, $\omega = \omega_0 - \omega_1$, and r_e is the classical electron radius. Decomposing the dynamical structure factor into different contributions, the free electron contribution can be analyzed to infer the temperature and density of the plasma.[5,7] An adequate treatment of interaction and collisions is needed to provide relations valid also in the non-ideal plasma region. In particular, at very high densities, Thomson scattering is one of the favoured methods for plasma diagnostics since X-rays can penetrate the plasma at condensed matter densities.[6]

3. Electrical Conductivity of Partially Ionized Argon Plasmas

As an example for recent progress in evaluating transport coefficients in non-ideal plasmas, we consider the electrical conductivity of partially ionized inert gas plasmas.[23,24] We report in particularly on argon. The conductivity of a fully ionized, non-ideal plasma has been investigated extensively, and reasonable agreement between analytical calculations, MD simulations and experiments with shock-induced dense plasmas has been obtained.[4] However, experiments are generally performed involving partially ionized plasmas. An important issue is the inclusion of the interaction with bound states when calculating the electrical conductivity.

The dc electrical conductivity is related to the static collision frequency according to

$$\sigma_{\rm dc} = \lim_{\omega \to 0} \sigma(\omega) = \frac{\epsilon_0\, \omega_{\rm pl}^2}{\nu_{\rm dc}} = \frac{e^2 n_{\rm e}}{m_e \nu_{\rm dc}}\,. \tag{9}$$

As shown in Eq. (5), the collision frequency can be decomposed into different parts determined by the different interactions present. In particular, the interaction between electrons and atoms will give a contribution to the collision frequency and lead to a considerable reduction of the electrical conductivity. Using empirical cross sections for the electron-atom scattering, it is possible to incorporate special effects such as the Ramsauer minimum which is characteristic for the electron scattering by inert gases.[24] Employing these empirical cross sections instead of the polarization potential approximation,[24] it was possible to develop a consistent description of the collision frequency and find excellent agreement with the corresponding conductivity. Results for argon are shown in Fig. 1.

Similar excellent agreement with experimental results has been obtained also for other inert gases.[24] Whereas the inclusion of bound states in evaluating the

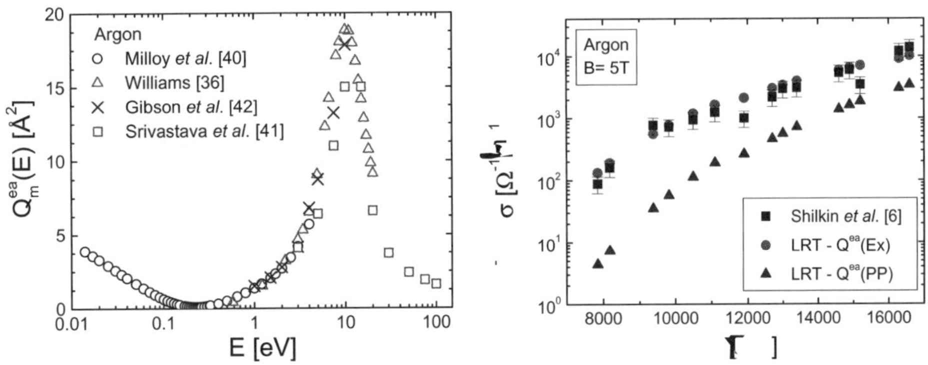

Fig. 1. Left: Experimental results of the ea momentum transfer cross section as a function of energy. Right: Electrical conductivity of argon as a function of temperature at different densities. Experimental results (Shilkin) are compared with calculations within linear response theory (LRT) with the ea interaction from polarization potential (PP) or experimental data (Ex), see left side of figure. For details and references see Adams *et al.*[24]

dynamical collision frequency is an open problem at present, this special example of static problems gives an idea how the transition from the fully ionized to a partially ionized non-ideal plasma can be performed.

4. Laser Excited Clusters

Recently, clusters at nearly solid densities are more accessible to experimental investigations[25] using laser intensities of $10^{13} - 10^{16}$ Wcm^{-2}. We are interested in the cluster size dependence of the collisional damping rates in comparison to bulk systems. Laser excited Na_{55}, Na_{147} and Na_{309} nano-clusters irradiated by a short pulse laser, are simulated using a MD code.[16] Initially, electrons are positioned on top of singly charged ions, see Fig. 2, for which isocahedral geometry is assumed. MD simulations were performed, using an error function pseudo-potential

$$V_{\rm ei}(r) = V_{\rm erf}(r) = -\frac{Z_i e^2}{4\pi\varepsilon_0 r}\,{\rm erf}\left(\frac{r}{\lambda}\right) \tag{10}$$

where $\lambda = 6.02\ a_B$ in order to reproduce the correct ionisation energy $I_P = V_{\rm erf}(r \to 0) = -5.1$ eV for sodium.

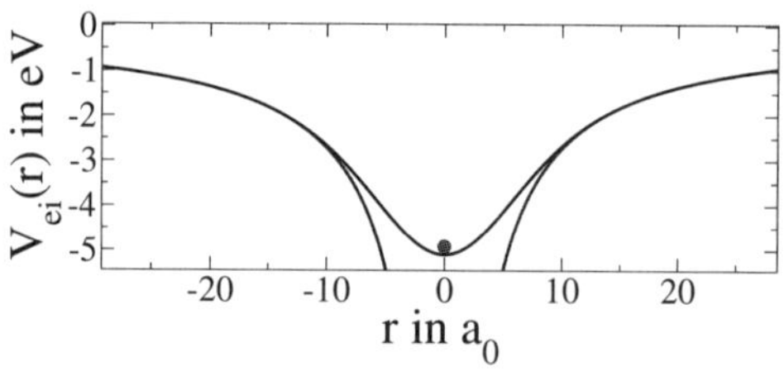

Fig. 2. Error function potential (full line) used for MD calculations of clusters consisting of sodium atoms in comparison to Coulomb potential (dashed line).

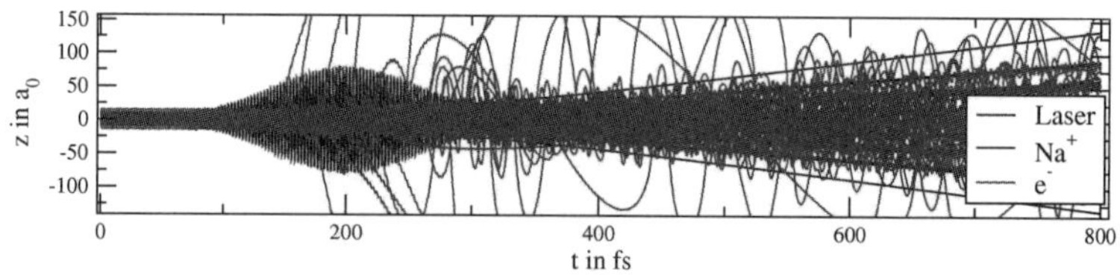

Fig. 3. Evolution of a Na_{55} cluster irradiated by a 100 fs laser pulse of $I = 0.5 \cdot 10^{12}$ Wcm^{-2} intensity. Electrons become delocalized and form a nanoplasma. The ion geometry (straight lines) expands.

Absorbing energy from the electromagnetic field, electrons become delocalized within the cluster, forming a nano-plasma. We consider a cosine square laser pulse with intensities of $I = 0.5;\ 1;\ 5 \cdot 10^{12}$ Wcm^{-2}. From the ion geometry, we define a cluster radius $r_c = \sqrt{5/3 \sum_i \langle \vec{r}_i^2 \rangle}$ which marks the spatial extension of the cluster. The scale of electron motion is in the fs region. On the time scale of 100 fs, some electrons escape this cluster extension r_c thus charging the cluster. Ion motion becomes essential on the timescale of ps, leading to the expansion of the cluster. Figure 3 illustrates the expansion of a Na_{55} cluster due to a 100 fs laser pulse of $I = 0.5 \cdot 10^{12}$ Wcm^{-2} intensity.

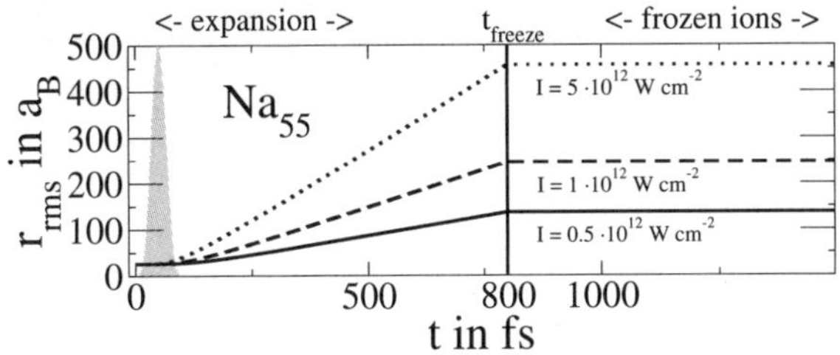

Fig. 4. Evolution of the Na_{55} cluster radius after irradiation with a 100 fs laser pulse of different intensities I. After 800 fs, the time of freezing the ions t_{freeze}, full MD simulations are continued by restricted MD simulations where the cluster radius does not expand any further.

During the expansion of the excited cluster, we assume local thermal equilibrium because of the small electron mass compared with the Na ions. In the following, we applied a restricted MD simulations scheme. For this, ion positions were frozen after an expansion time t_{freeze}, in order to analyze nano-plasmas at different thermodynamical parameters. Restricted MD simulations at those given ion positions allow to consider only electron dynamics. The evolution of cluster radius r_c is shown in Fig. 4 for three parameter sets. Different intensities at otherwise same conditions of pulse duration and starting geometry, leads to different densities and temperatures.

5. Single-Time Properties

After the freezing time, the electrons are allowed to equilibrate until a constant temperature T_e and cluster charge Z is reached. Here, we consider these single-time properties of the nanoplasma in quasi-equilibrium.

Equilibration of electron temperature, calculated as the mean kinetic energy of the electrons remaining within the cluster, is shown in Fig. 5 over a time period of 100 ns for a Na_{55}^{+11} cluster. The freezing time was $t_{\mathrm{freeze}} = 100$ fs and a radius of $r_c = 16$ a_B was obtained. The insert of Fig. 5 illustrates that the momentum distribution

$$f_e(p) = 4\pi p^2 \left(\frac{K}{\pi}\right)^{3/2} \exp\left(-Kp^2\right), \tag{11}$$

with $K = 1/(2m_e k_B T_e)$ taken from all times after 20 ns, has a Boltzmann like behaviour and the temperature can be deducted accordingly. An electron temperature of $T_e = 1.23$ eV was calculated as a time average of $N = 2 \cdot 10^8$ steps of $\Delta t = 0.1$ fs.

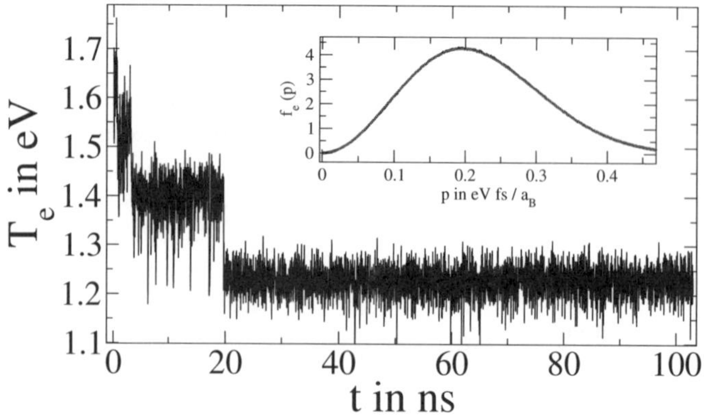

Fig. 5. Electron temperature at frozen ion geometry. After 20 ns, stable thermodynamical equilibrium is reached. From this moment, the time average $f_\mathrm{e}(p)$ of the momentum distribution function, seen in the insert, was calculated.

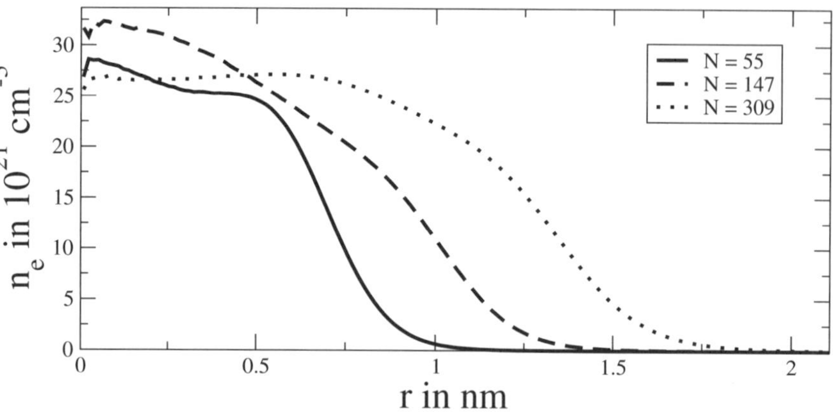

Fig. 6. Radial dependence of the electron density for different cluster sizes after irradiation with laser pulse of intensity $I = 0.5 \cdot 10^{12}$ W cm^{-2}, and equilibration after freezing time $t_\mathrm{freeze} = 100$ fs.

The density profile of the electrons remaining within the cluster can be deducted from the MD simulations. Figure 6 shows the result for a varying number of ions. A plateau is observed in the core which extends with increasing cluster size. Alternatively, the electron density can be calculated via a mean field potential ($n_\mathrm{e} \sim \exp(-U(r)/k_\mathrm{B}T_\mathrm{e})$, where $U(r)$ is extracted from MD simulations), see Fig. 7 for a Na_{55} cluster.

6. Two-Time Properties

Once local equilibrium is reached, we have also determined two-time properties. We calculated the momentum auto-correlation function (ACF) for the total momentum $\vec{P}_\mathrm{e}(t) = \sum_{i=1}^{N_\mathrm{e}} \vec{p}_i(t)$ of electrons in the equilibrated cluster within the restricted MD

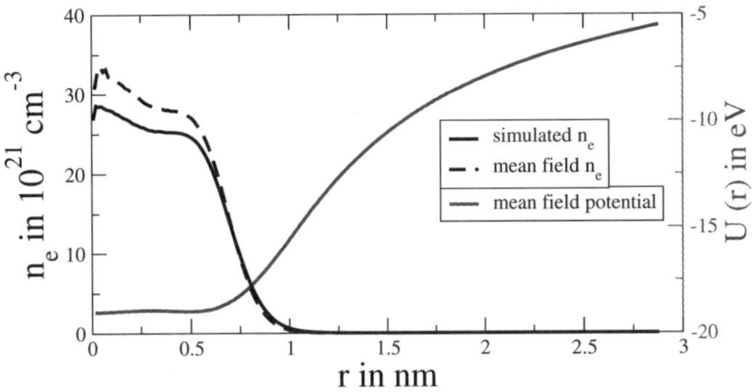

Fig. 7. Left axis: The electron density profile for a Na_{55}^{+11} cluster from same conditions as in Fig. 5 deducted directly from the simulation data (solid line) and calculated using an external mean field potential (dashed line). Right axis: Corresponding mean field potential.

simulations

$$K(t) = \langle \vec{P}_{\rm e}(t), \vec{P}_{\rm e}(0) \rangle \approx \frac{1}{N_\tau} \sum_{i=1}^{N_\tau} \vec{P}_{\rm e}(t + \tau i) \vec{P}_{\rm e}(\tau i) \tag{12}$$

with $N_\tau = 1 \cdot 10^6$ the number of averages. The normalized Laplace transformation of the momentum ACF Eq. (12) is

$$\tilde{K}(\omega) = \frac{1}{\langle P_{\rm e}^2 \rangle} \int_0^\infty {\rm d}t e^{{\rm i}\omega t} \langle \vec{P}_{\rm e}(t), \vec{P}_{\rm e}(0) \rangle \,. \tag{13}$$

The Laplace transforms of the momentum ACFs, Eq. (13), for three different cluster sizes with the same parameters as used in Fig. 6, are shown in Fig. 8. A double resonance structure is observed. A shift of the resonances is seen which depends not only on the cluster size but also on other cluster parameters as temperature, ionization degree and electron density.

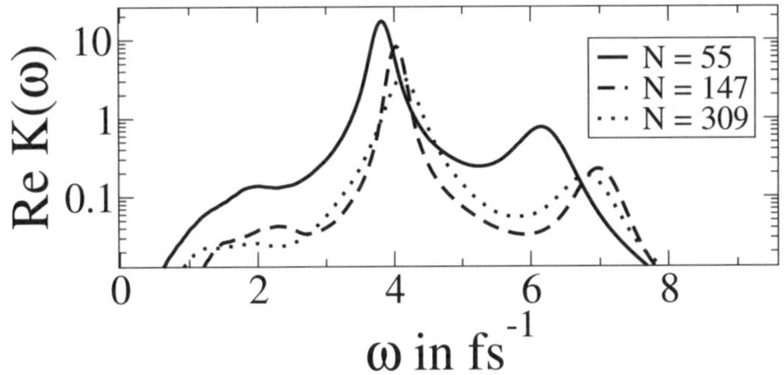

Fig. 8. The momentum ACF spectrum calculated from Laplace transform of the time dependent momentum ACF for different cluster sizes which correspond to the density profiles of Fig. 6.

In Fig. 9, the time evolution of the momentum ACF spectrum for the Na_{55} cluster is shown for different freezing times, thus following the expansion of the cluster and investigating different instants of the evolution. Considering a local equilibrium at each time step, the momentum ACF was calculated with restricted MD simulations. The freezing times considered were the time directly after the end of the laser interaction and four times later, taking a time step of $\Delta t = 50$ fs, till 200 fs afterwards. The electron density decreases, because the ion geometry expansion evolves with each time step, which also dilutes the electron gas. Collective modes can be seen for all electron densities, with changing positions and widths.

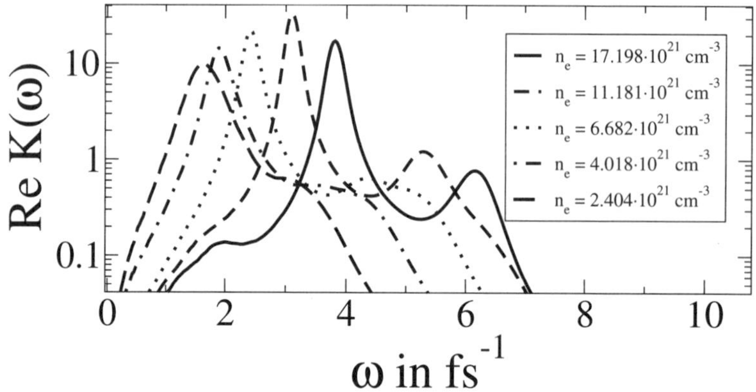

Fig. 9. The momentum ACF spectrum calculated from Laplace transform of the time dependent momentum ACF taking different freezing times. The electron density decreases with respect to the expansion of the ion geometry.

In a two-component bulk plasma, the dielectric function is given by the frequency dependent ACF $\tilde{K}(\omega)$ via

$$\frac{1}{\epsilon(\omega)} = 1 - \mathrm{i}\frac{\omega_{\mathrm{pl}}}{\omega}\tilde{K}(\omega) \tag{14}$$

with the plasma frequency $\omega_{\mathrm{pl}}^2 = \mathrm{e}^2 n_{\mathrm{e}}/(\varepsilon_0 m_{\mathrm{e}})$. Considering a generalized Drude approximation for the dielectric function,[4,8] Eq. (3), the momentum ACF can be directly related to the collision frequency $\nu(\omega)$

$$\tilde{K}(\omega) = \frac{\omega}{\nu(\omega)\omega - \mathrm{i}\left(\omega^2 - \omega_{\mathrm{pl}}^2\right)} \tag{15}$$

which has a Lorentzian form. Thus, information about the collisions in the systems can be gathered. We would like to apply this approach to finite systems using correlation functions to determine optical properties.

As can be seen from the structure of the ACFs in Fig. 8 and Fig. 9, a single Lorentzian fit as in bulk according to Eq. (15) is not possible in the case of clusters.

We will consider two resonance frequencies ω_0 and ω_1 and respective collision frequencies ν_0 and ν_1 which reproduce the width of the resonances. Both resonances are fitted assuming a double Lorentzian parametrizing positions and widths of the resonances. In Table 1 one can see the increase of the collision frequency with increasing size of charged cluster. This is in agreement with Ramunno *et al.*[26] By analyzing the damping rates of the ACF in larger clusters, they found collision frequencies lower compared to bulk systems.

Table 1. Resonance frequencies and damping rates for the first two modes and for two different number of ions.

Na_{55}^{11+} cluster						Na_{309}^{46+} cluster					
r_{rms} nm	T_e eV	ω_0 fs^{-1}	ν_0 fs^{-1}	ω_1 fs^{-1}	ν_1 fs^{-1}	r_{rms} nm	T_e eV	ω_0 fs^{-1}	ν_0 fs^{-1}	ω_1 fs^{-1}	ν_1 fs^{-1}
0.8	1.2	6.2	0.52	3.8	0.26	1.5	3.0	6.8	0.63	4.1	0.39
1.0	0.9	5.3	0.62	3.1	0.22	1.7	2.4	5.4	0.91	3.1	0.34
1.2	0.7	4.4	1.04	2.4	0.26	2.0	1.9	4.1	1.27	2.3	0.31

In general, the behavior of the dynamical collision frequency should be deducted when comparing the full spectrum with the resonance structure in the collisionless case. Restricted MD simulations of Na_{55} with suppressed electron collisions (mean field) taking 6 test particles per electron leads to sharper resonance peaks, as one can see in Fig. 10. This indicates that collisions should lead to broadening of the resonances.

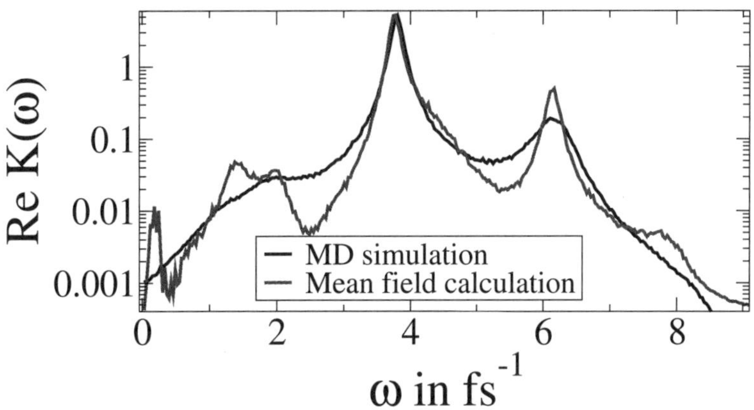

Fig. 10. The momentum ACF spectrum of the Na_{55} calculated via restricted MD simulations in comparison to a mean field calculation, taking 6 test particles per electron. Both calculations were done with parameters corresponding to Fig. 5.

7. Collective Excitations

One can calculate the collective modes of a spherically symmetric hydrodynamic electron system. Assuming harmonic oscillations $\vec{a}(\vec{r},t) = a(r)\mathrm{e}^{-\mathrm{i}\omega t}\vec{e}_\mathrm{z}$ in z-direction with a radially dependent amplitude $a(r)$, the eigenfrequencies can be calculated. The following equation of motion has to be solved:

$$m_\mathrm{e}\ddot{\vec{a}}(\vec{r},t) = -m_\mathrm{e}\omega^2 a(r)\mathrm{e}^{-\mathrm{i}\omega t}\vec{e}_\mathrm{z} = \Delta F_{\mathrm{U},z}(\vec{r},t)\vec{e}_\mathrm{z} \tag{16}$$

where the change of the force $\Delta\vec{F}_\mathrm{U}(\vec{r},t)$ is due to the mean field produced by the change of the surrounding electron cloud $\Delta n_\mathrm{e}(r)$ which can be expressed as

$$\Delta n_\mathrm{e}(r) = -\frac{\partial}{\partial z}\left[a(r)n_\mathrm{e}^\mathrm{eq}(r)\right]. \tag{17}$$

using the hydrodynamic equation of continuity. An additional term due to the changing pressure has been neglected so far. Then we find the following differential equation

$$m_\mathrm{e}\omega^2 a(r) = -\frac{\partial^2}{\partial z^2}\int \mathrm{d}^3\vec{r}_1\, a(r_1)n_\mathrm{e}^\mathrm{eq}(r_1)V_\mathrm{ee}(\vec{r}-\vec{r}_1). \tag{18}$$

In order to calculate the eigenfrequencies, this equation was symmetrized, using $\Psi(r) = a(r)\sqrt{n_\mathrm{e}^\mathrm{eq}(r)}$,

$$m_\mathrm{e}\omega^2\Psi(r) = -\frac{\partial^2}{\partial z^2}\int \mathrm{d}^3\vec{r}_1\,\Psi(r_1)\sqrt{n_\mathrm{e}^\mathrm{eq}(r_1)}V_\mathrm{ee}(\vec{r}-\vec{r}_1)\sqrt{n_\mathrm{e}^\mathrm{eq}(r)}. \tag{19}$$

Introducing a normalized set of basic functions $\int \Psi_i(r)\Psi_j(r)d^3\vec{r} = \delta_{ij}$, the eigenvalues of the resonance frequency can be calculated via diagonalization, similar to the Ritz variational approach in quantum mechanics. After transformation from variable z to the radial distance r, we find

$$\begin{aligned} m_\mathrm{e}\omega^2 = &-\frac{4\pi}{3}\int_0^\infty \mathrm{d}r\, r^2\Psi(r)\sqrt{n_\mathrm{e}^\mathrm{eq}(r)}\frac{\partial^2}{\partial r^2}\int \mathrm{d}^3\vec{r}_1\Psi(r_1)\sqrt{n_\mathrm{e}^\mathrm{eq}(r_1)}V_\mathrm{ee}(\vec{r}-\vec{r}_1) \\ &-\frac{8\pi}{3}\int_0^\infty \mathrm{d}r\, r\Psi(r)\sqrt{n_\mathrm{e}^\mathrm{eq}(r)}\frac{\partial}{\partial r}\int \mathrm{d}^3\vec{r}_1\Psi(r_1)\sqrt{n_\mathrm{e}^\mathrm{eq}(r_1)}V_\mathrm{ee}(\vec{r}-\vec{r}_1). \end{aligned} \tag{20}$$

In the following, we use Hermitean basis functions with a width parameter b as the normalized basis set,

$$\psi_{0,\mathrm{b}}(r) = \frac{\exp\left(-\frac{r^2}{b^2}\right)}{\left(\sqrt{\frac{\pi}{2}}b\right)^{3/2}}, \tag{21}$$

$$\psi_{1,\mathrm{b}}(r) = \frac{\exp\left(-\frac{r^2}{b^2}\right)}{\left(\sqrt{\frac{\pi}{2}}b\right)^{3/2}}\sqrt{\frac{8}{3}}\left(\frac{r^2}{b^2}-\frac{3}{4}\right), \tag{22}$$

$$\psi_{2,\mathrm{b}}(r) = \frac{\exp\left(-\frac{r^2}{b^2}\right)}{\left(\sqrt{\frac{\pi}{2}}b\right)^{3/2}}\sqrt{\frac{32}{15}}\left(\frac{r^4}{b^4}-\frac{5}{2}\frac{r^2}{b^2}+\frac{15}{16}\right). \tag{23}$$

For the case shown in Fig. 5, calculations of the resonance frequencies were done. The density profile from the restricted MD simulations was taken as the equilibrium density $n_e^{eq}(r)$. Analyzing the Laplace transfom of the electron momentum ACF, three collective modes were found at $\omega_0 = 6.16 \text{ fs}^{-1}$, $\omega_1 = 3.80 \text{ fs}^{-1}$ and $\omega_2 = 1.98 \text{ fs}^{-1}$. Using now the hydrodynamical approach, the eigenvalues for the resonance frequencies in dependence on the width parameter b have been calculated and illustrated in Fig. 11 . For $b = 6$ a_B, the resonance frequencies are maximal.

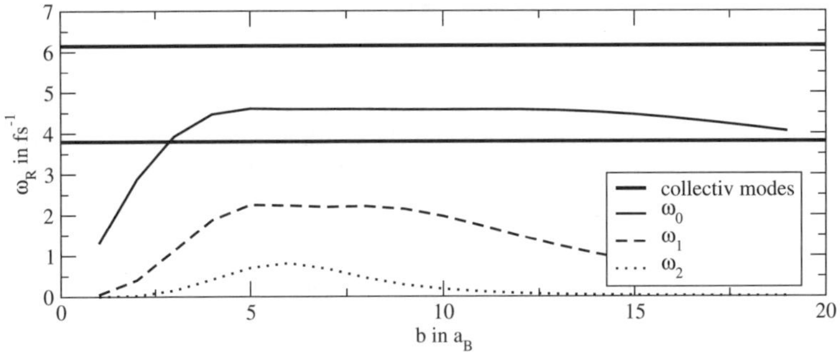

Fig. 11. The calculated eigenvalues of the resonance frequencies are shown in dependence on the width parameter b. The thick horizontal solid lines display the resonance frequencies as obtained from the Laplace transform of the ACF. The first (solid line), second (dashed line) and third (dotted line) eigenvalue were calculated taking the mean field interaction into account.

They were calculated as $\omega_0 = 4.59 \text{ fs}^{-1}$, $\omega_1 = 2.23 \text{ fs}^{-1}$ and $\omega_2 = 0.82 \text{ fs}^{-1}$. The ratio between the resonance frequencies is of the same order as for the results taken directly from the simulations. However, the position is too low. Further investigations are necessary to improve this result, e.g. using a more involved hydrodynamical approach taking into account corrections due to pressure gradients. The spatially dependent amplitude $a(r)$ was calculated via $\psi(r) = a(r)\sqrt{n_e(r)}$. The results are shown in Fig. 12.

8. Discussion and Conclusion

We have shown that the evaluation of equilibrium ACF for the electron momentum or the force allows to calculate the dynamical collision frequency which determines the dielectric function. Having the dielectric function to our disposal, different optical and transport properties of dense plasmas can be determined. Applications are dc conductivity, absorption, emission, reflection, bremsstrahlung, and Thomson scattering.

In particular, the methods developed to investigate homogenous non-ideal plasmas can also be applied to finite nano-plasmas, produced by laser irradiation of clusters. Using restricted MD simulations, we determined single-time properties (density and momentum distribution) as well as two-time properties, in particular

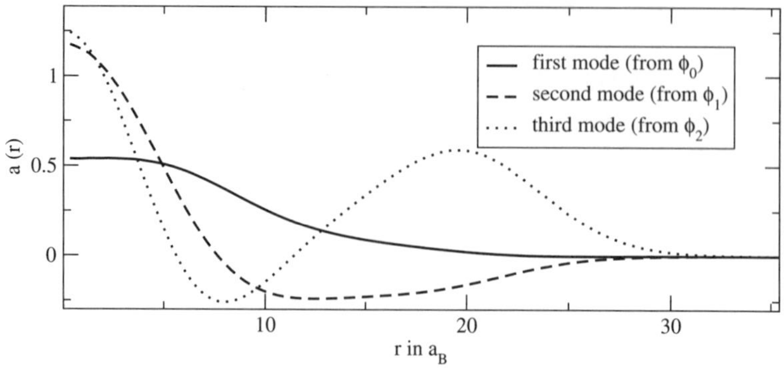

Fig. 12. Spatially dependent amplitudes $a(r)$ for the width parameter $b = 6\ a_B$, derived from the maximal resonance frequencies using the mean field contribution.

the momentum ACF spectrum. The single-time properties are well reproduced from equilibrium statistics. The momentum ACF shows a peak structure which is explained via collective excitations of the nano-plasma. Within exploratory calculations, collective modes are obtained, solving the self consistent mean field equation. The exact position of the peaks, however, needs further investigations.

Of interest are the damping rates which are related to the collision frequency. We found that damping rates obtained by Lorentzian fits to the excitation spectrum can be identified, which show a dependence on the size of the cluster. The collision frequency is increasing with increasing cluster size due to more frequent collisions with other particles in the average. A smooth transition to the bulk behavior is expected, see Ref. 26. More systematic investigations are needed to derive the size effects in the collision frequency, which will be the subject of future work.

Acknowledgments

We would like to thank E. Suraud for many fruitful discussions. The authors gratefully acknowledge financial support by the Deutsche Forschungsgesellschaft within the Sonderforschungsbereich SFB 652. I.V. Morozov acknowledges CRDF and Dynasty foundation.

References

1. R. Bonifacio, C. Pellegrini and L. Narducci, *Opt. Comm.* **50**, 373 (1984).
2. R. Brinkmann, K. F. B. Faatz, J. Rossbach, J. R. Schneider, H. Schulte-Schrepping, D. Trines, T. Tschentscher and H. Weise (eds.), *TESLA XFEL Technical Design Report (Supplement)* (DESY, Hamburg, Germany, DESY Rep. 2002-167, 2002).
3. *LCLS Design Study Report* (The LCLS Design Study Group, SLAC, Stanford, CA, 1998), SLAC-R-0521.
4. H. Reinholz, *Les Annales de Physique* **30**, 4–5 (2006).
5. R. Thiele, R. Redmer, H. Reinholz and G. Röpke, *J. Phys. A* **39**, 4365 (2006).
6. A. Höll, Th. Bornath, L. Cao, T. Döppner *et al.*, *HEDP* **3**, 120 (2007).

7. S. H. Glenzer, O. L. Landen, P. Neumayer, R. W. Lee, K. Widmann, S. W. Pollaine, R. J. Wallace, G. Gregori, A. Höll, T. Bornath, R. Thiele, V. Schwarz, W.-D. Kraeft and R. Redmer, *Phys. Rev. Lett.* **98** (2007).
8. H. Reinholz, Yu. Zaparoghets, V. Mintsev, V. Fortov, I. Morozov and G. Röpke, *Phys. Rev. E* **68**, 036403 (2003).
9. T. Raitza, H. Reinholz, G. Röpke, V. Mintsev and A. Wierling, *J. Phys. A* **39**, 4393 (2006).
10. A. Wierling, T. Millat, G. Röpke, R. Redmer and H. Reinholz, *Phys. Plasma* **8**, 3810 (2001).
11. C. Fortmann, R. Redmer, H. Reinholz, G. Röpke, A. Wierling and W. Rozmus, *HEDP* **2**, 57 (2006).
12. B. Omar, S. Günter, A. Wierling and G. Röpke, *Phys. Rev. E* **73**, 056405 (2006).
13. A. Sengebusch, S. H. Glenzer, H. Reinholz and G. Röpke, *Contrib. Plasma Phys.* **47**, 309 (2007).
14. J. R. Adams, H. Reinholz, R. Redmer and M. French, *J. Phys. A* **39**, 4329 (2006).
15. M. Belkacem, F. Megi, P.-G. Reinhard, E. Suraud and G. Zwicknagel, *Eur. Phys. J. D* **40**, 247 (2006).
16. H. Reinholz, T. Raitza and G. Röpke, *Int. J. Mod. Phys. B* **21**, 2460 (2007).
17. J. Köhn, R. Redmer, K.-H. Meiwes-Broer and T. Fennel, *Phys. Rev. A* **77**, 033202 (2008).
18. H. Reinholz, R. Redmer, G. Röpke and A. Wierling, *Phys. Rev. E* **62**, 5648 (2000).
19. R. Kubo, *J. Phys. Soc. Jpn.* **120**, 570 (1957).
20. I. Morozov, H. Reinholz, G. Röpke, A. Wierling and G. Zwicknagel, *Phys. Rev. E* **71**, 066408 (2005).
21. A. Selchow, G. Röpke, A. Wierling, H. Reinholz, T. Pschiwul and G. Zwicknagel, *Phys. Rev. E* **64**, 056410 (2001).
22. H. Reinholz, I. Morozov, G. Röpke and Th. Millat, *Phys. Rev. E* **69**, 066412 (2004).
23. J. R. Adams, N. S. Shilkin, V. E. Fortov, V. K. Gryaznov, V. B. Mintsev, R. Redmer, H. Reinholz and G. Röpke, *Phys. of Plasmas* **14**, 062303 (2007).
24. J. Adams, H. Reinholz, R. Redmer, V. B. Mintsev, N. S. Shilkin and V. K. Gryaznov, *Phys. Rev. E* **76**, 036405 (2007).
25. T. Döppner, Th. Diederich, A. Przystawik, N. X. Truong, Th. Fennel, J. Tiggesbäumker and K. -H. Meiwes-Broer, *Phys. Chem.* **9**, 4639 (2007).
26. L. Ramunno, C. Jungreuthmayer, H. Reinholz and T. Brabec, *J. Phys. B* **39**, 4923 (2006).

ON THE N-REPRESENTABILITY AND UNIVERSALITY OF $F[\rho]$ IN THE HOHENBERG-KOHN-SHAM VERSION OF DENSITY FUNCTIONAL THEORY

EDUARDO V. LUDEÑA

Centro de Química, Instituto Venezolano de Investigaciones Científicas, IVIC, Apartado 21827, Caracas 1020-A, Venezuela

eludena@ivic.ve

FRANCESC ILLAS

Department de Química Física & CeRQ, Universitat de Barcelona & PCB, Martí i Franques 1, E-08028 Barcelona, Spain

francesc.illas@ub.edu

ALEJANDRO RAMIREZ-SOLIS*

Department of Chemistry and Biochemistry, University of California at Santa Barbara, Santa Barbara, CA 93106-9510, USA

alex@buzon.uaem.mx

Received 31 July 2008

We discuss two basic problems in the Hohenberg-Kohn-Sham version of density functional theory, HKS-DFT: the first, the N-representability of the functional $F[\rho]$ and, the second, the universality of $F[\rho]$. In relation to the first, we show that $F[\rho]$ must satisfy N-representability conditions that follow from those on the 2-matrix $D^2(\mathbf{r}_1, \mathbf{r}_2; \mathbf{r}', \mathbf{r}'_2)$. In the case of the second, we provide arguments based on the equivalence between *ab initio* DFT and HKS-DFT to show that the functional $F[\rho]$ is not universal.

Keywords: Density functional theory; N-representability; Universality.

1. Introduction

Density functional theory (see Refs. 1–18) differs from ordinary quantum theory in that whereas the latter is based on the Schrödinger equation, or equivalently, on the variational principle or perturbation theory which, in general, involve N-particle wave functions, the former, is expressed in terms of an internal energy functional $F[\rho]$ which is a functional of the one-particle density $\rho(\mathbf{r})$. In the present work, we briefly comment on two aspects of the Hohenberg-Kohn-Sham formulation

*Permanent Address: Departamento de Física, Facultad de Ciencias, Universidad Autónoma del Estado de de Morelos, Cuernavaca, Morelos 62210, Mexico

of DFT, HKS-DFT,[19–21] (a more detailed version is given in Ref. 22) which in our opinion, are not sufficiently clarified. The first is whether there is a need to impose N-representability conditions on the energy functional $F[\rho]$ (not required in HKS-DFT) and the second, whether $F[\rho]$ may be truly regarded as a universal functional, a basic tenet of HKS-DFT. In Sec. 2, we show that there is an equivalence between the N-representability problem for the 2-matrix and that arising in the construction of the energy functional $F[\rho]$ in DFT.[23] We also argue that the N-representability conditions proposed by Weiner and Trickey[24] based on positivity, normalization and fixed density of N-electron matrices are not sufficient. In Sec. 3 we consider the problem of the universality of $F[\rho]$. We discuss this problem by exploring the connection between the *ab initio* DFT formalism and HKS-DFT. In Sec. 4 we present some conclusions.

2. *N*-Representability Conditions on *F*[*ρ*] in HKS-DFT

In the context of Levy's constrained-search formulation of DFT, $F[\rho]$ is defined by:[25]

$$\begin{aligned} F[\rho] = \quad & \inf \quad \left\{ < \Psi_\rho | \widehat{T} + \widehat{V}_{ee} | \Psi_\rho > \right\} \\ & \rho \in \mathcal{J}_N \equiv \{\rho : \rho \geq 0, \int \rho = N, \rho^{1/2} \in H^1(\mathbf{R}^3)\} \\ & \Psi_\rho \longrightarrow \rho \text{ (fixed)} \\ & \Psi_\rho \in \mathcal{L}_N \quad \text{(antisymmetric } N\text{-particle Hilbert space).} \end{aligned} \tag{1}$$

The wavefunction $\Psi_\rho\big((\mathbf{r}_1, \ldots, \mathbf{r}_N\big)$ is any N-particle wavefunction in $\mathcal{L}_N$ yielding the fixed density ρ. The variational principle for the energy is:

$$E_0[\rho] = \inf_{\rho \in \mathcal{J}_N} \left\{ F[\rho] + \int d\mathbf{r} v(\mathbf{r}) \rho(\mathbf{r}) \right\} \tag{2}$$

where clearly the infimum value of this functional coincides with the eigenvalue of the Schrödinger equation $\widehat{H}_v \Psi_0^v = E_0^v \Psi_0^v$ (where $\widehat{H}_v = \widehat{T} + \widehat{V}_{ee} + \sum_{i=1}^N v(\mathbf{r}_i)$ with $v(\mathbf{r}) \in \mathcal{V} \equiv L^\infty + L^{3/2}$) at the density $\rho(\mathbf{r}) = \rho_0^v(\mathbf{r})$, where

$$\rho_0^v(\mathbf{r}) = N \int d^3\mathbf{r}_3 \cdots \int d^3\mathbf{r}_N |\Psi_0^v(\mathbf{r}_1, \ldots, \mathbf{r}_N)|^2. \tag{3}$$

The connection of the above considerations with reduced 2-matrix theory is readily seen when we redefine $F[\rho]$ in terms only of the reduced 2-matrix, that is, without making any reference to the wavefunction.[26] Introducing the reduced internal two-particle operator:

$$\widehat{T} + \widehat{V}_{ee} = \sum_{i=1}^{N-1} \sum_{j=i+1}^{N} \Big[- \frac{\nabla^2_{\mathbf{r}_i} + \nabla^2_{\mathbf{r}_i}}{2(N-1)} + \frac{1}{|\mathbf{r}_i - \mathbf{r}_j|} \Big] = \sum_{i=1}^{N-1} \sum_{j=i+1}^{N} \widehat{K}_0^N(\mathbf{r}_i, \mathbf{r}_j) \tag{4}$$

we can rewrite the internal part of the energy as

$$< \Psi_\rho | \widehat{T} + \widehat{V}_{ee} | \Psi_\rho >= Tr[\widehat{K}_0^N D_\rho^2] \tag{5}$$

where $D_\rho^2(\mathbf{r}_1, \mathbf{r}_2; \mathbf{r}'_1, \mathbf{r}'_2)$ is the 2-matrix obtained from the wavefunction Ψ_ρ through

$$D_\rho^2 = \frac{N(N-1)}{2} \int d\mathbf{r}_3 \cdots \int d\mathbf{r}_N \Psi_\rho(\mathbf{r}_1, ..., \mathbf{r}_N) \Psi_\rho(\mathbf{r}'_1, \mathbf{r}'_2, \mathbf{r}_3, ..., \mathbf{r}_N). \tag{6}$$

In order to proceed further, we introduce the set $\mathcal{P}_N^2$ of N-representable 2-matrices. As the N-representability conditions on the 2-matrices are not completely known it is not possible to give a proper mathematical characterization of this set. Symbollically, however, we can define it as:

$$\mathcal{P}_N^2 \equiv \{ D_\rho^2 \quad | \quad \text{all necessary and sufficient } \textit{intrinsic} \text{ conditions on } D_\rho^2 \text{ to guarantee that it satisfies Eq. (6)} \} \tag{7}$$

We stress the adjective *intrinsic* for these conditions must be given entirely in terms of properties of the 2-matrices so as to eliminate all references to wave functions. The N-representability condition is then transferred to the requirement that $D_\rho^2 \in \mathcal{P}_N^2$.

Introducing Eq. (5) into Eq. (1) we are able to advance a definition of $F[\rho]$ that only involves *intrinsically N-representable* reduced 2-matrices:

$$F[\rho] = \inf_{\substack{\rho \in \mathcal{J}_N \\ D_\rho^2 \longrightarrow \rho \text{ (fixed)}, \quad D_\rho^2 \in \mathcal{P}_N^2}} \left\{ Tr[\widehat{K}_0^N D_\rho^2] \right\} \tag{8}$$

Thus, we see from the above definition that the *exact* functional $F[\rho]$ can be fully characterized in terms of the N-representability conditions on the reduced 2-matrix.[27] Hence, it follows that these conditions are required not only to characterize approximate functionals, but also the exact one.

We must stress here that the N-representability conditions on $F[\rho]$ are quite different from the N-representability conditions on the one-particle density. The latter guarantee that the one-particle density comes from a wavefunction. As has been shown by Gilbert[28] any $\rho \in \mathcal{J}_N$ is N-representable. Hence, this problem is solved. However, the problem of the N-representability of $F[\rho]$ requires a more careful handling. In this respect, let us mention some early warnings concerning the importance of this problem[4,29–31] as well as the explicit construction of v and N-representable functionals carried out in the context of the local-scaling transformation version of DFT, LST-DFT.[32–35] Quite recently, though, the N-representability of the *exact* $F[\rho]$ has been fully recognized as a true problem in HKS-DFT.[36] We emphasize the adjective *exact* because there is the widely held contention that these conditions are automatically fulfilled by the exact $F[\rho]$. In fact, it has been generally assumed in DFT that the N-representability conditions on the functional $F[\rho]$ are implicit in the representability conditions (particularly in the v-representability) of the one-particle density.[37,38] Thus, the usual assumption is that if the v- and N-representability of the one-particle density are guaranteed, then there is no N-representability problem for the exact $F[\rho]$.

Now, the need for a one-particle density to be v-representable arises from considerations regarding the existence of functional derivatives of particular instances of the functional $F[\rho]$ (this point has been thoroughly discussed by van Leeuwen[15] and Eschrig[14]). However, the existence of these functional derivatives is neither directly related to the problem of N-representability, nor to that of universality of the functional $F[\rho]$. In order to clarify this point, we refer here, to the work of Weiner and Trickey, who in a manner akin to that of Kryachko and Ludeña (see second paragraph of p. 225 of Ref. 24), define paths in N-particle operator space such that N-particle operators are labeled by densities in a 1-1 fashion. These paths are determined by optimization criteria which ensure that the resulting functionals have well-defined functional derivatives irrespective of the topology of the density.

The optimization (or minimization) of the functionals contains, according to Weiner and Trickey (p. 227, paragraph following Eq. (3.9) of Ref. 24) three types of constraints: positivity, normalization and a fixed density. These constraints would then guarantee the 1-1 correspondence between densities and N-matrices, which, in turn would automatically also guarantee fulfillment of the N-representability conditions on the exact $F[\rho]$. But it is easy to show (as has been done by Ludeña *et al.*[34]) that these conditions are necessary but not sufficient to guarantee N-representability. Ludeña *et al.* explicitly construct in the context of LS-DFT, for the case of $N = 2$, non-N-representable 2-matrices which are positive, normalized and yield the exact density of Hooke's atom. The non-N-representability of these 2-matrices is established by the fact that they lead to energy values lower than 2.000 hartrees, the exact energy value for Hooke's atom. Hence, through this explicit construction it is shown that these conditions do not suffice to guarantee the N-representability of the functional $F[\rho]$, as defined, for example in the work of Weiner and Trickey.[24] In consequence, as the N-representability problem for $F[\rho]$ is not solved by the imposition of these conditions, it remains as an open problem for the exact $F[\rho]$, and not only for approximations to this functional.

3. The Non-Universal Character of $F[\rho]$ in HKS-DFT

Hohenberg and Kohn[19] were the first ones to talk about $F[\rho]$ as a "universal "functional of the one-particle density. Rewriting their original definition in the present notation, this functional is defined as:

$$F_{HK}[\rho(\mathbf{r})] =< \Psi|\widehat{T} + \widehat{V}_{ee}|\Psi >; \rho \in \mathcal{A}_N, \Psi \in \mathcal{L}_N \tag{9}$$

where $\mathcal{A}$ is the set of one-particle densities corresponding to non-degenerate ground state wave functions $\Psi \in \mathcal{L}_N$ (with respect to an external potential $v \in \mathcal{V}$). As only the kinetic energy and electron-electron interaction appear in this expression (the external potential $v(\mathbf{r})$ which characterizes a given system is absent), they conclude that this expression must be the same for all N-electron systems. In addition, since this N-dependence can be eliminated by setting the condition that the one-particle density integrates to the number of electrons N, it is also claimed that $F_{HK}[\rho(\mathbf{r})]$ is N-independent. This N-independence has been questioned by Lieb.[13] More recently, Eschrig,[14] in the spirit of the Hohenberg-Kohn approach, has introduced a truly N-independent (but much more complex) functional.

In the constrained-search definition of Levy,[25] it is also asserted that $F[\rho]$ is universal in the sense that it is system independent: any v-representable ρ yields the same value of $F[\rho]$, regardless of what the external potential is. But actually, as pointed out by Kutzelnigg[39] the complicated relation between v and ρ that follows from the Hohenberg-Kohn theorem, makes it difficult to have functional of ρ which at the same time does not depend on v.

In fact, the belief that there exists a universal (but unattainable) functional $F[\rho]$ is one of the basic tenets of HKS-DFT. For this reason, a considerable effort has been devoted to the construction of energy functionals that approximate as much as possible the exact one. Some of the acomplishments and shortcomings of this assortement of approximate functionals has been recently reviewed.[18,40] We provide in what follows some arguments showing that the functional $F[\rho]$ is not universal.

3.1. *The non-universality of $F[\rho]$ in ab initio DFT*

As a consequence of the Hohenberg-Kohn theorem[41] the choice of the local potential $v_s(\mathbf{r})$ appearing in the Kohn-Sham equations

$$\left(\frac{1}{2}\nabla_{\mathbf{r}}^2 + v_s(\mathbf{r})\right)\phi_p^{KS}(\mathbf{r}) = \mathcal{E}_p\phi_p^{KS}(\mathbf{r}) \tag{10}$$

(where $v_s(\mathbf{r}) = v(\mathbf{r}) + v_H(\mathbf{r}) + v_{xc}(\mathbf{r})$ comprises the sum of the external, Coulomb and exchange-correlation potentials) defines the Kohn-Sham wavefunction, a single Slater determinant constructed from the first N orbitals:

$$\Phi^{KS}(\mathbf{r}_1, s_1, ..., \mathbf{r}_N, s_N) = \frac{det}{\sqrt{N!}}\{\phi_1^{KS}(\mathbf{r}_1), \sigma_1(s_1), ..., \phi_N^{KS}(\mathbf{r}_N), \sigma_N(s_N) \tag{11}$$

(where $\sigma(s)$ is the spin function), its associated non-interacting kinetic energy functional $T_s[\Phi^{KS}] =< \Phi^{KS}|\widehat{T}|\Phi^{KS} >$ and also its exchange energy functional $E_x[\Phi^{KS}]$. Moreover, through the definition of the exact one-electron density, namely, the density corresponding to the exact wave function $\Psi(\mathbf{r}_1, s_1, ..., \mathbf{r}_N, s_N)$ of the real N-electron system, solely in terms of the occupied Kohn-Sham orbitals.

The one-particle Kohn-Sham density $\rho_{\Phi^{KS}}(\mathbf{r})$ is:

$$\rho_{\Phi^{KS}}(\mathbf{r}) = \sum_{i=1}^{N} \phi_i^{KS}(\mathbf{r})\phi_i^{*KS}(\mathbf{r}). \tag{12}$$

However, a basic condition in Kohn-Sham theory is that the exact density $\rho(\mathbf{r})$ satisfies the condition:

$$\rho(\mathbf{r}) \equiv \rho_{\Phi^{KS}}(\mathbf{r}) = \rho_\Psi(\mathbf{r}) \tag{13}$$

This condition establishes a link between the Kohn-Sham and exact formulations as the the density corresponding to the exact wave function $\Psi(\mathbf{r}_1, s_1, ..., \mathbf{r}_N, s_N)$ of the real N-electron system, is described solely in terms of the occupied Kohn-Sham orbitals. But since, the Kohn-Sham orbitals are defined in terms of its potential $v_s(\mathbf{r})$, it follows that the potential $v_s(\mathbf{r})$ defines in turn the exact wavefunction Ψ, and, hence, its associated exact correlation functional. Because of this, it is possible to establish a connection between wave-function theory and DFT. The exact approach based on this connection is known as *ab initio* density functional theory.[42–46]

The existence of these connections allows us to analyze in the more transparent context of *ab initio* DFT some questions such as the universality of the functional $F[\rho]$, or more precisely, of the exchange-correlation functional E_{xc}^{KS}, whose examination at the DFT end is not so obvious. Elsewhere[22] we have discussed this issue in more detail and have presented the main aspects of the *ab initio* DFT method, following essentially the work of Goerling[41] and others.[47] The only novelty in our presentation is that, in order to avoid a circular argument, we neither introduce nor use the definition of the local potential

$$v_{xc}^{KS}(\mathbf{r}) = \frac{\delta E_{xc}^{KS}[\rho]}{\delta \rho(\mathbf{r})}. \tag{14}$$

This definition is basic in the derivation of HKS-DFT.

We proceed by writing the exact energy for the N-particle system (characterized by the exact wave function Ψ) as:

$$E[\Psi] = T[\Psi] + E_{ee}[\Psi] + E_{ext}[\Psi]. \tag{15}$$

This energy can be rewritten, resorting to the Kohn-Sham wave function Φ^{KS} as:

$$E[\Psi] = T_s[\Phi^{KS}] + E_{Coul}[\Phi^{KS}] + E_{ext}[\Psi] + E_{xc}^{KS}[\Phi^{KS}, \Psi] \tag{16}$$

where, the Kohn-Sham exchange-correlation functional is defined as:

$$E_{xc}^{KS}[\Phi^{KS}, \Psi] = T[\Psi] - T_s[\Phi^{KS}] + E_{ee}[\Psi] - E_{Coul}[\Phi^{KS}]. \tag{17}$$

However, since the exact density $\rho(\mathbf{r})$ satisfies Eq. (13), we can rewrite the exact energy as:

$$E[\Psi] = T_s[\Phi^{KS}] + E_{Coul}[\rho] + E_{ext}[\rho] + E_{xc}^{KS}[\Phi^{KS}, \Psi] \tag{18}$$

where we have used $E_{Coul}[\Phi^{KS}] \equiv E_{Coul}[\rho] = \int d^3\,\mathbf{r}_1 \int d^3\mathbf{r}_2 \frac{\rho(\mathbf{r}_1)\rho(\mathbf{r}_2)}{|\,\mathbf{r}_1 - \mathbf{r}_2|}$ and $E_{ext}[\Psi] \equiv E_{ext}[\rho] = \int d^3\mathbf{r} v(\mathbf{r})\rho(\mathbf{r})$. Because of the connection between $\rho(\mathbf{r})$ and the wavefunctions Φ^{KS} and Ψ it is customary in HKS-DFT to write also $E_{xc}^{KS}[\Phi^{KS}, \Psi]$ as $E_{xc}^{KS}[\rho]$, namely, as a functional of just the one-particle density $\rho(\mathbf{r})$. But as this way of writing the exchange-correlation functional obscures the issue at hand, we keep the former notation.

We start from the requirement that the variational derivative of the total energy with respect to the Kohn-Sham potential be equal to zero subject to orthonormalization condition on the occupied and unoccupied Kohn-Sham orbitals and to the observance of Eqs. (12) and (13):

$$\frac{\delta E[\Psi]}{\delta v_s(\mathbf{r})} = 0. \tag{19}$$

Note that for a given $E[\Psi]$, this variational condition **defines** the optimum Kohn-Sham potential $v_s(\mathbf{r})$ and, hence, the one-particle Kohn-Sham orbital set $\{\phi_p^{KS}(\mathbf{r})\}_{p=1}^{\infty}$ comprising occupied and virtual orbitals. We also assume, without loss of generality that $E[\Psi]$ can be expressed in terms of this complete orbital set. In fact, this is true when we consider either a CI expansion of the wavefunction Ψ, its Coupled-Cluster expression or its MBPT (many-body perturbation theory) expansion. However, an important condition that must be observed when writing Ψ in any one of these forms is that it must satisfy the Eq. (12), which states that its corresponding density is constructed only from the set of occupied Kohn-Sham orbitals $\{\phi_i^{KS}(\mathbf{r})\}_{i=1}^{N}$.

This follows from the fact that the exact wavefunction can be expanded as

$$\Psi = \sum_{I=1}^{\infty} C_I(\Psi)\Phi_I^{KS} \tag{20}$$

where the set $\{\Phi_I^{KS}\}$ is formed by the configuration state functions constructed from the complete set $\{\phi_p\}_{p=1}^{\infty}$ of occupied and virtual Kohn-Sham orbitals. We assume that the set $\{C_I(\Psi)\}_{I=1}^{\infty}$ is formed by optimal variational coefficients which minimize the energy subject to the condition on the one-particle density given by Eq. (13).

Thus, without loss of generality, we can write $E[\Psi] \equiv E[\{\phi_p\}_{p=1}^{\infty}, \{C_I(\Psi)\}_{I=1}^{\infty}]$, or, more concisely, as $E[\Psi] \equiv E[\{\phi_p\}_{p=1}^{\infty}, \mathbf{C}(\Psi)]$, a fact that implies that, similarly:

$$E_{xc}^{KS}[\Phi^{KS}, \Psi] = E_{xc}^{KS}[\{\phi_p^{KS}\}_{p=1}^{\infty}, \{C_I(\Psi)\}_{I=1}^{\infty}] \equiv E_{xc}^{KS}[\{\phi_p^{KS}\}_{p=1}^{\infty}, \mathbf{C}(\Psi)] \tag{21}$$

(note that the summation over the infinite set of Kohn-Sham orbitals comprises already the occupied ones making Φ^{KS}). Hence, the variational condition given by Eq. (19) can be rewritten as follows:

$$\frac{\delta E[\Psi]}{\delta v_s(\mathbf{r})} = \sum_{p=1}^{\infty} \int d^3\mathbf{r}' \left\{ \frac{\delta E[\{\phi_p^{KS}\}_{p=1}^{\infty}, \mathbf{C}(\Psi)]}{\delta \phi_p^{KS}(\mathbf{r}')} \frac{\delta \phi_p^{KS}(\mathbf{r}')}{\delta v_s(\mathbf{r})} \right\} = 0. \tag{22}$$

Introducing Eq. (18) into Eq. (22), we obtain:

$$\sum_{i=1}^{N} \int d^3\mathbf{r}' \left\{ \frac{\delta\Big(T_s[\Phi^{KS}] + E_{Coul}[\rho] + E_{ext}[\rho]\Big)}{\delta \phi_i^{KS}(\mathbf{r}')} \right\} \frac{\delta \phi_i^{KS}(\mathbf{r}')}{\delta v_s(\mathbf{r})} + \sum_{p=1}^{\infty} \int d^3\mathbf{r}' \frac{\delta E_{xc}^{KS}[\{\phi_p^{KS}\}_{p=1}^{\infty}, \mathbf{C}(\Psi)]}{\delta \phi_p^{KS}(\mathbf{r}')} \frac{\delta \phi_p^{KS}(\mathbf{r}')}{\delta v_s(\mathbf{r})} = 0. \tag{23}$$

Bearing in mind the well-known relations:

$$\frac{\delta T[\Phi^{KS}]}{\delta \phi_i(\mathbf{r})} = -\frac{1}{2}\nabla_{\mathbf{r}}^2 \phi_i(\mathbf{r}), \quad \frac{\delta E_{Coul}}{\delta \phi_i(\mathbf{r})} = v_H([\rho];\mathbf{r})\phi_i(\mathbf{r}), \quad \frac{\delta E_{ext}}{\delta \phi_i(\mathbf{r})} = v(\mathbf{r})\phi_i(\mathbf{r}) \tag{24}$$

and, in particular, the following one

$$\frac{\delta E_{xc}^{KS}}{\delta\phi_p(\mathbf{r})} = \hat{v}_{xc}^{NL}\phi_p(\mathbf{r}) \tag{25}$$

(where the **non-local** nature of the exchange-correlation potential $\hat{v}_{xc}^{NL}$ is stressed) it is easy to show (see, for example, Ref. 48 or the more explicit presentation of Ref. 22) that Eq. (23) can be rewritten as the following non-linear equation:

$$\int d^3\mathbf{r}' X(\mathbf{r},\mathbf{r}')v_{xc}^{KS}(\mathbf{r}') = \sum_{p=1}^{\infty}\phi_p(\mathbf{r})\int d^3\mathbf{r}' G_p(\mathbf{r},\mathbf{r}')\hat{v}_{xc}^{NL}\phi_i(\mathbf{r}') \tag{26}$$

where the Green's function and the linear response functional are, respectively,

$$G_p(\mathbf{r},\mathbf{r}') \equiv \sum_{q\neq p}\frac{\phi_q(\mathbf{r})\phi_q(\mathbf{r}')}{\epsilon_p - \epsilon_q} \tag{27}$$

and

$$X(\mathbf{r},\mathbf{r}') = \frac{\delta\rho(\mathbf{r})}{\delta v_s(\mathbf{r}')} = \sum_{i=1}^{N}\sum_{q>i}^{\infty}\frac{\phi_i(\mathbf{r})\phi_q(\mathbf{r})\phi_q(\mathbf{r}')\phi_i(\mathbf{r}')}{\epsilon_i - \epsilon_q}. \tag{28}$$

The above equation (26) establishes the basic relation between the **local** Kohn-Sham exchange correlation potential $v_{xc}^{KS}(\mathbf{r}')$ and the **non-local** exchange correlation potential $\hat{v}_{xc}^{NL}$ obtained in Eq. (25) as the functional derivative of the orbital-dependent exchange-correlation functional.[41,42,49] In fact, *ab-initio* DFT is essentially an application of Eq. (26), where E_{xc} is separated into its exchange and correlation parts $E_{xc} = E_x + E_c$ and, where, a perturbation theory expansion is used to write the correlation part as $E_c = \sum_{i=1}^{\infty} E_c^i$. The coupled equations defining the exchange potential $v_x(\mathbf{r})$ and each one of the components $v_c^i(\mathbf{r})$ of the correlation potential $v_c(\mathbf{r}) = \sum_{i=1}^{\infty} v_c^i(\mathbf{r}))$ are obtained then through application of Eq. (26). However, since the zeroth-order wavefunction (namely, the Kohn-Sham determinant) already yields the exact density, successive approximations to the local correlation potentials can be generated by requiring that all higher-order contributions to the density be zero.[42,44–46] On the other hand, in the Goerling-Levy perturbation theory,[50–52] it is assumed from the outset that the density is fixed. Hence, the potential is defined by the condition that the one-particle density remains fixed along an adiabatic path.[41,43]

Formally, we can solve for the exchange-correlation potential[53] by resorting to the inverse linear response function $X^{-1}(\mathbf{r},\mathbf{r}') = \delta v_s(\mathbf{r}'')/\delta\rho(\mathbf{r})$ which satisfies $\int d^3\mathbf{r}'' X(\mathbf{r}',\mathbf{r}'')X^{-1}(\mathbf{r}'',\mathbf{r}) = \delta(\mathbf{r}'-\mathbf{r})$

$$\begin{aligned} v_{xc}^{KS}(\mathbf{r}) &= \int d^3\mathbf{r}'' X^{-1}(\mathbf{r}'',\mathbf{r})\int d^3\mathbf{r}'\sum_p G_p(\mathbf{r}'',\mathbf{r}')\phi_p^{KS}(\mathbf{r}'')\frac{\delta E_{xc}^{KS}[\{\phi_p^{KS}\}_{p=1}^{\infty},\mathbf{C}(\Psi)]}{\delta\phi_p^{KS}(\mathbf{r}')} \\ &= \int d^3\mathbf{r}'\int d^3\mathbf{r}''\sum_p \frac{\delta E_{xc}^{KS}[\{\phi_p^{KS}\}_{p=1}^{\infty},\mathbf{C}(\Psi)]}{\delta\phi_p^{KS}(\mathbf{r}')}\frac{\delta\phi_p^{KS}(\mathbf{r}')}{\delta v_s(\mathbf{r}'')}\frac{\delta v_s(\mathbf{r}'')}{\delta\rho(\mathbf{r})}. \end{aligned} \tag{29}$$

An important aspect of Eq. (29) is that it is formally equivalent to Eq. (14) and, hence, the temptation arises to conclude from this relation that one can bypass the solution of the non-linear equation expressed either in terms of (26), or (29) by just calculating the variational derivative of the functional $E_{xc}^{KS}[\rho]$ as is indicated in Eq. (14). Moreover, as in HKS-DFT it is assumed that $E_{xc}^{KS}[\rho]$ is a universal functional, all that would be needed is a universal expression for $E_{xc}^{KS}[\rho]$ as an explicit functional of the one-particle density. This alluring feature of HKS-DFT is, however, not based on fact, as we show below.

Let us consider now certain aspects of HKS-DFT that become clearer when viewed from the perspective of *ab initio* DFT and which lend support to the contention that $F[\rho]$ is not universal. From Eqs. (1) and (18), it follows that this analysis can be carried out by considering E_{xc}^{KS}.

1. The functional $E_{xc}^{KS}[\rho]$ which is basic to HKS-DFT does not appear in *ab initio* DFT. Instead, we have, according to Eq. (21) $E_{xc}^{KS}[\Phi^{KS}, \Psi] \equiv E_{xc}^{KS}[\{\phi_p^{KS}\}_{p=1}^{\infty}, \mathbf{C}(\Psi)]$.
2. The functional $E_{xc}^{KS}[\{\phi_p^{KS}\}_{p=1}^{\infty}, \mathbf{C}(\Psi)]$ is not universal because it depends on the wavefunction Ψ, which is the particular ground-state wave function of a given system.
3. However, when it is assumed that the functional satisfies the equivalence $E_{xc}^{KS}[\Phi^{KS}, \Psi] \equiv E_{xc}^{KS}[\rho]$, then the system-dependent wave function appearing in the left-hand-side is absent in the right-hand side. Thus, this unwarranted equivalence gives the impression that in the right-hand side we are dealing with a functional $E_{xc}^{KS}[\rho]$, that is, with a functional that only depends upon the one-particle density and which, in addition, is universal (i.e., valid for all systems).
4. Since the wavefunction-dependence must be necessarily included, the proper way to express this functional is: $E_{xc}^{KS}[\Phi^{KS}, \Psi] \equiv E_{xc}^{KS}[\rho, \Psi]$. But then, the system-dependence (or, equivalently, its non-universality) becomes evident.
5. It is far from trivial to find an explicit expression for the system-dependent (i.e., non-universal) functional $E_{xc}^{KS}[\rho, \Psi]$ (see Refs. 54–69 for some progress along this line within the context of the local-scaling transformation version of DFT, LST-DFT). Since, by and large, only the energy expression $E_{xc}^{KS}[\{\phi_p^{KS}\}_{p=1}^{\infty}, \mathbf{C}(\Psi)]$ is readily available, As pointed out by van Leeuwen (see, for example, Eqs. (111) and (113) of Ref. 15) and since, by and large, only the energy expression $E_{xc}^{KS}[\{\phi_p^{KS}\}_{p=1}^{\infty}, \mathbf{C}(\Psi)]$ is readily available, the formal equivalence given by Eq. (14) implies that the functional derivative must be calculated through application of the chain rule of Eq. (29). The particular interpretation of functional derivatives through response functions, turns out to be a general trait of the various types of realizations of the functionals $F[\rho]$.[15]
6. In addition, since, for each particular many-body systems there is either a minimizing wave function, or a minimizing ensemble, one can obtain an

explicit expression for the energy, which is particular to that system. Therefore, in *ab initio* DFT, the notion of a universal functional is replaced by the particular orbital-dependent energy expression. This fact, however, is not in contradiction with the basic definitions of DFT, as the theorems establishing the existence of a functional derivative proper, or of a tangent derivative (see, for example Refs. 15, 38 for a discussion on this topic) neither imply nor establish the existence of a universal functional. Moreover, these theorems remain equally valid for system-dependent functionals.

7. Finally, in the words of Bartlett *et al.*[46]: "A negative aspect of *ab initio* DFT is that we sacrifice the universality of the functionals $E_{XC}[\rho]$ and its potential $V_{XC}[\rho]$ by using orbital dependent forms. The hope of being able to "guess" $E_{XC}[\rho]$ and its potential and have the universal treatment of everything is seductive but its complexity [...] points to the difficulty. Not only must the density dependent potential and functional reflect the shell structure as a function of $\mathbf{r}$, and account for the self-interaction, but it must also reflect the discontinuities with a change in particle number. Such a quantity will be highly difficult to obtain from model problems and consistency conditions."To this quotation we must add that it is not a practical inconvenience that bars the way to an exact universal functional $F[\rho]$, but that simply, such a functional does not exist. What we have, instead, is the set $\{F[\rho, \Psi]\}$ of system-dependent functionals.

4. Conclusion

Based on the premise that there exists a universal functional $F[\rho]$, an impressive amount of work has been carried out in HKS-DFT in order to obtain explicit expressions for the functional $F[\rho]$. The approximate forms of these functionals include ρ and its derivatives ($\nabla\rho$,$\nabla^2\rho$, etc.). Some more recent versions include also the kinetic energy term τ and the paramagnetic current density $\mathbf{j}_\sigma$ (in both of these cases, the Kohn-Sham orbitals are indirectly incorporated). There are other versions that partially include the Hartree-Fock exchange, and still others that satisfy various types of constraints (asymptotic behavior, density scaling transformations, reduction to low and high-density limits, etc.) For up-to-date reviews, see Refs. 18, 70. The families of functionals resulting from the application of these series of modifications have been assigned to particular rungs of what has been called "Jacob's ladder".[71] Among the functionals belonging to these families, we have the LDA, DGE, GGA meta-GGA, hyper-GGA, hybrid functionals, etc. In addition, let us mention "non-empirical" functionals constructed by a systematic satisfaction of constraints.[72]

The surprising fact about these approximate (and, purportedly, universal) functionals is that they lead to sometimes quite accurate prediction of molecular properties.[18] This fact indicates the existence of underlying common properties of many-electron systems which the approximate functionals of HKS-DFT describe quite well. It is quite possible, for example, that exchange and correlation holes belong to this category (the description of these holes by ordinary quantum chemical methods is highly complicated and usually requires the incorporation of very large radial and angular harmonic expansions).

But, we must stress the fact that HKS-DFT functionals present the serious difficulty of not being susceptible to systematic improvements allowing convergence of the calculated results to the experimental values. As the functionals are not variational, they can lead to energy values below the experimental ones. Moreover, there are fine-tuning effects which demand the use more and more specialized functionals if one wishes to obtain results with physical or chemical significance.

Based on our present analysis showing the non-universality of density functionals, and also taking into account the fact that the description of fine-tuning effects of electronic systems demands that the particular aspects of the system-dependent functionals be incorporated as modifications or improvements to the HKS-DFT functionals, we advocate here an inversion concerning the actual way of handling this problem. Thus, instead of trying to refine functionals in order to make them more general, or universal, precisely the opposite course of action should be adopted. Through these particular system-dependent modifications, it would then be easier to attain any desired accuracy in the calculated results.

Acknowledgments

E.V.L. gratefully acknowledges FONACIT of Venezuela for support through Group Project G-97000741 and the Basque government for support during his stays at EHU, Donosti, Euskadi where part of this work was carried out; he also acknowledges Profs. J.M. Ugalde and X. Lopez for stimulating discussions and for their kind hospitality and Prof. P.W. Ayers for his enlightening and helpful comments. F.I. acknowledges financial support from the Spanish Ministry of Education and Science (Projects CTQ2005-08459-CO2-01 and UNBA05-33-001) and, in part, by *Generalitat de Catalunya* (Projects 2005SGR-00697, 2005 PEIR 0051/69 and *Distinció per a la Promoció de la Recerca Universitaria de la Generalitat de Catalunya* granted to F.I.). A.R.S. acknowledges financial support from CONACYT (México) through project 45986, a sabbatical research grant from UCMEXUS/CONACYT at UC Santa Barbara, unlimited amounts of CPU time on the IBM-p690 supercomputer at UAEM through project FOMES2000-SEP(Cómputo Científico) and on the CNSI at UC Santa Barbara.

References

1. R.G. Parr and W. Yang, *Density Functional Theory of Atoms and Molecules*, (Oxford University Press, Oxford, 1989).
2. R.M. Dreizler and E.K.U. Gross, *Density Functional Theory*, (Springer-Verlag, Berlin, 1990).
3. E.S. Kryachko and E.V. Ludeña, *Energy Density Functional Theory of Many Electron Systems*, (Kluwer Academic Publishers, Dordrecht, 1990).

4. N.H. March, *Electron Density Theory of Atoms and Molecules*, (Academic Press, New York, 1992).
5. J. Cioslowski (ed.), *Many-Electron Densities and Reduced Density Matrices*, (Kluwer Academic/Plenum Publishers, New York, 2000).
6. E.K.U. Gross and R.M. Dreizler (eds.), *Density Functional Theory*, NATO ASI Series, Vol. B337, (Plenum, New York, 1995).
7. J.M. Seminario and P. Politzer (eds.), *Modern Density Functional Theory: A Tool for Chemistry*, (Elsevier, Amsterdam, 1995).
8. J.F. Dobson, G. Vignale and M.P. Das, *Electronic Density Functional Theory. Recent Progress and New Directions*, (Plenum Press, New York, 1998).
9. D.P. Chong (ed.), *Recent Advances in Density Functional Methods*, (World Scientific, Singapore, 1995).
10. P. Geerlings, F. de Proft and W. Langenaeker (eds.), *Density Functional Theory. A Bridge between Chemistry and Physics*, (VUB University Press, Brussels, 1999).
11. R.F. Nalewajski (ed.), *Density Functional Theory*, in *Topics in Current Chemistry, Vols. 180-183* (Springer-Verlag, Berlin, 1996).
12. K.D. Sen, (ed.), *Reviews in Modern Quantum Chemistry: A Celebration of the Contribution of Robert G. Parr*, (World Scientific, 2002).
13. E.H. Lieb, *Int. J. Quantum Chem.* **24**, 243 (1983).
14. H. Eschrig, *The Fundamentals of Density Functional Theory*, (Teubner, Sttutgart, 1996), Section 6.3.
15. R. van Leeuwen, *Adv. Quantum Chem.* **43**, 24 (2003).
16. W. Koch and M.C. Holthausen, *A Chemist's Guide to Density Functional Theory*, (Wiley-VCH, Weinheim, 2000).
17. J. Kohanoff and N.I. Gidopoulos, in *Handbook of Molecular Physics and Quantum Chemistry*, Vol. 2: *Molecular Electronic Structure*, (Wiley, Chichester, 2003).
18. S.F. Sousa, P.A. Fernandes and M.J. Ramos, *J. Phys. Chem. A* **111**, 10439 (2007)
19. P. Hohenberg and W. Kohn, *Phys. Rev.* **136**, B864 (1964)
20. W. Kohn and L.J. Sham, *Phys. Rev.* **140**, A1133 (1965)
21. P.C. Hohenberg, W. Kohn and L. Sham, *Adv. Quantum Chem.* **21**, 7 (1990).
22. E.V. Ludeña, F. Illas, and A. Ramírez-Solís, in *New Developments in Quantum Chemistry*, J.L. Paz and A. Hernández, (Eds.), Research Signpost, 2008, (in press).
23. E.V. Ludeña, *J. Mol. Struct. (Theochem)* **709**, 25 (2004).
24. B. Weiner and S.B. Trickey, *Adv. Quantum Chem.* **35**, 217 (1999).
25. M. Levy, *Proc. Natl. Acad. Sci. USA* **76**, 6062 (1979).
26. D.A. Mazziotti, *J. Chem. Phys.* **112**, 10125 (2000).
27. A.J. Coleman and V.I. Yukalov, *Reduced Density Matrices: Coulson's Challenge* (Springer, New York, 2000).
28. T.L. Gilbert, *Phys. Rev. B* **12**, 2111 (1975).
29. P.-O. Löwdin, in *Density Matrices and Density Functionals*, (eds.) R. Erdahl and V.H. Smith, Jr. (Reidel, Dordrecht, 1987), p. 21.
30. R. McWeeny, *Phil. Mag. B* **69**, 727 (1994).
31. E.V. Ludeña and J. Keller, *Adv. Quantum Chem.* **21**, 46 (1990).
32. E.S. Kryachko and E.V. Ludeña, *Phys. Rev. A* **43**, 2179 (1991).
33. E.S. Kryachko and E.V. Ludeña, Cond. Matt. Theor. Vol. 7, (eds.) J. Aliaga and A. Proto (Plenum, New York, 1992), p. 229.
34. E.V. Ludeña, V.V. Karasiev, A. Artemyev, and D. Gómez, in *Many-Electron Densities and Reduced Density Matrices*, ed. J. Cioslowski (Kluwer, New York, 2000), p. 209.
35. O. Bokanowski, *J. Math. Chem.* **26**, 271 (1999).
36. P.W. Ayers and S. Liu, *Phys. Rev. A* **75**, 022514 (2007).

37. H. Englisch and R. Englisch, *Physica A* **121**, 253 (1983).
38. H. Englisch and R. Englisch, *Phys. Stat. Sol. B* **123**, 711 (1984).
39. W. Kutzelnigg, *J. Mol. Struct. (Theochem)* **768**, 163 (2006).
40. J. Katriel, S. Roy and M. Springborg, *J. Chem. Phys.* **124**, 234111 (2006).
41. A. Goerling, *J. Chem. Phys.* **123**, 062203 (2005).
42. I. Grabowski, S. Hirata, S. Ivanov and R.J. Bartlett, *J. Chem. Phys.* **116**, 4415 (2002).
43. S. Ivanov, S. Hirata, I. Grabowski and R.J. Bartlett, *J. Chem. Phys.* **118**, 461 (2003).
44. A. Beste and R.J. Bartlett, *J. Chem. Phys.* **120**, 8395 (2004).
45. R.J. Bartlett, I. Grabowski, S. Hirata, and S. Ivanov, *J. Chem. Phys.* **122** 034104 (2005).
46. R.J. Bartlett, V.F. Lotrich, and I.V. Schweigert, *J. Chem. Phys.* **123**, 062205 (2005).
47. E. Engel and R.M. Dreizler, *J. Comput. Chem.* **20**, 31 (1999).
48. J.D. Talman and W.F. Shadwick, *Phys. Rev. A* **14**, 36 (1976).
49. E.J. Baerends, *Phys. Rev. Lett.* **87**, 133004 (2001).
50. A. Goerling and M. Levy, *Phys. Rev. A* **50**, 196 (1994)
51. A. Goerling and M. Levy, *Int. J. Quantum Chem. Symp.* **29**, 93 (1995).
52. M. Levy, *Phys. Rev. A* **43**, 4637 (1991).
53. S. Hirata, S. Ivanov, I. Grabowski, R.J. Bartlett, K. Burke and J.A. Talman, *J. Chem. Phys.* **115**, 1635 (2001)
54. E.V. Ludeña, R. López-Boada, *Top. Curr. Chem.* **180**, 169 (1996) (see references therein for pioneer work on LST).
55. E.V. Ludeña, V. Karasiev, R. López-Boada, E. Valderrama and J. Maldonado, *J. Comput. Chem.* **20**, 155 (1999)
56. O. Bokanowski and B. Grébert, *J. Math. Phys.* **37**, 1553 (1996).
57. O. Bokanowski and B. Grébert, *Int. J. Quantum Chem.* **68**, 221 (1998).
58. O. Bokanowski, *J. Math. Phys.* **41**, 2568 (2000).
59. T. Koga, Y. Yamamoto and E.V. Ludeña, *J. Chem. Phys.* **94**, 3805 (1991).
60. T. Koga, Y. Yamamoto and E,V. Ludeña, *Phys. Rev. A* **53**, 5814 (1991).
61. E.V. Ludeña, R. López-Boada and R. Pino, *Can. J. Chem.* **74**, 1097 (1996)
62. R. López-Boada, R. Pino and E.V. Ludeña, *Int. J. Quantum Chem.* **63**, 1025 (1997).
63. R. López-Boada, E.V. Ludeña and R. Pino, *J. Chem. Phys.* **107**, 6722 (1997).
64. V.V. Karasiev, E.V. Ludeña and A.N. Artemyev, *Phys. Rev. A* **62**, 062510 (2000).
65. E.V. Ludeña, V.V. Karasiev and L. Echevarría, *Int. J. Quantum Chem.* **91**, 94 (2003).
66. E.V. Ludeña, V. Karasiev and P. Nieto, *Theor. Chem. Acc.* **110**, 395 (2003).
67. A. Artemyev, E.V. Ludeña and V. Karasiev, *J. Mol. Struct. (Theochem)* **580**, 47 (2002).
68. E.V. Ludeña, D. Gómez, V. Karasiev and P. Nieto, *Int. J. Quantum Chem.* **99**, 297 (2004).
69. D. Gómez, E.V. Ludeña, V.V. Karasiev and P. Nieto, *Theor. Chem. Acc.* **116**, 608 (2006).
70. G.E. Scuseria and V.N. Staroverov, Ch. 12 in *Theory and Application of Computational Chemistry: The First 40 Years (A Volume of Technical and Historical Perspectives)*, eds. C.E. Dykstra, G. Frenkling, K.S. Kim, and G.E. Scuseria (Elsevier, Amsterdam, 2005).
71. J.P. Perdew and K. Schmidt, in *Density Functional Theory and its Applications to Materials*, eds. V. Van Doren, C. Van Alsenoy, and P. Geerlings (AIP Melville, New York, 2001).
72. V.N. Staroverov, G.E. Scuseria, J. Tao and J.P. Perdew, *Phys. Rev. B* **69**, 075102 (2004).

DYNAMIC PAIR EXCITATIONS IN ALUMINUM

HELGA M. BÖHM*, ROBERT HOLLER,
ECKHARD KROTSCHECK and MARTIN PANHOLZER
*Institute of Theoretical Physics, Johannes Kepler University, Altenbergerstr. 69
A-4040 Linz, Austria*
**helga.boehm@jku.at*

Received 2 October 2008

We present a calculation of the excitation spectrum of the electron liquid that includes time-dependent pair correlations. For the charged boson fluid these correlations provide a major mechanism for lowering the plasmon energy; here we extend that study to the much more demanding fermionic case. Based on the formalism of correlated basis functions we derive coupled equations of motion for time-dependent 1- and 2-particle correlation amplitudes. Our solution strategy for these equations ensures the fulfillment of the first two energy–weighted sum rules and, in the appropriate limit, is consistent with the bosonic version. Results are presented for the dynamic structure factor with special emphasis being put on studying the double plasmon.

Keywords: Electron liquid; pair correlations; vertex corrections.

1. Introduction

Over decades the electron liquid has been the subject of intensive studies[1] due to its relevance to a broad variety of systems, including (three-dimensional) metals and (lower dimensional) semiconductor hetero–structures.

Despite the many efforts that have gone into the investigation of the electron liquid, an unambiguous explanation for some of its basic features is still lacking. One of these properties is the experimentally observed decrease[2] of the plasmon dispersion coefficient over the series of alkali metals with r_s (related to the volume density n of the electrons via $r_s a_{\rm B} = (3/4\pi n)^{1/3}$ where $a_{\rm B}$ is Bohr's radius). A possible explanation attributes this effect to *dynamic* pair excitations, as has been argued on a mode–mode coupling basis.[3,4] However, as can be seen from Fig. 1, the influence of *static* correlations is of the same order of magnitude and none of the two mechanisms fully accounts for the steep drop of the plasmon dispersion coefficient at $r_s = 6$ to a negative value. This is consistent with findings for the charged Bose gas,[5] where static and dynamic correlations contribute equally to a negative dispersion of the plasmon.

This example demonstrates that correlations are important even in the long wavelength limit, where the random phase approximation (RPA) already gives

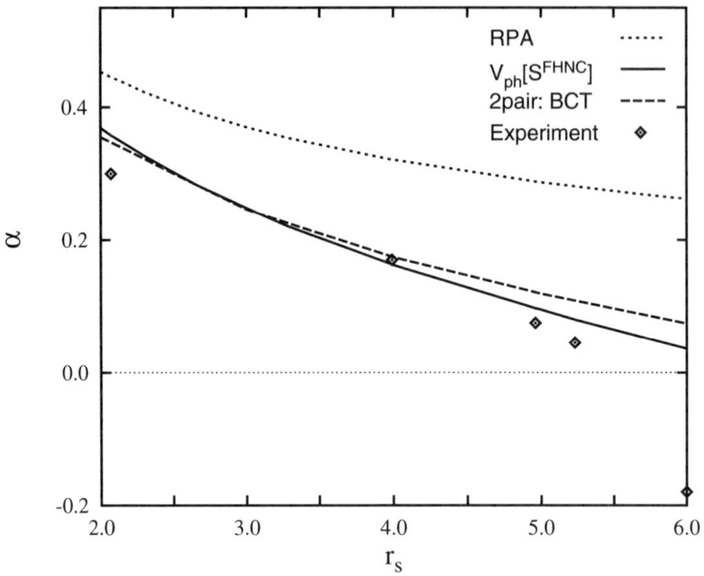

Fig. 1. Plasmon dispersion coefficient α defined as $\omega_{\rm P}(q) = \omega_{\rm P}(0) + \alpha\hbar q^2/2m$ (where q is the plasmon's wave vector and $\omega_{\rm P}$ its frequency). BCT (long-dashed line) is the theory of Ref. 3, whereas the full line was calculated with a static effective interaction $V_{\rm ph}(q)$ as described in Sec. 2.1.

a reasonable description of the dominant excitation. For short wavelengths, in particular those of the order of the mean inter-particle spacing or below, correlation effects become even more pronounced. Recent inelastic X-ray scattering (IXS) measurements[6] on Al have revealed a shoulder on the high energy side of the dynamic structure factor $S(q,\omega)$ that can be identified as a correlation–induced (i.e., "intrinsic") double-plasmon excitation. A remarkable agreement was found with the theoretical results of Sturm and Gusarov,[7] who computed the leading Feynman diagrams correcting the RPA polarizability. We here address the question whether the inclusion of higher order correlations, both of static and dynamic nature, significantly influences these results.

Our studies are performed within the framework of correlated basis functions (CBF) theory invoking dynamic pair correlations. This approach has proved successful[8] in yielding an accurate phonon dispersion curve in ^{4}He. An extension of the formalism to the much more demanding fermionic case is highly non-trivial; preliminary studies[9] on two-dimensional ^{3}He have shown that, again, the inclusion of dynamic pair correlations in the density response function leads to a significant improvement of the phonon dispersion. In this work we first improve on some of the approximations in Ref. 9 and then apply the theory to the electronic case.

2. Theoretical Framework

2.1. *CBF theory*

The starting point of our investigations is the following ansatz for the wave function of an excited state

$$\Psi = \frac{1}{\sqrt{\mathcal{N}}} e^{-iE_0 t/\hbar} \widehat{F} e^{\widehat{U}(t)} |\Phi_0\rangle , \tag{1}$$

where E_0 is the energy of the (correlated) ground state, $\mathcal{N}$ the normalization integral, $\widehat{F}$ a Jastrow correlation factor, and $|\Phi_0\rangle$ a Slater determinant; the excitation operator $\widehat{U}$ invokes dynamic single-particle and two-particle amplitudes $u^{(1)}_{ph}(t)$ and $u^{(2)}_{pp',hh'}(t)$ in the form

$$\widehat{U}(t) = \sum_{ph} u^{(1)}_{ph}(t)\, a^\dagger_p a_h + \tfrac{1}{2} \sum_{pp'h'h} u^{(2)}_{pp',hh'}(t)\, a^\dagger_p a^\dagger_{p'} a_{h'} a_h \,. \tag{2}$$

Here, the indices p and h denote "particle" and "hole" states, respectively. The correlation amplitudes $u^{(1)}_{ph}$ and $u^{(2)}_{pp',hh'}$ are determined from minimizing the action corresponding to the time-dependent Schrödinger equation.[9] Assuming that the Hamiltonian H contains a small external perturbation $H^{\rm ext}$, in the spirit of linear response theory it is sufficient to keep only terms linear in $u^{(1)}_{ph}$ and $u^{(2)}_{pp',hh'}$ in the equations of motion (EOMs) that result from this procedure; their explicit form is given in appendix A.

In a fermionic system these EOMs are still quite intricate and further approximations are necessary to bring them into a numerically tractable form. As a first step we retain only those terms, that have proved to be essential for a quantitatively accurate description of bosonic systems.[5,8,10] In particular, this implies neglecting exchange contributions and certain topologically similar (ladder-type) diagrams. Since exchange contributions have not been considered by Sturm and Gusarov[7] either, this will allow a direct comparison with their work. Also, where technically unavoidable, the correlation amplitudes are replaced by their Fermi sea averages, which only depend on the momentum transfers $\mathbf{p}$–$\mathbf{h}$ and $\mathbf{p}'$–$\mathbf{h}'$, respectively. Finally, higher order distribution functions and matrix elements are treated in the so-called "modified convolution approximation".[11]

Once the EOMs are solved, the deviation $\delta\rho$ of the system's density from its unperturbed expectation value is obtained, again within linear response, from

$$\delta\rho_{\mathbf{q}} = \frac{\langle\Psi|\delta\hat{\rho}_{\mathbf{q}}|\Psi\rangle}{\langle\Psi|\Psi\rangle} =: \chi H^{\rm ext} , \tag{3}$$

thus yielding an expression for the density–density response function χ. This leads to our final result for χ, which has the following structure

$$\chi^{\rm 2pair}(q,\omega) = \frac{\chi^s(q,\omega)}{1 - V_{\rm ph}(q)\chi^s(q,\omega) - \Lambda(q,\omega)} \,. \tag{4}$$

Before giving the detailed expressions for χ^s, $V_{\rm ph}$ and Λ in Sec. 2.3, we compare this response function with other forms reported in the literature. First we state that omitting the dynamic pair correlations $u^{(2)}_{pp',hh'}$ altogether, but solving the fully correlated EOMs for the $u^{(1)}_{ph}$ results in

$$\chi^{\rm cRPA}(q,\omega) \;=\; \frac{\chi^0(q,\omega)}{1 - V_{\rm ph}(q)\,\chi^0(q,\omega)}\,, \tag{5}$$

where $\chi^0 \equiv \chi^{0+} + \chi^{0-}$ is the Lindhard function (for N particles with excitation energies $e_k = \hbar^2k^2/2m$ and free occupation numbers $n_{\mathbf{k}}$)

$$\chi^{0\pm}(q,\omega) \;=\; \tfrac{1}{N}\sum_{\mathbf{h}} \frac{n_{\mathbf{h}}\,(1-n_{\mathbf{h}+\mathbf{q}})}{\pm\hbar[\omega + i0^+] - (e_{\mathbf{h}+\mathbf{q}} - e_{\mathbf{h}})}\,, \tag{6}$$

and

$$V_{\rm ph}(q) \;=\; \frac{q^2}{4m}\left[\frac{1}{S(q)^2} - \frac{1}{S_{\rm F}(q)^2}\right]. \tag{7}$$

Here $S(q)$ and $S_{\rm F}(q)$ denote the static structure factor and its free counterpart, respectively. The form (5) is often termed "generalized RPA" and expressed in terms of the Coulomb potential $v^{\rm C}$ and a "local field correction" $\mathcal{G}$

$$\chi^{\rm GRPA}(q,\omega) \;=\; \frac{\chi^0(q,\omega)}{1 - v^{\rm C}(q)[1-\mathcal{G}]\chi^0(q,\omega)}\,. \tag{8}$$

It has been pointed out[1,12] that the asymptotic behavior of $\mathcal{G}$ depends crucially on whether χ^0 in Eqs. (8) and (6) is defined with the free or the exact occupation numbers $n_{\mathbf{h}}$: Whereas $\mathcal{G}$ must diverge for large q if the free Fermi functions are used, it tends towards a finite value otherwise. Though not directly comparable to expressions obtained from Feynman diagrams, our result (4) takes contributions of the interacting $n_{\mathbf{h}}$ into account in the single-particle function χ^s, consistent with the vanishing of $V_{\rm ph}$ for large q.

Although a clear-cut separation of many-body properties into definite n-particle contributions is normally impossible, it has been argued that certain physical single-particle effects (e.g., arising from "backflow") are described more clearly[13] if not packed into a single dynamic interaction $v^{\rm C}(q)\mathcal{G}(q,\omega)$. A more intuitive understanding of the system's response function is achieved by accounting for these effects in a better one-particle function χ^s instead of the Lindhard function χ^0. That Eq. (8) is not a natural way to describe the dynamics of the system was also demonstrated by Neilson *et al.*[14]

2.2. *Generalized time-dependent Hartree–Fock (HF) theory*

Whenever the Fourier transform of the interaction exists (as it is the case for the Coulomb potential), the RPA can be derived by a wealth of different routes; one of them being time-dependent HF. For all other systems (such as helium, which interacts via a hard core potential) one can derive a density response function of the form (5) by using a (yet unspecified) effective interaction $W(q)$ in the HF analysis. In this section we sketch how our theory connects to such an approach.

Let $\widehat{H}' = \widehat{H} - E_0$ denote the deviation of the Hamiltonian from the ground state energy. Then the EOMs for the *single-pair correlation amplitudes* $u^{(1)}_{ph}$ *alone* can be cast into the matrix form (cf. Appendix A, $(ph) \neq (p'h')$)

$$\underline{\underline{\mathbf{M}}} \begin{pmatrix} u^{(1)}_{p'h'} \\ u^{(1)*}_{p'h'} \end{pmatrix} = \begin{pmatrix} H^{\rm ext}_{ph,0} \\ H^{\rm ext}_{0,ph} \end{pmatrix} + \begin{pmatrix} H'_{ph,p'h'} & H'_{php'h',0} \\ H'_{0,php'h'} & H'_{p'h',ph} \end{pmatrix} \begin{pmatrix} u^{(1)}_{p'h'} \\ u^{(1)*}_{p'h'} \end{pmatrix} . \tag{9}$$

The matrix elements of the Hamiltonian in a weakly interacting system simply are $(\mathbf{q} \equiv \mathbf{p} - \mathbf{h} = \pm(\mathbf{p}' - \mathbf{h}'))$

$$\left.\begin{matrix} H'_{ph,p'h'} \\ H'_{0,php'h'} \end{matrix}\right\} = W(q) \,, \tag{10}$$

yielding the RPA with the weak potential $W(q)$. For strong interactions one has

$$\left.\begin{matrix} H'_{ph,p'h'} \\ H'_{0,php'h'} \end{matrix}\right\} = \frac{e_p - e_h + e_{p'} - e_{h'}}{2} \Gamma(q) + W(q) \,, \tag{11a}$$

$$W(q) = \left(\frac{S(q)}{S_{\rm F}(q)}\right)^2 V_{\rm ph}(q) + \frac{\hbar^2 q^2}{4m} \Gamma^2(q) S_{\rm F}(q) \,, \tag{11b}$$

where the function $\Gamma(q)$ arises from the non-orthogonality of the CBF wave functions; it is related to the structure factor via $\Gamma S_{\rm F} = (S/S_{\rm F} - 1)$. This leads to the cRPA result (5) with the interaction $V_{\rm ph}(q)$ from (7).

Finally, the effect of the two-pair correlation amplitudes $u^{(2)}_{pp',hh'}$ can be mimicked by adding a dynamic operator into the EOMs for $u^{(1)}_{ph}$

$$H'_{ph,p'h'} \to H'_{ph,p'h'} + \frac{S(q)}{S_{\rm F}(q)} \mathcal{A}(q, i\hbar\tfrac{\partial}{\partial t}) \tag{12}$$

(but no correction to $H'_{0,php'h'}$). This procedure reproduces our result for $\chi(q,\omega)$ as derived in Sec. 2.1. Thus CBF theory provides a means to map strong correlations onto weak effective static (as $V_{\rm ph}(q)$) and dynamic (as $\mathcal{A}(q,\omega)$) interactions.

2.3. *Density response function with pair correlations*

We now give the explicit expression for the dynamic interaction entering the density response function (in the following $\mathbf{q}''$ denotes $-(\mathbf{q}+\mathbf{q}')$)

$$\mathcal{A}^{\pm}(q,\omega) = \tfrac{1}{2N} \sum_{\mathbf{q}'} \left| W_3(\mathbf{q},\mathbf{q}',\mathbf{q}'') \right|^2 \mu^{\rm Pair}(\mathbf{q}',\mathbf{q}'',\pm\omega) \,, \tag{13}$$

where $\mu^{\rm Pair}$ corresponds to taking the two-particle Lindhard function

$$\mu^0(\mathbf{q},\mathbf{q}',\pm\omega) = \tfrac{1}{N^2} \sum_{\mathbf{h},\mathbf{h}'} \frac{n_{\mathbf{h}}(1-n_{\mathbf{h}+\mathbf{q}})\, n_{\mathbf{h}'}(1-n_{\mathbf{h}'+\mathbf{q}'})}{\pm\hbar[\omega + i0^+] - (e_{\mathbf{h}+\mathbf{q}} - e_{\mathbf{h}}) - (e_{\mathbf{h}'+\mathbf{q}'} - e_{\mathbf{h}'})} \tag{14}$$

at the shifted frequency argument

$$\pm\hbar\omega + \frac{\hbar^2 q'^2}{2m} X(q') + \frac{\hbar^2 q''^2}{2m} X(q'') \,, \tag{15}$$

where $X \equiv 1/S_{\rm F} - 1/S$ is the direct correlation function. In the bosonic limit, $n_{\mathbf{h}} = N\delta_{\mathbf{h},\mathbf{0}}$ and $S_{\rm F} = 1$, consequently this becomes

$$\mu^{\rm Pair}(\mathbf{q}', \mathbf{q}'', \pm\omega) = \frac{1}{\pm\hbar\omega - \varepsilon(q') - \varepsilon(q'')} . \tag{16}$$

Here, $\varepsilon(q') = \hbar^2 q'^2/(2m\, S(q'))$ are the Bijl–Feynman excitation energies. It is obvious that in the long wave length limit this form has poles at twice the plasmon energy; and in general at frequencies built up from two collective modes with wave vectors q' and q''.

Finally, the vertex in Eq. (13) is the fermion analogue of the expression derived by Campbell[11] (here without triplet correlations)

$$W_3 = \tfrac{\hbar^2}{2m} \sqrt{S(q')\, S(q'')/S(q)}\, \left[\mathbf{q}\cdot\mathbf{q}' X(q') + \mathbf{q}\cdot\mathbf{q}'' X(q'')\right] . \tag{17}$$

(The full form of the fermionic W_3 is given in appendix B.) We stress that W_3, μ and thus $\mathcal{A}$ are functionals of the static structure factor $S(q)$ only, so that our theory contains no free parameter. Finally, the single-particle part of the response function is given by

$$\chi^s = \chi^0 + \chi^{0+}\chi^{0-}(\mathcal{A}^+ + \mathcal{A}^-) , \tag{18}$$

and Λ in Eq. (4) is to be taken as

$$\Lambda = \frac{S^2 + S_{\rm F}^2}{4SS_{\rm F}}\, \chi^0(\mathcal{A}^+ + \mathcal{A}^-) + \frac{1}{2}(\chi^{0+} - \chi^{0-})(\mathcal{A}^+ - \mathcal{A}^-) - \chi^{0+}\chi^{0-}\mathcal{A}^+\mathcal{A}^- . \tag{19}$$

We next apply these expressions to the electron liquid.

3. Results

For the actual computations we use the static structure factor obtained from Fermi hypernetted chain (FHNC) calculations in Ref. 15. In Fig. 2 we compare this $S(q)$ with the most recent Monte Carlo (MC) data; it is seen that the FHNC reproduces the static structure extremely well. The particle–hole interaction $V_{\rm ph}$ obtained from this structure factor according to Eq. (7) shows somewhat larger differences (right part of Fig. 2). We have found, however, that the influence of these differences on $\Im m\, \chi(q, \omega)$ is insignificant. Results shown here were obtained with the FHNC input.

Figure 3 shows the dynamic structure factor $S(q, \omega) = -\frac{1}{\pi}\Im m \chi(q, \omega)$ of Al ($r_s = 2.07$) for 4 typical wave vectors q. For low q values the high–energy shoulder is clearly visible; from the analysis of Eqs. (13)–(15) it can be attributed to a double-plasmon excitation. Once q is large enough for the contributing plasmons to become Landau damped, the shoulder gets less and less pronounced (lower right part of Fig. 3).

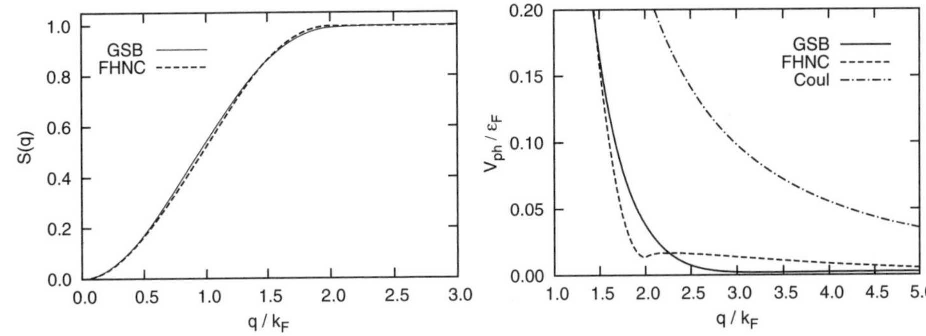

Fig. 2. Left part: Aluminum static structure factor $S(q)$ from the FHNC/EL calculations[15] (dashed line) in comparison with MC data[16] (solid line). Right part: Particle-hole interaction $V_{\mathrm{ph}}(q)$ obtained with these static structure factors from Eq. (7) in comparison with the bare Coulomb potential (dash-dotted line).

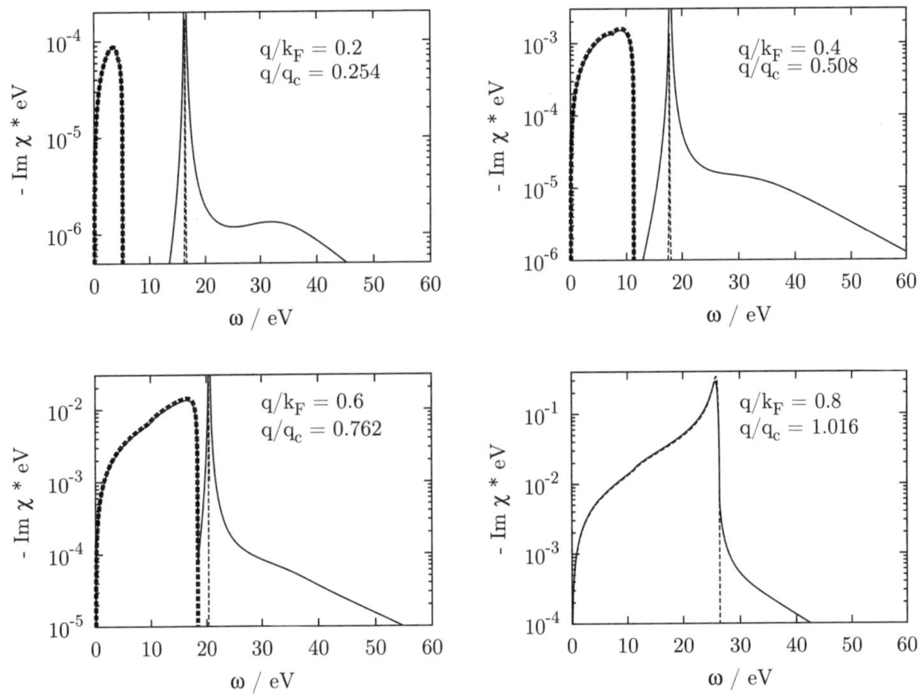

Fig. 3. Imaginary part of the density-density response function $\chi(q,\omega)$ of Al for 4 typical wave vectors as denoted in the plots (k_F is the Fermi wave vector and q_c the value where Landau damping sets in for the single-plasmon). Dashed line: RPA; full line: present theory.

Finally, in Fig. 4 the dynamic structure factor $S(q,\omega)$ is shown for the full (q,ω)-plane. The position of the double-plasmon was obtained *without* subtracting any fit for the high frequency tail (as the specific form used for this tail can modify the peak positions). Our results (full line in Fig. 4) lies between the experimental data[6] and the theoretical ones of Sturm.[7] Experimentally a plasmon–plasmon dispersion coefficient α_2 was obtained by fitting the data to

$$\hbar\omega_{\rm pl-pl}(q) = 2\omega_p(0) + \alpha_2 \frac{\hbar^2 q^2}{m} , \tag{20}$$

resulting in $\alpha_2 = 0.27$ and $\omega_p(0) = 15.2$eV, in contrast to the theoretical value of 15.8eV for the plasma frequency of Al. We adopt the view that the discrepancy in $\omega_p(0)$ arises from effects going beyond a jellium model description (such as band structure or core polarizability) but that α_2 is nevertheless dominated by correlation effects. Accordingly, in Fig. 4, in addition to the experimental data we show the double-plasmon positions obtained from Eq. (20) with the theoretical value of $\omega_p(0)$ at the experimentally investigated wave vectors. The dispersion is clearly positive but rather flat. For the large q-values of the IXS experiments the structure of $\Im m\chi$ is quite weak both in theory and in experiment. Consequently, a precise determination of the collective excitations is very difficult and the general agreement between the experimental data and and our predictions is very good.

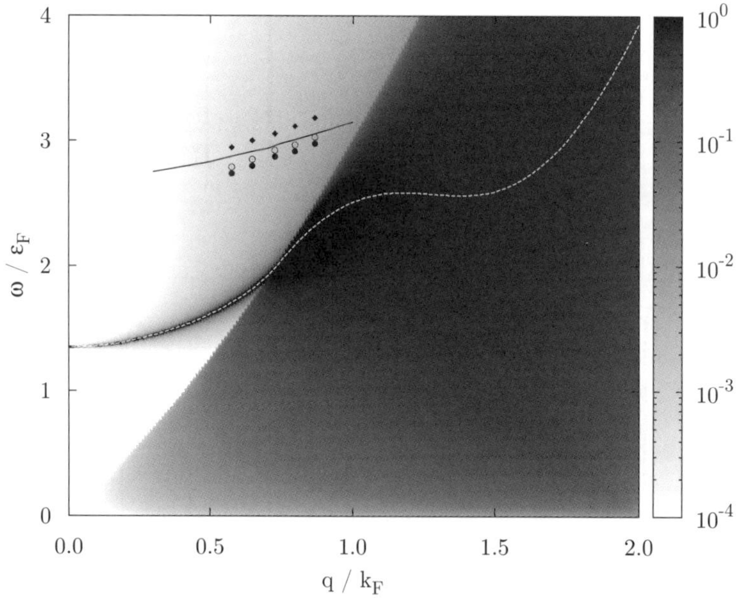

Fig. 4. Dynamic structure function of Al including time-dependent pair correlations. The gray scale ranges from white for zero to black for the highest scattering probability (the logarithmic scale was cut-off at 10^{-4}). The dashed white line gives the position of the main peak, the solid black line that of our theoretical double-plasmon. Full circles mark the experimental data[6] and open circles the latter shifted by $\omega_p^{\rm th}(0) - \omega_p^{\rm exp}(0)$. The full diamonds are the results of Sturm/Gusarov.[7]

4. Conclusion

In summary, the inclusion of dynamic pair correlations leads to a novel structure of the density–density response function which is a functional of the static structure function of the ground state. Building upon the most accurate available data for $S(q)$ of the electron liquid we obtain a result for the excitation spectrum that clearly shows a double-plasmon excitation consistent with experiments.

Acknowledgments

This work was supported by the Austrian research fund FWF under project P18134-N08. We thank S. Huotari for providing us with valuable information about the experiments. One of us (H.M.B.) also acknowledges helpful discussions with C.E. Campbell.

Appendix A. Explicit Form of the Equations of Motion

Abbreviating the particle-hole excitation energies between different excited states with $e_{ph} \equiv e_p - e_h$, denoting $\widehat{H}' = \widehat{H} - E_0$, and using Einstein's summation convention, the correlated single-particle EOMs read explicitly

$$\begin{aligned}\left[i\hbar\frac{\partial}{\partial t} - e_{ph}\right] u^{(1)}_{ph} + i\hbar\widetilde{N}_{ph,p'h'}\frac{\partial}{\partial t}u^{(1)}_{p'h'} &= H^{\rm ext}_{ph,0} \\ &+ \left[H'_{ph,p'h'}\,u^{(1)}_{p'h'} + H'_{php'h',0}\,u^{(1)*}_{p'h'}\right] \\ &+ \left[H'_{ph,p'h'p''h''}\,u^{(2)}_{p'p'',h'h''} + H'_{php'h'p''h'',0}\,u^{(2)*}_{p'p'',h'h''}\right] .\end{aligned} \tag{A.1}$$

The summations are understood as *not* to include the diagonal contributions to the matrix elements; in particular $\widetilde{N}_{ph,p'h'}$ describes the overlap of the non-orthogonal CBF wave functions for $(ph)\neq(p'h')$

$$\begin{aligned}\widetilde{N}_{ph,p'h'} &\equiv \left\langle \widehat{F}a^\dagger_p a_h\Phi_0 \left|1\right| \widehat{F}a^\dagger_{p'}a_{h'}\Phi_0\right\rangle \equiv \left\langle\psi_{ph}\left|1\right|\psi_{p'h'}\right\rangle \\ \widetilde{N}_{php'h',0} &\equiv \left\langle \widehat{F}a^\dagger_p a_h a^\dagger_{p'}a_{h'}\Phi_0\left|1\right|\Phi_0\right\rangle \equiv \left\langle\psi_{php'h'}\left|1\right|\psi_0\right\rangle ,\end{aligned} \tag{A.2}$$

etc. In the matrix elements of the Hamiltonian and the external perturbation unity is replaced by $\widehat{H}'$ and $\widehat{H}^{\rm ext}$, respectively. The EOMs for the correlated two-particle amplitudes $u^{(2)}_{pp',\{hh'\}_a} \equiv u^{(2)}_{pp',hh'} - u^{(2)}_{pp',h'h}$ are derived as

$$\begin{aligned}&\left[i\hbar\,\frac{\partial}{\partial t} - e_{ph} - e_{p'h'}\right] u^{(2)}_{pp',\{hh'\}_a} \\ &-\tfrac{1}{2}i\hbar\left[\widetilde{N}_{php'h',p''h''p'''h'''} - \widetilde{N}_{php'h',0}\widetilde{N}_{0,p''h''p'''h'''}\right]\frac{\partial}{\partial t}u^{(2)}_{p''p''',h''h'''} = \\ &\widetilde{H}'_{php'h',p''h''}u^{(1)}_{p''h''} + \widetilde{H}'_{php'h'p''h'',0}u^{(1)*}_{p''h''} + \tfrac{1}{2}\Big\{\widetilde{H}'_{php'h',p''h''p'''h'''}u^{(2)}_{p''p''',h''h'''} + \\ &\left[\widetilde{H}'_{php'h'p''h''p'''h''',0} - \widetilde{N}_{php'h',0}\widetilde{H}'_{p''h''p'''h''',0}\right] u^{(2)*}_{p''p''',h''h'''}\Big\} .\end{aligned} \tag{A.3}$$

In order to be able to solve these equations we assume that the excitations are from an optimized ground state, implying that the Brillouin conditions

$$\widetilde{H}'_{php'h',0} = 0 \qquad \text{and} \qquad \widetilde{H}'_{php'h'p''h'',0} = 0 \tag{A.4}$$

are fulfilled (only) in the Fermi sea average. Momentum conservation ensures that $\widetilde{H}'_{php'h',0} \propto \delta_{p-h+p'-h',0}$, $\widetilde{H}'_{ph,p'h'} \propto \delta_{p-h,p'-h'}$, and analogously for $\widetilde{N}$ and higher order matrix elements.

The overlap elements $\widetilde{N}_{ph,p'h'}$ and $\widetilde{N}_{php'h',0}$ are approximated as functions of the momentum transfer $\mathbf{q} = \mathbf{p}–\mathbf{h}$ only, so that they can be replaced by $\frac{1}{N}\Gamma(q)$ (implicitly including $\delta_{\mathbf{p}-\mathbf{h},\pm\mathbf{p}'-\mathbf{h}'}$ and Fermi functions $n_{h'}(1-n_{p'})$). The function $\Gamma(q)$ can be identified as the sum of diagrams $\Gamma_{\rm dd}(q)$ of the Fermi-hypernetted chain theory that contribute to $h(r)$ (pair distribution function minus 1) having no exchange lines at either of the external points. This implies the relation to the static structure factor $\Gamma(q)S_{\rm F}(q) = S(q)/S_{\rm F}(q) - 1$.

The matrix elements of the Hamiltonian are treated in a similar manner. First the non-local contribution due to the kinetic energy is split off, leading to the introduction of an effective interaction $\widetilde{W}$

$$\left.\begin{matrix} \widetilde{H}'_{ph,p'h'} \\ \widetilde{H}'_{0,php'h'} \end{matrix}\right\} =: \tfrac{1}{2}\left(e_{ph} + e_{p'h'}\right) \left\{\begin{matrix} \widetilde{N}_{ph,p'h'} \\ \widetilde{N}_{0,php'h'} \end{matrix}\right\} + \left\{\begin{matrix} \widetilde{W}_{ph,p'h'} \\ \widetilde{W}_{0,php'h'} \end{matrix}\right\} \tag{A.5}$$

(again the $\widetilde{}$ denotes that we deal only with off-diagonal terms). Next $\widetilde{W}$ is approximated as a local function of the momentum transfer. Making use of the Brillouin condition (A.4) this function can again be expressed by $\Gamma(q)$

$$\tfrac{1}{N}\sum_{hh'} \widetilde{W}_{0,php'h'} = W(q)\, S_{\rm F}^2(q) = -\Gamma(q)\, S_{\rm F}(q)\, \frac{\hbar^2 q^2}{2m}\,. \tag{A.6}$$

This procedure is easily generalized to the higher order matrix elements.

The EOMs for the pair amplitudes $u^{(2)}_{pp',hh'}$ are treated in the same way with additionally replacing the 4–particle overlap elements $\widetilde{N}_{php'h',p''h''p'''h'''}$ (and their counterparts arising from $\widetilde{W}_{php'h',p''h''p'''h'''}$) by a product decoupling ("uniform limit approximation" or "modified convolution approximation"[11]).

Neglecting the pair contributions in Eq. (A.1) gives the correlated EOMs for $u^{(1)}$

$$\begin{aligned} &\begin{pmatrix} i\hbar\frac{\partial}{\partial t} - e_{ph} + i\hbar\widetilde{N}_{ph,p'h'}\frac{\partial}{\partial t} & 0 \\ 0 & -i\hbar\frac{\partial}{\partial t} - e_{ph} - i\hbar\widetilde{N}_{p'h',ph}\frac{\partial}{\partial t} \end{pmatrix} \begin{pmatrix} u^{(1)}_{p'h'} \\ u^{(1)*}_{p'h'} \end{pmatrix} = \begin{pmatrix} H^{\rm ext}_{ph,0} \\ H^{\rm ext}_{0,ph} \end{pmatrix} \\ &+ \frac{e_{ph} + e_{p'h'}}{2} \begin{pmatrix} \widetilde{N}_{ph,p'h'} & \widetilde{N}_{php'h',0} \\ \widetilde{N}_{0,php'h'} & \widetilde{N}_{p'h',ph} \end{pmatrix} \begin{pmatrix} u^{(1)}_{p'h'} \\ u^{(1)*}_{p'h'} \end{pmatrix} + \begin{pmatrix} \widetilde{W}_{ph,p'h'} & \widetilde{W}_{php'h',0} \\ \widetilde{W}_{0,php'h'} & \widetilde{W}_{p'h',ph} \end{pmatrix} \begin{pmatrix} u^{(1)}_{p'h'} \\ u^{(1)*}_{p'h'} \end{pmatrix}, \end{aligned} \tag{A.7}$$

which leads to the cRPA result, Eq. (5).

Appendix B. Fermion vertex function

For the function W_3 to be used in the dynamic potential $\mathcal{A}$ in Eq. (13) we derive

$$-\tfrac{4m}{\hbar^2} W_3(\mathbf{q},\mathbf{q}',\mathbf{q}'')S_{\rm F}(q)\sqrt{S(q)\frac{S(q')S(q'')}{S_{\rm F}(q')S_{\rm F}(q'')}} = \left[q'^2\Gamma(q') + q''^2\Gamma(q'')\right] S_{\rm F}^{(3)}(\mathbf{q},\mathbf{q}',\mathbf{q}'')$$
$$+\,\Gamma(q')S_{\rm F}(q')\,\mathbf{q}'\cdot[\mathbf{q}''S_{\rm F}(q) - \mathbf{q}S_{\rm F}(q'')] + \Gamma(q'')S_{\rm F}(q'')\,\mathbf{q}''\cdot[\mathbf{q}'S_{\rm F}(q) - \mathbf{q}S_{\rm F}(q')]$$
$$-\,2\Gamma(q')S_{\rm F}(q')\Gamma(q'')S_{\rm F}(q'')\,\mathbf{q}\cdot\left[\mathbf{q}''S_{\rm F}(q') + \mathbf{q}'S_{\rm F}(q'')\right]. \tag{B.1}$$

The free three-particle structure factor was evaluated in Ref. 17 and is defined by

$$S_{\rm F}^{(3)}(\mathbf{q},\mathbf{q}',\mathbf{q}'') = \frac{1}{N}\sum_{\mathbf{k}} n_{\mathbf{k}}(1-n_{\mathbf{k}-\mathbf{q}''})[1 - n_{\mathbf{k}+\mathbf{q}} - n_{\mathbf{k}+\mathbf{q}'}]\,. \tag{B.2}$$

References

1. G. Giuliani and G. Vignale, *Quantum Theory of the Electron Liquid* (Cambridge University Press, Cambridge, 2005).
2. A. vom Felde, J. Sprösser-Prou and J. Fink, *Phys. Rev. B* **40**, 10181 (1989).
3. H. M. Böhm, S. Conti and M. P. Tosi, *J. Phys.: Condens. Matter* **8**, 781 (1996).
4. R. Nifosì, S. Conti and M. P. Tosi, *Phys. Rev. B* **58**, 12758 (1998).
5. V. Apaja, J. Halinen, V. Halonen, E. Krotscheck and M. Saarela, *Phys. Rev. B* **55**, 12925 (1997).
6. C. Sternemann, S. Huotari, G. Vankó, M. Volmer, G. Monaco, A. Gusarov, H. L. K. Sturm and W. Schülke, *Phys. Rev. Lett.* **95**, 157401 (2005).
7. K. Sturm and A. Gusarov, *Phys. Rev. B* **62**, 16474 (2000).
8. B. E. Clements, E. Krotscheck and C. J. Tymczak, *Phys. Rev. B* **53**, 12253 (1996).
9. H. M. Böhm, H. Godfrin, E. Krotscheck, H. J. Lauter, M. Meschke and M. Panholzer, *International Journal of Modern Physics B* **21**, 2055 (2007).
10. M. Saarela, *Phys. Rev. B* **33**, 4596 (1986).
11. C. C. Chang and C. E. Campbell, *Phys. Rev. B* **13**, 3779 (1976).
12. A. Holas, in *Strongly Coupled Plasma Physics*, eds. F. J. Rogers and H. E. DeWitt (Plenum Press, New York, 1986), pp. 463–482.
13. N. H. March and M. P. Tosi, *Coulomb Liquids* (Academic Press Inc., London, 1984).
14. D. Neilson, L. Swierkowski, A. Sjölander and J. Szymanski, *Phys. Rev. B* **44**, 6291 (1991).
15. E. Krotscheck, *Ann. Phys. (NY)* **155**, 1 (1984).
16. P. Gori-Giorgi, F. Sacchetti and G. B. Bachelet, *Phys. Rev. B* **61**, 7353 (2000).
17. M. J. D. Powell, *Mol. Phys.* **7**, 591 (1963/64).

ON TIME DEPENDENT DFT WITH SIC

J. MESSUD, P. M. DINH and E. SURAUD
Laboratoire de Physique Théorique, IRSAMC, CNRS,
Université de Toulouse, F-31062 Toulouse, France

P.-G. REINHARD
Institut für Theoretische Physik, Universität Erlangen,
D-91058 Erlangen, Germany

Received 31 July 2008

We discuss an extension of time dependent density functional theory by a self-interaction correction (SIC). A strictly variational formulation is given taking care of the necessary constraints. A manageable and transparent propagation scheme using two sets of wave-functions is proposed and applied to laser excitation with subsequent ionization of a dimer molecule.

Keywords: Density functional theory; time dependent theory; self-interaction correction.

Static Density Functional Theory (DFT)[1–5] and its Time Dependent counterpart (TDDFT)[6] is a tool of choice to describe exchange and correlation effects in sizable electronic systems. Practically, TDDFT methods require approximations, the most widely used being the Local Density Approximation (LDA).[4] But LDA does not cancel exactly the self-interaction, i.e., the interaction of an electron with itself. As a consequence, observables as the Coulomb asymptotics, the ionization potential, the potential energy surface are wrong. This is very embarassing for the description of TD processes like ionization.

A correct description of those observables could be obtained by introduction of a self-interaction correction (SIC).[7] In the static case, SIC methods were applied in various domains of physics (atomic, molecular, cluster and solid state physics[8–11]). But the original SIC scheme leads to an orbital dependent (thus non hermitian) mean-field which, as a consequence, violates orthonormality, which is dramatic in time dependent cases. Approximations which allow to recover hermiticity were proposed through the OEP formalism[6,12–16] but no exact formulation preserving orthonormality has been derived yes. Here we propose an exact time dependent manageable SIC scheme called Diagonal SIC (DSIC), which guarantees orthonormality.

We work in the Kohn-Sham scheme of DFT with SIC which is composed of a set $\{\psi_\alpha,\, \alpha = 1, \ldots, N\}$ of orthonormalized single particle orbitals. The starting point is the SIC energy functional

$$E_{\mathrm{SIC}} = E_{\mathrm{kin}} + E_{\mathrm{ion}} + E_{\mathrm{LDA}}[\rho] - \sum_{\beta=1}^{N} E_{\mathrm{LDA}}[\rho_\beta] \tag{1}$$

where the single particle densities are $\rho_\alpha = |\psi_\alpha|^2$ and the total density is $\rho = \sum_\alpha \rho_\alpha$ (all summations run over occupied states only). The total SIC energy E_{SIC} is then varied with respect to ψ^*_α, leading to

$$\frac{\delta E_{\mathrm{SIC}}}{\delta \psi^*_\alpha} = \hat{h}_\alpha \psi_\alpha,\; \hat{h}_\alpha = \hat{h}_{\mathrm{LDA}} - U_\alpha\,, \tag{2}$$

$$\hat{h}_{\mathrm{LDA}} = -\frac{\hbar^2 \Delta}{2m} + U_{\mathrm{LDA}}[\rho]\,, \tag{3}$$

$$U_\alpha = U_{\mathrm{LDA}}[|\psi_\alpha|^2]\,, \tag{4}$$

$$U_{\mathrm{LDA}}[\varrho] = \left.\frac{\delta E_{\mathrm{LDA}}}{\delta \rho}\right|_{\rho=\varrho}\,. \tag{5}$$

The emerging Kohn-Sham Hamiltonian $\hat{h}_\alpha$ depends on the state ψ_α on which it acts through the SIC term U_α, thus it is not unitary invariant.

1. TDSIC with Explicit Orthonormalization

1.1. *The formalism*

For the time dependent SIC (TDSIC), one has to stationarize the quantum action. This is done by imposing explicitly the orthonormality :

$$0 = \delta \int_{t_0}^{t} dt' \Big(E_{\mathrm{SIC}} - \sum_\alpha (\psi_\alpha | \mathrm{i}\hbar\partial_t | \psi_\alpha) - \sum_{\beta,\gamma} (\psi_\beta|\psi_\gamma)\lambda_{\gamma\beta} \Big)\,. \tag{6}$$

To our knowledge, no explicit orthonormalization constraint was added up to now on action stationarization, because calculations were generally done for hermitian schemes. For TDSIC, this additionnal constraint is crucial because we deal with state dependent Hamiltonians which are non-linear operators and thus are not hermitian.

After variation with respect to ψ^*_α, one is left with two equations (7)–(8) to be fulfilled for the $\{\psi\}$:

$$\big(\hat{h}_\alpha - \mathrm{i}\hbar\partial_t\big)|\psi_\alpha) = \sum_\beta |\psi_\beta)\lambda_{\beta\alpha}\,, \tag{7}$$

$$\lambda_{\beta\alpha} = (\psi_\beta | h_\alpha - \mathrm{i}\hbar\partial_t | \psi_\alpha)$$

$$\lambda_{\alpha\beta} = \lambda^*_{\beta\alpha} \Rightarrow (\psi_\beta|\hat{h}_\alpha|\psi_\alpha) = (\psi_\alpha|\hat{h}_\beta|\psi_\beta)^*\,. \tag{8}$$

Equation (7) is easy to obtain. The way the other one, Eq. (8), appears in the TD case is less obvious. This is what we are going to discuss now. The overlap matrix

$$s_{\alpha\beta} = (\psi_\alpha|\psi_\beta)$$

is hermitian. Let us separate $\lambda_{\beta\alpha}$ into hermitian and anti-hermitian part as

$$\lambda_{\beta\alpha} = \mu_{\beta\alpha} + \mathrm{i}\kappa_{\beta\alpha} \quad , \quad \mu^*_{\alpha\beta} = \mu_{\beta\alpha} \quad , \quad \kappa^*_{\alpha\beta} = \kappa_{\beta\alpha} \,. \tag{9}$$

The double summation in the constraint in Eq. (6) thus has the following features

$$\sum_{\alpha,\beta} s_{\alpha\beta}\mu_{\beta\alpha} \in \mathbb{R} \quad , \quad \mathrm{i}\sum_{\alpha,\beta} s_{\alpha\beta}\kappa_{\beta\alpha} \in \mathbb{I}$$

where $\mathbb{I}$ is the set of purely imaginary numbers. The two other terms in Eq. (6) are purely real. The energy functional is real by definition. The term $\sum_\alpha(\psi_\alpha|\mathrm{i}\partial_t|\psi_\alpha)$ is real because of the hermiticity of $(\psi_\alpha|\mathrm{i}\partial_t|\psi_\beta)$, the relation $(\psi_\alpha|\mathrm{i}\partial_t|\psi_\beta) = (\psi_\beta|\mathrm{i}\partial_t|\psi_\alpha)^*$ being trivial since $\partial_t(\psi_\alpha|\psi_\beta) = 0$ as we imposed orthonormality. There is thus no term which can compensate a purely imaginary contribution from the constraint. We then have to require that

$$\kappa_{\beta\alpha} = 0 \Rightarrow \lambda_{\beta\alpha} = \mu_{\beta\alpha}$$

or directly

$$\lambda^*_{\alpha\beta} = \lambda_{\beta\alpha} \,. \tag{10}$$

Using Eq. (7), we see that the condition (10) directly leads to Eq. (8). We will call this equation the 'symmetry condition'. This is not a hermiticity condition because the Hamiltonian $\hat{h}_\alpha$ depends of the state on which it acts (thus it is not unitary invariant and non-linear). As a consequence, it is not possible to diagonalize the $\lambda_{\beta\alpha}$ matrix while preserving orthonormality of the $\{\psi\}$. The symmetry condition fixes a particular solution and should be realized at all times.

It is instructive to introduce the projector on the unoccupied space

$$\hat{\Pi}_\perp = \hat{1} - \sum_\beta |\psi_\beta)(\psi_\beta| \,. \tag{11}$$

A compact formulation of Eq. (7) then shows the key importance of the symmetry condition (8):

$$\hat{\Pi}_\perp \left(\hat{h}_\alpha - \mathrm{i}\hbar\partial_t\right)|\psi_\alpha) = 0 \,, \tag{12}$$

This equation determines the propagation of the occupied subspace as a whole[a]. The time evolution within the occupied space is given by the symmetry condition Eq. (8).

[a] Or more exactly of the unoccupied subspace as the whole, thus as a consequence of the occupied subspace as a whole.

2. The Diagonal Version of TDSIC: The Way to Propagate TDSIC

The above TDSIC equations (7) and (8) are quite involved. Hence the resulting scheme is highly non-trivial to propagate. An equivalent formulation, which we will call Diagonal SIC (DSIC), comes to our help.

2.1. *The formalism*

In order to overcome the propagation difficulties, we note that there is always the freedom of unitary transformations amongst the set of occupied orbitals $\{\psi_\alpha\}$, which preserve orthonormality and do not change the total density ρ. Thus TDSIC could be formulated in terms of another set $\{\varphi_i\}$ connected to the $\{\psi_\alpha\}$ by a unitary transformation:

$$|\varphi_i(t)) = \sum_\beta |\psi_\beta(t))\ v^*_{i\beta}(t) \tag{13}$$

with

$$\sum_\beta v^*_{i\beta}(t) v^*_{j\beta}(t) = \delta_{ij} \tag{14}$$

so that the unitary transformation conserves the total density $\rho = \sum_\beta |\psi_\beta|^2 = \sum_i |\varphi_i|^2$.

Let us demonstrate the interest of this degree of freedom. We multiply the TDSIC equation (7), written for the state α, by $v^*_{i\alpha}$ and sum it over α. Keeping in mind that the symmetry condition (8) should always be realized, one obtains:

$$(\hat{h}_{\text{SIC}} - \mathrm{i}\partial_t)|\varphi_i) = \sum_j \theta_{ij}|\varphi_j) \tag{15}$$

with $\theta_{ij} = \sum_{\alpha,\beta} v^*_{i\alpha}\lambda_{\alpha\beta}v_{\beta j} = (\varphi_j|\hat{h}_{\text{SIC}} - \mathrm{i}\partial_t|\varphi_i)$ and with the resulting Hamiltonian:

$$\hat{h}_{\text{SIC}} = \sum_\alpha \hat{h}_\alpha|\psi_\alpha)(\psi_\alpha| \quad . \tag{16}$$

We have used in the last equation the property $v^*_{i\alpha} = (\varphi_i|\psi_\alpha)$, see Eq. (13). This new Hamiltonian is now state independent and it should be calculated with the $\{\psi_\alpha\}$ set, while it is applied on the $\{\varphi_i\}$ set. One could also note that it is non-hermitian in the general case. However, the symmetry condition (8) imposes that $(\varphi_j|\hat{h}_{\text{SIC}}|\varphi_i) = (\varphi_i|\hat{h}_{\text{SIC}}|\varphi_j)^*$. Since this Hamiltonian is now identical for all states φ_α, the symmetry condition becomes a hermiticity condition for $\hat{h}_{\text{SIC}}$ (in the occupied subspace only). This allows us to claim that there exists a particular orthonormal set of occupied $\{\varphi_i\}$ which diagonalizes the $\hat{\theta}$ matrix while conserving orthonormality. Thus, instead of Eq. (15), it is sufficient to solve the Schrödinger-like equation:

$$(\hat{h}_{\text{SIC}} - \mathrm{i}\partial_t)|\varphi_i) = \theta_{ii}|\varphi_i)\,.$$

The solution of this equation reads:

$$|\varphi(t)) = \exp\left(-\mathrm{i}\int_{t_0}^{t} \mathrm{d}t'\ \theta_{ii}\right) \exp\left(-\mathrm{i}\int_{t_0}^{t} \mathrm{d}t'\ \hat{h}_{\mathrm{SIC}}\right) |\varphi(t_0)) \tag{17}$$

which is physically the same as:

$$|\varphi(t)) = \exp\left(-\mathrm{i}\int_{t_0}^{t} \mathrm{d}t'\ \hat{h}_{\mathrm{SIC}}\right) |\varphi(t_0)) \tag{18}$$

because Eqs. (17) and (18) differ only by the global unphysical phase factor $\exp\left(-\mathrm{i}\int_{t_0}^{t} \mathrm{d}t'\ \theta_{ii}\right)$. Differentiating Eq. (18) shows that it is finally sufficient to solve only:

$$\begin{aligned} &(\hat{h}_{\mathrm{SIC}} - \mathrm{i}\partial_t)|\varphi_i) = 0 \\ &(\psi_\beta|\hat{h}_\alpha|\psi_\alpha) = (\psi_\alpha|\hat{h}_\beta|\psi_\beta)^* . \end{aligned} \tag{19}$$

with the symmetry condition explicitely written again. We call this scheme TD Diagonal SIC (TD-DSIC).

Since we go from TDSIC to TD-DSIC by a unitary transformation, which is reversible (see Eq. (13)), we deduce that the two schemes are equivalent. One can either find the $\{\psi_\alpha\}$ which verify the symmetry condition and the TDSIC equation (7), or find the $\{\psi_\alpha\}$ which satisfy the symmetry condition and whose particular unitary transformation verify the TD-DSIC equation (19). In both cases, we would obtain the same $\{\psi_\alpha\}$ and the same total energy. One advantage in TD-DSIC is that we can give a physical meaning to the $\{\varphi_i\}$ because they verify a Shrödinger-type equation. However the real gain is that we now can write a manageable time propagation scheme:

$$\begin{aligned} &|\varphi(t)) = \exp\left(-\mathrm{i}\int_{t_0}^{t} \mathrm{d}t'\ \hat{h}_{\mathrm{SIC}}\right) |\varphi(t_0)) , \\ &(\psi_\beta|\hat{h}_\alpha|\psi_\alpha) = (\psi_\alpha|\hat{h}_\beta|\psi_\beta)^* . \end{aligned} \tag{20}$$

We have to perform a joined propagation of $\{\varphi_i\}$ and $\{\psi_\alpha\}$ such that each of the two sets of orbitals (connected by a unitary transformation) contribute either the propagation or the symmetry condition. The latter should be achieved thanks to a symmetry unitary transformation on the propagated $\{\varphi_i\}$ at each step, which gives the good propagated $\{\psi_\alpha\}$ and thus the good mean-field. We call them symmetrized orbitals. The $\{\varphi_i\}$ could be interpreted as single electron orbitals. We call them diagonalized orbitals. It is nevertheless crucial to note that while the diagonal orbitals φ_i can be interpreted as single particle states, the mean-field $\hat{h}_{\mathrm{SIC}}$ which describes their evolution should be calculated with the set of (symmetrized) orbitals $\{\psi_\alpha\}$.

2.2. *Existence of a solution*

The symmetry condition (8) is a highly non-linear relation. Thus one could question the possibility to find a solution which satisfied this condition at all t. One can show that there exist at all t a unitary transformation which verifies this condition, thas it, which stationarizes the energy. As the energy is bound in the unitary transformed subspace, at a given t, it will always be possible to find a stationary point, and thus to satisfy the symmetry condition.

3. Numerical Results

We use a simple one-dimensional model for a molecule, without taking into account the spin, in the spirit of.[17] The 1D test case have the advantage to be much more sensible to orthonormalization than the 3D realistic calculations, and orthonormalization is the crucial point of TDSIC.

As an interaction, we use the following smoothed Coulomb potential (in Hartree units)

$$w(x, x') = \frac{1}{\sqrt{(x - x')^2 + a_i}} \tag{21}$$

where the parameters a_i for electron-electron, electron-ion and ion-ion interactions are tuned to reproduce typical molecular energies. The interest of that interaction is that there is no singularity in $x = x'$.

Taking that interaction, we develop with LDA an energy functional for the exchange term only. As a result, we obtain the LDA exchange potential (γ is the possible degeneracy number, equal to 1 in the next results):

$$\begin{aligned} U^{\mathrm{x}}_{\mathrm{LDA}}[\rho](x) &= -\frac{1}{2\pi}\int_{-\infty}^{+\infty} \mathrm{d}y \, \frac{\sin[2\pi\rho(x)y/\gamma]}{y\sqrt{y^2+a}} \\ U^{\mathrm{x}}_{\mathrm{LDA}}[\rho_i](x) &= -\frac{1}{2\pi}\int_{-\infty}^{+\infty} \mathrm{d}y \, \frac{\sin[2\pi\rho_i(x)y/\gamma]}{y\sqrt{y^2+a}} \, . \end{aligned} \tag{22}$$

Working at the level of exchange only makes time dependent HF (TDHF) calculations the benchmark.

Numerical tests of the propagation scheme, Eqs. (20), were performed on a diatomic system of 2 electrons. One starts from the stationary states and applies an initial instantaneous boost. This simulates a very short laser pulse[18] and has the advantage that energy conservation can be used as a test for the calculations. We also checked conservation of orthonormality. The result was fully satisfying.

We also checked $\langle X \rangle$, $\langle X^2 \rangle$ and $\langle X^3 \rangle$. The results are shown in Fig. 1 comparing the TDHF benchmark with TDLDA and TD-DSIC. It is obvious that TDLDA overshoots the TDHF curves, while TD-DSIC (which is the practical way to propagate FSIC) comes much closer to TDHF.

4. Conclusion

We have proposed an exact variational formulation of time-dependent SIC with preservation of orthonormalization, together with a manageable propagation scheme

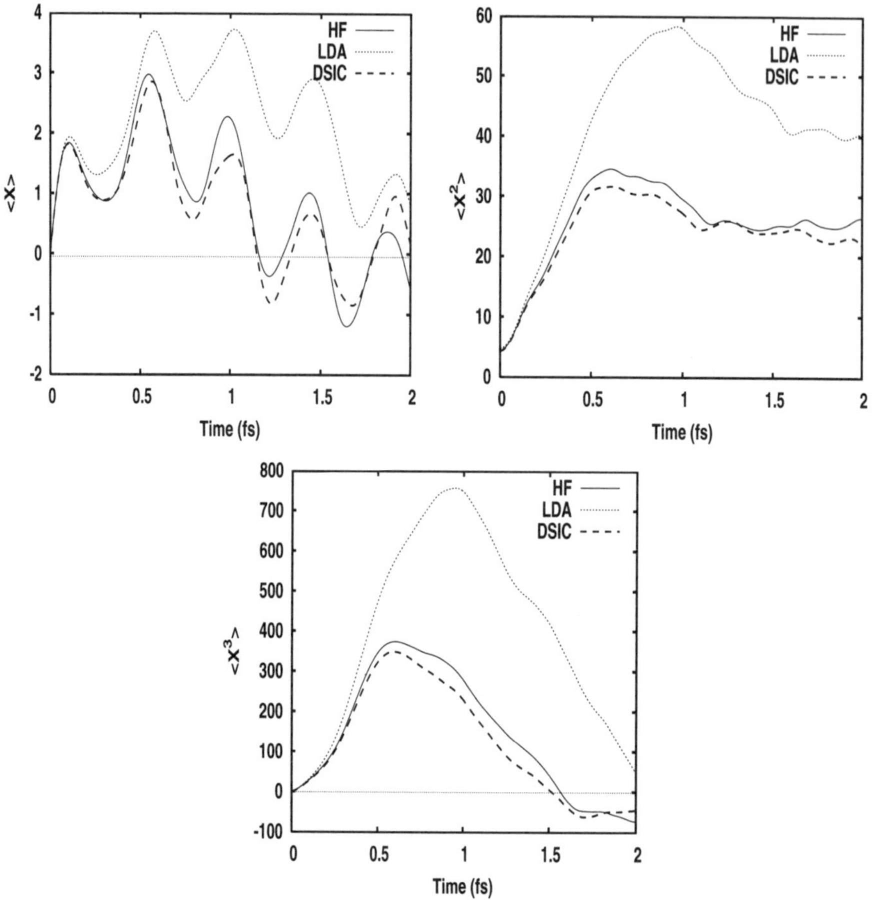

Fig. 1. Time evolution of $\langle X \rangle$, $\langle X^2 \rangle$ and $\langle X^3 \rangle$ for a dimer molecule with two electrons. Results from full TDHF, TDLDA and TD-DSIC are compared.

called TD-DSIC. The formalism also conserves the energy. We applied it to a one-dimensional bi-atomic molecule with a laser-induced electronic motion. We checked the evolution of the first three electronic moments and found good agreement of the TD-DSIC results with TDHF benchmarks.

This work was supported by Agence Nationale de la Recherche (ANR-06-BLAN-0319-02), the Deutsche Forschungsgemeinschaft (RE 322/10-1), and the Humboldt foundation.

References

1. P. Hohenberg and W. Kohn, *Phys. Rev.* **136**, 864 (1964).
2. R.G. Parr and W. Yang, *Density-Functional Theory of Atoms and Molecules* (Oxford University Press, 1989).

3. R.M. Dreizler and E.K.U. Gross, *Density Functional Theory* (Springer, Berlin, 1990).
4. W. Kohn, *Rev. Mod. Phys.* **71**, 1253 (1999).
5. E. Runge and E. K. U. Gross, *Phys. Rev. Lett.* **52**, 997 (1984).
6. C. A. Ullrich, U. J. Gossmann and E. K. U. Gross, *Phys. Rev. Lett.* **74**, 872 (1995).
7. J. P. Perdew and A. Zunger, *Phys. Rev. B* **23**, 5048 (1981).
8. M. R. Pederson, R. A. Heaton and C. C. Lin, *J. Chem. Phys.* **80**, 1972 (1984).
9. S. Goedecker and C.J. Umrigar, *Phys. Rev. A* **55**, 1765 (1997).
10. V. Polo, E. Kraka and D. Cremer, *Mol. Phys.* **100**, 1771 (2002).
11. O. A. Vydrov and G. E. Scuseria, *J. Chem. Phys.* **121**, 8187 (2004).
12. S. Kuemmel and L. Kronik, *Rev. Mod. Phys.* (2007) in press.
13. X. M. Tong and Shih-I. Chu, *Phys. Rev. A* **55**, 3406 (1997).
14. X. M. Tong and Shih-I. Chu, *Phys. Rev. A* **57**, 452 (1998).
15. X. M. Tong and Shih-I. Chu, *Phys. Rev. A* **64**, 013417 (2001).
16. X. Chu and Shih-I. Chu, *Phys. Rev. A* **63**, 023411 (2001).
17. H. Yu, T. Zuo and A. D. Bandrauk, *Phys. Rev. A* **54**, 3290 (1996).
18. F. Calvayrac, P.–G. Reinhard and E. Suraud, *Ann. Phys. (N.Y.)* **255**, 125 (1997).

DYNAMICAL PHASE TRANSITIONS IN OPINION NETWORKS: COEXISTENCE OF OPPORTUNISTS AND CONTRARIANS

KARL E. KÜRTEN
Fakultät für Physik, Universität Wien, Boltzmanngasse 5, A-1090 Wien, Austria
and
Department of Physics, Loughborough University, LE11 3TU, UK
karl.kuerten@univie.ac.at

Received 31 July 2008

We study a model for the emergence of collective decision making, consisting of N interacting agents, whose opinions are described by Ising spin variables. In particular, we present dynamical phase transitions from ordered to chaotic behavior in the space-time evolution of the binary choice network. One focus of this study is the determination of critical parameters, where the network is placed "at the edge of chaos," i.e., at a subtle compromise between stability and flexibility, where the system has both, the necessary stability and the potential for "evolutionary" improvements.

1. Introduction

There has always been substantial interest of theoretical physicists in complex phenomena departing from the traditional avenue of physics research. In particular, the application of statistical physics methods to social phenomena such as opinion formation, socio- and economic dynamics has been rather fruitful during the last few years.[1–8] The present study extends the analysis of the dynamics of randomly assembled threshold networks to social phenomena first introduced by Galam,[3] to damage spreading analysis successfully applied earlier to the theory of genetic networks, neural networks and spin models.[9–11] This earlier work established the existence of two dynamical phases: a stable phase, where the system is resistent to damage spreading and a chaotic phase, where an initially small damage might spread all over the system. In the context of socio physics this may help to understand under which conditions special shocking events or propaganda are able to influence the results of elections. In this model, individuals gather in groups of fixed size K, where K randomly chosen agents discuss a topic in order to arrive at a final decision, in favor or against. The dynamical decision making process is based on the majority- and minority rule such that each individual may adopt the opinion of the majority or minority, respectively. The model thus describes how people convince each other, in either their or the opposite direction expressing their desire to identify with certain social groups or to differentiate themselves from them. In particular, we study the effects of opportunists and contrarians in Galam's opinion dynamics model with interactions based on only two behavior schemes reminiscent of the theory of cellular automata. In the absence of opportunists (contrarians), the dynamics is ordered and leads to stable states, whereas the coupled system exhibits a dynamical phase transition to chaotic dynamics. One focus of this study is the determination of critical network parameters, where the network is placed at the "borderline" between order and chaos,[9,12] where the system satisfies necessary requirements of order, evolvability and stability found in any living organism.

2. A General Model of Discrete Automata

The model network consists of a population of N agents of a democratic community who are characterized by a specific opinion on a particular subject. The opinion of agent i is described by an Ising spin variable σ_i only capable to take the value $+1$ or -1, i.e., to say YES and NO, to buy or not to buy, or to vote Party A or Party B, respectively. We assume that each agent i can be influenced by K other agents of the community with $1 \leq K \leq N$. These so-called "neighbors" can be chosen randomly (infinite range) according to a given probability distribution, i.e., by constructing a completely random or a Barábasi network[13] associated with a power-law probability distribution. In the case of a lattice version one could choose a square lattice based on nearest neighbor interactions, or a one-dimensional spin chain with periodic boundary conditions. The latter variant would represent an opinion exchange at a roundtable, where only nearest- and next nearest neighbors are involved. The dynamical rule governing the time evolution of the network that is, whether agent i changes its opinion at the next time step or not, is given by the linear threshold rule

$$\sigma_i(t+1) = \operatorname{sgn}\left(\sum_{j=1}^{N} c_{ij}\sigma_j(t) + h_i\right) \qquad i = 1, ..., N\,. \tag{1}$$

The coupling coefficient c_{ij}, usually not symmetric with respect to interchange of the subscripts, defines the interaction strength between agent i and agent j. It takes a positive (negative) value, if agent j has an excitatory (inhibitory) effect on agent i; its magnitude quantifies the efficacy or interaction strength of agent j upon the target agent i. It is zero, if agent j does not influence agent i at all. Let us now distinguish between the nature of the efficiencies c_{ij}: **"Activating"** influences (**A**) ($c_{ij} > 0$) can be associated with convincible behavior (ferromagnetic), where agent i is stimulated to adopt the opinion of agent j. On the other hand, **"Inhibiting"** influences (**I**) ($c_{ij} < 0$) can be associated with contrarian behavior (anti-ferromagnetic), where agent i is stimulated to adopt the choice opposite to the one of the actual choice of agent j. The dynamics is discrete in time, deterministic, and the opinion of agent i depends only on the value of its K "neighbors" at the previous time step. The bracket term in Eq. (1) specifies the net internal stimulus felt by agent i, and is given in terms of a weighted sum over those input states which influence agent i. The quantity h_i represents a threshold which can be thought of as an external field on agent i associated with mass media, public relations, targeted propaganda, or personal preferences toward either orientation. A large positive threshold h_i favours a YES state, while a negative threshold h_i favours a NO state of agent i. The agents form their opinions at the next time step via the interactions with their neighbors and the influence of the external stimuli. Note however that, in principle, the interactions as well as the thresholds are time dependent. Due to the binary nature of the binary variables there exist only 2^N different configurations in phase-space. Consequently, the motion driven by Eq. (1) will inevitably evolve through a transient phase to an attractor, either a fixed point or a limit cycle, characterized by periodic changes of the opinions of specific agents. In this model the time evolution of the N-dimensional opinion vector $\overline{\sigma}(t) = (\sigma_1(t), \sigma_2(t), ..., \sigma_N(t))$, where each agent expresses his opinion at time t, could describe the time evolution of a discussion which — due to the deterministic nature of the dynamics Eq. (1) — might be periodic or reach a fixed point. The macroscopic variable, the public opinion at time t, which defines the degree of acceptance of the YES or NO state, can be defined by the total magnetization

$$m(t) = \frac{1}{2N} \sum_{i=1}^{N} (\sigma_j(t) + 1) \tag{2}$$

such that the system relaxes to $m = 1$ or $m = 0$ in the case of consensus, or to $m = \frac{1}{2}$ in the case of a tie. The sensitivity to slight changes of the initial conditions is expected to depend strongly on the connectivity structure of the network, the couplings as well as on the individual thresholds h_i associated with the external influence on the agents. One important aspect is the question how local perturbations influence the time evolution of the opinion vector $\overline{\sigma}(t)$. Such a perturbation could consist of changing the input variable of only one or a few randomly selected agents in the initial configuration. In order to study these effects one considers an identical replica of the system and compares the time evolution of the two replicas with slightly different initial conditions, where one or a few agents start with different opinions. The so-called Hamming distance between the two opinion vectors $\sigma^{(1)}(t)$ and $\sigma^{(2)}(t)$

$$d(t) = \frac{1}{2N} \sum_{\nu=1}^{N} |\sigma_\nu^{(1)}(t) - \sigma_\nu^{(2)}(t)| \tag{3}$$

specifies the fraction of agents who differ in their opinions when we compare the opinion vectors $\sigma^{(1)}(t)$ and $\sigma^{(2)}(t)$ of the two replicas at time t. This quantity is of crucial importance, since its long-time behavior determines whether the system is placed into the ordered or chaotic phase. In the ordered phase an initially small Hamming distance eventually vanishes for large times, whereas in the chaotic phase a small initial distance will evolve into a finite distance and the natural fixed point $d^* = 0$ becomes unstable. One can then study how such a perturbation spreads over the system as a function of time and eventually affects a big fraction of the agents. Damage spreading processes have originally introduced by Kauffman,[9] who studied the stability of biological gene regulatory networks with respect to external influences. Later this concept has been successfully applied in the theory of Ising models, spin glasses and neural network models.[11]

3. Mean-Field Approach

In the thermodynamic limit of asymptotically large N, statistical predictions for the time evolution of the magnetization $m(t)$ Eq. (2) as well as for the time evolution of the distance $d(t)$ Eq. (3) can be derived analytically, provided that:

(i) Every agent has exactly K incoming connections from K distinct input agents j chosen with uniform probability among the other $N-1$ agents.
(ii) The weights c_{ij} of input connections are chosen randomly according to a given probability density distribution $\rho(c_{ij})$.
(iii) The network is sparsely connected ($K \sim \log N$).[10]

The 2^K distinct input configurations can be distributed into $K+1$ equivalence classes of size $\binom{K}{k}$ with k with $k = 0, 1, 2, ..., K$, where k of the agents are in the YES state and the remaining $(K-k)$ agents are in the NO state. Since the K input agents are chosen with equal probability, the probability distribution $M(k)$ of these input configurations is binomial

$$M(k) = \binom{K}{k} m(t)^k (1-m(t))^{K-k} \qquad k = 0, 1, ..., K. \tag{4}$$

Note that the distribution is identical for any arbitrary network with indegree K independent of the dynamical rule provided that the "neighbors" are chosen democratically at random. In contrast, the output probabilities that k agents are in state YES and $(K-k)$ agents are in state NO, depend crucially on the dynamical rule and are given by the K-dimensional probability integrals[10,14,15]

$$a_k^{(K)}(h) = \int \cdots \int dx_1 \cdots dx_K \rho(x_1) \cdots \rho(x_K) g(x_1, x_2, ..., x_K, h) \tag{5}$$

with

$$g(x_1, x_2, ..., x_K, h) = \theta\left((x_1 + \cdots + x_k) - (x_{k+1} + \cdots + x_K + h)\right), \tag{6}$$

where the set of dynamical "rules" is incorporated in the probability distribution $\rho(x)$ of the couplings and the threshold h. The quantities $x_i \in \{-\infty, \infty\}$ are random variables representing the $c_{ij}\sigma_j$ and $\theta(x)$ is the Heaviside step function. Eventually, the magnetization, i.e., the probability that a randomly chosen agent is in the YES state in the **next** time step is given by the iterative mapping

$$m(t+1) = \sum_{k=0}^{K} a_j^{(K)}(h) \binom{K}{k} m(t)^k (1-m(t))^{K-k}. \tag{7}$$

With suitable choices of the probability coefficients $a_k^{(K)}$, i.e., the probability distribution of the couplings and suitable threshold parameters h, the dynamical behavior of the magnetization can be predicted for all known models based on the linear threshold model Eq. (1). For a symmetric probability distribution of the couplings ($\rho(c_{ij}) = \rho(-c_{ij})$) Eq. (7) collapses to $m(t+1) = \frac{1}{2}$. The corresponding time evolution for the distance $d(t)$ is

$$d(t+1) = \sum_{k=0}^{K} I_k^{(K)} \binom{K}{k} d(t)^k (1-d(t))^{K-k} \tag{8}$$

where the so-called sensitivity integrals $I_k^{(K)}$specify the probability that a sign reversal of j input variables at time t will not affect the output state of the system at time $t+1$.[10,11,14,15] We note that for a non-symmetric distribution of the coupling coefficients ($\rho(c_{ij} \neq \rho(-c_{ij})$) the situation is more complicated and the distance $d(t)$ might also depend on the actual magnetization $m(t)$.[16]

4. Specific Voting Rules: Majority and Minority

The basic principle of the majority/minority rule model often applicable in the real life context was originally proposed by Galam[5,6] to describe public debates, where at each time step a discussion group of K agents is selected at random and all agents take the opinion of the majority inside the group. If K is odd, there is always a majority in favour of either opinion. If K is even, instead, there is the possibility of a tie. For simplicity we will further concentrate on K odd, although the formalism for K even will be similar. Agents which follow the majority rule can be considered as opportunists who always adopt the opinion of the majority in their neighborhood. For the majority rule we have $c_{ij} = +1$ for **all** values of j and $h = 0$ (ferromagnetic case). According to Eq. (1) the dynamical evolution takes the form

$$\sigma_i(t+1) = \operatorname{sgn}\left(\sum_{(j)} \sigma_j(t)\right) \qquad i = 1, ..., N \quad . \tag{9}$$

Furthermore, the probability coefficents $a_k^{(K)}(h)$ specified in Eq. (5) can only take the values zero or one, depending on whether the number of "ones" in the input configuration is dominant or not. We have

$$a_k^{(K)}(h) = \begin{cases} 1 & \text{for } k \geq \frac{K+1}{2}, \\ 0 & \text{for } k < \frac{K+1}{2}. \end{cases} \tag{10}$$

The time evolution of the magnetization of the opportunists is then

$$m_O(t+1) = \sum_{k=\frac{K+1}{2}}^{K} \binom{K}{k} m_O^k(t)(1-m_O(t))^{K-k}. \tag{11}$$

The natural stable fixed points are $m_O^* = 0$ and $m_O^* = 1$, while the fixed point $m_O^* = \frac{1}{2}$ is always unstable. Figure 1 (left) depicts the time evolution of the opportunists, where depending on the initial condition, which could describe the degree of support of the YES state before a discussion, one of the competing states YES or NO are monotonically approched after a short transient.

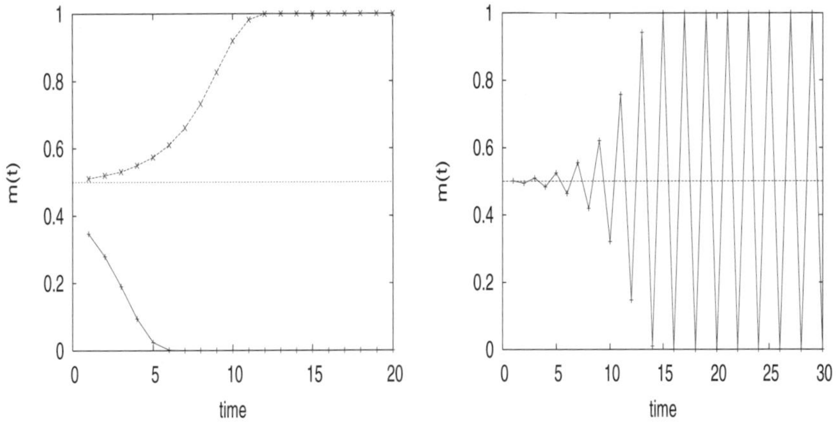

Fig. 1. Time dependence of the magnetization $m(t)$ for the majority (left) and the minority rule (right).

Another group of individuals the so-called contrarians cannot be convinced at all by agents in their neighborhood. They always adopt the opposite opinion of the majority. Accordingly, for the minority rule we have $c_{ij} = -1$ for **all** values of j and $h = 0$ (anti-ferromagnetic case) and for the magnetization of the contrarians we have the complementary time evolution

$$m_C(t+1) = 1 - m_O(t+1). \tag{12}$$

Now the fixed points $m^* = 0$, $m^* = 1$, and $m^* = \frac{1}{2}$ are unstable, and the long-time behavior is oscillatory between the values one and zero specifying a stable cycle of period two. Figure 1 (right) depicts the oscillatory time evolution between the competing states YES or NO independent on the initial condition. In both cases, for the the majority- as well as the minority interaction rule, damage spreading remains confined and it is easy to show that an arbitrary initial distance $d(t)$ decreases monotonically to zero.

5. Binary Mixtures of Opportunists (O) and Contrarians (C)

Mixtures of opportunistic and contrary behavior are of special interest since these systems do not have any obvious fixed points. The time evolution of the magnetization $m(t)$ for a mixture with a fraction p of opportunists and a fraction

$1-p$ of contrarians is a linear superposition of the pure systems $m(t+1) = p\, m_O(t) + (1-p)\, m_C(t)$ and we have

$$m(t+1) = (2p-1) \sum_{i=\frac{K+1}{2}}^{K} \binom{K}{k} m^k(t)(1-m(t))^{K-k} + (1-p). \tag{13}$$

For $p = \frac{1}{2}$ we have $m(t+1) = \frac{1}{2}$ after the first time step. Moreover the fixed point $m^* = \frac{1}{2}$ exists for arbitrary values of p, while the trivial fixed points $m^* = 0$ and $m^* = 1$ only exist for $p = 0$ and as $p = 1$. Since the stability of an arbitrary fixed point m^* is governed by

$$\frac{\partial m_{t+1}}{\partial m_t}|_{m=m^*} = (2p-1)\frac{K!}{(\frac{K-1}{2})!2}\,(m^*(1-m^*))^{\frac{K-1}{2}}\ , \tag{14}$$

the fixed point $m^* = \frac{1}{2}$ changes its stability, when the slope equals one such that we find for the critical p_c^m

$$p_c^m(K) = \frac{1}{2} \pm \frac{(\frac{K-1}{2})!^2}{K!} 2^{K-2}\ . \tag{15}$$

For $K = 3$ and $K = 5$ the critical values are given by $p_c^m{}_{\pm}(3) = \frac{1}{2} \pm \frac{1}{3} = 0.5 \pm 0.333...$ and $p_c^m{}_{\pm}(5) = \frac{1}{2} \pm \frac{4}{15} = 0.5 \pm 0.266...$, respectively. In the large $K-$ limit $p_c^m(K)$ tends to $\frac{1}{2}$. For $p < p_{c_-}^m$, where contrarians are in the majority, we find oscillatory behavior. Here the system undergoes a sea-saw motion between a YES and NO intention, respectively. On the other hand, for $p > p_{c_+}^m$, where opportunists are in the majority, the system is bistable in eather the YES or NO intention. The outcome depends uniquely on the initial condition. In between for $p_{c_-}^m < p < p_{c_+}^m$, where the fixed point $m^* = \frac{1}{2}$ is stable, the outcome is effectively random characterized by gaussian fluctuations around the "tie" with amplitudes depending on the system size N. In marked contrast to the pure systems the mixed systems show a remarkable increase of the length of the transients.[5] Since the distribution of the couplings is only symmetric for $p = \frac{1}{2}$ the time evolution of the distance for arbitrary values of p depends also on the magnetization $m(t)$. For $K = 3$ we find for the distance

$$d(t+1) = 6m(t)(1-m(t))\ d(t) - \frac{3}{2}d^2(t) + d^3(t) \tag{16}$$

with the corresponding magnetization

$$m(t+1) = (2p-1)(3m^2(t) - 2m^3(t)) + (1-p) \tag{17}$$

according to Eq. (13). For arbitrary values of K one can show that for our system the time evolution of the distance has a Taylor expansion in powers of $d(t)$ with the first order coefficient

$$\frac{\partial m_{O_{t+1}}}{\partial m_{O_t}} = \frac{K!}{(\frac{K-1}{2})!2}\,(m(t)(1-m(t)))^{\frac{K-1}{2}}\ . \tag{18}$$

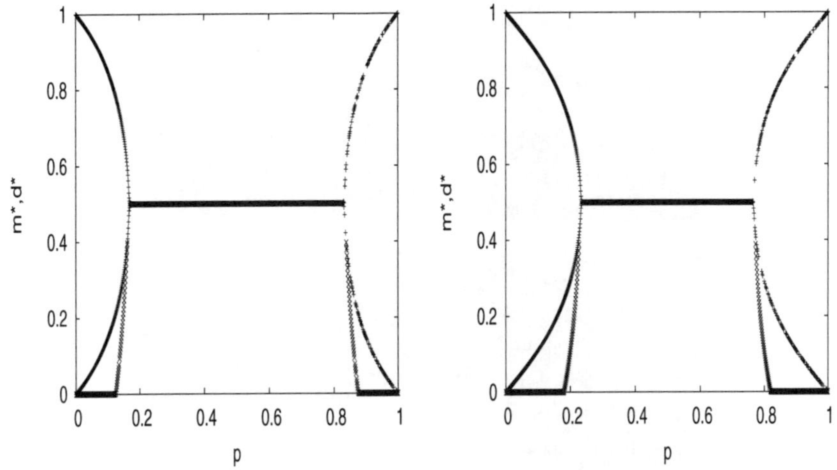

Fig. 2. Magnetization m^* (upper curves) and distance d^* (lower curves) as a function of the concentration p for $K = 3$ (left) and $K = 5$ (right).

Since critical behavior of the distance $d(t)$, where the fixed point $d^* = 0$ becomes unstable, is uniquely determined by the first order coefficient of the expansion Eq. (18), the transition to chaotic behavior is given by the relation

$$\frac{K!}{(\frac{K-1}{2})!2}\,(m^*(1-m^*))^{\frac{K-1}{2}} = 1\ . \tag{19}$$

Inserting this critical fixed point of the magnetization

$$m^* = \frac{1}{2} \pm \sqrt{\frac{1}{2} - \left(\frac{(\frac{K-1}{2})!2}{K!}\right)^{\frac{2}{K-1}}} \tag{20}$$

into Eq. (13) we can directly calculate the critical values for p_c^d. For $K = 3$ and $K = 5$ we find $p_{c\pm}^d(3) = 0.5 \pm 0.375$ and $p_{c\pm}^d(5) = 0.5 \pm 0.319...$, respectively. Note that the critical concentrations for the distance are slightly above (below) the corresponding critical values for the magnetization. It is the forward and backward bifurcation of the magnetization (see Fig. 2) which gives rise to a phase transition from chaotic to ordered dynamics when we approach the two extremes of the parameter range p, where the opportunists and contrarians are present in higher concentrations. In the large $K-$ limit $p_c^d(K)$ also tends to 0.5 such that with increasing values of K the ordered regime will be extended at expense of the chaotic regime. Note that as long as the the fixed point $m^* = \frac{1}{2}$ is stable, also the distance tends to $d^* = \frac{1}{2}$, such that independent of the initial distance, the time evolution of the individual agents is essentially random. Figure 3 depicts the phase-time portraits in the frozen regime for $p = 0.95$ with $d^* = 0$ (upper figure), in the supercritical regime for $p = 0.86$ with $d^* = 0.1$ (middle figure), and in the chaotic regime for $p = 0.5$ with $d^* = 0.5$ (lower figure). In the frozen regime we find fixed point behavior after a short transient. In

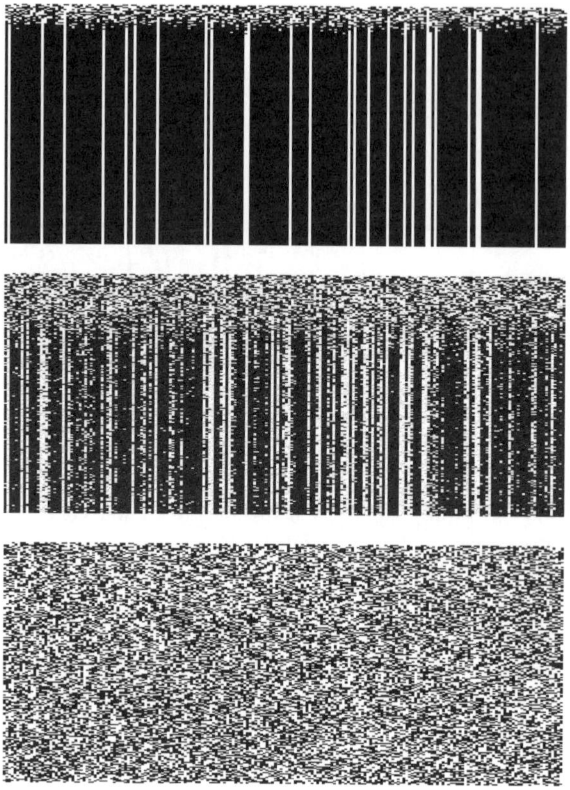

Fig. 3. Space time portraits (horizontal-vertical) in the frozen- in the supercritical- and in the chaotic regime for $p = 0.95$, $p = 0.86$, and $p = 0.5$, respectively. (from above)

the supercritical regime we find cycling behavior of moderate length and observe that a substantial fraction of the agents do not change their state at all. These agents establish the frozen core of the system which is necessary for the existence of the frozen phase. We clearly observe that in the chaotic regime the agents change their opinion more or less randomly. Computer experiments reveal that after the transient phase the fraction of agents who change their opinion from one time step to the next, who can be identified with (unentschlossen), coincides with the value of the distance $d(t)$. If this feature is desirable in a realistic model or not might give rise to further discussions and possible improvements of the model.

6. Discussion

We have presented a two-agent model to study the subtle balance of opportunistic and contrarian behavior in the dynamics of opinion spreading. At high concentrations of the opportunists we find two stable fixed point configurations with a clear YES or NO majority. In contrast, at low concentrations of the opportunists, where the contrarians are present in high concentrations, we find stable periodic behavior with alternationg Yes and NO preferences. In between, an election would result in a random outcome due to statistical fluctuations around the tie. These results are reminiscent of the hung elections in the US (2000) and Germany (2002). Galam suggested that those were not chance driven, but possibly arised due to the coexistence of opportunists and contrarians.[5] On the other hand, damage spreading analysis reveals that this situation is associated with deterministic **chaotic** behavior, where 50% of the agents change their opinion essentially randomly from on time step to the next. It would be of interest to study this model on a scale free topology, where the individual agents are not constrained to have the same fixed number K of neighbors. Here one could explore the extent to which the stable regime can be expanded by adjusting the powerlaw coefficient of the scale free probability distribution. For this more realistic model a substantial increase of the ordered regime at the expense of the chaotic regime is to be expected.[15]

Acknowledgments

The author thanks J.W. Clark, P. Klimek, F.V. Kusmartsev and D. Stauffer for numerous valuable discussions. Thanks are also due to P. Martin, S. Charpentier, and O. Machon of the Université du Littoral de Dunkerque for computational assistance. This work has been supported by a grant of the U.S. Army Research Office.

References

1. C. Castellano, S. Fortunato and V. Loreto, Statistical Physics of Social Dynamics, arXiv:0710.3256v1 (2007).
2. D. Stauffer, Opinion Dynamics and Sociophysics, arXiv:0705.0891 (2008).
3. S. Galam, *J. Math. Psychol.* **30**, 426 (1986).
4. S. Galam, *Europhys. Lett.* **70**, 705 (2005).
5. S. Galam, *Physica A* **333**, 453 (2004).
6. S. Galam, *J. Stat. Phys.* **61**, 943 (1990).
7. S. Galam, *Europhys. Lett.* **70**, 705 (2005).
8. J. J. Schneider, *Int. J. Mod. Phys. C* **5**, 659 (2004).
9. S. A. Kauffman, *Origins of Order: Self-Organization and Selection in Evolution* (Oxford University Press, Oxford) (1993).
10. K. E. Kürten, *Phys. Lett. A* **129**, 157 (1988).
11. B. Derrida, *J. Phys. A* **20**, L721 (1987).
12. K. E. Kürten and H. Beer, *J. Stat. Phys.* **87**, 929 (1997).
13. A. L. Barabási and R. Albert, *Science* **286**, 509 (1999).
14. K. E. Kürten, *J. Phys. A: Math. Gen.* **21**, L615 (1988).
15. K. E. Kürten and J. W. Clark, *Phys. Rev. E* **77**, 1 (2008).
16. A. A. Moreira and L. A. Nunes Amaral, *Phys. Rev. Lett.* **94**, 218702 (2005).

PROBING THE EQUATION OF STATE OF NUCLEAR MATTER IN THE NUCLEAR RAINBOW SCATTERING

DAO T. KHOA

Institute for Nuclear Science and Technique, VAEC
P.O. Box 5T-160, Nghia Do, Hanoi, Vietnam
khoa@vaec.gov.vn

W. VON OERTZEN and H.G. BOHLEN

Hahn-Meitner-Institut Berlin, Glienicker Str.100, D-14109 Berlin, Germany

Received 31 July 2008

We present a brief overview of the light wave interference in the atmospheric rainbow and how a similar mechanism can be observed in the elastic nucleus-nucleus scattering which gives rise to the nuclear rainbow. The latter phenomenon, observed at energies of around few tens MeV/nucleon, has been well investigated based on the basic concepts of the nuclear optical model. Given a weak absorption associated with the nuclear rainbow scattering, the observed data can be used to probe the density dependence of the effective nucleon-nucleon (NN) interaction based on the folding model study of elastic scattering. Most of the rainbow scattering data were found to be best described by a density dependent NN interaction which gives a nuclear incompressibility $K \approx 230-260$ MeV in the Hartree-Fock calculation of nuclear matter. This result implies a rather *soft* equation of state of nuclear matter.

Keywords: Nuclear rainbow; double-folding model; nuclear incompressibility.

1. Introduction

1.1. *Atmospheric rainbow*

The commonly known atmospheric rainbow is observed whenever there are water droplets illuminated by the sunlight. It can be seen during the rain with the sunshine not completely covered by the clouds (see Fig. 1) or from a fountain, when the sunlight enters from behind the point of observation. Besides the fascinating effect of color splitting due to the dependence of the refraction on the wavelength, the more interesting physics effect is the *increased* light intensity around the rainbow angle Θ_R and the *shadow* region lying beyond Θ_R.

The first modern explanation of the atmospheric rainbow was given by Descartes in 1637 in his book "Les Meteores". An illustration of the atmospheric rainbow according to Descartes' interpretation is shown in Fig. 1. Since Descartes' time up to the present, the physics of the atmospheric rainbow has been described by different

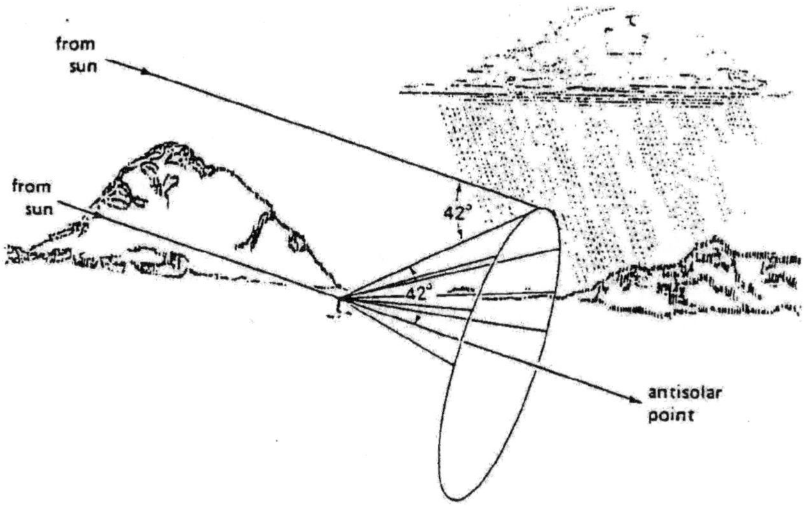

Fig. 1. Descartes traced the light ray reflected from a uniform rain drop and found a "critical" ray which has a minimal deflection angle of about 138°, whose supplementary $\Theta_R \simeq 42°$ is the largest and known nowadays as the "rainbow" angle. The atmospheric rainbow is produced by the piling up of the light rays near Θ_R which is slightly larger than 42° for red light and smaller for blue, and hence the color splitting of the "white" sunlight. Illustration taken from Ref. 1.

models, ranging from a simple classical geometrical ray optics to the quantum mechanical complex angular momentum theory for the scattering of electromagnetic waves[2,3] with higher and higher mathematical sophistication. For interested readers there exists a number of excellent monographs and reviews on the physics of the atmospheric rainbow, like the book by Greenler[1] on rainbows, halos and glories or the review articles by Nussenzweig[2] and Adam.[3] In this work we will concentrate on the refractive nucleus-nucleus scattering which gives rise to the nuclear rainbow.

In terms of the ray optics, the atmospheric rainbow is formed as a result of a refraction-reflection-refraction process by the light rays in water drops as illustrated in Fig. 2. One can see in Fig. 2 an interesting variation of the "deflection angle" as function of the impact parameter b, from the "head-on" ray with a maximal deflection at 180° to the "rainbow" ray with a minimal deflection at about 138° (whose supplementary known as the rainbow angle $\Theta_R \simeq 42°$). The interesting physics effect here is the enhanced light intensity (concentration of many light rays, e.g., those numbered 6 to 12 in Fig. 2) near the rainbow angle Θ_R, which is followed by a "shadow" region. Classically,[4] this shadow is produced by the maximum of the deflection function $\Theta(b)$ because the intensity of the scattered light is proportional to the inverse of the first derivative of $\Theta(b)$

$$\frac{d\sigma}{d\Omega} = \sum_b \frac{b}{\sin\Theta(b)|d\Theta(b)/db|}. \tag{1}$$

Here the sum is taken over the light rays entering the water drop at different impact parameters b, some of them scatter to the same angle Θ, like those numbered 7, 8

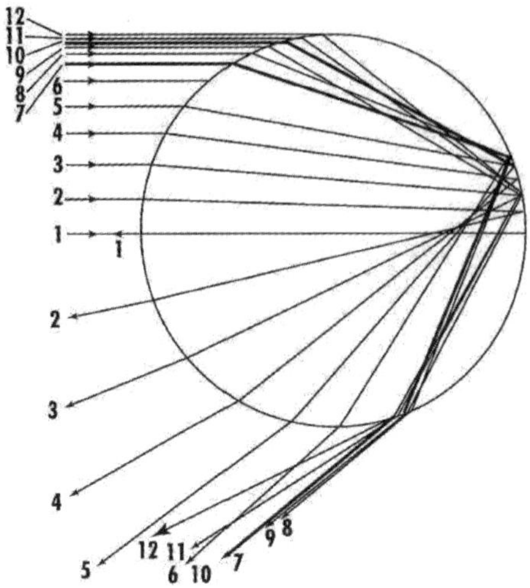

Fig. 2. The paths of light rays entering a spherical water drop at different impact parameters and leaving it after a refraction-reflection-refraction sequence. The ray numbered 7 is the rainbow ray, at which light is deflected to a maximal (negative) scattering angle of around 42°. Illustration taken from Ref. 3.

and 9 in Fig. 2, with the maximal (negative) deflection occurring near the rainbow angle Θ_R. This simple expression of light intensity has a *divergence* at Θ_R (with $d\Theta(b)/db{=}0$), which is also known as caustic in the optics. Such a divergence of the light intensity at the rainbow angle could not be remedied in either Descartes' theory or more advanced ray optics by Newton and Young. Moreover, the observation of the supernumerary bows inside the primary rainbow (see Fig. 4 below) persistently pointed to the inadequacy of Descartes' and Newton's models which could not explain the origin of the supernumeraries. All this remained a challenge until Airy, in the 19th century, provided the first mathematical model of the rainbow based on the light wave diffraction and interference.[5]

By using the standard Huygens' concept of the light wavefront rather than the optical light rays, Airy has shown the self-interference of such a wavefront as it becomes folded onto itself during the refraction and reflection within the rain drop. As a result, the primary rainbow is the first interference maximum, the second and third maxima being the first and second supernumerary bows, respectively. Airy

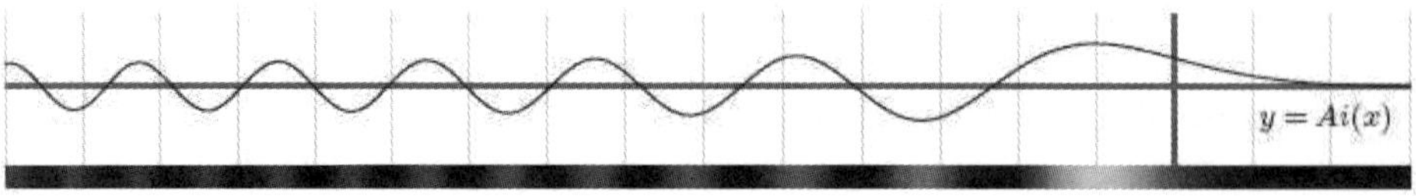

Fig. 3. Graph of the Airy rainbow integral $Ai(x)$. The argument x is proportional to $\Theta - \Theta_R$ so that $x = 0$ corresponds to $\Theta = \Theta_R$ and positive x is on the dark side of the rainbow. Below, according to Airy's theory, illumination is proportional to $Ai(x)^2$ which gives rise to the *primary* bow along with several *supernumerary* bows. Illustration taken from Ref. 3.

described the local intensity of scattered light wave with the help of a "rainbow integral", which is now known as the Airy function $Ai(x)$ (see Fig. 3). Although more sophisticated approaches to describe the atmospheric rainbow have been developed later in the 20th century,[2,3] Airy's model remains a simple but very realistic approximate description of the rainbow, which has been used widely to identify the rainbow patterns observed in molecular, atomic and nuclear scattering.[4,6]

1.2. *Nuclear rainbow*

Nuclei are known to have wave properties and they can be diffracted, refracted and suffer interference just like the sunlight. Consequently, the nucleus-nucleus scattering may also display rainbow features depending on the scattering conditions and binding structure of the projectile and target. In terms of the wave scattering theory, the nuclear rainbow should be strongest in the elastic scattering channel if a nucleus-nucleus system with weak absorption is chosen. Indeed, the nuclear rainbow pattern has been first observed during the 70's of the last century in the elastic α-nucleus scattering,[8–10] and later on in the elastic scattering measured for several

Fig. 4. Photographic image of the atmospheric rainbow where both the primary and secondary rainbows are seen. The faint bows located inside the primary rainbow are the supernumeraries which were first explained in 1838 by Airy. The secondary rainbow is formed by light rays undergoing a second reflection in the rain drops and hence is fainter and has a reversed sequence of colors. The inset along the primary bow shows the elastic $^{16}O+^{16}O$ scattering data measured at different laboratory energies, where the most pronounced rainbow pattern associated with the first Airy maximum has been observed at 350 MeV. Illustration taken from Ref. 7.

light heavy-ion (HI) systems like $^{12}C+^{12}C$[11–13] or $^{16}O+^{16}O$.[14–16] For these systems the absorption due to nonelastic reactions was sufficiently low for the rainbow effect to appear. The concept of the wave *refraction* implies that the wavelength changes whenever the wave penetrates from one medium into another. In case of the elastic nucleus-nucleus scattering, de Broglie wavelength λ_B of the scattered wave is changed as the projectile penetrates the target nucleus at the internuclear distance R where the strong projectile-target interaction occurs, due to the local strength $V(R)$ of the (real) nucleus-nucleus optical potential (OP)

$$\lambda_B(R) = h/\sqrt{2\mu[E - V(R)]}. \tag{2}$$

Here E is the energy of the relative motion (in the center-of-mass system) and μ is the reduced mass. It is obvious that the nuclear OP, used in the optical model (OM) calculation to describe the elastic scattering, is the most important physics input in the study of nuclear rainbow scattering.

2. Theoretical Basis of the Nucleus-Nucleus Optical Potential

In general, the nuclear OP is an effective interaction $U(R)$ between the two nuclei (separated by the distance R) used in the Schrödinger equation for elastic scattering

$$\left[-\frac{\hbar^2}{2\mu}\nabla^2 + U(R) + V_C(R) - E\right]\chi(\boldsymbol{R}) = 0, \tag{3}$$

where $V_C(R)$ is the Coulomb potential. It is assumed in the OM calculation that the effects from nonelastic reaction channels are taken into account by the imaginary part W of the OP which describes the loss of incident flux (*absorption*) into the nonelastic channels. The simplest procedure of an OM analysis is to adopt a phenomenological functional form for $U(R)$ and adjust its parameters until the calculated elastic cross section agrees with the measurement. However, if one uses some microscopic approach to calculate the nucleus-nucleus OP for the OM study of elastic *refractive* nucleus-nucleus scattering, valuable information on the effective nucleon-nucleon (NN) interaction and/or nuclear wave functions of the two nuclei can be obtained.

2.1. *Feshbach's formalism for the nucleus-nucleus optical potential*

A microscopic model of the nuclear OP must be an approach to predict the OP starting essentially from the in-medium NN interaction and realistic nuclear wave functions of the projectile and target nuclei. In such a formulation, the rigorous microscopic formalism can be established only for the nucleon-nucleus OP based on the G-matrix studies of nuclear matter.[19] The interaction between two composite nuclei is a much more complicated many-body problem due to the HI collision dynamics, and there is no truly microscopic theory for the nucleus-nucleus OP like that for the nucleon-nucleus OP. However, an approximate approach can be formulated[4] based on the reaction theory by Feshbach.[20]

Given the total wave function of the dinuclear system expanded over the internal wave functions of the projectile (a) and target (A) nuclei as

$$\Psi = \sum_{mn} \chi_{mn}(\boldsymbol{R})\psi_m^{(a)}(\xi_a)\psi_n^{(A)}(\xi_A), \tag{4}$$

where $\chi_{mn}(\boldsymbol{R})$ describes the relative motion of the system (with the projectile and target being in states labeled by m and n, respectively) the OP should generate $\chi_{00}(\boldsymbol{R})$ for the elastic scattering when used in Eq. (3). In general, (4) must be inserted into the many-body Schrödinger equation which gives an infinite set of coupled equations for $\chi_{mn}(\boldsymbol{R})$. By using Feshbach's projection operator,[18,20] one can obtain the effective interaction between the two nuclei U which acts in the elastic channel only and, hence, can be used in Eq. (3) to determine $\chi_{00}(\boldsymbol{R})$.

$$U = V_{00} + \lim_{\epsilon \to 0} \sum_{\alpha\alpha'}{}' V_{0\alpha} \left(\frac{1}{E - H + i\epsilon} \right)_{\alpha\alpha'} V_{\alpha' 0}. \tag{5}$$

Here $V_{\alpha\alpha'}$ is the first-order interaction between the two nuclei and $\alpha = mn$ stands for a pair of internal states of the projectile and target. The primed sum runs over all the pair states excluding $\alpha = 00$. The first term of (5) is real and can be evaluated within the double-folding approach[17,24,25]

$$V_{00} = V_{\rm F} \equiv (\psi_0^{(a)}\psi_0^{(A)}|V|\psi_0^{(a)}\psi_0^{(A)}), \tag{6}$$

where the round brackets denote integration over the internal coordinates ξ_a and ξ_A of the two nuclei being in their ground states (g.s.) $\psi_0^{(a)}$ and $\psi_0^{(A)}$, respectively. We can further rewrite Eq. (5) as

$$U = V_{\rm F} + \Delta U, \tag{7}$$

where ΔU is the so-called *dynamic polarization potential* (DPP) which arises from couplings to all open nonelastic channels. While ImΔU is the main source of the imaginary part of the OP due to transitions to the nonelastic channels, ReΔU is at least one order of magnitude smaller than V_F.[4] Since the direct (one-step) elastic scattering occurs via $V_{\rm F}$, the weaker the absorption caused by the DPP the stronger the refractive scattering which can lead to the appearance of the nuclear rainbow at appropriate incident energies.

2.2. *Double-folding model*

It is clear that $V_{\rm F}$ is the dominant part of the nucleus-nucleus interaction when elastic scattering proceeds directly in one step. Among various models for the nucleus-nucleus OP, the double-folding model (DFM) has been used widely as a microscopic method to calculate V_F starting from an appropriately chosen effective NN interaction between nucleons in the system and nuclear density distributions of the projectile and target nuclei. The established success of the DFM in describing elastic scattering measured for many HI systems[4,17,21–25] suggests that it indeed produces the dominant part of the real nucleus-nucleus OP.

In the folding approach, the real nucleus-nucleus interaction V is evaluated as a sum of effective NN interactions v_{ij} between nucleon i in the projectile a and nucleon j in the target A

$$V = \sum_{i \in a, j \in A} v_{ij}. \tag{8}$$

Although the individual g.s. wave functions $\psi_0^{(a)}(\xi_a)$ and $\psi_0^{(A)}(\xi_A)$ in Eq. (4) are each taken to be antisymmetrized, the Pauli principle still requires the total wave function Ψ to be antisymmetric under interchange of nucleons between the two nuclei. Therefore, v_{ij} in Eq. (8) should be replaced by

$$v_{ij}(1 - P_{ij}) = v_{\rm D} + v_{\rm EX} P_{ij}^x, \tag{9}$$

$$\text{where } v_{\rm D} \equiv v_{ij}^{(\rm D)} = v_{ij} \text{ and } v_{\rm EX} \equiv v_{ij}^{(\rm EX)} = -v_{ij} P_{ij}^\sigma P_{ij}^\tau. \tag{10}$$

Here P_{ij}^x, P_{ij}^σ and P_{ij}^τ are the exchange operators for the spatial, spin and isospin coordinates of the nucleon pair, respectively. Due to the exchange of the spatial coordinates, the first term of Eq. (5) becomes a Hartree-Fock-type potential consisting of the local direct- and *nonlocal* exchange parts

$$V_{00} = V_{\rm F} \equiv \left(\psi_0^{(a)} \psi_0^{(A)} \middle| \sum_{i \in a, j \in A} v_{ij}(1 - P_{ij}) \middle| \psi_0^{(a)} \psi_0^{(A)} \right) = V_{\rm F}^{(\rm D)} + V_{\rm F}^{(\rm EX)}. \tag{11}$$

As can be seen from Eq. (5) the imaginary part of the OP should be constructed from an appropriate theory for ΔU. This is, however, a complicated task and lies well beyond the DFM. Therefore, it is common to resort to a hybrid approach by using the DFM to generate the real part of the OP but to use a phenomenological (local) Woods-Saxon (WS) form factor for the imaginary part.[21–25]

In general, the real folded potential (11) supplemented by a local WS imaginary potential should be inserted into Eq. (3) and, given a nonlocal exchange potential, one needs to solve an integro-differential equation for the elastic scattering.[26] However, from a practical point of view a reliable local approximation for $V_{\rm F}^{(\rm EX)}$ is highly desirable. The direct part of the folded potential is local and obtained from a double-folding integral over $v_{\rm D}$ and the g.s. densities of the two nuclei as

$$V_{\rm F}^{(\rm D)}(R) = \int \rho_a(\boldsymbol{r}_a) \rho_A(\boldsymbol{r}_A) v_{\rm D}(s) d^3r_a d^3r_A, \quad \boldsymbol{s} = \boldsymbol{r}_A - \boldsymbol{r}_a + \boldsymbol{R}. \tag{12}$$

If one uses a local WKB approximation[18] for the change in the relative motion wave function induced by P_{ij}^x, then a following local expression for the exchange folded potential can be obtained

$$V_{\rm F}^{(\rm EX)}(R) = \int \rho_a(\boldsymbol{r}_a, \boldsymbol{r}_a + \boldsymbol{s}) \rho_A(\boldsymbol{r}_A, \boldsymbol{r}_A - \boldsymbol{s}) v_{\rm EX}(s) \exp\left(\frac{i\boldsymbol{K}(R)\boldsymbol{s}}{M} \right) d^3r_a d^3r_A. \tag{13}$$

The local momentum of relative motion must be determined self-consistently through the total *real* OP as

$$K^2(R) = \frac{2\mu}{\hbar^2}[E - V_{\rm F}^{(\rm D)}(R) - V_{\rm F}^{(\rm EX)}(R) - V_C(R)]. \tag{14}$$

The calculation of $V_F^{(EX)}(R)$ still contains a self-consistency problem and involves an explicit integration over the *nonlocal* nuclear density matrices of the projectile and target (see Ref. 27 and references therein for more details on the treatment of the exchange term).

3. Rainbow Scattering as a Probe of the Nuclear Incompressibility

Beside the nuclear densities, it is necessary to have an appropriate in-medium NN interaction as input for the folding calculation of the nucleus-nucleus OP. To evaluate in-medium NN interaction starting from first principles (using free NN interaction) still remains a challenge for the nuclear many-body theory. Therefore, most of the folding calculations so far still use different kinds of effective NN interaction. Such interactions can be roughly divided into two groups.

In the first group one parameterizes the interaction directly as a whole, like the Skyrme forces, without any connection with the realistic free NN interaction. In the second group one parameterizes the interaction in a functional form amenable to the folding calculation, based on results of the many-body calculation using free NN potential. A popular choice in the second group is the so-called M3Y interaction which was designed in terms of three Yukawas to reproduce the G-matrix elements of the Reid[28] and Paris[29] free NN potentials (see their explicit expressions

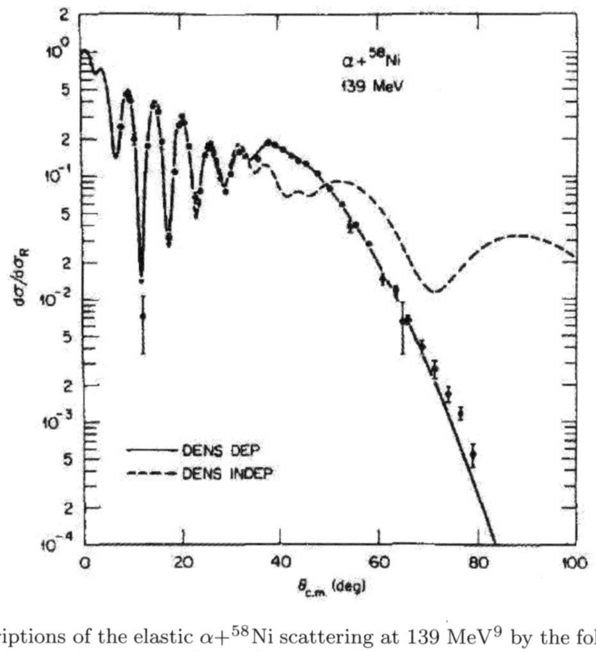

Fig. 5. OM descriptions of the elastic $\alpha+^{58}$Ni scattering at 139 MeV[9] by the folded OP obtained with (solid curve) and without (dashed curve) density dependence of the effective NN interaction. The density dependence was empirically introduced[18] to reduce the depth of the real OP at small distances while leaving the potential at the surface nearly unchanged. Illustration taken from Ref. 18.

in, e.g., Ref. 27). The original M3Y interaction has been used with some success in the DFM calculation of the nucleus-nucleus OP at low energies,[17] with elastic data usually limited to the forward angles and, thus, sensitive to the OP at the surface only. However, in cases of rainbow nucleus-nucleus scattering where the data are sensitive to the OP over a much wider radial domain, these *density independent* M3Y interactions failed to give a good description of the data. Namely, the real folded OP is too deep at small distances R to reproduce the elastic cross sections at the large angles. A typical example is shown in Fig. 5 where the inclusion of a density dependence into the effective NN interaction was found essential to describe the observed rainbow shoulder (the first Airy maximum) at large angles. In terms of medium effects, the inclusion of an explicit density dependence was needed to account for a reduction in the OP strength that occurs at small R where the overlap density of the nuclear collision increases.

The physical origin of the density dependence of effective NN interaction can be very well illustrated[30] in a Hartree-Fock (HF) calculation of the nuclear matter (NM) binding energy

$$E = \frac{1}{2}\sum_{k\sigma\tau}\sum_{k'\sigma'\tau'}[\ <\boldsymbol{k}\boldsymbol{\sigma}\boldsymbol{\tau}, \boldsymbol{k}'\boldsymbol{\sigma}'\boldsymbol{\tau}'|v_{\rm D}|\boldsymbol{k}\boldsymbol{\sigma}\boldsymbol{\tau}, \boldsymbol{k}'\boldsymbol{\sigma}'\boldsymbol{\tau}'> \\ + <\boldsymbol{k}\boldsymbol{\sigma}\boldsymbol{\tau}, \boldsymbol{k}'\boldsymbol{\sigma}'\boldsymbol{\tau}'|v_{\rm EX}|\boldsymbol{k}'\boldsymbol{\sigma}\boldsymbol{\tau}, \boldsymbol{k}\boldsymbol{\sigma}'\boldsymbol{\tau}'>] \tag{15}$$

using plane waves for $|\boldsymbol{k}\boldsymbol{\sigma}\boldsymbol{\tau}>$. Our HF calculation[30] of the NM energy (15) has shown that the original density independent M3Y interaction failed to saturate NM, leading to a collapse. Only the introduction of a density dependence into the original M3Y interaction, which accounts effectively for higher-order NN correlations (see Fig. 6), can lead to a proper HF description of the NM saturation as shown in Fig. 7. Although different versions of the density dependence saturate NM at the same density $\rho = \rho_0$, they give different curvatures of the NM energy curve, i.e., they are associated with different values of the NM incompressibility determined as

$$K = 9\rho^2\frac{d^2[E/A]}{d\rho^2}\Big|_{\rho=\rho_0}. \tag{16}$$

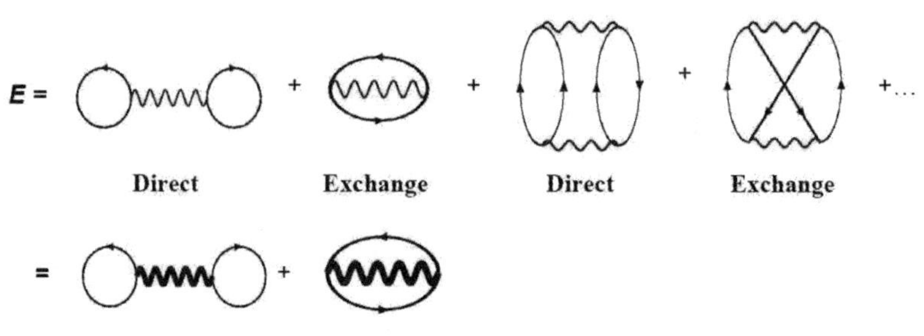

Fig. 6. The inclusion of a density dependence into the M3Y interaction effectively accounts for the contributions from higher-order ladder diagrams of the direct and exchange NN correlation in the first-order HF scheme.

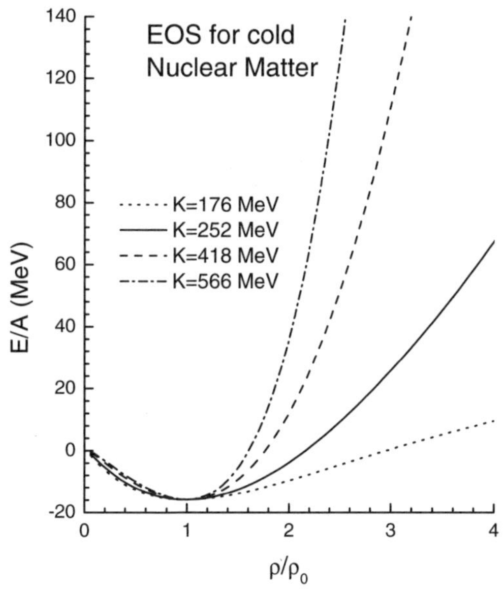

Fig. 7. NM binding energy as function of the density given by the HF calculation (15) using 4 choices of the density dependent M3Y interaction. $\rho_0 \approx 0.17$ fm^{-3} is the NM saturation density. These density dependent interactions are associated with the nuclear incompressibility K ranging from 176 to 566 MeV.[22,23]

Since the equation of state (EOS) of NM, which is important in both nuclear physics and astrophysics, usually is identified with a given K value,[27] it is of high interest to have some sensitivity of nucleus-nucleus scattering data to the nuclear incompressibility K. Indeed, we found out that the folding model analysis of the nuclear rainbow scattering data is an efficient method to determine K.

An interesting case is illustrated in Fig. 8, where different density dependent M3Y interactions were used in the DFM to calculate the real OP for the OM analysis of the elastic ^{16}O+^{16}O data at 350 MeV. These data have been shown[25] to be strongly refractive, with a broad primary-rainbow maximum observed at large angles following the first Airy minimum at $\Theta_{c.m.} \approx 43°$. Given a big difference in the OP's at small distances where the overlap density of the ^{16}O+^{16}O system is large, the rainbow pattern in the elastic ^{16}O+^{16}O data at 350 MeV can be used to probe the real OP's given by different density dependent interactions. The OM results shown in Fig. 8 confirm that the CDM3Y6 interaction is the most favorable interaction. This version of the density dependent M3Y interaction gives $K \approx 252$ MeV in the HF calculation of nuclear matter.[23] Similar (and quite unambiguous) conclusion about the nuclear incompressibility K has also been reached in our OM analysis[23] of the elastic α-nucleus scattering data. The weak absorption observed in the refractive α-nucleus scattering at medium energies, with the appearance of the nuclear rainbow pattern, offers a unique opportunity to probe the density dependence of the effective NN interaction. The results of our OM analysis of the

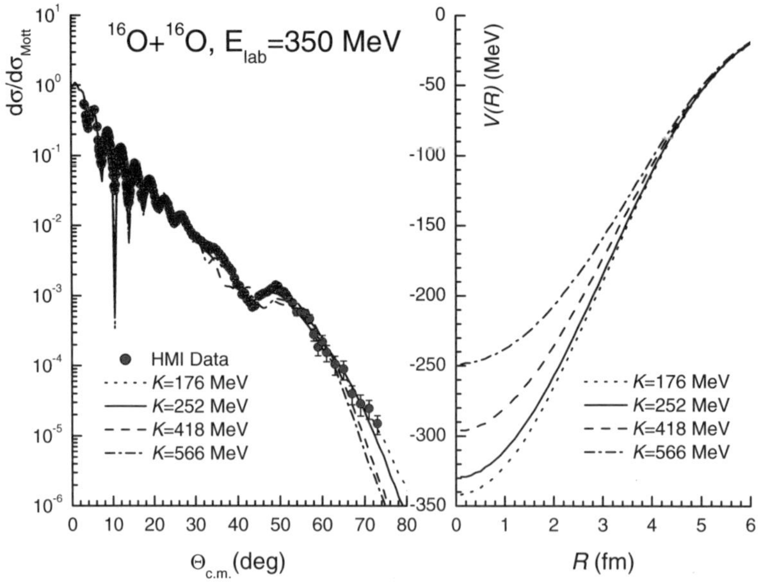

Fig. 8. OM description (left panel) of the elastic $^{16}O+^{16}O$ data at $E_{\rm lab} = 350$ MeV[14] given by the real OP's (right panel) predicted by the DFM using the same density dependent M3Y interactions as those used in the HF calculation of NM shown in Fig. 7 and an absorptive WS imaginary OP taken from Ref. 25. The best OM fit is associated with $K \approx 252$ MeV.

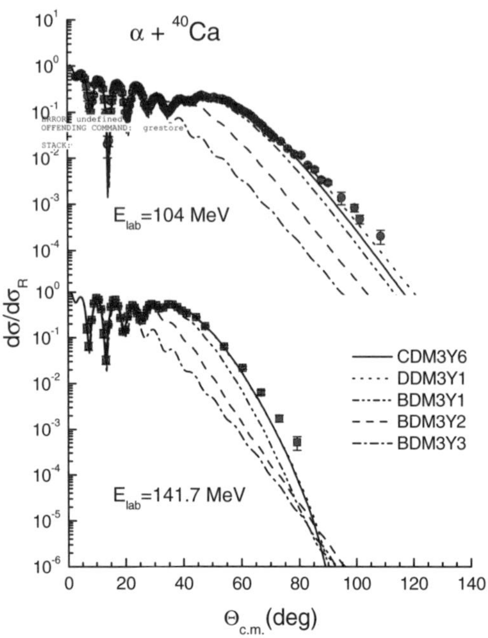

Fig. 9. OM description of the elastic $\alpha+^{40}$Ca scattering data at $E_{\rm lab} = 104$[31] and 147 MeV[10] given by the different real folded potentials.

elastic $\alpha+^{40}$Ca data[10,31] at $E_{\rm lab} = 104$ and 147 MeV are shown in Fig. 9 where the broad rainbow shoulders observed at large angles confirm unambiguously again the CDM3Y6 interaction as the most appropriate density dependent interaction.

4. Summary

We have presented a brief overview of the nuclear rainbow, a fascinating phenomenon observed in the elastic α- and light HI scattering at medium energies which can well be understood based on the basic concepts of the OM description of the elastic scattering. Despite the striking similarity in the interference structure between the nuclear rainbow and atmospheric rainbow, the former has proven to be much harder to observe experimentally. It is important to stress that the occurrence of the rainbow pattern in the α-nucleus and light HI elastic scattering is due to a *strong* mean field caused by the two nuclei overlapping each other. As a result, the rainbow scattering data can be used to probe the density dependence of the in-medium NN interaction and most of the rainbow scattering data were found to be best described by a *deep* real OP given by the DFM using a density dependent M3Y interaction which gives a nuclear incompressibility $K \approx 250$ MeV in the HF calculation of nuclear matter. This result indicates that the EOS of nuclear matter should be rather soft.

Acknowledgments

This work has been supported, in part, by Vietnam Natural Science Council, Alexander von Humboldt Stiftung of Germany and Julian Schwinger Foundation.

References

1. R. Greenler, *Rainbows, Halos and Glories* (Cambridge University Press, 1980).
2. H.M. Nussenzweig, *Scientific American* **236**, 116 (1977).
3. J.A. Adam, *Phys. Rep.* **356**, 229 (2002); *Notices of the AMS* **49** 1360.
4. M.E. Brandan and G.R. Satchler, *Phys. Rep.* **285**, 143 (1997).
5. G.B. Airy, *Trans. Camb. Phil. Soc.* **6**, 379 (1838).
6. M.E. Brandan, M.S. Hussein, K.W. McVoy, G.R. Satchler, *Comments Nucl. Part. Phys.* **22**, 77 (1996).
7. W. von Oertzen, D.T. Khoa, H.G. Bohlen, *Europhysicsnews* **31**, 5 (2000); W. von Oertzen, H.G. Bohlen, V. Subotin, D.T. Khoa, *Acta Physica Polonica B* **33**, 93 (2002).
8. D.A. Goldberg and S.M. Smith, *Phys. Rev. Lett.* **29**, 500 (1972).
9. D.A. Goldberg et al., *Phys. Rev. C* **7**, 1938 (1973).
10. D.A. Goldberg, S.M. Smith, G.F. Burdzik, *Phys. Rev. C* **10**, 1362 (1974).
11. R.G. Stokstad et al., *Phys. Rev. C* **20**, 655 (1976).
12. H.G. Bohlen et al., *Z. Phys. A* **308**, 121 (1982).
13. H.G. Bohlen et al., *Z. Phys. A* **322**, 241 (1985).
14. E. Stiliaris et al., *Phys. Lett. B* **223**, 291 (1989).
15. H.G. Bohlen et al., *Z. Phys. A* **346**, 189 (1993).
16. G. Bartnitzky et al., *Phys. Lett. B* **365**, 23 (1996).
17. G.R. Satchler and W.G. Love, *Phys. Rep.* **55**, 183 (1979).

18. G.R. Satchler, *Direct Nuclear Reactions* (Oxford: Oxford University Press, 1983).
19. C. Mahaux and R. Sartor, *Adv. Nucl. Phys.* **20**, 1 (1991) and references therein.
20. H. Feshbach, *Theoretical Nuclear Physics* Volume II (Wiley-Interscience: New York, 1992).
21. D.T. Khoa, W. von Oertzen, H.G. Bohlen, *Phys. Rev. C* **49**, 1652 (1994).
22. D.T. Khoa et al., *Phys. Rev. Lett.* **74**, 34 (1995).
23. D.T. Khoa, G.R. Satchler, W. von Oertzen, *Phys. Rev. C* **56**, 954 (1997).
24. D.T. Khoa and G.R. Satchler, *Nucl. Phys. A* **668**, 3 (2000).
25. D.T. Khoa, W. von Oertzen, H.G. Bohlen, F. Nuoffer, *Nucl. Phys. A* **672**, 387 (2000).
26. W.G. Love, *Nucl. Phys. A* **312**, 160 (1978).
27. D.T. Khoa, W. von Oertzen, H.G. Bohlen, S. Ohkubo, *J. Phys. G* **34**, R111 (2007).
28. G. Bertsch, J. Borysowicz, H. McManus, W.G. Love, *Nucl. Phys. A* **284**, 399 (1977).
29. N. Anantaraman, H. Toki, G.F. Bertsch, *Nucl. Phys. A* **398**, 269 (1983).
30. D.T. Khoa and W. von Oertzen, *Phys. Lett. B* **304**, 8 (1993); *Phys. Lett. B* **342**, 6 (1995).
31. H.J. Gils et al., *Phys. Rev. C* **12** 1239 (1980).

NON-MONOTONIC ALPHA- AND ^{6}Li-POTENTIALS FROM ENERGY DENSITY FUNCTIONAL FORMALISM

S. HOSSAIN

Department of Physics, Shahjalal University of Science & Technology, Sylhet 3114, Bangladesh

A. K. BASAK* and M. A. UDDIN

Department of Physics, University of Rajshahi, Rajshahi 6205, Bangladesh
akbasak2001@yahoo.com

M. N. A. ABDULLAH

Department of Physics, Rajshahi University of Engineering & Technology, Rajshahi 6204, Bangladesh

I. REICHSTEIN

School of Computer Science, Carleton University, Ottawa, Ontario K1S 5B6, Canada

F. B. MALIK

Department of Physics, Southern Illinois University, Carbondale, IL 62901, USA & Department of Physics, Washington University, St. Louis, Mo 63130, USA

Received 31 July 2008

The present status of the α-nucleus potential, generated from the energy density functional (EDF) formalism using a realistic two-nucleon potential, which incorporates the Pauli principle, is discussed. The EDF potentials, calculated using a density distribution of α-particle that yields a binding energy of 20 MeV with a reasonable root-mean-squared radius and observed density distributions of ^{6}Li and various target nuclei, are found to be shallow and non-monotonic in character. This non-monotonic EDF potential reproduces satisfactorily the experimental elastic scattering data, particularly at energies above the Coulomb barrier. Since the elastic scattering data and the binding energies of all nuclei considered herein are well reproduced using the mean field generated from a realistic two-nucleon potential for nuclear and nucleonic matter, one may conclude to have reasonable information on the equation of states of nuclear and nucleonic matter from a very low to the saturation density from the present investigation.

Keywords: Nucleus-nucleus potentials from energy density functional; non-monotonic α-^{58}Ni,^{90}Zr,^{208}Pb and ^{6}Li-^{28}Si,^{40}Ca,^{90}Zr potentials; analyses of α and ^{6}Li elastic scattering data.

1. Introduction

A knowledge of α and ^{6}Li-nucleus potential is pre-requisite for understanding many processes involving these projectiles, some of which are of astrophysical interest.[1,2] A first step toward this is the understanding of elastic scattering data of α-particle and ^{6}Li by various nuclei. Some of these data exhibit anomalous large angle scattering[3] at incident energies slightly larger than the Coulomb barrier heights as well as the refractive structure of nuclear rainbow[4,5] at higher energies, both of which are reasonably accounted for by non-monotonic[6] and squared Woods-Saxon (SWS)[3] empirical complex potentials. However, the phenomenological potentials, including the SWS ones, lack adequate predicting powers as these suffer from continuous and discrete (family) ambiguities[7,8] which sometimes impede consistency in terms of energy and mass variations of their volume integrals. Attempts have, therefore, been made to develop these potentials, particularly their real parts, from two-nucleon potentials. Widely used theory of the double-folding (DF) model, which attempts to derive the real part of the potential using various two-nucleon including M3Y potentials,[9,10] does not properly consider the Pauli principle among incident and target nucleons of the same type. The derived potential, therefore, quite often, requires an empirical adjustment of its strength or renormalization.[11–14]

An alternative approach has been to derive the real part of the complex potential from a realistic two-nucleon potential describing all the properties of deuteron and observed phase shifts up to about pion-production energy using the energy- density functional (EDF) theory,[15,16] the details of which are outlined in the next section. In the EDF theory, the Pauli principle is considered in determining the mean-field in nuclear and nucleonic matter approximation. The first application of the EDF theory for α-scattering was done for ^{28}Si by Manngård *et al.*[6] yielding a non-monotonic real part very close to the one needed to fit the elastic scattering data for 14.0 to 28.0 MeV incident energy. Since then the procedure has led to the determination of non-monotonic α-nucleus potentials[17,18] for ^{24}Mg, ^{27}Al, ^{29,30}Si, ^{40,44,48}Ca.

Recent realization that the best fits to α elastic scattering by ^{4}He at low incident energy is the actual potential calculated by the EDF theory[19] and that by Ni-isotopes, very close to the potential generated by the EDF theory, has led us to revisit the work on Billah *et al.*[20] for the Ni-isotopes and analyze α and ^{6}Li elastic scattering data by other nuclei in terms of the actual potentials calculated directly by the EDF theory. In this investigation, we report our initial results.

2. Energy-Density Functional (EDF) Formalism

In the EDF theory, the energy of a nucleus for a given density distribution $\rho(r)$ is expressed[19,20] as

$$E = \int \epsilon\,[\rho(\mathbf{r})]\, d^3\mathbf{r}, \tag{1}$$

where

$$\epsilon[\rho(\mathbf{r})] = 0.3\left(\frac{\hbar^2}{2M}\right)\left(\frac{3\pi^2}{2}\right)^{2/3}\left[(1-\xi)^{5/3} + (1+\xi)^{5/3}\right] \times \rho^{5/3}$$
$$+\nu(\rho,\xi)\rho + \frac{e}{2}\Phi_C(\mathbf{r})\rho_P - 0.739e^2\rho_P^{4/3} + \left(\frac{\hbar^2}{8M}\right)\eta(\nabla\rho)^2. \quad (2)$$

Here, M is the nucleon mass and $\xi = (N-Z)/A$ is the neutron excess parameter. The first term arises from the nucleon kinetic energy in nuclear matter. $\nu(\rho,\xi)$, the nucleonic mean field, is determined from the realistic nucleon-nucleon potential[21] in the Brueckner-Hartree-Fock theory, which relates the matrix elements of the potential[22] to those of scattering operator K with full consideration of the Pauli principle among the nucleons of the same type in the nuclear and nucleonic matter approximations, i.e., using plane wave for nucleonic wave functions. The density dependence of energy per nucleon E/A in this nuclear matter has been calculated by Brueckner *et al.*[22] using this potential with full consideration of the Pauli principle. The calculated $\nu(\rho,\xi)$ has been parametrized analytically as

$$\nu(\rho,\xi) = b_1(1+a_1\xi^2)\rho + b_2(1+a_2\xi^2)\rho^{4/3} + b_3(1+a_3\xi^2)\rho^{5/3}. \quad (3)$$

The coefficients, $a_1 = -0.2$, $a_2 = 0.316$, $a_3 = 1.646$, $b_1 = -741.28$, $b_2 = 1179.89$ and $b_3 = -467.54$, are derived[23] from fitting the calculated curves for the density distribution of the energy per nucleon, E/A. Hence the nucleon potential in the mean field incorporates the exchange effect. Φ_C in the third term is the Coulomb energy on a single proton from a charge distribution ρ_P. The fourth term represents the exchange correction among the protons due to the Coulomb energy. The fifth term is the inhomogeneity correction, which incorporates the correlation part of the nuclear interaction not included in the mean field. This term includes a parameter η. An optimum value of $\eta = 8.0$ has been found to reproduce the correct masses[23] and hence, binding energies, of the nuclei in the range from ^{12}C to ^{238}U.

The potential $V(R)$ between projectile I and target T in the EDF theory is then given[24,25] by

$$V(R) = E[\rho(\mathbf{r},R)] - E[\rho_I(\mathbf{r},R=\infty)] - E[\rho_T(\mathbf{r},R=\infty)]. \quad (4)$$

Here, ρ is the density function of the composite system of projectile and target. ρ_I and ρ_T are the density functions of the projectile and target when they are at $R=\infty$. In the sudden approximation,

$$\rho(r) = \rho_I(r) + \rho_T(r). \quad (5)$$

3. Parametrization of the Projectile-Target Potentials

The EDF generated potential, which serves as the real part of the projectile-target potential, is parametrized in terms of the following analytic expression:[20]

$$V(R) = -V_0\left[1+\exp\left(\frac{R-R_0}{a_0}\right)\right]^{-1} + V_1\exp\left[-\left(\frac{R-D_1}{R_1}\right)^2\right] + V_C(R). \quad (6)$$

The first term is attractive with the Woods-Saxon geometry. The real part is made non-monotonic with the inclusion of the second term, which is repulsive with the shifted Gaussian form-factor. $V_C(R)$ is the Coulomb potential of a uniformly charged sphere of radius R_C given by

$$\begin{aligned} V_C(R) &= \left[\frac{Z_1 Z_2 e^2}{2R_C}\right]\left[3 - \frac{R^2}{R_C^2}\right] \quad \text{for } R \le R_C, \\ &= \frac{Z_1 Z_2 e^2}{R} \qquad\qquad\quad \text{for } R > R_C. \end{aligned} \tag{7}$$

The imaginary part of the potential is taken to be composed of a volume term and a surface one as

$$W_m(R) = -W_0 \exp\left[-\left(\frac{R}{R_W}\right)^2\right] - W_S \exp\left[-\left(\frac{R - D_S}{R_S}\right)^2\right]. \tag{8}$$

4. Analyses and Results

In the EDF calculations, the α density distribution function[26] is taken as

$$\rho_\alpha(r) = 4\left(\frac{\gamma}{\pi}\right)^{3/2} \exp(-\gamma r^2), \tag{9}$$

with the width parameter $\gamma = 0.45$ which yields BE of 20.0 MeV using $\eta = 8.0$ and R_{rms} of 1.825 fm, which are close to the observed values of $BE = 28$ MeV

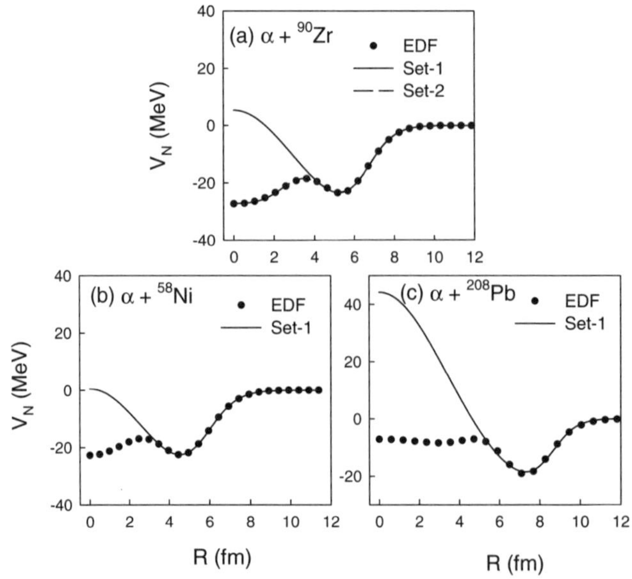

Fig. 1. Parametrization of the nuclear parts of the α-nucleus potentials from the EDF calculations (solid dots) in terms of expression in (6) excluding V_C.

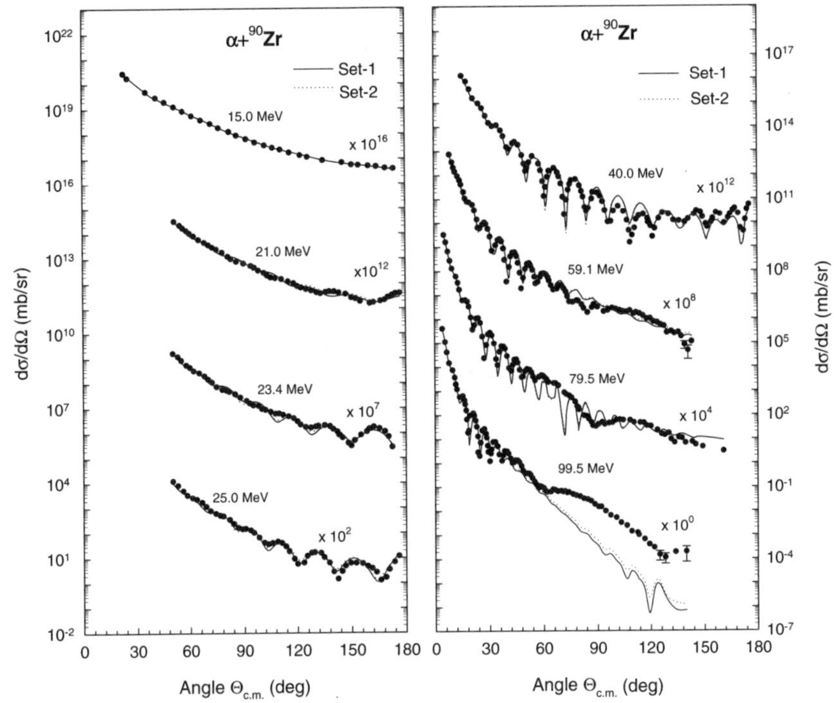

Fig. 2. Experimental differential cross sections (solid dots) for the α+^{90}Zr elastic scattering at different energies are compared to the predictions using the set-1 (solid lines) and set-2 (dotted lines) potential parameters in Tables 1–2.

Table 1. Parameters of the real part of the projectile-target potentials. The incident energy E_I, V_0 and V_1 are in MeV; R_0, a_0, R_1, D_1 and R_C in fm; and the volume integral $J_R/(A_I A_T)$, in MeV.fm^3. Set-1 and set-2 parameters are generated from the EDF theory. ENM represents the parameters of the empirical non-monotonic potential.

Proj-Targ.	Pot-set	E_I	V_0	R_0	a_0	V_1	R_1	D_1	R_C	$J_R/(A_I A_T)$
α-^{90}Zr	Set-1	15.0-59.1	37.62	6.65	0.700	42.94	4.32	0.0	10.2	89.5
	Set-2	15.0-59.1	27.24	6.78	0.614	8.63	1.51	3.44	10.2	96.4
	ENM	79.5-99.5	45.00	6.98	0.470	150.0	3.67	0.0	10.2	71.4
		118.0-141.7	40.50	6.96	0.570	101.0	4.00	0.0	10.2	69.4
α-^{58}Ni	Set-1	23.4-42.0	41.60	4.80	0.760	42.0	4.20	0.0	9.5	103.2
α-^{208}Pb	Set-1	27.0-104.0	29.40	8.60	0.639	73.6	4.84	0.0	10.5	43.5
^{6}Li-^{28}Si	Set-1	11.0-99.0	60.80	4.90	0.852	60.0	3.93	0.0	8.6	110.9
	ENM	135.0-210.0	50.00	5.54	0.600	96.0	3.43	0.0	8.6	108.1
		318.0	55.00	5.06	0.750	135.0	2.90	0.0	8.6	107.1
^{6}Li-^{40}Ca	Set-1	26.0-88.0	53.30	5.39	0.868	45.0	4.14	0.0	9.0	108.9
^{6}Li-^{90}Zr	Set-1	70.0-156.0	45.50	6.80	0.810	36.0	4.50	0.0	10.0	92.7

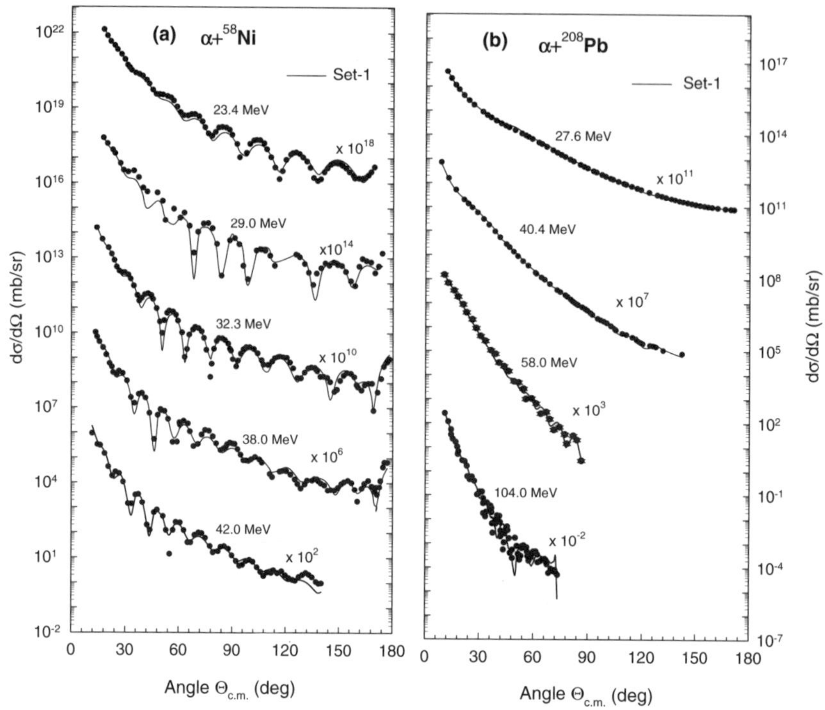

Fig. 3. Experimental differential cross sections (solid dots) for (a) $\alpha+^{58}$Ni and (b) $\alpha+^{208}$Pb elastic scattering at different energies are compared to the predictions (solid lines) using the set-1 parameters.

and $R_{rms} = 1.67 - 1.70$ fm.[27] This is a considerable improvement over the density distribution functions quite often used in DF calculations which, although yielding good R_{rms} for α-particle, lead to *unbound* α-particle states, as noted in Abdullah *et al.*[19]

The sources of other density functions are Bray *et al.*[28] for ^{6}Li, Stanley *et al.*[29] for ^{28}Si, and de Vries *et al.*[27] for ^{40}Ca, ^{58}Ni, ^{90}Zr and ^{208}Pb. These density functions are parametrized in terms of two-parameter Fermi (2pF) function given by

$$\rho(r) = \rho_0 \left[1 + \exp\left(\frac{r-c}{z}\right)\right]^{-1} \tag{10}$$

for use in the EDF calculations.

Figure 1 shows the nuclear parts of the EDF generated α potentials in solid dots for ^{90}Zr, ^{58}Ni and ^{208}Pb. Set-1 (solid lines) and set-2 (dashed line) denote the analytical forms of the EDF potentials using, respectively, $D_1 = 0$ and $D_1 \neq 0.0$ in Eq. (6). For the α-^{90}Zr interaction, the set-1 and set-2 potentials, although widely different up to about $R = 4$ fm, have the similar volume integrals per nucleon pair $J_R/(4A) = 89.5$ and 96.4 MeV.fm^3, respectively, as noted in Table 1.

Table 2. Parameters of the imaginary part of the α-nucleus potentials along with the χ^2 per point for the fits. E_I, W_0 and W_S are in MeV; R_W, R_S and D_S, in fm; and the volume integral $J_I/(4A_T)$, in MeV.fm^3. Real parameters of the set-1 and set-2 parameters are generated from the EDF theory. ENM represents the parameters of the empirical non-monotonic potential.

Target	Pot-set	E_I	W_0	R_W	W_S	R_S	D_S	$J_I/(4A_T)$	χ^2
^{90}Zr	Set-1	15.0	6.0	5.00	1.31	1.16	7.23	16.5	0.17
		21.0			3.45			24.2	2.9
		23.4			4.00		6.97	25.7	3.7
		25.0	7.0					27.7	5.7
		40.0	10.0	6.0	1.75	0.90	8.00	39.5	64.4
		59.1	10.5		3.45	1.10	8.20	43.0	40.0
		79.5	84.0	4.25	1.00	0.40	7.80	101.3	21.8
		99.5	86.0		0.00	-	-	102.1	67.0
	Set-2	15.0	6.0	5.00	2.00	1.16	7.23	19.2	0.26
		21.0			3.45			24.7	3.7
		23.4			3.80		6.97	25.0	4.7
		25.0	7.0					27.7	10.3
		40.0	10.0	6.0	1.85	0.90	8.00	39.8	59.2
		59.1	10.5			1.10	8.20	43.2	45.1
		79.5	84.0	4.25	1.00	0.40	7.80	101.3	26.2
		99.5	86.0		0.00	-	-	102.1	78.6
	ENM	79.5	82.0	4.50	0.0			115.6	18.6
		99.5	89.0					125.5	17.7
		118.0	250.0	3.70				195.9	51.3
		141.7	252.0					197.5	27.2
^{58}Ni	Set-1	23.4	13.0	4.80	8.0	0.58	5.32	47.0	6.2
		29.0			13.0	0.70	5.24	52.6	11.2
		32.3	19.0		6.7			54.6	17.5
		38.0	23.0		4.0			57.8	7.5
		42.0	25.0		3.0			60.3	19.4
^{208}Pb	Set-1	27.6	15.0	6.25	7.2	0.62	7.20	30.5	0.45
		40.4	24.5		8.0			46.6	1.55
		58.0	30.0		6.0			53.8	5.72
		104.0	181.0	5.00	0.0	-	-	151.4	135.2

Figure 2 presents the predicted cross sections for the set-1 and set-2 parameters and the comparison of experimental data of the $\alpha+^{90}$Zr elastic scattering with the predictions. Two significant points emerge from the figure. Firstly, the almost identical results from set-1 and set-2 suggest that the elastic scattering is not sensitive to the details of potential in the nuclear interior. This is because, at these incident energies, the elastic scattering is primarily determined by the surface part of the potential. Secondly, both the parameter sets, representing the EDF potential, describe the wide angular distribution data well up to the 59.1 MeV without any need for adjustment of parameters or renormalization of the potential. The parameters for the imaginary part of the set-1 and set-2 potentials are given in Table 2.

Figure 3 displays the fits to the scattering data on ^{58}Ni and ^{208}Pb. Here again the parameters of the EDF generated potential describe the data well excepting the case of ^{208}Pb at 104.0 MeV. The parameters of the real and imaginary parts of the α-potentials are noted in Tables 1 and 2.

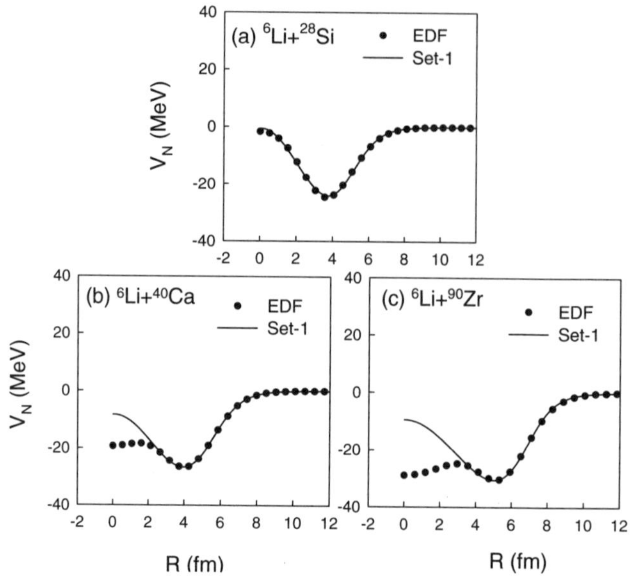

Fig. 4. Same as in Fig. 1 for ^{6}Li-nucleus potentials.

Table 3. Same as Table 2 for ^{6}Li-nucleus potentials with the volume integral $J_I/(6A_T)$.

Target	Pot-set	E_I	W_0	R_W	W_S	R_S	D_S	$J_I/(6A_T)$	χ^2
^{28}Si	Set-1	11.0	5.2	4.4	14.4	0.30	4.8	27.9	1.2
		13.0	14.8		15.5			56.0	4.4
		25.0	28.0		8.0			86.4	14.3
		27.0	34.5	4.2	2.4	0.47	5.2	88.8	13.5
		30.0	40.0		2.1			101.8	81.1
		34.0	41.0			0.45	5.0	103.8	16.3
		46.0	43.0		2.0			108.6	4.6
		75.6	46.0		0.0	-	-	113.0	6.2
		99.0	52.0					127.7	20.3
	ENM	135.0	32.0	5.1				139.8	21.8
		154.0						139.8	14.0
		210.0	46.0	4.50				140.5	6.6
		318.0	80.0	3.8				145.5	27.2
^{40}Ca	Set-1	26.0	100.0	3.82	5.5	0.52	5.4	137.1	4.7
		30.0			7.7	0.66		143.2	12.1
		34.0	102.0		8.0			146.3	10.4
		88.0	135.0		0.0			174.6	32.3
^{90}Zr	Set-1	70.0	158.0	4.50	10.0	0.26	6.54	153.1	0.60
		99.0	162.0		8.0			155.9	2.6
		156.0	185.0	4.40	0.0	-	-	162.5	9.5

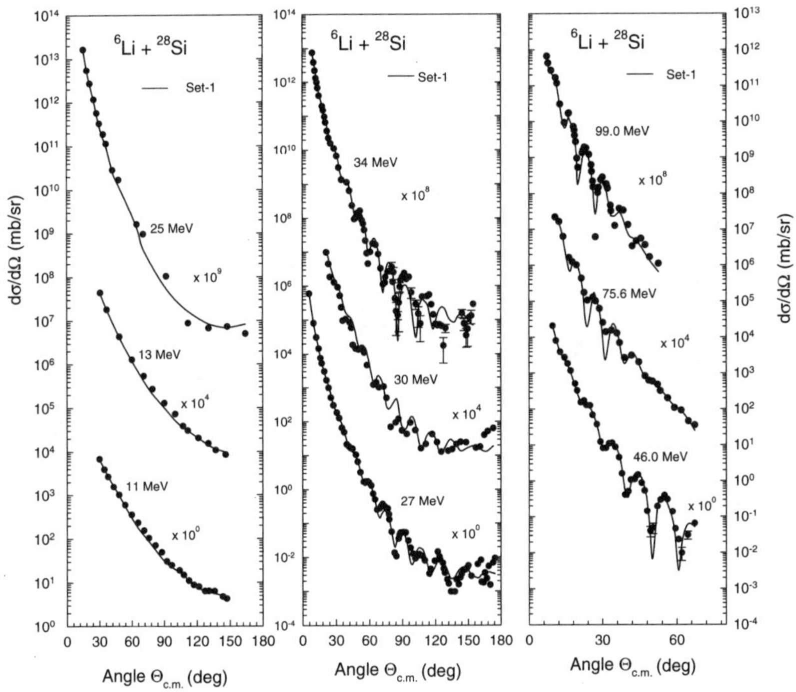

Fig. 5. Same as in Fig. 3 for ^{6}Li+^{28}Si elastic scattering using the set-1 potential parameters.

The nuclear parts of the EDF generated ^{6}Li potentials are shown in Fig. 4. The solid lines represent the analytical fit to the potentials with the parameters listed in Table 1. Figure 5 presents the comparison between the cross sections, predicted by the EDF generated parameters, and the experimental data on ^{28}Si. The data, here again, are reproduced well by the calculations up to 99.0 MeV. The associated imaginary parameters are given in Table 3.

Figure 6 shows the comparison of the experimental cross sections of the ^{6}Li elastic scattering on ^{40}Ca and ^{90}Zr with the predictions from the EDF generated potential parameters. The wide angular distributions for ^{40}Ca in energy interval of 26.0-34.0 MeV and the narrow angular distributions for ^{90}Zr are reproduced well by the predictions. The parameters of the real and imaginary parts of the potentials are noted in Tables 1 and 3.

Figure 7 presents the fits to (α+^{90}Zr) and (^{6}Li+^{28}Si) elastic scattering data, which have the refractive structure of nuclear rainbow. The fits are produced by the empirical non-monotonic (ENM) sets of real parameters in Table 1, generated from the EDF potential parameters as the starting ones. The corresponding imaginary parameters are given in Tables 2 and 3.

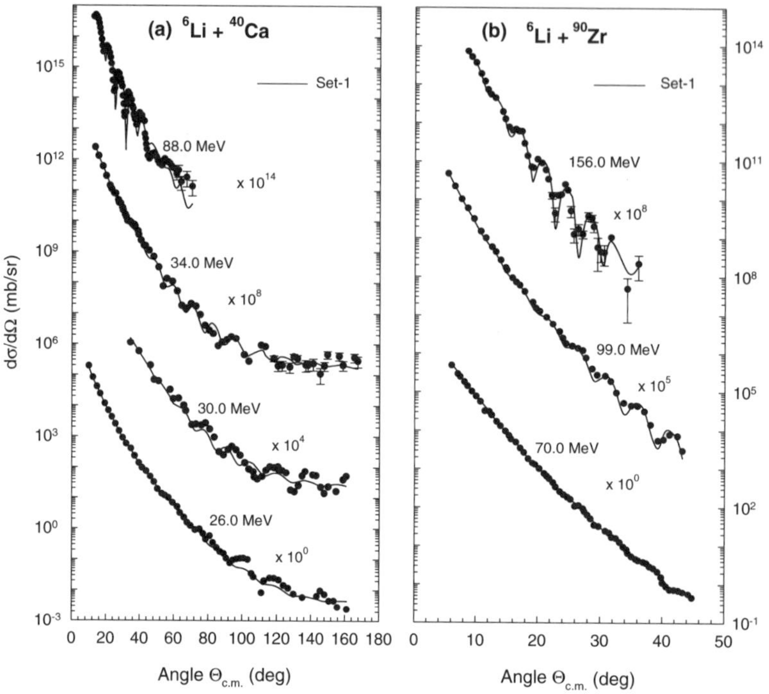

Fig. 6. Same as in Fig. 3 for ^{6}Li+^{28}Si elastic scattering using the set-1 potential parameters.

5. Conclusion

The real part of the potential, calculated directly using the EDF theory, can account for the elastic scattering data of α- particle and ^{6}Li by nuclei considered herein for incident energies of a few times the value of Coulomb barrier heights. The binding energies of all but ^{4}He are also reproduced very well with the observed density distribution functions using the EDF theory.[15] Hence, the mean field used in the EDF theory provides reasonably accurate information of the equation of states of nuclear and nucleonic matter (i.e., asymmetric nuclear matter) from very small to slightly greater than the saturation density. The calculated potentials are non-monotonic with shallow attractive part leading to smaller values (compared to potentials obtained in the DF approach) of the volume integral $J_R/(A_I A_T)$, A_I and A_T being the mass numbers of the projectile and target nuclei, respectively. At higher incident energies, the real part of the EDF generated potential needs revision due to the onset of strong absorption in accordance with the dispersion relations[30] between the real and imaginary parts. However, the functional form of the real part remains non-monotonic and volume integrals stay consistent in their values.

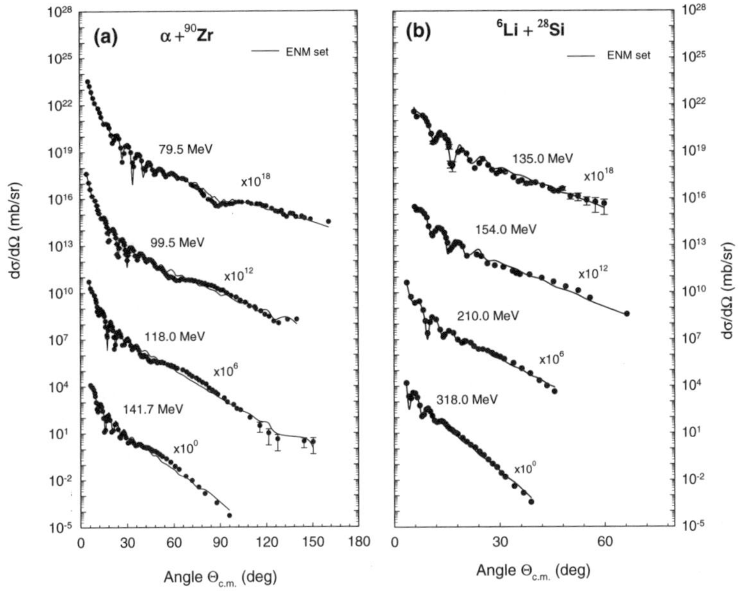

Fig. 7. Same as in Fig. 3 for (a) α+^{90}Zr and (b) ^{6}Li+^{28}Si elastic scattering using the ENM parameters in Tables 1–2.

Acknowledgments

The authors thankfully acknowledge grants from the U.S. National Science Foundation (Grant No. INT-0209584) and the Bangladesh University Grant Commission in support of this work. In addition, Basak and Malik express their thanks, respectively, to the Julian Schwinger Foundation and to the U.S. Army Research Office for grants to attend the 31st Condensed Matter Theories workshop held in Bangkok in 2007.

References

1. P. Demetrious, C. Grama and S. Goriely, *Nucl. Phys. A* **707**, 253 (2002).
2. F.D. Becchetti, D. Overway, J. Jänecke and W.W. Jacobs, *Nucl. Phys. A* **344**, 336 (1980).
3. Th. Delbar, Gh. Grégoire, G. Paic, R. Ceuleneer, F. Michel, R. Vanderpoorten, A. Budzanskowski, H. Dabrowski, L. Freindl, K. Grotowski, S. Micek, R. Planeta and A. Strzalkowski, *Phys. Rev. C* **18**, 1237 (1978).
4. M.E. Brandan and G.R. Satchler, *Phys. Rep.* **285**, 143 (1997).
5. D.T. Khoa, W. von Oertzen, H.G. Bohlen and F. Nuoffer, *Nucl. Phys. A* **672**, 387 (2000).
6. P. Manngård, M. Brenner, M.M. Alam, I. Reichstein and F.B. Malik, *Nucl. Phys. A* **504**, 130 (1989).
7. G. R. Satchler, *Direct Nuclear Reactions* (Clarendron Press, Oxford, 1983).
8. P. Mohr, T. Rauscher, H. Oberhummer, Z. Máté, Zs. Fülöp, E. Somorjai, M. Jaeger and G. Staudt, *Phys. Rev. C* **55**, 1523 (1997).

9. G. Bertsch, J. Borysowicz, H. McManus and W.G. Love, *Nucl. Phys. A* **284**, 399 (1977),
10. G.R. Satchler and W.G. Love, *Phys. Rep.* **55**, 183 (1979).
11. D. T. Khoa, *Phys. Rev. C* **63**, 034007 (2001).
12. D. T. Khoa, *Nucl. Phys. A* **484**, 376 (1988).
13. U. Atzrott, P. Mohr, H. Abele, C. Hillenmayer and G. Staudt, *Phys. Rev. C* **53**, 1336 (1996) .
14. D. T. Khoa. G.R. Satchler and W. von Oertzen, *Phys. Rev. C* **56**, 954 (1997).
15. K. Brueckner, J.R. Buchler and M.M. Kelly, *Phys. Rev.* **173**, 944 (1968).
16. R.J. Munn, B. Block and F.B. Malik, *Phys. Rev. Lett.* **21**, 159 (1968).
17. A.S.B. Tariq, A.F.M.M. Rahman, S.K. Das, A.S. Mondal, M.A. Uddin, A.K. Basak, H.M. Sen Gupta and F.B. Malik, *Phys. Rev. C* **59**, 2556 (1999).
18. M. N. A. Abdullah, M. S. Mahbub, S. K. Das, A. S. B. Tariq, M. A. Uddin, A. K. Basak, H. M. Sen Gupta and F. B. Malik, *Eur. Phys. J. A* **15**, 477 (2002); M. N. A. Abdullah, A. B. Idris, A. S. B. Tariq, M. S. Islam, S. K. Das, M. A. Uddin, A. K. Basak, I. Reichstein, H. M. Sen Gupta and F. B. Malik, *Nucl. Phys. A* **760**, 40 (2005).
19. M.N.A. Abdullah, M.S. Sabra, M.M. Rashid, Z. Shehadeh, M.M. Billah, S.K. Das, M.A. Uddin, A.K. Basak, I. Reichstein, H.M. Sen Gupta and F.B. Malik, *Nucl. Phys. A* **775**, 1 (2006).
20. M. M. Billah, M.N.A. Abdullah, S.K. Das, M.A. Uddin, A.K. Basak, I. Reichstein, H.M. Sen Gupta and F.B. Malik, *Nucl. Phys. A* **762**, 50 (2005).
21. K.A. Brueckner and J.L. Gammel, *Phys. Rev.* **109**, 1023 (1958).
22. K.A. Brueckner, S.A. Coon and J. Dabrowski, *Phys. Rev.* **168**, 1184 (1968).
23. M.A. Hooshyar, I. Reichstein and F.B. Malik, *Nuclear Fission and Cluster Radioactivity* (Springer-Verlag, Berlin, 2005).
24. L. Rickertsen, B. Block, J.W. Clark and F.B. Malik, *Phys. Rev. Lett.* **22**, 951 (1969).
25. H. Ngô and Ch. Ngô, *Nucl. Phys. A* **348**, 140 (1980).
26. L.R.B. Elton, *Nuclear Sizes* (Oxford University Press, London, 1961).
27. H. de Vries, C.W. de Jagers and C. de Vries, *At. Data Nucl. Data Tables* **36**, 495 (1987) .
28. K.H. Bray, M. Jain, K.S. Jayaraman, G. Lobianco, G.A. Moss, W.T.H. va Oers and D.O. Wells, F. Petrovich, *Nucl. Phys. A* **189**, 35 (1972).
29. D.P. Stanley, F. Petrovich and P. Schwandt, *Phys. Rev. C* **22**, 1357 (1980).
30. C. Mahaux, H. Ngô and G.R. Satchler, *Nucl. Phys. A* **449**, 354 (1986).

RESONANCES AND ANGULAR DISTRIBUTION IN α DECAY

F. F. KARPESHIN

Fock Institute of Physics, Physical Department
St. Petersburg State University, RU-198504 St. Petersburg, Russia
karpechine@teor.fis.uc.pt

Received 31 July 2008

Tunneling of α particles through the Coulomb barrier is consecutively treated. The effect of sharp peaks arising in the case of coincidence of the α energy with that of a quasistationary state within the barrier is elucidated. Peaks' energy depend on the α-nucleus potential. They can give rise to "anomalous" properties of some neutron resonances. The peaks can also be observed in the incoming α-nucleus channel. The method is also applied for calculation of the angular distribution of the emitted α particles from the α decay of a compound nucleus of ^{135}Pr. Next fundamental test of the theory is proposed to be the α decay of superheavies.

Keywords: Angular distribution; giant resonances in alpha decay.

1. Introduction

α decay is an exciting problem. Classical α-decay, as well as α-decay from the compound nuclei formed in fusion-fission reactions and neutron resonances, comprises a significant element of the nuclear study. In the fusion-fission reaction, prefission emission of alphas and other light charged particles provides us with a source of information about the timescale of the process.[1] However, some details of the theoretical description still remain not enough elucidated in the literature. Thus, emission of alphas and light charged particles from hot compound sources produced in heavy-ion or energetic proton-nucleus collisions is usually made in terms of the inverse cross section.[2] Such an approach assuming the time reversibility of the process leaves out of scope a possibility of its profound experimental check. This is in contrast with a number of suggestions that violations of the reversibility may arise, e.g., due to the back-transparency of the inner slope of the potential barrier in the incoming channel,[3] or different response of the nuclear surface on the interaction with the emitted or an incoming particle of the same type. For description of classical α decay, usually the Gamov's theory of decay of quasistationary states is used. The theory was proposed by Gamov as an extension of the theory of quasistationary states by introducing the complex energy as an eigenvalue.[4] In this theory, the real part of the energy has the physical sense of the eigenvalue, and the imaginary part gives the probability of tunneling through the barrier. There is, however, one point in this way, namely, that there are *no* quasistationary α orbits in the nuclei. They are only formed virtually, mainly on the nuclear surface (e.g. Ref. 5). Then they have a chance to escape through the barrier, otherwise they are immediately destroyed and absorbed by the bulk of the nuclear matter. The same also refers to the decay of atomic clusters. According to Mang,[6,7] the probability to find an α particle on the surface is approximately 1/100.

This simple circumstance has important consequences on the physics of the decay. Were alphas on the stationary orbits, their eigenenergies would coincide with the separation energies, in accordance with the theorem. Positive eigenvalues would mean the quasistationary states, while the energies of the emitted alphas would coincide with the eigenvalues. Therefore, the parameters of the α-nucleus potential could be fitted in the way that the energy of the emitted particle exactly coincides with a quasistationary state inside the barrier, and the width of the α line — with the inverse lifetime.[8] Contrary, if a created virtual alpha does exist for only nuclear time of approximately $\tau \approx 10^{-22}$ s, its energy is uncertain within $\Delta E \approx \hbar/\tau \approx 10$ MeV, which approximately equals the energy separation of the neighbor eigenvalues. Therefore, the α decay energy has nothing to do with the eigenenergies of the α-nucleus potential, being determined by the mother and daughter nuclear masses.[9]

One tries to overcome this circumstance. The formation probability of the α particle is defined in a certain model. This probability is then multiplied by the tunneling probability which is given by the classical action. However, the latter is only true if the alpha energy coincides with the eigenvalue, then its wavefunction smears down under the barrier. But how to proceed if the wave function *increases* under the barrier? The latter is just the case if the alpha energy is not an eigenvalue.

An appropriate way of calculating the decay width at any energy of the emitted particle has been proposed in Ref. 9. Strong resonance effects are, specifically, predicted in alpha spectra from subbarrier decay of compound systems. The resonances can be also observed in the incoming channel in the subbarrier alpha-nucleus reactions. Such resonances were observed in proton-nucleus reaction.[11,12] However, in the case of α-nucleus reactions, to the best of our knowledge, we cannot refer to any data. Note that account of such resonances can manifest themselves in unusual, "anomalous" properties of α decay of some neutron resonances.[10]

Note that the predicted quasi-collective resonances appear at certain energy of the emitted particle, *independent* of the energies of the initial and final nuclear states. In this attitude they are similar to the giant dipole resonance (GDR) in nuclei. The latter manifests itself in γ spectrum of highly excited nuclei as a characteristic bump with exactly the same parameters as in reverse processes of photonuclear reactions. The spectrum depends on the transition energy, and not on the initial nuclear energy as a consequence of the Axel–Brink hypothesis. Analogous picture is thus predicted in the α decay. Discovery of these resonances will make clear the concept of the α-nucleus potential and will provide us with direct information about its parameters.

In the present paper, we apply the method to calculation of the angular distribution. As a next fundamental test of the theory, it is supposed to apply the method for calculation of the α decay of superheavies. In this case, all the peculiarities of the methods used should manifest themselves especially distinctively.

2. Reminder of the Model

Let $\Psi(\mathbf{r}_1, \ldots, \mathbf{r}_A)$ be a wave function of the source nucleus with the mass number A. Let then the nucleus make a transition $i \to f$. As a result, in the exit channel we observe the system in a state which is described by a wavefunction as a superposition of the plane wave and ingoing spherical wave[13] at large α-nucleus distances R:

$$\psi_{f\mathbf{p}}(\mathbf{r}_1, \ldots, \mathbf{r}_A) \underset{R\to\infty}{\sim} \mathcal{A}\,\varphi_f(\mathbf{r}_1, \ldots, \mathbf{r}_{A-4}) g_{\mathbf{p}}(\mathbf{r}_{A-3}, \ldots, \mathbf{r}_A)\ , \tag{1}$$

$$g_p(\mathbf{r}_{A-3}, \ldots, \mathbf{r}_A) \underset{R\to\infty}{\sim} \mathcal{F}_{\mathbf{p}}(\mathbf{R}) \xi(\mathbf{r}_{A-3} - \mathbf{R}, \ldots, \mathbf{r}_A - \mathbf{R})\ ; \tag{2}$$

$$\mathcal{F}_{\mathbf{p}}(\mathbf{R}) \equiv \left[e^{i\mathbf{pR}} + \frac{A(\vartheta, \varphi)}{R} e^{-ipR} \right]\ . \tag{3}$$

Here $g_{\mathbf{p}}$ is the channel wavefunction, which is the eigen function of the α-nucleus Hamiltonian with an appropriate mean-field single particle potential $U_\alpha(R)$:

$$(H - \varepsilon_{\mathbf{p}})\mathcal{F}_{\mathbf{p}} = 0\,, \tag{4}$$

$$\varepsilon_p = p^2/2M_\alpha\,. \tag{5}$$

Symbol $\mathcal{A}$ in Eqs. (1)–(3) designates antisymmetrization over all the identical nucleons whose variables enter different wavefunctions, ξ is the intrinsic α wavefunction. All the wave functions are supposed to be properly antisymmetrized.

The recipe can therefore be proposed as follows. Firstly, we project the wavefunction of the mother nucleus onto the wavefunction of the daughter one $\varphi_f(\mathbf{r}_1, \ldots, \mathbf{r}_{A-4})$. Resulting amplitude is a function of the four coordinates of the nucleons constituting the α particle:

$$\langle \varphi_f(\mathbf{r}_1, \ldots, \mathbf{r}_{A-4}) | \Psi(\mathbf{r}_1, \ldots, \mathbf{r}_A) \rangle = \eta(\mathbf{r}_{A-3}, \ldots, \mathbf{r}_A)\,. \tag{6}$$

The resulting product can be called α cluster in the adiabatic representation. Within the framework of the shell model, its wavefunction vanishes outside the nuclear surface.

Therefore, we have to relate the shell-model wavefunction of the source nucleus to the channel wavefunction (1). Then the re-expansion coefficients give the transition amplitude under consideration. This way is similar to that found by Migdal when solving his classical problem of shake of an atom in β decay.[13] To find the transition amplitude, one has to project the right part of Eq. (1) onto the primary wavefunction Ψ:

$$M_{f\mathbf{p}} = \langle g_{\mathbf{p}}(\mathbf{r}_{A-3} - \mathbf{R}, \ldots, \mathbf{r}_A - \mathbf{R}) | \eta(\mathbf{r}_{A-3}, \ldots, \mathbf{r}_A) \rangle\,. \tag{7}$$

According to Ref. 9, the α decay amplitude is thus determined by the formation-penetration factor (7), which includes both the formation and penetration amplitudes. One can further break this factor into these conventional amplitudes by inserting the set of the α wavefunctions, reduced to the single ground state ξ of an α particle which is in the point R, and then integrate over R. As a result, we get the following expression:

$$M_{f\mathbf{p}} = \langle g_{\mathbf{p}}(\mathbf{r}_{A-3} - \mathbf{R}, \ldots, \mathbf{r}_A - \mathbf{R}) | \xi(\mathbf{r}_{A-3} - \mathbf{R}, \ldots, \mathbf{r}_A - \mathbf{R})\rangle \times \\ \times \langle \xi(\mathbf{r}_{A-3} - \mathbf{R}, \ldots, \mathbf{r}_A - \mathbf{R}) \eta(\mathbf{r}_{A-3}, \ldots, \mathbf{r}_A)\rangle \equiv \\ \equiv \int_0^\infty \mathcal{F}_p(R) f(R)\, dR\,. \quad (8)$$

In (8), $f(R)$ can be considered as a wavefunction of an α particle inside the nucleus. The preformation probability P is therefore included in this factor:

$$P = |\int_0^\infty f(R)\, dR|^2\,. \quad (9)$$

The conventional spectroscopic factor s for α decay is given by the overlapping integral

$$s = \int_0^\infty \langle \xi(\mathbf{r}_{A-3} - \mathbf{R}, \ldots, \mathbf{r}_A - \mathbf{R}) | \eta(\mathbf{r}_{A-3}, \ldots, \mathbf{r}_A)\rangle\, dR = \int_0^\infty f(R)\, dR\,. \quad (10)$$

On the other hand, the other overlapping integral in (8) gives the tunneling probability t:

$$t = |\int_0^\infty \mathcal{F}_p(R) f(R)\, dR\, |^2. \quad (11)$$

Furthermore, taking into account the asymptotics (3), the wavefunction $\mathcal{F}_{\mathbf{p}}(\mathbf{R})$ can be expressed in terms of the spherical harmonics in a usual way:

$$\mathcal{F}_p(\mathbf{R}) = \sum_{\ell=0}^{\infty} i^\ell (2\ell+1) e^{-i\delta_\ell} R_{p\ell}(R) Y_{\ell m}(\theta, \varphi)\,. \quad (12)$$

Taking into account that the wave functions $\mathcal{F}_{\mathbf{p}}$ are normalized at one particle in a unit volume, with the flux $\mathbf{v} \equiv \mathbf{p}/M_\alpha$, we obtain from (8) the following expression for the decay probability per unit time:

$$\Gamma_{\mathbf{p}} \equiv \frac{d^3W}{d^3p} = |M_{f\mathbf{p}}|^2\, v\delta(E_f - E_i)\,. \quad (13)$$

$\delta(E_f - E_i)$ in (13) expresses conservation of energy.

Under reasonable supposition of the utmost fragmentation[9] of the eigen α levels, the spectroscopic factor for the α decay as well as the preformation probability are approximately constant, independent of energy. According to Mang, that is $P \approx 0.01$ for all the nuclei. Then the probability of emission is determined by the penetration probability. Let us study this dependence.

3. Surface Cluster Model

Let the wavefunctions of the initial and final states be $\psi_i(r)$ and $\psi_{\mathbf{k}}^{(-)}(\mathbf{r})$:

$$\psi_{\mathbf{k}}^{(-)}(\mathbf{r}) = \frac{2\pi}{k} \sum_{l=0}^{\infty} \sum_{m=-l}^{l} i^l e^{-i\delta_l} R_{kl}(r) Y_{lm}^{\star}(\hat{k}) Y_{lm}(\hat{r}) \,. \tag{14}$$

Here $R_{kl}(r)$ is the radial wavefunction, normalized at

$$R_{kl}(r) \sim \frac{2}{r} \sin(kr + \frac{1}{k} \log 2kr - l\frac{\pi}{2} + \delta_l) \,. \tag{15}$$

According to what is said in the previous section, the transition amplitude is

$$C_{\mathbf{k}i} = \langle \psi_{\mathbf{k}}^{(-)}(\mathbf{r}) | \psi_i \rangle \tag{16}$$

where $\psi_i(\mathbf{r})$ is the initial alpha wavefunction. Eqs. (14), (16) allow us to calculate the angular distribution of the emitted alphas. Very instructive physical results can be obtained within the framework of the surface cluster model SCM.[6,14,15] Its physical justification is that α clusters are most often formed on the nuclear surface, as it is explained in the Introduction. Let us put down the initial wavefunction in the form

$$\psi_i(\mathbf{r}) = D\delta(\mathbf{r} - \mathbf{R}_0) \,, \tag{17}$$

where D is the normalization, and R_0 is a point on the nuclear surface, where an α cluster is formed. Substituting (17) into (16), one obtains that the transition amplitude is given by the following expression:

$$M_{\mathbf{k}i} = \frac{2\pi}{k} \sum_{l=0}^{\infty} \sum_{m=-l}^{l} i^l e^{-i\delta_l} R_{kl}(\mathbf{R}_0) Y_{lm}(\hat{k}) \tag{18}$$

$$\frac{dN}{d\,\cos(\theta)} = |M_{\mathbf{k}i}|^2 \,. \tag{19}$$

Another kind of the initial wavefunction can be represented by the wavefunction with a definite angular momentum:

$$\psi_i(\mathbf{r}) = D_l \delta(r - R_0) Y_{l0}(\hat{r}) \,. \tag{20}$$

Then the angular distribution is given by the same definite angular harmonic:

$$\frac{dN}{d\,\cos(\theta)} \sim |C_{\mathbf{k}i}^{(l)}|^2 \,, \tag{21}$$

$$C_{\mathbf{k}i}^{(l)} = D_l \frac{2\pi}{k} \sum_{l=0}^{\infty} \sum_{m=-l}^{l} i^l e^{-i\delta_l} R_{kl}(R_0) Y_{l0}(\hat{k}) \,. \tag{22}$$

4. Method of Solution

Usually, the angular momentum of the α particle is more or less fixed. This is the case e.g. if the final (or initial) state has zero spin. In this case, the SCM with a definite angular momentum is appropriate for the description. The angular momentum is given by the square of the corresponding spherical harmonic (22). However, if the initial spin of the mother nucleus I_i is high enough, then many spherical harmonics can contribute to the angular distribution. In the latter case, the SCM (18) can be appropriate. Let us see the results it provides with.

The calculations were performed for the same tentative nucleus as in Ref. 9 emitting α particles with the energy $E_\alpha = 12.5$ MeV. SCM (18) has been used. Numerical solutions $R_{kl}(r)$ of Eq. (4) were obtained by sewing solutions which are regular at the origin with the regular and irregular Coulomb wavefunctions, F_{Coul} and G_{Coul}, respectively. In finer detail, the Schrödinger equation for an α particle in the field of a nucleus reads as follows:

$$\left\{-\frac{1}{2m}\left[\frac{d^2}{dr^2} - \frac{l(l+1)}{r^2}\right] + V(r)\right\} R_{kl}(r) = \varepsilon R_{kl}(r)\,, \tag{23}$$

with the potential

$$V(r) = V_{SW}(r) + V_C(r)\,, \tag{24}$$

$$V_{SW} = \frac{-V_0}{1+\exp\left(\frac{r-c}{a}\right)}\,. \tag{25}$$

Calculations were performed with the Saxon–Woods potential (25), with the parameters $V_0 = 100$ MeV, $s = 2.3$ Fm, $c = 1.2A^{1/3}$ Fm. The imaginary part of the potential is usually omitted for the energies below 12 MeV. Furthermore, the Coulomb potential was taken into account as due to the sharp-edge charge distribution:

$$V_C(r) = \begin{cases} \frac{\alpha Z}{2R_0}\left[3 - \left(\frac{r}{R_0}\right)^2\right] & \text{for } r < R_0\,, \\ \frac{\alpha Z}{r} & \text{for } r \geq R_0\,, \end{cases} \tag{26}$$

with R_0 being the nuclear radius.

On the radius segment between the origin $r = 0$ and the first turning point c_1: $0 < r \leq c_1$, equation (23) was integrated numerically with the initial condition

$$\Psi(r) \underset{r\to 0}{\sim} r^l\,. \tag{27}$$

The general solution of the Schrödinger equation under the barrier is a linear combination of the two linearly independent solutions. One of them exponentially vanishes, and the other exponentially increases with increasing r. The coefficients were obtained by sewing the functions at some internal point R_s under the barrier. The solution was checked for self-consistency, that is, the coefficients do not depend on the position of R_s. The fundamental set was hence obtained by numerical integration from c_2 to c_1 with two different initial conditions:

$$y_1(c_2) = F_{Coul}(c_2), \qquad y'(c_2) = F'_{Coul}(c_2) \tag{28}$$

$$y_2(c_2) = G_{Coul}(c_2), \qquad y'(c_2) = G'_{Coul}(c_2) \,. \tag{29}$$

We note that after sewing at $r = R_s$, the resulting solution increases under the barrier if not an eigenstate, in contrast with the behavior of each of the fundamental solutions. This demonstrates mathematical correctness of the method. For the numerical integration, the Runge–Kutta–Nyström method was used. The Shtermer method was also tried, with the same results. Behind the barrier, the both solutions oscillate.

5. Numerical Results

The efficient number of the spherical harmonics contributing to the decay can be estimated for this energy as follows:

$$kR_0 \sim R_0\sqrt{8M_pE_\alpha} \sim 8 \,. \tag{30}$$

The calculations have been performed with the account of up to 10 spherical harmonics in the wavefunction in the continuum (18). The results of the calculation are presented in Fig. 1. As it was expected the angular distribution is given by superposition of several spherical harmonics. A distinctive feature of Fig. 1 is that the probability of forward and backward emission is completely different. As it

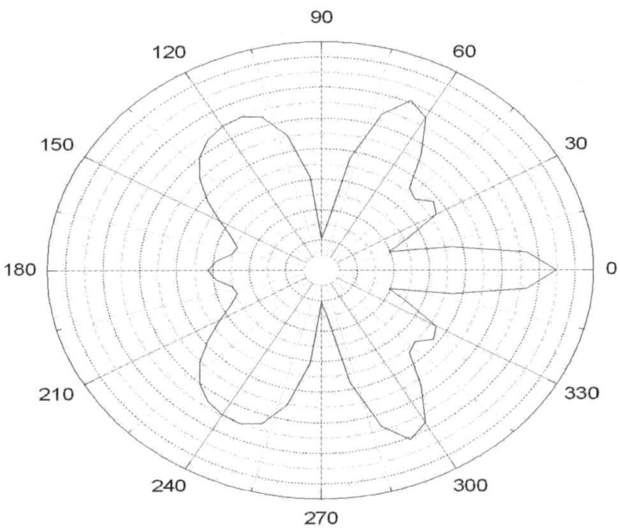

Fig. 1. Angular distribution of alpha particles emitted from a compound nucleus of ^{135}Pr with the energy of 12.5 MeV.

should be expected on the physical grounds, the α particles are emitted in the forward direction. Backward emission is suppressed by the shadow effect of the nucleus. Analogous picture was obtained in Ref. 16 in prompt fission of muonic atoms for the prompt muons, which are emitted as a result of shake off effect brought about by the rupture of the neck. In that case, the angular asymmetry was caused by more probable position of the α particle near the heavier fragment.

Another example where a similar picture of the angular emission is observed is in fission, where the ternary α particles are emitted perpendicular to the fission axis. This result is qualitatively clear in view of the above consideration: the probability to find an α particle in the vicinity of the fragments which repel the α particle is suppressed. Usually, the angular distribution is calculated by means of Monte-Carlo method, using classical trajectories. Our method gives an opportunity of quantum-mechanical calculation. The question of the angular distribution becomes actual in view of the study of the effect of the left-right asymmetry in emission of ternary α particles.[17]

References

1. E. Vardaci, A. Brondi, G. LaRana et al. *Yad. Fiz.* **66**, 1218 (2003). *Engl. Transl. Phys. Atom. Nucl.* **66**, 1182 (2003).
2. F.F. Karpeshin, G.Ye. Belovitsky, V.N. Baranov and O.M. Steingrad, in *Proc. of the XXXIV-th Intern. Winter Meeting on Nucl. Phys., Bormio, Italy, January 23-27 1996*, ed. I. Iori (Ricerca Scientifica ed Educazione Permanente, Supplemento Nr. 102), pp. 139–148; *Yad. Fiz.* **62**, 37 (1999). (*Engl. transl. Phys. Atom. Nucl.* (USA), **62**, 32 (1999)).
3. M. Kildir, G. La Rana, R. Moro *et al.*, *Phys. Rev. C* **46**, 2264 (1992); **51**, 1873 (1995).
4. A.I. Baz, Ya.B. Zeldovich and A.M. Perelomov, *Scattering, reactions and decays in quantum mechanics.* M.: Nauka, 1981.
5. L.A.Sliv, in *Proc. PNPI Winter School on Nucl. Phys.* St. Petersburg: PNPI, 1982. Vol. 1.
6. M. Ali Hooshyar, I. Reichstein and F. B. Malik, *Nuclear Fission and Cluster Radioactivity* (Springer, Berlin, Heidelberg, New York, 2005).
7. H. J. Mang, *Phys. Rev.* **119**, 1069 (1960); *Z. Phys.* **149**, 572 (1957).
8. M. Rizea, A. Sandulescu and W. Scheid. *Rev. Roum. Phys.* **26**, 14 (1981).
9. F. F. Karpeshin, G. LaRana, E. Vardaci, A. Brondi, R. Moro, S. N. Abramovich and V. I. Serov. *J. Phys. G: Nucl. Part. Phys.* **34**, 587 (2007).
10. Yu. P. Gangrsky, F. F. Karpeshin, Yu. P. Popov and M. B. Trzhaskovskaya, *Particles and Nuclei Lett.* **3**, 90 (2006).
11. D. P. Lindstrom *et al.*, *Nucl. Phys. A* **168**, 37 (1971).
12. M. G. Bowler. *Nuclear Physics* (Oxford, Pergamon, 1973).
13. L. D. Landau and E. M. Livshitz, *Quantum Mechanics. Nonrelativistic Theory* (Pergamon, London, 1958).
14. G. H. Winslow, *Phys. Rev.* **96**, 1032 (1954).
15. F. B. Malik, *Invited talk at the XXX Int. Workshop on CMT* (2006).
16. F. F. Karpeshin, *Nucl. Phys. A* **617**, 211 (1997).
17. F. F. Karpeshin, Right-left asymmetry of radiation from fission, nucl-th/0710.1799.

GIANT DIELECTRIC BEHAVIOR OF $BaFe_{0.5}Nb_{0.5}O_3$ PEROVSKITE CERAMIC

URAIWAN INTATHA
School of Science, Mae Fah Luang University,
Chiang Rai, 57100, Thailand

SUKUM EITSSAYEAM and TAWEE TUNKASIRI*
Department of Physics, Faculty of Science, Chiang Mai University,
Chiang Mai, 50202, Thailand
**tawee@chiangmai.ac.th*

Received 31 July 2008

Single-phase cubic $Ba(Fe,Nb)_{0.5}O_3$ (BFN) powder was synthesized by solid-state reaction at 1443 K for 4 hour in air. X-ray diffraction indicated that the BFN oxide mixture calcined at 1200°C crystallizes to the pure cubic perovskite phase. BFN ceramics were produced from this powder by sintering at 1623–1673 K for 4 hrs in air. Samples prepared under these conditions achieved up to 94.7% of the theoretical density. The temperature dependence of their dielectric constant and loss tangent, measured at difference frequencies, shows an increase in the dielectric constant with temperature which is probably due to disorder on the B site ion of the perovskite. Non-Debye type of relaxation phenomena has been observed in the BFN ceramics as confirmed by Cole–Cole plots. The higher value of ε' at the lower frequency is explained on the basis of the Maxwell–Wagner (MW) polarization model.

Keywords: Dielectric properties; Barium iron niobate; solid state reaction.

1. Introduction

The development of a new dynamic memory generation implies downscaling of the device dimensions, while maintaining the storage charge at an adequate level to retain data against leakage current. The data, representing one of the two binary logical states is stored as an electrical charge. Depending on the circuit associated with the memory cell, such as the sensitivity of the sense amplifier and the parasitic capacitance. With the ever continuing downscaling, this minimum charge was ensured through the compensation of the capacitance by increasing the area (A) or using thinner dielectric (thickness: d) having relatively high dielectric constant (ε_r):

$$C = \frac{\varepsilon_0 \varepsilon_r A}{d}\,.$$

---

*Corresponding author.

Ceramic type perovskites of the general formula ABO_3 are widely used in various electronics and microelectronic devices. Among these electroceramics, those of complex perovskite $A(B'B'')O_3$, are particularly attractive for various applications such as in microwave frequency resonators, multilayer capacitors, sensors, detectors and actuators. The high relative permittivity over a very wide temperature interval that has been demonstrated by some of these complex perovskites is related to the disorder in B-site cations of the perovskite unit lattice. Indeed, composition fluctuation can occur in B-site of these perovskites, and as a consequence, the different local Curie temperatures give rise to frequency dispersion and relaxor behavior.[1] Recently, high dielectric permittivity (ε_r) has been reported for certain ceramic compositions of ternary perovskite $Ba(Fe,Nb)_{0.5}O_3$ (BFN). On the other hand, several previous investigators, including Yokosuka,[2] Tezuka *et al.*,[3] Raevskite,[4] Saha and Sinha,[5,6] have reported that the BFN-based electroceramics exhibits very attractive dielectric and electrical properties over a wide temperature range, together with relaxor behavior. On the other hand, there still exist considerable doubt concerning the physical mechanisms governing their electrical behavior.[4] We will focus on the fabrication of BFN electroceramics where the effect of preparation condition on the dielectric and physical properties, as well as their microstructure, will be investigated.

2. Experimental

Barium iron niobate ($BaFe_{0.5}Nb_{0.5}O_3$: BFN) was synthesized by solid-state reaction of appropriate mixtures of barium carbonate, $BaCO_3$ (Fluka, 99.0% purity), iron oxide, Fe_2O_3 (Riedel — deHaën, 99.0% purity) and niobium oxide, Nb_2O_5 (Aldrich, 99.9% purity). The powders were milled together in ethanol for 24 hrs in a polyethylene container with zirconia balls, dried at 393 K and calcined at 1473 K, with a dwell time of 4 hrs and a heating/cooling rate of 5 K/min. The BFN powder was pressed into discs using polyvinyl alcohol as a binder and sintered at 1623 K, with a dwell time of 4 hrs at a heating and cooling rate of 5 K/min. The sintered ceramics were subsequently examined at room temperature by X-ray diffraction using CuK_α radiation (XRD; Siemens D500/D501) to identify the phase composition. Scanning electron microscopy (SEM) (Model: JEOL JSM-840A) performed at 10 kV was used to determine the as-sintered surface morphology of the ceramics. The sintered samples were polished to obtain plane and parallel surfaces and silver paste electrodes were applied for the purpose of the impedance spectroscopy measurements which were made at 100 Hz to 1 MHz and temperatures of 293 to 600 K using a impedance/gain - phase analyzer (Solartron model SI 1260).

3. Results and Discussion

The density for a sample sintered at 1623 K is 6.2 g/cm^3, or 94.7% of the theoretical density (reported in ICSD[7]). It is known that the relative density is the most important factor for a dielectric material since the dielectric constant are in good correlation with relative density. The X-ray diffraction (XRD) patterns of the BFN ceramics (Fig. 1) indicate that pure cubic $BaFe_{0.5}Nb_{0.5}O_3$ (BFN) is formed in the BFN ceramics sintered at 1623 K. The XRD phase analysis was based on the Inorganic Crystal Structure Database (ICSD) code. 52835. The cubic cell parameter (a) of this compound is 4.080 Å, in the space group Pm3m. Our results are in agreement with Rama *et al.*[8] and Tezaka *et al.*[3] who also prepared this compound by solid state reaction. However, Saha and Sinha[5,6] deduced a monoclinic-type BFN structure by using a standard computer program (POWD) analysis. Microstructure development during sintering was investigated by scanning electron microscopy (SEM). The surface micrographs of BFN ceramics are shown Fig. 2. The SEM micrographs of as-sintered ceramics show a few pores at grain boundaries, but cracks or micro cracks were not detected. The average grain size of BFN was around ~10 μm.

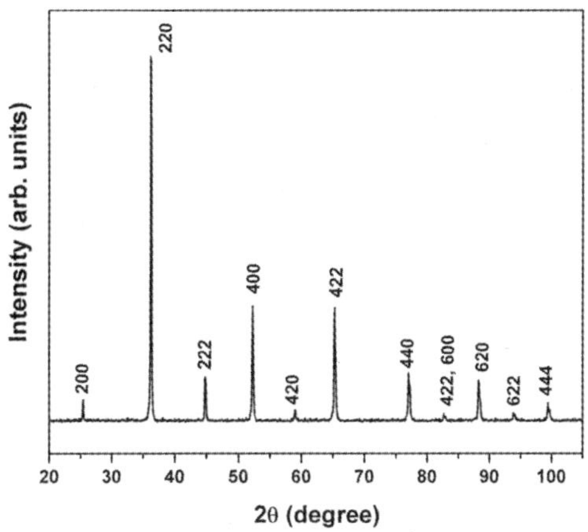

Fig. 1. X-ray pattern of BFN ceramic sintered at 1623 K.

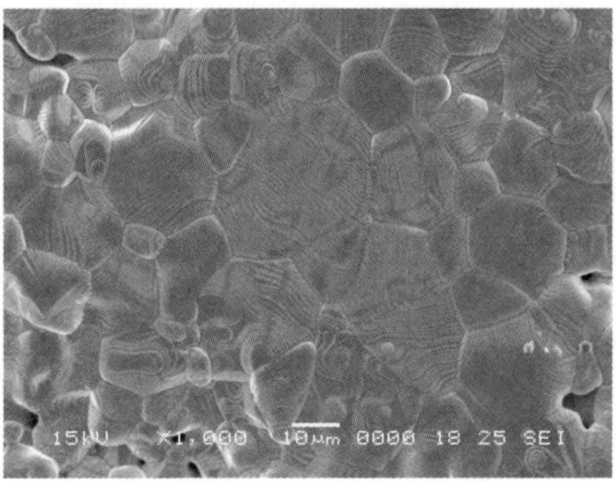

Fig. 2. SEM micrograph of BFN ceramic sintered at 1623 K.

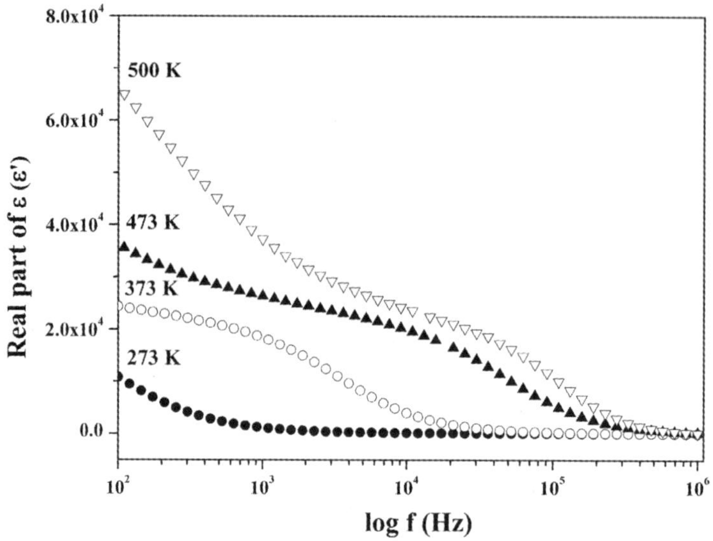

Fig. 3. Frequency dependence of ε' of $BaFe_{0.5}Nb_{0.5}O_3$ ceramics at different temperature.

The frequency dependence of the dielectric permittivity and the dielectric loss of BFN ceramic as a function of temperature are plotted in Fig. 3. At 293 K, the real part dielectric permittivity (ε') gradually decrease with increasing frequency, with a value of around 10,000 at 100 Hz. With increasing temperature apparently, which become more significant at low frequency (below 1 kHz). This nature of variation is very much consistent with that of a normal ferroelectrics, because higher values of nature of ε' at lower frequencies are due to the presence of all different types of polarizations (i.e., interfacial, dipolar, atomic, ionic, electronic) in the material. At high frequencies some of the above polarizations may have less contribution in ε'. The high value of ε' in the low frequency region has been explained using Maxwell–Wagner (MW) polarization effect sometimes called as interfacial polarization. Thus high values of permittivity are not usually intrinsic but rather are associated with a heterogeneous conduction in the grain and grain boundary structure of the compounds, which arises due to the materials of grains separated by more insulating intergrain barriers as in boundary layer capacitor.[9]

The variation of dielectric permittivity with temperature is shown in Fig. 4. The real part dielectric permittivity (ε') was found to decrease for a given frequency. There are two possible explanations for the systematic increase in dielectric constant with sintering temperature. One possibility is that it is due to increased conductivity in the samples, suggested by Ananta and Thomas[10] to result from the presence of Fe^{2+} in sintered BFN ceramics. The concentration of such species is known to be highly densitive to the sintering temperature, increasing with increasing temperature.[11] It is known that the co-existence of Fe^{2+} and Fe^{3+} on equivalent crystallographic sites can give rise to an electron-hopping conduction

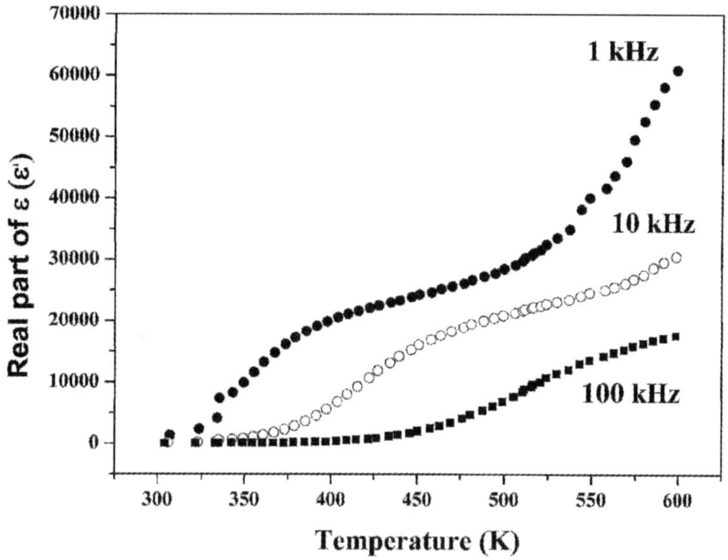

Fig. 4. Temperature dependence of ε' of $BaFe_{0.5}Nb_{0.5}O_3$ ceramics at different temperature.

mechanism, which, owing to its finite hopping (or jump) probability, tends to come into effect at lower frequencies. An alternative explanation is related to disorder in the B site cations, as suggested by Majumder *et al.*[12] to occur in complex perovskites. In $A(B_IB_{II})O_3$-type perovskites such as $Pb(Fe^{+3}_{0.5}Nb^{+5}_{0.5})O_3$, the Fe^{+3} (B_I) and Nb^{+5}(B_{II}) ions randomly occupy the octahedral B sites surrounded by O^{-2} anions. Due to the presence of larger Fe^{+3} (B_I) cations, a larger "rattling space" is available for the relatively smaller Nb^{+5} (B_{II}) cations. When an oscillating ac signal is applied to such disordered systems, the smaller B_{II} cations (with a large rattling space) can readily move without distorting the oxygen framework. In an ordered perovskite, a comparatively smaller rattling space is available for B site cations. Therefore a larger dielectric constant is expected in disordered complex perovskites such as PFN compared with ordered perovskites (e.g., $PbTiO_3$).

It is the most convenient way to check the nature of dielectric relaxation through complex argand plane of ε' versus ε'', usually called Cole–Cole plots.[13,14] For monodispersive Debye process, one expects a semicircle in Cole–Cole plot with the center located on the ε'-axis, whereas for polydispersive relaxation, the argand plane plots are closer to circular arc permittivity in such situations is known to be described by an empirical relation

$$\varepsilon^*(\omega) = \varepsilon' - i\varepsilon'' = \frac{\varepsilon_\infty + (\Delta\varepsilon)}{1 + (i\omega\tau)^{1-\alpha}} \tag{1}$$

where $\Delta\varepsilon = \varepsilon_s - \varepsilon_\infty$ is the contribution of the relaxation to static permittivity ε_s; ε_∞ is the contribution of higher frequency polarization and τ is the mean relaxation time of the relaxors. The parameter α characterizes the distribution of relaxation time,

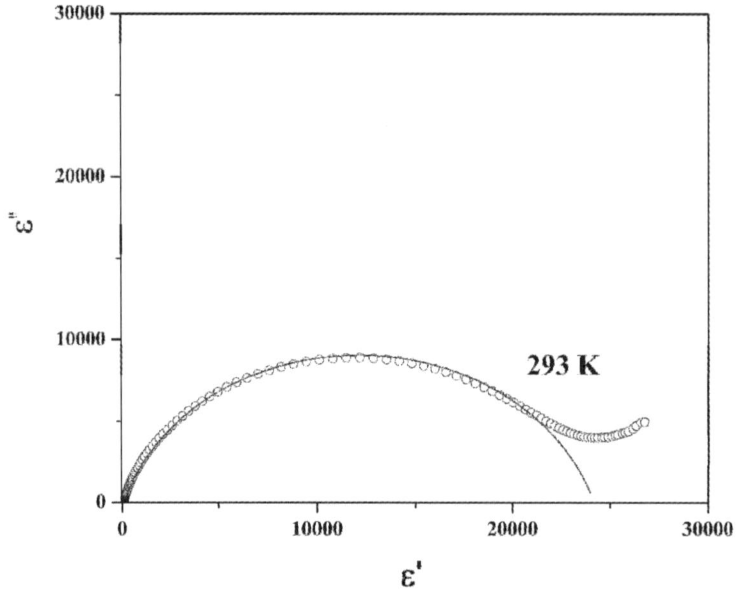

Fig. 5. Cole–Cole plot of $BaFe_{0.5}Nb_{0.5}O_3$ ceramics at room temperature.

which increases with increasing internal degrees of freedom of relaxors, indicating the departure of electrical response from an ideal condition. It can, however, be determined from the location of the center of the Cole–Cole circles. Figure 5 is a representative plot of BFN at different temperature. When α goes to zero $(1-\alpha \rightarrow 1)$ the above equation reduces to classical Debye's formalism. It can be seen from this plot that the relaxation process of the BFN ceramics is quite different from that of monodispersive Debye-type (for which $\alpha = 0$). The parameter α, as determined from the angle subtended by the radius of the circle with the ε'-axis passing through the origin of the ε''-axis, is always greater than zero. Therefore, the non-zero value of α confirms the polydispersive nature of dielectric relaxation of BFN ceramics. Further, it is observed that the value of α increases with the increase in temperature. The value of parameter α is about 0.283 for 293 K. Our results are in good agreement with Saha and Sinha who studied at low temperature.[5] Therefore, the polydispersive nature of the ceramics changes with the temperature.

4. Conclusion

$BaFe_{0.5}Nb_{0.5}O_3$ ceramics were prepared by solid-state reaction, producing single-phase cubic $BaFe_{0.5}Nb_{0.5}O_3$ powders when calcined at 1473 K. The sintering temperature affects the density and dielectric properties of $BaFe_{0.5}Nb_{0.5}O_3$, densities 94.7% of theoretical being achieved upon sintering at 1623 K. The high value of ε' in the low frequency region has been explained using Maxwell–Wagner polarization effect, which is associated with a heterogeneous conduction in the grain and grain boundary structure of the compounds. The non-zero value of α from the Cole–Cole plot confirms the polydispersive nature of dielectric relaxation of BFN ceramics.

Acknowledgments

The authors would like to express their sincere thanks to the Thailand Research Fund, the Commission on Higher Education and Graduate School, Chiang Mai University for financial support throughout the project.

References

1. B. Jaffe and W. R. Cook, *Piezoelectric Ceramic* (R.A.N. Publisher, 1971).
2. M. Yokosuka, *Jpn. J. Appl. Phys.* **34**, 5338 (1995).
3. K. Tezuka, K. Henmi, Y. Hinatsu and N. M. Masaki, *J. Solid State Chem.* **154**, 591 (2000).
4. I. P. Raevski, S. A. Prosandeev, A. S. Bogatin, M. A. Malitskaya and L. Jastrabik, *J. Appl. Phys.* **93**, 4130 (2003).
5. S. Saha and T. P. Sinha, *Phys. Rev. B* **65**, 134103 (2002).
6. S. Saha and T. P. Sinha, *J. Phys.: Condens. Matter* **14**, 249 (2002).
7. The Inorganic Crystal Structure Database (ICSD) No. 43622.
8. N. Rama, J. B. Philipp, M. Opel, K. Chandrasekaran, V. Sankaranarayanan, R. Gross and M. S. R. Rao, *J. Appl. Phys.* **95**, 7528 (2004).
9. Y. J. Li, X. M. Chen, R. Z. Hou and Y. H. Tang, *Solid State Comm.* **137**, 120 (2006).
10. S. Ananta and N. W. Thomas, *J. Eur. Ceram. Soc.* **19**, 1873 (1999).
11. K. Singh, S. A. Band and W. K. Kinge, *Ferroelectrics* **306**, 179 (2004).
12. S. B. Majumdar, D. Bhattacharyya, R. S. Katiyar, A. Manivannan, P. Dutta and M. S. Seehra, *J. Appl. Phys.* **99**, 024108 (2006).
13. K. S. Cole and R. H. Cole, *J. Chem. Phys.* **9**, 341 (1941).
14. K. S. Cole and R. H. Cole, *J. Chem. Phys.* **10**, 98 (1942).

STRUCTURAL AND PIEZOELECTRIC PROPERTIES OF $(1-x)$PZT-xBFN ($x = 0.1 - 0.2$) SOLID SOLUTION

KRIT SUTJARITTANGTHAM, SUKUM EITSSAYEAM,* KAMONPAN PENGPAT, GOBWUTE RUJIJANAGUL and TAWEE TUNKASIRI

Department of Physics, Faculty of Science, Chiang Mai University, Chiang Mai, 50202, Thailand

**sukum99@yahoo.com*

GANNAGA SATITTADA and URAIWAN INTATHA†

School of Science, Mae Fha Luang University, Chiang Rai, 57100, Thailand

†i_uraiwan@yahoo.com

Received 31 July 2008

The structural, dielectric and piezoelectric properties of $(1-x)PbZr_{0.52}Ti_{0.48}O_3$–$xBaFe_{0.5}Nb_{0.5}O_3$ ceramic system with the composition near the morphotropic phase boundary were investigated as a function of the $BaFe_{0.5}Nb_{0.5}O_3$ content by X-ray diffraction (XRD), dielectric measurement and piezoelectric measurement techniques. Studies were performed on the samples prepared by solid state reaction for $x = 0.10$, 0.12, 0.14, 0.16, 0.18 and 0.20. The XRD analysis demonstrated that with increasing BFN content in $(1-x)PZT_{-x}BFN$, the structural change occurred from tetragonal to the mixture of tetragonal and cubic phase. Changes in the dielectric behavior and piezoelectric properties were found to relate with these structural changes depending on the BFN contents.

Keywords: Piezoelectric properties; solid solution; PZT-BFN.

1. Introduction

Actuators make use of the strain response of ferroelectric ceramics in high electric fields. Usable physical effects are piezoelectricity, electrostriction and field induced phase switching between antiferroelectric and ferroelectric states.[1] Among them, only piezoelectricity operates in a wide temperature ranges. Undoubtedly nowadays, ferroelectric lead zirconate titanate (PZT) compositions belong to the materials destined besides the classical piezoelectric using also for micro electromechanical systems (MEMS) applications, especially for micro-actuator applications because their excellent piezoelectric and pyroelectric properties are well known.[2]

Due to the rapid development of functional ceramics, many kinds of piezoelectric materials have been employed for ceramic actuators to find out the advantages of

*Corresponding author.

fast response, high generating force, and low power consumption. $BaFe_{0.5}Nb_{0.5}O_3$ is also another interesting perovskite compound which possesses a considerably high dielectric constant. This work then aimed at developing the dielectric and piezoelectric properties of PZT by adding BFN. For this purpose, we first prepared the PZT-BFN ceramics by conventional mixed oxide method to find out the suitable condition for actuator applications. In earlier work,[3] $(1-x)PbZr_{0.52}Ti_{0.48}O_3$–$xBaFe_{0.5}Nb_{0.5}O_3$ ($x = 0.1$, 0.2, 0.3, 0.4 and 0.5) ceramic system was investigated. It was reported that the $(1-x)PZT_{-x}BFN$ ceramics of $x = 0.1$ and 0.2 have interesting dielectric and piezoelectric properties and phase evolution of tetragonal phase become cubic phase as $x = 0.3$. Then, in this study we have emphasized where $0.1 \leq x \leq 0.2$ on sub set of $(1-x)PZT_{-x}BFN$ ceramic system. We varied the ratio of adding BFN in PZT from 10 to 20% by mole denoted as P90, P88, P86, P84, P82 and P80 respectively. The phase evolutions of the samples were investigated by an X-ray diffractometer. The piezoelectric and dielectric properties of PZT-BFN ceramics were determined.

2. Experimental

The PZT–BFN ceramics used in this study were prepared from powders using the conventional mixed-oxide method. The $(1-x)PbZr_{0.52}Ti_{0.48}O_3$–$xBaFe_{0.5}Nb_{0.5}O_3$ ($1-x$PZT–xBFN) powders were first prepared by mixing the starting materials PbO (> 99%), ZrO_2 (> 99%), TiO_2 (> 99%), $BaCO_3$ (> 99%), Fe_2O_3 (99.9%) and Nb_2O_5 (99.9%) powders in the desired mole ratios, ($x = 0.1$, 0.12, 0.14, 0.16, 0.18 and 0.2). These powders were ball-milled for 24 h with zirconia balls. Ethanol was used as a milling medium. After drying at 120°C, the mixed powders were then calcined at 800–1100°C for 2 h with heating and cooling rate of 5°C/min. Subsequently, the most appropriate calcined samples were pressed into disc shape and sintered at various temperatures ranging from 1150 to 1300°C depending upon the compositions. The samples were heated for 2 h with constant heating and cooling rates of 5°C/min. The phase formations of the calcined powders and sintered specimens were studied by an X-ray diffractometer (Philips model X-pert) at 40 kV and 30 mA in the 2θ range from 10 to 60 degrees with step scan of 0.01°. The microstructure was examined by a scanning electron microscopy (Jeol model 6335F). Prior to electrical properties characterizations, the sintered samples were ground to obtain parallel faces, and the faces were then coated with silver paste as electrodes. The dielectric constant and loss tangent of the sintered ceramics were measured as a function of temperature at 1 KHz using a precision impedance analyzer (Agilent 4294A). The capacitance and the dielectric loss tangent were determined over the temperature range of 30 and 300°C at the frequency 1 KHz. The samples were poled at 150°C in silicone oil bath with DC field of 25–30 kV/cm for 30 min before piezoelectric measurement. The piezoelectric coefficient (d_{33}) of the samples was measured using a piezoelectric-d_{33}-meter (PM3001, KFC TECH).

3. Results and Discussion

The XRD results of calcined $(1-x)$PZT–xBFN powders showed that the perovskite structure of PZT started to form at 800°C. It showed clearer at higher calcining temperature and most pronounced at 1100°C. No trace of pyrochlore phase was detected. Therefore, the powder calcined at 1100°C was employed to prepare ceramic samples. After that the $(1-x)$PZT–xBFN ceramics were prepared from optimum condition of calcined powder. The sintering temperature was varied from 1150 to 1300°C. Maximum density of $(1-x)$PZT–xBFN was found at 1250°C. They are between 7.68 to 7.50 g/cm^3, comparable to the theoretical value of PZT 8.006 g/cm^3 and the density of BFN is 6.51 g/cm^3.[4,5] The density was decreased with amount of BFN in ceramic system. Figure 1 shows the X-ray diffractograms of the highest density ceramics in $(1-x)$PZT–xBFN system $x = 0.10$, 0.12, 0.146, 0.16, 0.18 and 0.20, denoted as P90, P88, P86, P84, P82 and P80 respectively.

The phase evolution of $(1-x)$PZT–xBFN ceramics of each sintered temperature was investigated by XRD techniques. The results indicated that change of crystal structure occurred as a function of PZT–BFN compositions. Mixed ferroelectric tetragonal and cubic phases began to occur in all samples, as can be seen from (001) and (100) peaks. Similar result was also found by Vittayakorn *et al.*[6] who studied the $(1-x)$PZT–xPZN. Neither BFN nor other impurities were detected.

The SEM micrographs of the $(1-x)$PZT–xBFN ceramics were compared with that of the pure PZT ceramic (Fig. 2). It can be seen that only single phase appeared in all micrographs, and no other phase was observed. The average grain size of PZT,

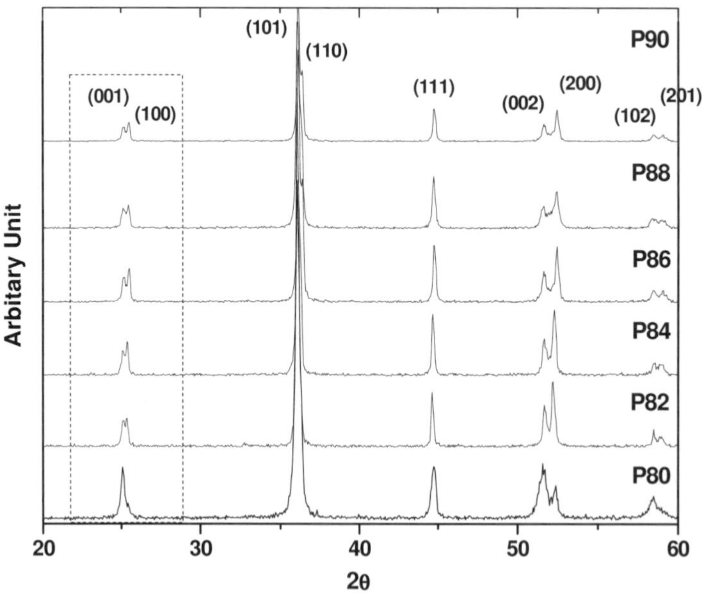

Fig. 1. X-ray patterns of ceramics at maximize bulk density condition.

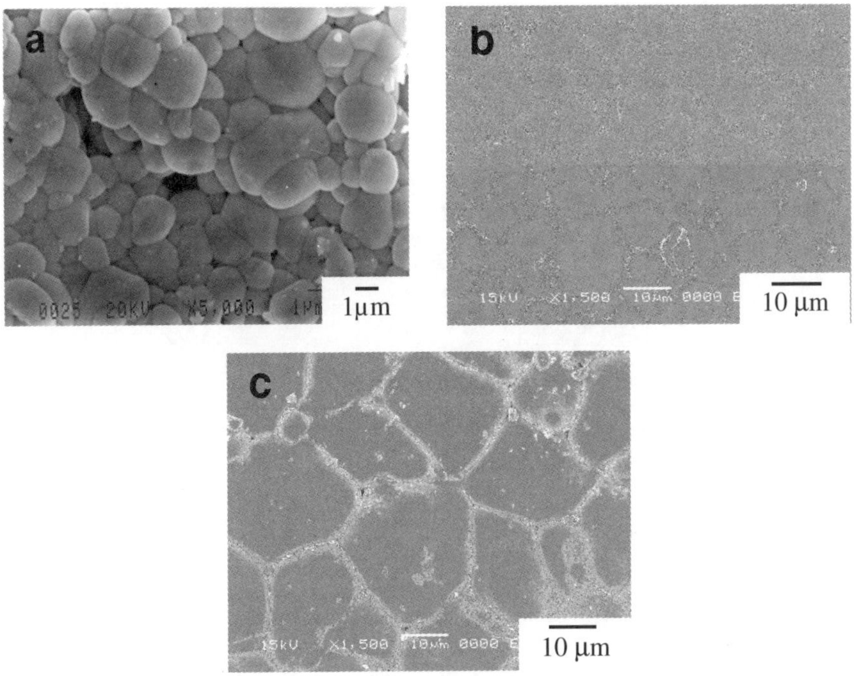

Fig. 2. SEM micrograph of PZT and $(1-x)$PZT–xBFN ceramics. (a) PZT, (b) P90 and (c) P80.

P90 and P80 are 1.66 μm, 5.03 μm and 15.68 μm respectively. The small addition of BFN of about $x = 0.10$ causes significant increase on grain size [Fig. 2(b)] and reaches maximum of about 15 μm at $x = 0.2$ [Fig. 2(c)].

The relationships between dielectric properties and temperature are shown in Figs. 3 and 4. It was clearly shown that the relative dielectric constant depends on temperature measured at 1 KHz on the ceramic sample of $(1 - x)$PZT–xBFN

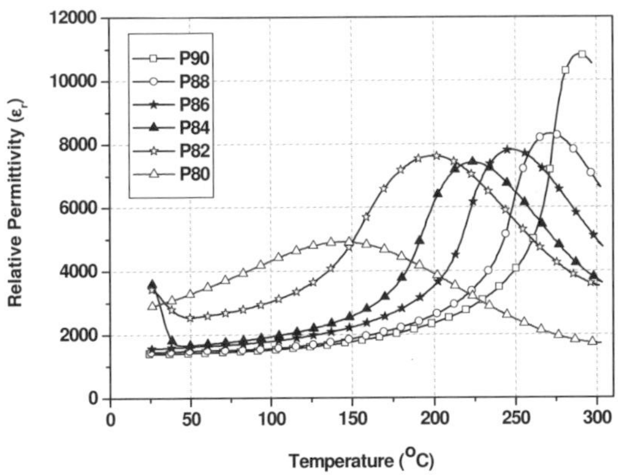

Fig. 3. The relationship between dielectric permittivity and temperature.

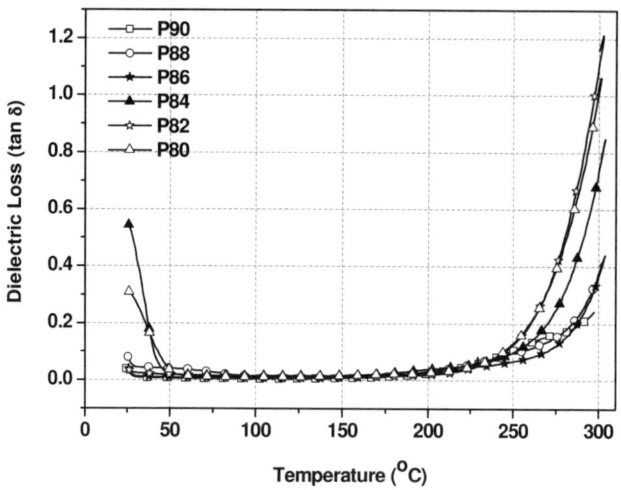

Fig. 4. The relationship between dielectric loss and temperature.

system. The broadness of the dielectric constant peak increased with increasing BFN content. Since the Curie temperature for all of investigated compositions showed a tendency to decrease from 280°C to below 150°C. Not only the Curie temperature decreased, but also the maximum dielectric constant reduced from 10,800 to 4900 for P9B1 and P8B2 respectively. Considerable broadening of dielectric constant curves was found to increase, with the dielectric loss lower than 0.05 at working temperature (room temperature to 200°C). Furthermore, it was seen that the role of BFN affected the dielectric properties of samples.

The piezoelectric coefficient (d_{33}) and electro- mechanical coupling factors (k_p) of $(1-x)$PZT–xBFN ceramics were shown in Table 1. The value of d_{33} was not apparent in the same trend. The maximum value of d_{33} (290 pC/N) was observed at P80 sample. The value of k_p increase from 26% (P90) to 71% (P86), and then decrease slightly afterward. The maximum value of k_p (71%) was observed at P86 sample. The trends of d_{33} value are just like the trend of corresponding properties

Table 1. The summary of piezoelectric and dielectric properties of PZT-BFN ceramics compared with soft and hard PZT from other works.

Sample	k_p (%)	d_{33} (pC/N)	ε_r at room temp	tan δ at 25°C
Soft – PZT [1]	60	580	2100	0.02
Hard – PZT [1]	57	230	900	0.002
P90	26	269	1390	0.03
P88	37	222	1400	0.07
P86	71	229	1550	0.03
P84	45	262	3590	0.5
P82	35	220	3430	0.03
P80	30	290	2900	0.3

of hard — PZT.[1] However, the value of k_p are not similar with hard and soft PZT but depend on the concentration of BFN in $(1-x)$PZT–xBFN ceramics.

4. Conclusion

The effect of BFN on the structure and dielectric of $(1-x)$PZT–xBFN system was investigated for various chemical compositions. The $(1-x)$PZT–xBFN (when $x = 0.10, 0.12, 0.14, 0.16, 0.18$ and 0.20) ceramics were prepared by the mixed oxide method. Lattice parameters of the tetragonal phase and cubic phases were found to vary with chemical composition. The evolution of the tetragonal phase, (100)/(001) transformed to a single peak (100) which indicating cubic symmetry, while optimum sintering temperature was standing at 1250°C. They were identified as a single PZT phase material with a perovskite structure having the symmetry from tetragonal to cubic when the ratio of BFN increased. The broadness of dielectric constant peak increased and Curie temperatures showed a tendency to decrease below 150°C while the dielectric loss was still less than 0.05. The SEM micrographs also revealed single phase of the materials. The maximum value of d_{33} (290 pC/N) was observed at P80 sample and the maximum value of k_p (71%) was observed at P86 sample. According to the results, it can be concluded that adding BFN can improve the piezoelectric and dielectric properties of PZT ceramics, pointing the way to further developments for actuator applications.

Acknowledgments

The authors would like to express their sincere thanks to the Thailand Research Fund, the Commission on Higher Education and Graduate School, Chiang Mai University for financial support throughout the project.

References

1. A. Schnecker, H.-J. Gesemann and L. Seffner, *IEEE* 263 (1996).
2. Y. Jing and J. Luo, *J. Mats Sci.: Mats in Elec.* **16**, 287 (2005).
3. S. Eitssayeam, U. Intatha, G. Rugiganagul, K. Pengpat and T. Tunkasiri, *Appl. Phys. A* **83**, 295 (2006).
4. Powder diffraction file. Joint committee on Powder Diffraction Standards. No. 33–784.
5. The Inorganic Crystal Structure database (ICSD) No. 43622.
6. N. Vittayakorn, G. Rujijanagul, T. Tunkasiri, X. Tan and D. Cann, *Mat. Sci. Eng. B* **108**, 258 (2004).

EFFECTS OF HEAT TREATMENT ON SPIN HAMILTONIAN PARAMETERS OF Cr^{3+} IONS IN NATURAL PINK SAPPHIRE

T. KITTIAUCHAWAL* and P. LIMSUWAN

Department of Physics, Faculty of Science,
King Mongkut's University of Technology Thonburi,
Bangkok 10140, Thailand
** opticslaser@yahoo.com*

SUKUM EITSSAYEAM and TAWEE TUNKASIRI†

Department of Physics, Faculty of Science, Chiang Mai University,
Chiang Mai, 50202, Thailand
† tawee@chiangmai.ac.th

Received 31 July 2008

The purpose of this research work is to investigate the effects of heat treatment on spin Hamiltonian parameters of Cr^{3+} ions in natural pink sapphire sample. The pink sapphire were heated at different temperatures of 1200, 1300, 1400, 1500 and 1600°C in oxygen atmosphere for 12 h, respectively. Electron Spin Resonance (ESR) spectra of pink sapphire samples before and after heat treatment were recorded in X-band frequency by mounting the crystal sample with the *c*-axis perpendicular to the applied magnetic field direction. The spectra were recorded in the range of 0-180 degrees for every 15 degrees of rotation angle (ϕ). It was found that, five strong ESR absorption peaks at the magnetic fields of 97, 163, 312, 519 and 724 mT were obtained. The magnetic fields of 163 and 519 mT corresponding to Cr^{3+} ions replacing the Al^{3+} ions site of corundum structure (α-Al_2O_3), and those of 97, 312, and 724 mT were assigned to Fe^{3+} ions. Some conspicuous peaks of Cr^{3+} ion at various rotation angles were used to calculate the spin Hamiltonian parameters by least-squares fit method with the help of a computer program. The optimum spin Hamiltonian parameters of Cr^{3+} ions were then used to simulate the energy levels diagram of Cr^{3+} ions in sample before and after heat treatment.

Keywords: Spin Hamiltonian parameters; Cr^{3+} ion; natural pink sapphires; electron spin resonance.

1. Introduction

A corundum crystal composed of pure aluminum oxide (α-Al_2O_3) is rather an uninteresting appearing material also known as white or colorless sapphire.[1] When it contains 1% or little more of chromium oxide (Cr_2O_3), so that about 1 of every 100

*Corresponding author.
†Corresponding author.

aluminum atoms is replaced by chromium atom, the material acquires a beautiful luminous red or pink color.[2]

The objective of this research is not only to study impurity ion in sapphire but also the interaction of an applied magnetic field with unpaired electron in solids. These unpaired electrons are most conveniently investigated using an Electron Spin Resonance (ESR) spectrometer. Actually the magnetic resonance can be employed in different experimental techniques but ESR, in which the electron absorbs energy, is the most prominent technique. In the case of charge misfit ion substitution such as in the case where the impurity ions carry greater charge than the host ion they replace, the electrostatic attractive force acting on the impurity ions by the nearest neighboring O^{2-} ions becomes larger than that on the replaced host ion. The impurity ions with excessive charge at the trigonal-distorted octahedral sites in corundum structure will therefore move towards the larger O^{2-} triangles.[3,4] The effect of impurity on its displacement in the sapphire-type structure has not been clearly understood. Therefore, we need to investigate the effects of impurity displacement of Cr^{3+} ions in Al_2O_3 crystals from the ESR spectra. The valence of impurity especially 3d transition paramagnetic ions usually depends on heat treatment which plays an important role in coloring sapphires.[5,6] In this research, the ESR spectra of Cr^{3+} ions in pink sapphire crystal before and after heating at the temperatures of 1200, 1300, 1400, 1500 and 1600°C in oxygen atmosphere were carried out at various rotation angles. Some conspicuous peaks of Cr^{3+} ions at various rotating angles were used to calculate the spin Hamiltonian parameters by least-squares fit method with the help of an EPR-NMR computer program. The optimum spin Hamiltonian parameters of Cr^{3+} ions were then used to simulate the energy levels diagram.

1.1. *Spin Hamiltonian of Cr^{3+} ions in sapphire*

The configuration of Cr atom (^{24}Cr) is [Ar] $3d^5$ $4s^1$ as shown in Fig. 1(a). In a single crystal sapphire, the Cr atoms replace the Al atoms in Al_2O_3 crystal and lose three electrons. The electron configuration of Cr atom that lose three electrons is changed from [Ar] $3d^5$ $4s^1$ to [Ar] $3d^3$ as shown in Fig. 1(b). Furthermore, Cr is converted to Cr^{3+} ion. Therefore, Cr^{3+} ion has three valence electrons in 3d subshell.[2] The electronic spin state of Cr^{3+} ion is $S = 3/2$ and the ground state of Cr^{3+} ion is $^4F_{3/2}$. For the ground state, the orbital angular momentum quantum number, $L = 0$, and spin magnetic quantum number, $M_s = +3/2, +1/2, -1/2$ and $-3/2$. In an applied magnetic field, B the Cr^{3+} ion ground-state energy level splits into $2S + 1 = 4$ energy states, then two allowed transitions according to selection rule, $\Delta M_s = \pm 1$, are obtained as shown in Fig. 2.

The ESR spectra of Cr^{3+} ions can be described by a spin Hamiltonian (H) incorporating with Zeeman interaction, hyperfine structure and crystal field operators. It is given by equation

$$\mathrm{H} = \mathrm{H}_{\mathrm{Zeeman}} + \mathrm{H}_{\mathrm{HF}} + \mathrm{H}_{\mathrm{CF}}\,. \tag{1}$$

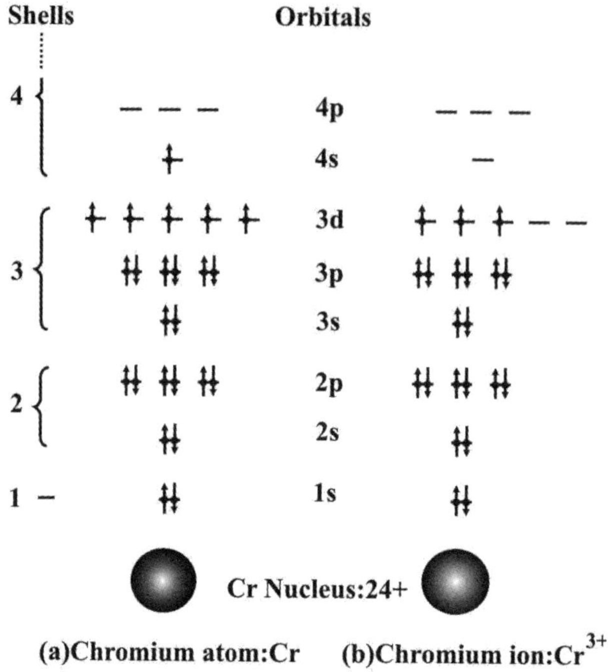

Fig. 1. Electron distribution in the ground state of: (a) chromium atom and (b) Cr^{3+} ion.

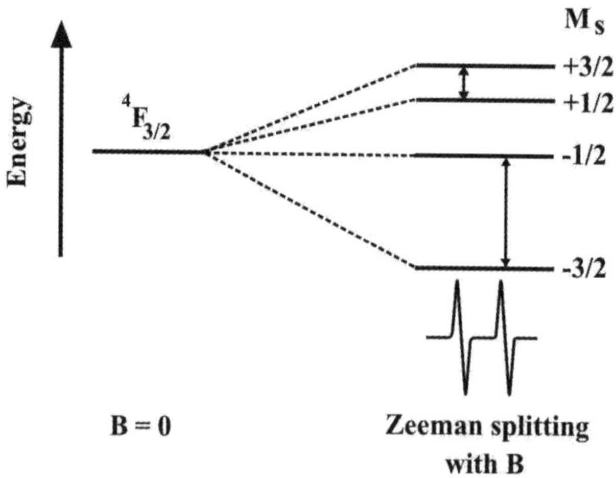

Fig. 2. Energy level diagram of Cr^{3+} ions in sapphire.

The first term corresponds to the Zeeman interaction arises from the interaction between electron spin angular momentum and external magnetic field. The Zeeman term is given by equation

$$\mathrm{H}_{\mathrm{Zeeman}} = \beta\mathbf{S}\cdot\mathbf{g}\cdot\mathbf{B}$$

$$= \beta[bfS_x, \mathbf{S}_y, \mathbf{S}_z]\cdot\begin{bmatrix} g_{xx} & g_{xy} & g_{xz} \\ g_{yx} & g_{yy} & g_{yz} \\ g_{zx} & g_{zy} & g_{zz}\end{bmatrix}\cdot\begin{bmatrix}\mathrm{B}_x \\ \mathrm{B}_y \\ \mathrm{B}_z\end{bmatrix} \tag{2}$$

where β, $\mathbf{S}$, $\mathbf{g}$ and $\mathbf{B}$ are the Bohr magnetron; spin operators $\mathbf{S}_x, \mathbf{S}_y, \mathbf{S}_z$; gyromagnetic tensor, and magnetic field, respectively.

The second term is the hyperfine interaction term arises from the interaction between the electron spin and nuclear spin. The hyperfine term is written in the form

$$\mathrm{H}_{\mathrm{HF}} = \mathbf{S}\cdot\mathbf{A}\cdot\mathbf{I} \tag{3}$$

where $\mathbf{I}$ and $\mathbf{A}$ are nuclear spin operator and hyperfine coupling tensor, respectively.

The third term is the crystal field term arises from the crystal field potential generated from the surroundings of the paramagnetic ion in sapphire. This term depends on the local symmetry of paramagnetic ion site and the electronic configuration of ion. It is the sum of spin angular momentum operators called Stevens' operators or equivalent operators ($\mathrm{O}_2^0, \mathrm{O}_4^0$ and O_4^3) with their coefficients ($\mathrm{B}_2^0, \mathrm{B}_4^0$ and B_4^3) as the crystal field parameters. The crystal field Hamiltonian is usually written in the spherical-tensor notation as the following relation

$$\mathrm{H}_{\mathrm{CF}} = \mathrm{B}_2^0\mathrm{O}_2^0 + \mathrm{B}_4^0\mathrm{O}_4^0 + \mathrm{B}_4^3\mathrm{O}_4^3\,. \tag{4}$$

When the fourth orders of the crystal field terms are sufficiently small to be neglected, Eq. (4) can be written in the form

$$\mathrm{H}_{\mathrm{CF}} = \mathrm{B}_2^0\mathrm{O}_2^0\,. \tag{5}$$

Therefore, the spin Hamiltonian is given by equation

$$\mathrm{H} = \beta\mathbf{S}\cdot\mathbf{g}\cdot\mathbf{B} + \mathbf{S}\cdot\mathbf{A}\cdot\mathbf{I} + \mathrm{B}_2^0\mathrm{O}_2^0\,. \tag{6}$$

However, the hyperfine term is small compared with the Zeeman term so that it can be neglected. Then, the spin Hamiltonian in Eq. (6) becomes

$$\mathrm{H} = \beta\mathbf{S}\cdot\mathbf{g}\cdot\mathbf{B} + B_2^0\mathrm{O}_2^0\,. \tag{7}$$

The parameters in Eq. (7) were calculated from the resonance magnetic field positions in the ESR spectra.[7–10]

2. Experimental

The natural pink sapphires obtained from a Phu marple mine, Luc Yen district, Yenbai province, Vietnam were used in this research work. They were first cleaned with acid and solvent to remove all stains and other impurities on the surfaces. Then, the pink sapphires were heated at the temperatures of 1200, 1300, 1400, 1500

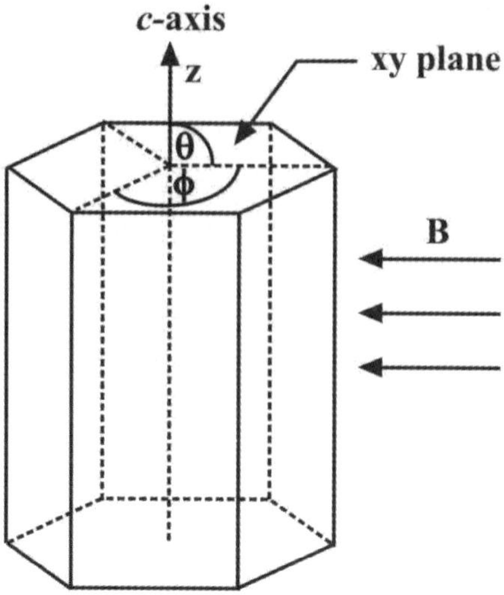

Fig. 3. The direction of magnetic field, B and the rotation angle (ϕ).

and 1600°C in oxygen atmosphere for 12 h, respectively. The pink sapphire crystals before and after heat treatment at various temperatures were investigated by ESR spectrometer (JEOL JES-RE2X). The crystal sample was attached to a goniometer with the *c*-axis perpendicular to the applied magnetic field direction as shown in Fig. 3.

The theta (θ) is the angle between magnetic field direction and *c*-axis, and the phi (ϕ) is angle between projection of magnetic field on xy plane and x-axis. The spectra were recorded in the range of 0–180 degrees for every 15 degrees of rotation angle (ϕ). From the positions of absorption peak of ESR spectra, we used an EPR-NMR computer program to calculate the g-factor values and crystal field parameters (B_2^0). The optimum spin Hamiltonian parameters of Cr^{3+} ions were then used to simulate the energy levels diagram.

3. Results and Discussion

Figure 4 shows ESR spectra of natural pink sapphire crystal at 9.446 GHz before and after heat treatment at the temperatures of 1200–1600°C. It was found that, there were five main ESR absorption peaks at magnetic fields of 97, 163, 312, 519 and 724 mT. For the magnetic field positions of 163 and 519 mT correspond to Cr3+ ions replacing the Al^{3+} ions site of corundum structure, and those of 97, 312, and 724 mT are assigned to Fe^{3+} ions.

In our experiments, the ESR spectra were also observed at different rotation angles (ϕ). Figure 5 shows experimental and calculated resonance magnetic field positions as a function of rotation angles for unheated and heat-treated sapphires from 1200–1600°C with the applied magnetic field perpendicular to the *c*-axis.

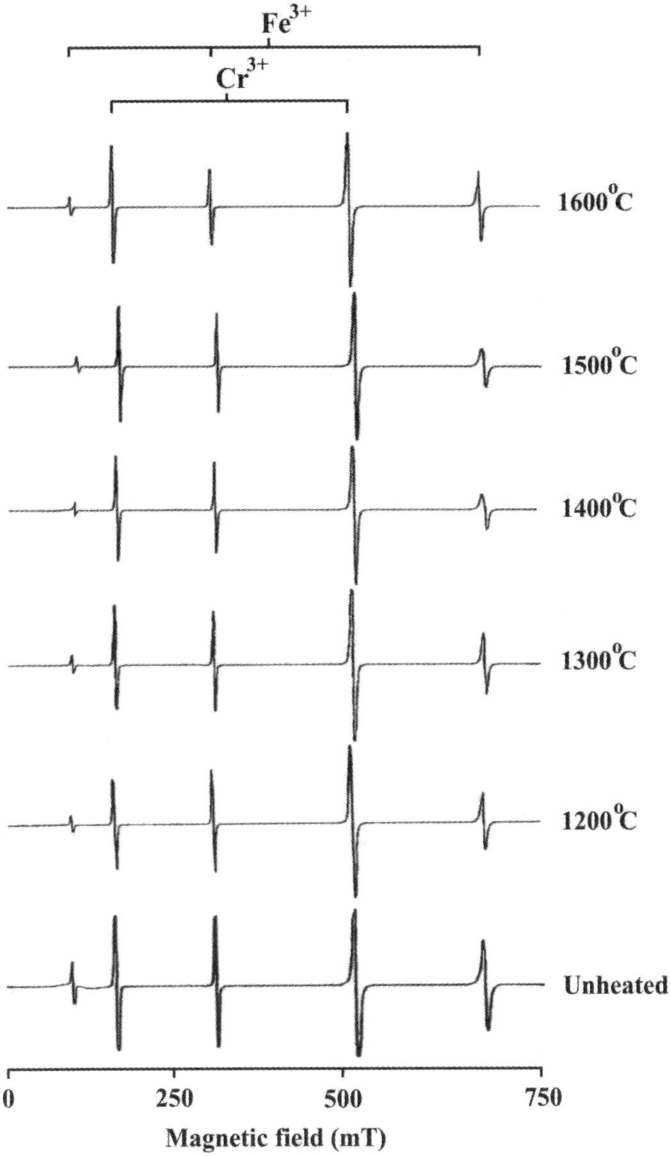

Fig. 4. ESR spectra of a natural pink sapphire crystal at 9.446 GHz before and after heat treatment at different temperatures, when applied magnetic field direction is perpendicular to the c-axis of crystal, $\theta = 90°$ and $\phi = 0°$.

The prominent peaks of Cr^{3+} ions at various rotating angles were used to calculate the spin Hamiltonian parameters by least-squares fit method with the help of an EPR-NMR computer program. The results of calculated spin Hamiltonian parameters of Cr^{3+} ions in natural pink sapphire crystal before and after heat treatment at various temperatures are shown in Table 1.

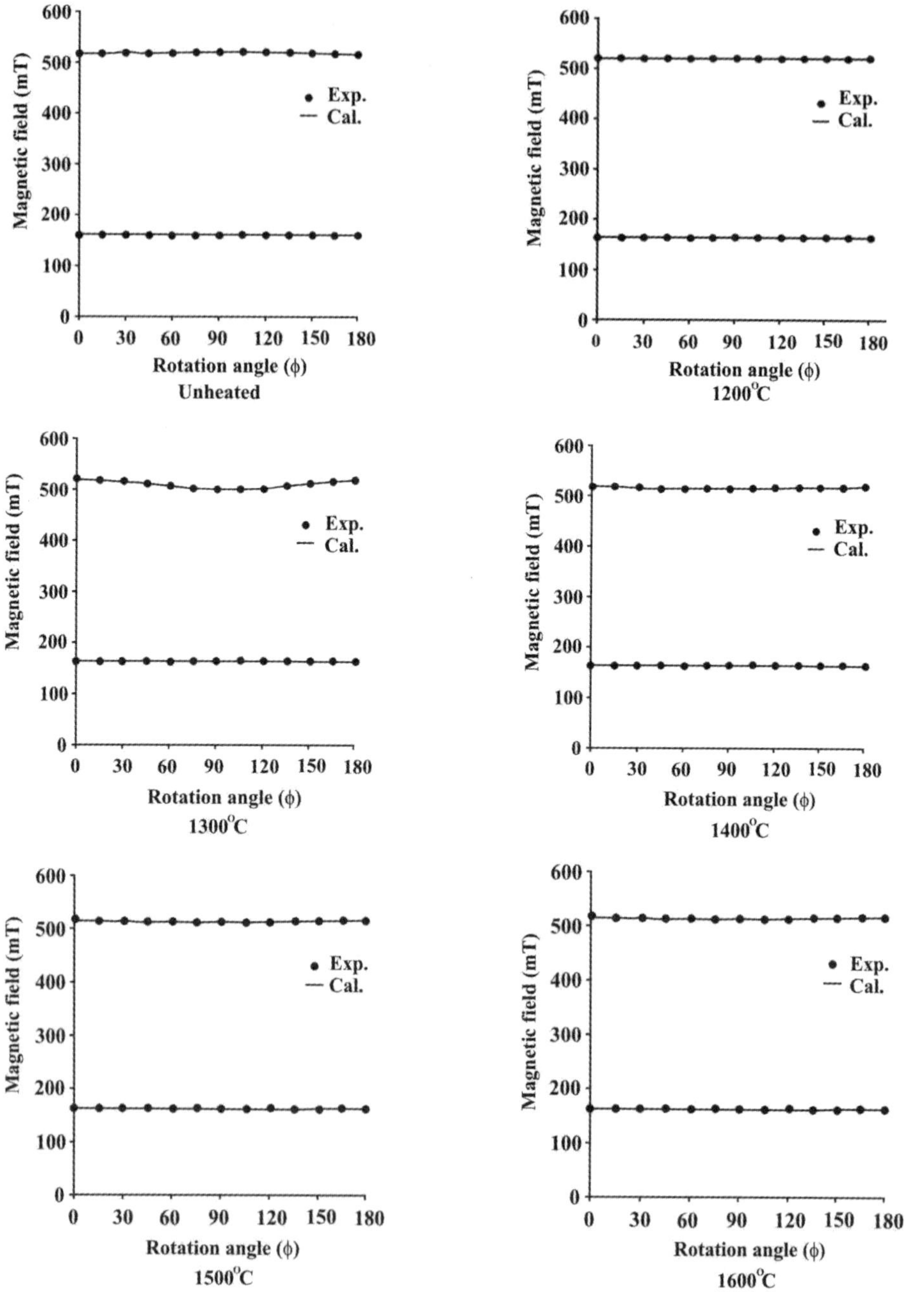

Fig. 5. Experimental (solid circles) and calculated (solid line) resonance magnetic field positions in the ESR spectra of Cr^{3+} ions in a natural pink sapphire crystal before and after heat treatment at various temperatures.

Table 1. Calculated spin Hamiltonian parameters of Cr^{3+} ions in a natural pink sapphire crystal before and after heat treatment at various temperatures.

Temperatures (°C)	g		B_2^0 (mT)	RMSD* (mT)
	$g_{xx} = g_{yy}$	g_{zz}		
Unheated	1.976(7)	1.967(7)	92.099(1)	0.3904
1200	1.979(3)	1.975(4)	92.648(3)	0.2334
1300	1.985(6)	1.984(5)	91.445(3)	0.3261
1400	1.983(2)	1.971(0)	88.215(1)	0.3886
1500	1.987(5)	1.983(1)	91.845(6)	0.3182
1600	1.967(1)	1.963(8)	94.527(9)	0.3545

From Table 1, the spin Hamiltonian parameters of Cr^{3+} ions were used for the simulation of the energy levels diagram. The typical values of computed energy levels of Cr^{3+} ions, for $B = 0$ and $B = 100$ mT, before and after heat treatment at various temperatures are shown in Table 2. From data in Table 2, we can plot the energy levels diagram, and the results are shown in Fig. 6. The ground state energy level of the Cr^{3+} ions in natural pink sapphire, without magnetic field, was split into two doublets, denoted by $M_s = \pm 3/2$ and $M_s = \pm 1/2$, due to crystal field or zero field splitting as shown in Fig. 6(a). When the magnetic field was applied to the sapphire, the magnitude of splitting becomes larger due to Zeeman interaction as shown in Fig. 6(b).

Table 2. The energy levels of Cr^{3+} ions in a natural pink sapphire crystal before and after heat treatment at various temperatures.

Heat treatment temperature (°C)	Energy at magnetic field = 0 (GHz)				Energy at magnetic field = 100 mT (GHz)			
	M_S=+3/2	M_S=+1/2	M_S=-1/2	M_S=-3/2	M_S=+3/2	M_S=+1/2	M_S=-1/2	M_S=-3/2
Unheated	0.55(7)	-0.55(7)	-0.55(7)	0.55(7)	3.80(8)	2.01(3)	-0.89(7)	-4.92(4)
1200	0.77(8)	-0.77(8)	-0.77(8)	0.77(8)	3.34(7)	2.15(4)	-0.59(6)	-4.90(5)
1300	0.76(8)	-0.76(8)	-0.76(8)	0.76(8)	3.39(7)	2.15(7)	-0.61(9)	-4.93(5)
1400	0.74(1)	-0.74(1)	-0.74(1)	0.74(1)	3.28(2)	2.08(3)	-0.59(9)	-4.76(6)
1500	0.77(2)	-0.77(2)	-0.77(2)	0.77(2)	3.39(1)	2.15(9)	-0.61(5)	-4.93(5)
1600	0.79(4)	-0.79(4)	-0.79(4)	0.79(4)	3.32(8)	2.16(9)	-0.57(9)	-4.91(7)

From Table 2 and Fig. 6, it is seen that after heat treatment at different temperatures, the energy levels of Cr^{3+} ions in sapphire were changed in comparison with the unheated sapphire.

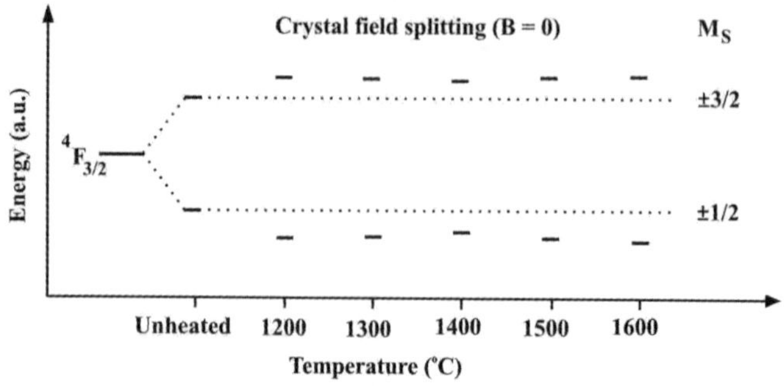

(a)

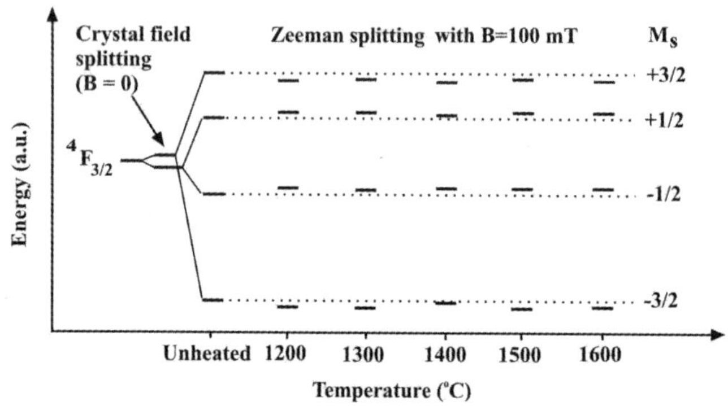

(b)

Fig. 6. Energy level diagrams of Cr^{3+} ions in sapphire before and after heat treatment at different temperatures, in cases of: (a) crystal field splitting (B = 0) and (b) Zeeman splitting (B = 100 mT).

4. Conclusion

The effects of heat treatment on spin Hamiltonian parameters of Cr^{3+} ions in natural Vietnamese pink sapphire crystals were investigated. The heat treatments were carried out in an electric furnace at 1200, 1300, 1400, 1500 and 1600°C in oxygen atmosphere for 12 hours, respectively. It can be concluded that before and after heat treatments the Cr^{3+} ions replace the Al^{3+} ions sites on *c*-axis. This can be clearly seen that the resonant peak positions in ESR spectra at rotation angle (ϕ) of 120° are exactly the same as those of original position ($\phi = 0$). After heat treatments at different temperatures, the energy levels of Cr^{3+} ions in sapphire were changed in comparison with the unheated sample. This resulted from the displacement of Cr^{3+} ions during the heat treatment as well as the distortion of crystal structure.

Acknowledgments

The authors wish to thanks Prof. Dr. John A. Weil, the University of Saskachewan, Canada for EPR-NMR computer program. This work is financially supported by the

Strategic Scholarship Fellowships Frontier Research Networks of Thailand Commission on Higher Education and the National Research Council of Thailand (NRCT).

References

1. W. H. Richard, *Corundum* (Courier International Ltd., London, 1990).
2. K. Nassau, *The Physics and Chemistry of Color*, 2nd edn. (John Wiley & Sons, New York, 2001).
3. W. C. Zheng, *Physica B* **245**, 119 (1998).
4. R. Hughes, *Ruby and Sapphire* (White Lotus, Bangkok, 1990).
5. P. Winotai, S. Saiseng and T. Sudyoadsuk, *Mod. Phys. Lett. B* **15**, 873 (2001).
6. P. Winotai, T. Wichan, I. M. Tang and J. Yaokulbodee, *Int. J. Mod. Phys. B* **14**, 1693 (2000).
7. G. Morin and D. Bomin, *J. Magn. Resonance* **136**, 176 (1999).
8. V. Y. Nagy, P. N. Komozin and M. F. Desrosiers, *Anal. Chim. Acta* **339**, 31 (1997).
9. C. P. Poll, Jr. and H. A. Farach, *The Theory of Magnetic Resonnance* (John Wiley & Sons, New York, 1972).
10. T. F. Yen, *Electron Spin Resonance of Metal Complexes* (Plenam Press, New York, 1969).

THE SPIN HAMILTONIAN PARAMETERS CALCULATION OF ^{14}N AND ^{15}N IN NATURAL TYPE I DIAMOND

CHITTRA KEDKAEW* and PICHET LIMSUWAN

Department of Physics, Faculty of Science,
King Mongkut's University of Technology Thonburi,
Bagnkok, 10140, Thailand
** chittra193@hotmail.com*

KANPHOT THONGCHAM and SIWAPORN MEEJOO

Department of Chemistry, Faculty of Science, Mahidol University,
Bangkok, 10400, Thailand

Received 31 July 2008

The main purpose of this work is to present the ESR spectra and calculate the spin Hamiltonian parameters of ^{14}N and ^{15}N impurities in natural diamond. The ESR spectra of diamond crystal were measured on ESR spectrometer operating at X-band microwave frequency. The results of ESR spectra show that the diamond has a P1 center. This center gives rise to three strong resonance absorption peaks at $\theta = 90°$, $\phi = 0°$ due to hyperfine interaction between electron spin and nuclear spin of ^{14}N. The ESR spectra of ^{15}N impurity consist of two satellites at the same rotation angle (ϕ). The effects of isolated substitution nitrogen on carbon atom produced a symmetric distortion from T_d to C_{3V} symmetry. According to this symmetry, the resonance magnetic field positions of ESR spectra for the rotation angles of 0°, 90° and 180° are almost overlap. The g-factor values and spin Hamiltonian parameters of ^{14}N and ^{15}N are: $g = 2.0019$, $A_\perp = 29.73$, $A_\| = 40.24$ and $g = 2.0019$, $A_\perp = -39.90$, $A_\| = -57.05$, respectively.

Keywords: Spin Hamiltonian; ESR; diamond; nitrogen impurity.

1. Introduction

Diamond crystals are very attractive because of their rarity and beauty. The diamond cost usually depends on the colors of diamond. However, the crystal structure of beautiful diamonds may not always be perfect. The color of diamond partly arises from the impurities and defects in the crystal. The nitrogen atom is a major impurity found in natural diamond and the size of nitrogen atom is about the same as carbon atom. The Electron Spin Resonance (ESR) spectroscopy can be used to investigate the impurities and the defects in the diamond.[1–6] The use of ESR

*Corresponding author.

spectroscopy technique to study the detailed microscopic nature of impurities and that of point defects, especially nitrogen substitution was firstly reported by Smith, *et al.*[7]

In this work, we used ESR spectroscopy technique to observe the ESR spectra of nitrogen impurities in natural type I diamond and then to calculate the spin Hamiltonian parameters due to the substitution of nitrogen atoms at the site of carbon atoms and finally the crystal symmetry of diamond will be obtained.

2. Experimental

The natural type I diamond samples with a cubic shape were cleaned with acid and solvent to remove all stains and other impurities on the surfaces. The ESR spectra of diamond crystals were measured at room temperature on a Bruker E500 CW ESR spectrometer operating at X-band microwave frequency using 0.5 mT field (100 kHz) modulation amplitude with a time constant of 0.03 s.

The diamond crystal was mounted on a sample holder with the crystallographic axis (c-axis) perpendicular to the magnetic field direction and then it was inserted into a goniometer to control the orientation of the c-axis with respect to the external applied magnetic field direction as shown in Fig. 1. The theta (θ) is the angle between magnetic field direction and c-axis and the phi (ϕ) is the angle between projection of magnetic field on xy plane and x axis. The magnetic field was varied from 340 to 360 mT. The ESR spectra were measured at every 15° of rotation angle (ϕ) about z-axis from 0°–180°. From the positions of absorption peak of ESR spectra, we used EPR-NMR program to calculate the g-factor values and hyperfine coupling constant parameters, A.

3. Results and Discussion

Figure 2 shows a typical result of ESR spectra of one diamond sample with the applied magnetic field perpendicular to the c-axis ($\theta = 90°$) and the rotation angle

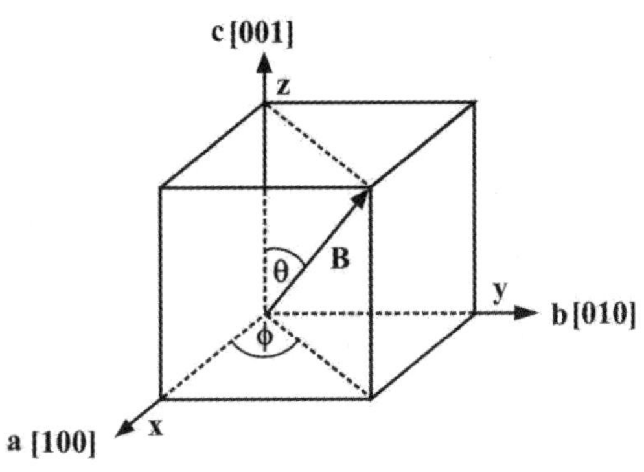

Fig. 1. Direction of magnetic field, B and the rotation angles θ and ϕ.

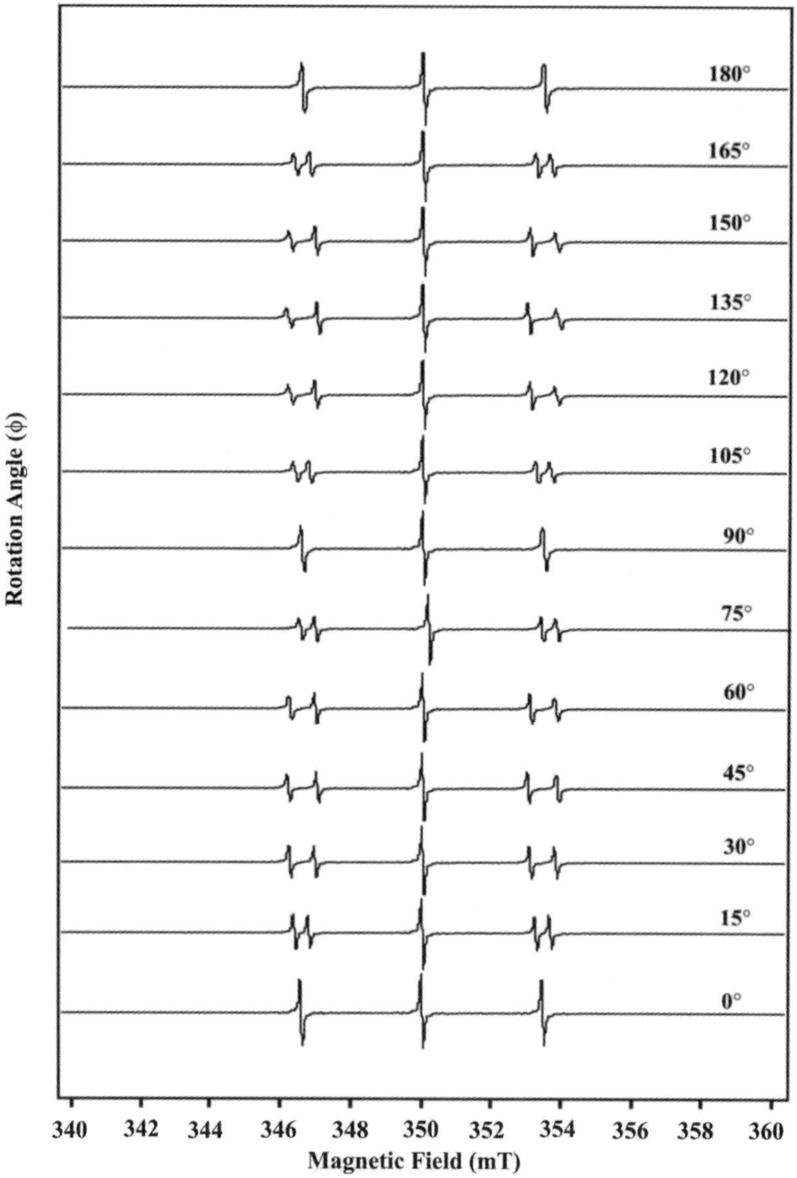

Fig. 2. X-band ESR spectra observed in a natural type I diamond crystal at $\theta = 90°$ and rotation angles $\phi = 0°, 15°, \ldots 180°$.

ϕ was varied every 15° from 0°–180°. The resonance absorption peaks occurred in the magnetic field range of 340 to 360 mT.

Each carbon atom has four valence electrons and the diamond structure is formed with covalence bond of four carbon atoms. When one nitrogen atom substitutes one carbon atom, an unpaired electron was remained in C–N bond, and

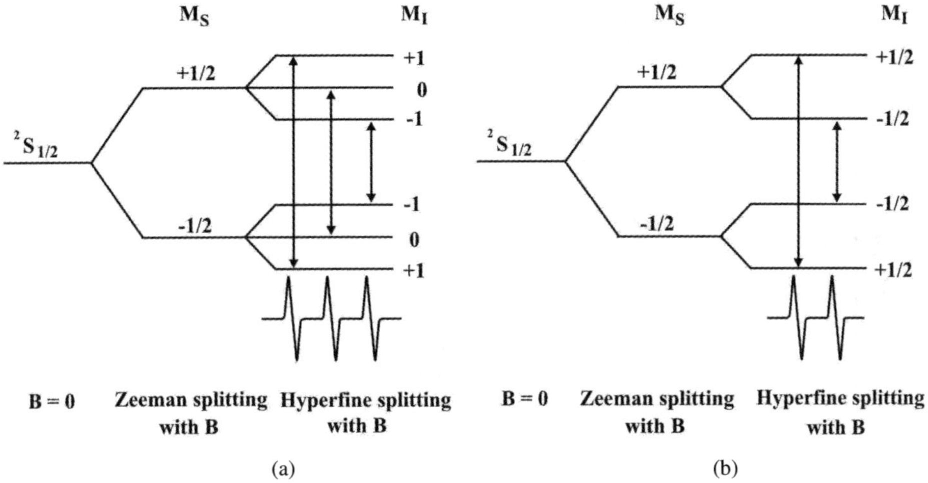

Fig. 3. (a) Energy level of ^{14}N atom splits into six energy sublevels in magnetic field B and gives rise to three allowed transitions and (b) four energy sublevels and two transitions are observed for ^{15}N atom in magnetic field B.

hence the ground state of unpaired electron of nitrogen atom is $^2S_{1/2}$. This single unpaired electron with spin S (spin quantum number, S = 1/2) interact with a single nitrogen impurity atom. In the case of ^{14}N impurity, the nuclear spin I (nuclear spin quantum number, I = 1) whereas those of ^{15}N, I = 1/2. The interaction between electron spin (S) and nuclear spin (I) causes the electron energy level splits into (2I + 1) sublevels.[3,8,9] In an applied magnetic field, for the energy level of ^{14}N ($\text{I} = 1, M_s = \pm 1/2$) splits into $(2\text{S}+1)(2\text{I}+1) = 6$ energy sublevels and for the energy level of ^{15}N ($\text{I} = 1/2, M_s = \pm 1/2$) splits into $(2\text{S}+1)(2\text{I}+1) = 4$ energy sublevels as shown in Fig. 3. According to the selection rules, $\Delta M_s = \pm 1$ and $\Delta M_\text{I} = 0$, three and two allowed transitions for ^{14}N and ^{15}N, respectively are obtained [Figs. 3(a) and 3(b)].

From the ESR spectra in Fig. 2, three strong resonance absorption peaks are observed at $\theta = 90°$, $\phi = 0°$, hence these three sharp peaks arise from ^{14}N impurity. This defect is due to single nitrogen substitutions in diamond and it is called P1 center. The three resonance peaks were also observed at $\phi = 90°$ and 180°, but five resonance peaks were observed at the remainder rotation angles, $\phi = 15°$, 30°, 45°, 60°, 75°, 105°, 120°, 135°, 150° and 165°. The resonance peaks due to ^{15}N impurity cannot be seen in Fig. 2, however for a closed-up ESR spectrum at $\theta = 90°$, $\phi = 0°$, we can observe two small sharp peaks due to ^{15}N impurity as shown in Fig. 4. Many other satellites (small sharp peaks) were also observed according to four ^{13}C atoms. Each of three ^{13}C atom gives rise to six resonance peaks and the rest one of ^{13}C atom gives only two peaks.

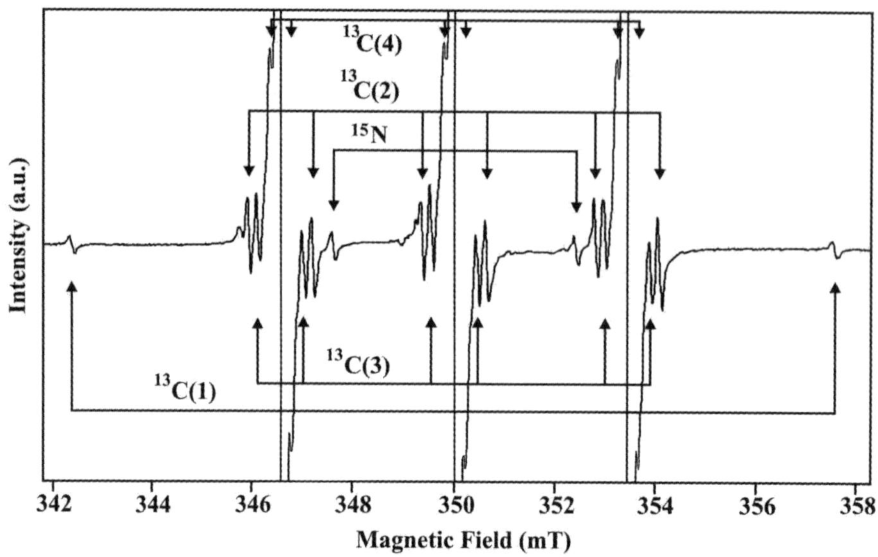

Fig. 4. The closed-up ESR spectrum at $\theta = 90°$, $\phi = 0°$.

3.1. *Spin Hamiltonian of nitrogen in diamond*

The ESR spectra of nitrogen in diamond can be described by a spin Hamiltonian (H) incorporating with Zeeman interaction and hyperfine structure.[10] It is given by equation

$$\mathrm{H} = \mathrm{H}_{\mathrm{Zeeman}} + \mathrm{H}_{\mathrm{HF}}\,. \tag{1}$$

The first term corresponds to the Zeeman interaction arises from the interaction between electron spin angular momentum and external magnetic field. The Zeeman term is given by equation

$$\begin{aligned}\mathrm{H}_{\mathrm{Zeeman}} &= \beta \mathbf{S}\cdot\mathbf{g}\cdot\mathbf{B}\\ &= \beta[\mathrm{S}_x, \mathrm{S}_y, \mathrm{S}_z]\cdot\begin{bmatrix} g_{xx} & g_{xy} & g_{xz}\\ g_{yx} & g_{yy} & g_{yz}\\ g_{zx} & g_{zy} & g_{zz}\end{bmatrix}\cdot\begin{bmatrix}\mathrm{B}_x\\ \mathrm{B}_y\\ \mathrm{B}_z\end{bmatrix}\end{aligned} \tag{2}$$

where β, $\mathbf{S}$, $\mathbf{g}$ and $\mathbf{B}$ are the Bohr magnetron; spin operators $\mathrm{S}x, \mathrm{S}y, \mathrm{S}z$; gyromagnetic tensor, and magnetic field, respectively.

The second term is the hyperfine interaction term arises from the interaction between the electron spin and nuclear spin. The hyperfine term is written in the form

$$\begin{aligned}\mathrm{H}_{\mathrm{HF}} &= \mathbf{S}\cdot\mathbf{A}\cdot\mathbf{I}\\ &= \mathbf{S}\cdot\mathbf{A}\cdot\mathbf{I}_{14N} + \mathbf{S}\cdot\mathbf{A}\cdot\mathbf{I}_{15N}\end{aligned} \tag{3}$$

where **I** and **A** are nuclear spin operator and hyperfine coupling tensor, respectively. For tensors **g** and **A**, the cross terms can be eliminated by a suitable choice for the x, y, z axes (known as the principal axes). Then g and A can be written in matrix form

$$g = \begin{bmatrix} g_\perp & 0 & 0 \\ 0 & g_\perp & 0 \\ 0 & 0 & g_\| \end{bmatrix} \tag{4}$$

and

$$A = \begin{bmatrix} A_\perp & 0 & 0 \\ 0 & A_\perp & 0 \\ 0 & 0 & A_\| \end{bmatrix} \tag{5}$$

where $g_\perp = g_{xx} = g_{yy}, g_\| = g_{zz}$ and $A_\perp = A_{xx} = A_{yy}$, $A_\| = A_{zz}$.

For cubic symmetry, $g_{xx} = g_{yy} = g_{zz}$ and $A_{xx} = A_{yy} = A_{zz}$ hence $g_\perp = g_\|$ and $A_\perp = A_\|$. Using the resonance peak positions in the ESR spectra from Fig. 2, we can calculate the spin Hamiltonian parameters, g and A with the help of EPR-NMR program. The results are shown in Table 1.

From Table 1, the g-tensor of P1 center has isotropic symmetry while the tensor of hyperfine interaction reveals a well-defined axial symmetry. The calculated spin Hamiltonian parameters of ^{14}N and ^{15}N indicated that the symmetry is trigonal.

3.2. *Angular dependence of resonance magnetic field position of ^{14}N and ^{15}N spectra*

From the ESR spectra in Fig. 2, we can find the relation between the positions of resonance magnetic and the rotation angles (ϕ) field for ^{14}N and ^{15}N, and the results are shown by solid circles for ^{14}N and open circles for ^{15}N in Fig. 5. It is seen that three peaks of ESR spectra, corresponding to $\phi = 0°$, 90°, 180°, were not split, but both of two neighboring peaks split into two peaks with lower intensity at $\phi = 15°$, 30°, 45°, 60°, 75°, 105°, 120°, 135°, 150° and 165°. This splitting is due to the nitrogen (^{14}N and ^{15}N) substitute at one of the four C–N bonds and this C–N bond, called the hyperfine axis, makes an arbitrary angle with applied magnetic field.

We used essential parameters from Ref. 11, $g = 2.0024$, $A_\| = 40.73$, $A_\perp = 29.04$ for ^{14}N and $g = 2.0024$, $A_\| = -57.05$, $A_\perp = -39.90$ for ^{15}N, to simulate the

Table 1. Spin Hamiltonian parameters of ^{14}N and ^{15}N.

g	A_14N	A_{14N} [ref. 11]	A_{15N}	A_{15N} [ref. 11]	RMSD* (G)
$g_\| = g_\perp = 2.0019$	$A_\| = 40.24$ $A_\perp = 29.73$	$A_\| = 40.73$ $A_\perp = 29.04$	$A_\| = -57.60$ $A_\perp = -40.50$	$A_\| = -57.05$ $A_\perp = -39.90$	0.4447

*Root mean sum of squares of weighted differences between observed and calculated transition frequencies.

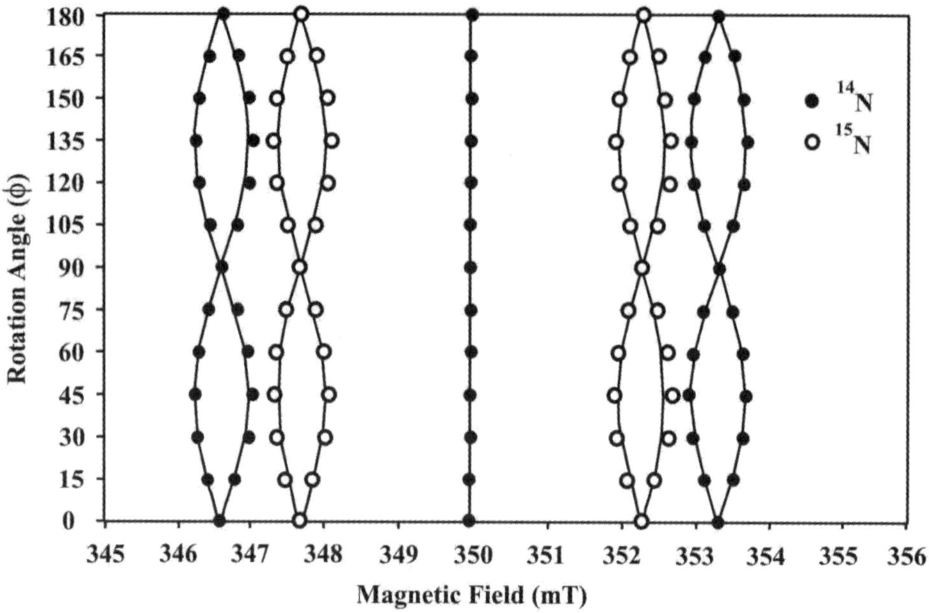

Fig. 5. The angular dependence of resonance magnetic field positions in the ESR spectra of ^{14}N and ^{15}N in diamond: solid and open circles represent observed values and solid lines represent simulated values.

characteristic of the angular dependence of the resonance magnetic field position. The results of the angular dependence of resonance magnetic field positions of the observed and simulated values are shown in Fig. 5. It can be concluded from Fig. 5 that the substitution of single ^{14}N in diamond gives rise to trigonal symmetry which agrees with that of spin Hamiltonian parameters calculation. This symmetry was also classified as C_{3v} symmetry. On the other hand, the resonance magnetic field position are not split for all rotation angles, ϕ as shown in Fig. 6(a) due to g factor and A tensor are isotropic symmetry. In the case of g-factor and A tensor are all axial symmetry, the resonance magnetic field positions are split for all rotation angles except $\phi = 0°$, $90°$ and $180°$ as shown in Fig. 6(c).

In general, the symmetric crystal describes the resonance absorption peaks data of the unpaired electrons associated with a particular defect, and give sensitive and qualitative data of any distortion from the regular arrangement of the surrounding diamagnetic ions. In diamond, each carbon atom is tetrahedrally surrounded by four other carbon atoms. If an impurity atom as nitrogen substitutes one of four carbon atoms, tetrahedral symmetry need no longer be preserved due to the Jahn–Teller effect. The symmetry of the surroundings in this case then drops from T_d to C_{3v} symmetry.

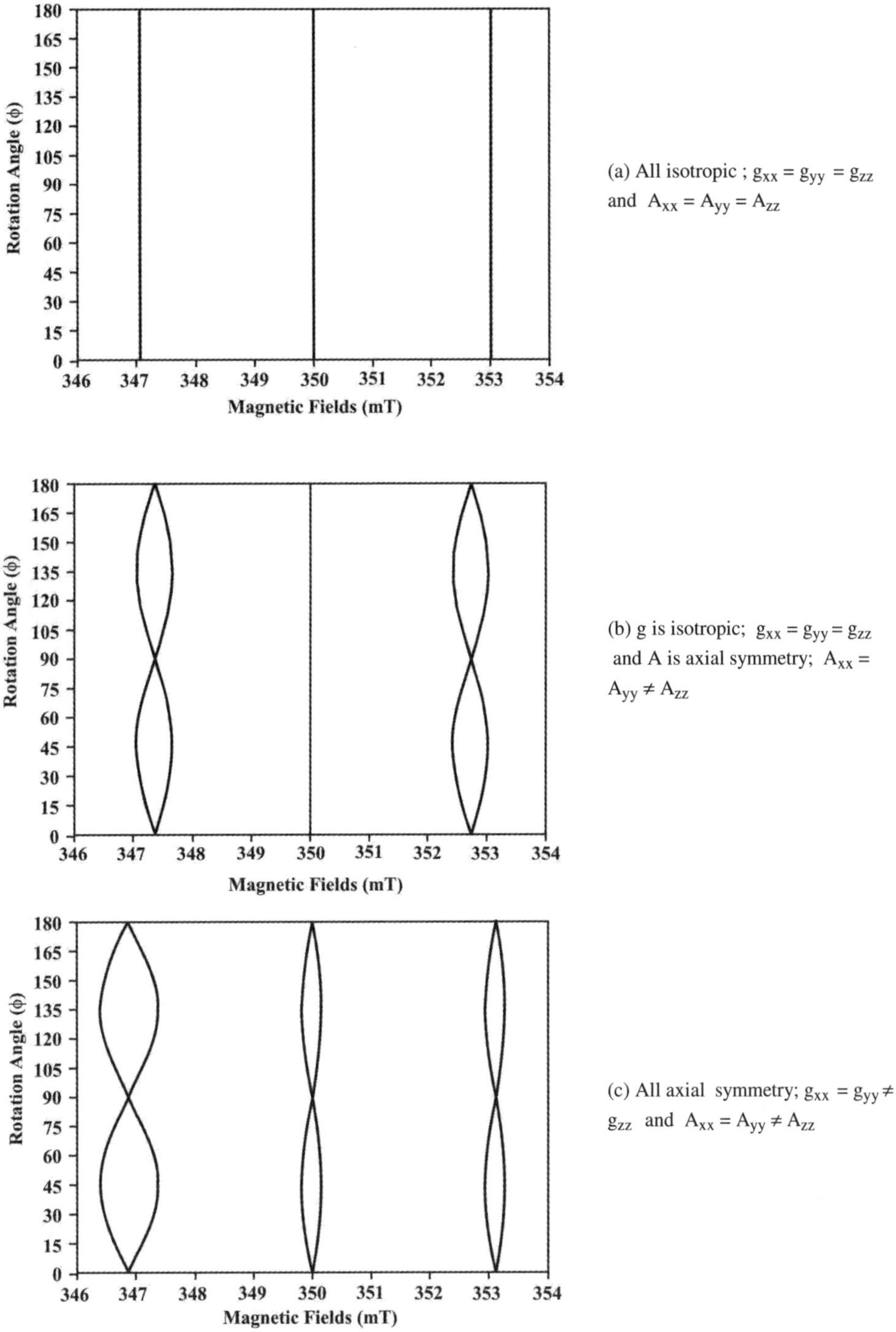

Fig. 6. The simulated angular dependence for: (a) g and A are isotropic, (b) g is isotropic and A is axial symmetry and (c) g and A are axial symmetry.

4. Conclusion

The characterization of paramagnetic center in diamond was studied by ESR spectroscopy. The results of ESR spectra show that three strong lines were obtained at $\phi = 0°, 90°$ and $180°$. This lines generated from a coupling between the electron spin (S = 1/2) and the nuclear spin (I = 1 for ^{14}N), of the isolated substitutional nitrogen atom, called P1 center. For two satellites that occurred between strong resonant absorption lines are generated from a hyperfine interaction of ^{15}N (I = 1/2). The calculated spin Hamiltonian parameters are $g = 2.0019$ and for ^{14}N ; $A_{\parallel} = 40.24$, $A_{\perp} = 29.73$ and for ^{15}N ; $A_{\parallel} = -57.60$, $A_{\perp} = -40.50$ agree with these of Cox *et al.*[11] The g-tensor is isotropic symmetry while the A-tensor is axial symmetry. The resonance magnetic field positions repeat itself for every rotation angles of 90°. Due to P1 center, the symmetry of diamond used in this work is trigonal symmetry (C_{3v}).

Acknowledgments

The authors wish to thank Prof. Dr. John A. Weil, the University of Saskachewan, Canada for EPR-NMR computer program, PERCH-CIC and Department of Chemistry, Faculty of Science, Mahidol University for ESR spectrometer. This work is financially supported by the Department of Physics, Faculty of Science, KMUTT and the National Research Council of Thailand (NRCT).

References

1. P. R. Briddond and R. Jones, *Physica B* **185**, 179 (1993).
2. M. E. Newton and J. M. Baker, *J. Phys: Condens. Matter* **1**, 10549 (1989).
3. J. H. N. Loubser and J. A. van Wyk, *Rep. Prog. Phys.* **41**, 1201 (1978).
4. S. Nokhrin, J. Rosa, M. Vanecek, A. G. Badalyan and M. Nesladek, *Diamond Relat. Mater.* **10**, 480 (2001).
5. J. A. van Wyk, J. H. N. Loubser, M. E. Newton and J. M. Baker, *J. Phys: Condens. Matter* **4**, 2651(1992).
6. J. Isoya, H. Kanda, I Sakaguchi, Y. Morita and T. Ohshima, *Radiat. Phys. Chem.* **50**, 321 (1997).
7. W. V. Smith, P. P. Sorokin, I. L. Gelles and G. J. Lasher, *Phys. Rev.* **115**, 1546 (1959).
8. G. E. Pake and T. L. Estle, *The Physical Principles of Electron Paramagnetic Resonance* (W. A. Benjamin, Inc., 1973), p. 110.
9. J. A. Weil, J. R. Bolton and J. E. Wertz, *Electron Paramagnetic Resonance* (John Wiley & Sons. Inc., 1994), p. 310.
10. P. E. Klingsporn, M. D. Bell and W. J. Leivo, *J. Appl. Phys.* **41**, 2977 (1970).
11. A. Cox, M. E. Newton and J. M. Baker, *J. Phys.: Condens. Matter* **6**, 551 (1994).

LOW SINTERING TEMPERATURE OF LEAD MAGNESIUM NIOBATE-LEAD TITANATE (0.9PMN–0.1PT) BY ADDING OXIDE ADDITIVES

NUTTAPON PISITPIPATHSIN, KAMONPAN PENGPAT*,
SUKUM EITSSAYEAM and TAWEE TUNKASIRI
Department of Physics, Faculty of Science, Chiang Mai University,
Chiang Mai, Thailand 50200
**kpengpat@gmail.com*

SOMNUK SIRISOONTHORN and SAVITTREE BUDCHAN
National Metal and Materials Technology Center (MTEC), 114 Paholyothin Rd.,
Klong Luang, Pathumthani, Thailand 12120

URAIWAN INTATHA
School of Science, Mae Fah Luang University,
Chiang Rai, 57100, Thailand

Received 31 July 2008

This research work is aimed at lowering the sintering temperature of 0.9PMN–0.1PT ceramics by adding oxide additives. The oxides used for this purpose were Bi_2O_3 and Li_2CO_3 with various amounts, following the formula of $x\mathrm{Bi_2O_{3+y}Li_2CO_3}$, where $x+y=$ 10 mol% and $x = 1, 3, 5$ and the mixed oxide additive powders were then added to the dried powder of 0.9PMN–0.1PT with 1 wt% concentration. An excess PbO content of approximately 3 wtadded to all compositions to compensate the lead loss. After that, the mixed powders were pressed into pellets and subsequently sintered to form the ceramic samples. The results showed that the sintering temperature of 0.9PMN–0.1PT ceramics could be lowered down from 1250°C to 900–1000°C by the addition of small amount of the oxide additives, where the optimum composition was found in the sample with $x : y = 1 : 9$ at sintering temperature of 1000°C. Moreover, the densification, dielectric and ferroelectric properties of this sample remain acceptable in particular uses, promising positive future for reduction of lead loss from electrical and electronic industries.

Keywords: PMN–PT; low sintering temperature; conventional mixed oxide method.

1. Introduction

Lead-based perovskite-type solid solutions, consisting of normal and relaxor ferroelectric materials, have been extensively studied due to their high dielectric and piezoelectric properties, which may be applied to microactuators and miniaturized transducers.[1–3] Among the lead-based complex perovskites, lead magnesium niobate ($Pb(Mg_{1/3}Nb_{2/3})O_3$, PMN) and lead titanate (PT) have been the subject of much research, particularly for the 0.9PMN–0.1PT composition as it has high dielectricity and piezoelectricity over broad range of temperature at about 25–30°C.[4–6] Normally, a sintering temperature in the range of 1200–1250°C is generally required for densifying PMN-PT ceramics. However, high-temperature sintering tends to cause a significant increase in the grain size, and volatilization of PbO, which degrades the properties of the ceramics. There is also the additional problem dealing with the environmental effects of lead vapor in the atmosphere. Lowering the sintering temperature may be one of the possible solutions to these problems.

Recently, several kinds of fluxing agents have been used to decrease sintering temperatures of ceramic materials.[3,8] The addition of V_2O_5 to $Pb(Zr_{0.53}Ti_{0.47})O_3$ (PZT) ceramics by Wittmer and Buchanan,[9] was found to reduce the formation temperature of PZT from 1280°C to below 975°C. Chen and Fu[6] reported that the sintering temperature of PMN–PT could be reduced by addition of Bi_2O_3/Li_2O. Lithium-containing compounds, e.g., Li_2CO_3, have been used as sintering aids in dielectric ceramic systems. Tunkasiri[11] employed B_2O_3–Bi_2O_3–CdO as low melting frit to lower the sintering temperature of PZT ceramics, and found that the sintering temperature could be reduced from 1250° to about 800°C by the addition of a small amount of the low melting frit.

Therefore, it is interesting to study the effect of the liquid phase on electrical properties of the 0.9PMN–0.1PT ceramics prepared by various sintering conditions and compositions of a low melting frit series. In this work, a Bi_2O_3/Li_2CO_3 series was employed.

2. Experimental

The samples were prepared using the conventional solid state reaction methods. The commercially available 0.9PMN–0.1PT, PbO, Bi_2O_3 and Li_2CO_3 (Fluka chemika), all having at least 98.0% purity were used as raw materials. The Bi_2O_3 and Li_2CO_3 were pulverized and 1 wt% of these additive powders with the mole ratios of Bi_2O_3:Li_2CO_3 (x:y) equal to 1:9, 3:7 and 5:5 were added into the 0.9PMN–0.1PT powder, designated as A1, A2 and A3 respectively. Non-oxide additive powder (0.9PMN–0.1PT) was denoted as A0. All mixtures contained an excess of 3.0 wt% PbO, to compensate the vaporization of lead oxide. The batch powders were ball-milled using ZrO_2 balls, grinding media in acetone for 24 h. Subsequently, the resulting slurry was filtered in a rotary evaporator and dried at 100°C for 6 h in an oven. After sieving, a binder of 3.0 wt% PVA was added and the powder was then again granulated prior to pressing. The mixed powder was uniaxially pressed at 196 MPa into circular disks of approximately 2.5 cm in diameter and 1 mm in thickness. The pressed disks were sintered at various sintering temperatures.

After polishing, silver paste was applied onto both sides of the samples to make electrical contact. Dielectric properties were measured with a LF impedance analyzer (4192A, Hewlett-Packard) and impedance-capacitance-resistance (LCR) meter (Model HP 4284A, Hewlett-Packard) in a temperature control chamber from

$-100°C$ to $100°C$, at various frequencies between 100 Hz and 100 kHz. A conventional strain gauge method was used to detect the change of strains with an electric field. The ferroelectric properties were evaluated from the $P - E$ hysteresis curves, using a high voltage test system (Model RT66A, Radiant Technologies, Albuquerque, NM), in a silicone oil bath from 0°C up to 150°C. A strain gauge system was connected to a Wheatstone bridge. The electric field, generated by a digital-dialog (DA) converter assisted by computer interfaces, was amplified by a high voltage amplifier (Standford Research System SRS 830 Lock-In Amplifier) and applied to the sample continuously between −2 kV and +2 kV per mm. A minimum of three samples was measured for each composition and the data were reported as the average values.

3. Results and Discussion

The density results of the sintered specimens are listed in Table 1. The average density for the A0 samples (0.9PMN–0.1PT) was found to be 7.95 g/cm^3, which is approximately 98% of the theoretical density of the 0.9PMN–0.1PT ceramic (8.12 g/cm^3). At low sintering temperature of 800°C, a decrease of Li_2CO_3 content did not significantly affect the density of all specimens, while at high sintering temperature of about 1000°C; the decrease in Li_2CO_3 content resulted in the reduction of density values. The maximum density of 7.7 g/cm^3 with relative density of about 95% of the theoretical density, was obtained in the sample A1 (x:y = 1:9) with maximum Li_2CO_3 content at the sintering temperature of 1000°C. This may be due to the help of the liquid phase sintering which normally occurs when high amount of alkaline oxide was added into ceramic samples.

Figure 1 shows the X-ray diffraction (XRD) patterns of the A1 samples, as a function of sintering temperature, comparing to the simulated XRD peaks of PMN and PT phase from the Joint Committee on Powder Diffraction Standards (JCPDS).[12] It was found that, in A1 samples, the rhombohedral phase of PMN started to appear at 800°C. Traces of unidentified phase were detected, however, they disappeared at 900°C onward. The similar XRD patterns (not shown here), indicating the formation of single rhombohedral phase of PMN, were also observed

Table 1. Sintering density and sintering conditions for each specimen.

Sintering Temperatures	Density (g/cm^3)			
	Oxide added samples			0.9PMN–0.1PT sample
	A1	A2	A3	A0
800°C	7.4	7.4	7.4	
900°C	7.5	7.5	7.4	
1000°C	7.7	7.6	7.5	
1250°C				7.95

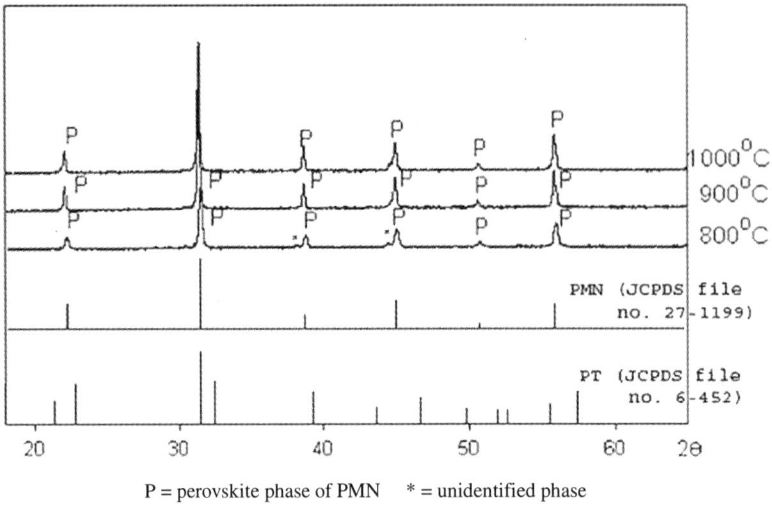

Fig. 1. XRD patterns of samples A1 sintered at 800–1000°C for 1 h.

Table 2. Values of $\varepsilon_{\max}$, $\tan\delta$ and T_m for each sintered specimen at 1000°C, 1 kHz.

Sample Code	0.9MN–0.1PT				PMN	PMN	0.9PMN–0.1PT	PMN
	A0	A1	A2	A3	(16)	(17)	(18)	(19)
$\varepsilon_{\max}$ 12300	8700	5800	4300	18000	14000	18500	12000	
$\tan\delta$	0.006	0.040	0.003	0.002	0.003	0.66		0.0036
T_m(°C)	15	−5	−10	−10	−5	—	40	−20
$\bar{P}_s$ (μC/cm^2)	65.50	36.71	28.41	25.9				
$\bar{P}_r$ (μC/cm^2)	4.0	3.8	3.7	3.6	—	4.6		
$\bar{x}_{\max}$ ($\times 10^{-4}$)	6.0	1.9	1.0	0.8	—	—		

in A2 and A3 samples. No traces of PT[13] phase were detected. It may be assumed that the 10 mol% of PT phase is in a solubility limit of PMN-PT binary system, retaining the rhombohedral structure of PMN phase in this solid solution sample. This agrees pretty well with other previous works.[14,15] Taking into account of the density of the sintered samples obtained in this study, the specimens sintered at 900°C and 1000°C were chosen for further dielectric measurement.

The temperatures of maximum dielectric constant (T_m) exhibited relaxor type frequency dependency. $T_m(s)$ were found to increase with increasing frequency, which is consistent with that of typical PMN relaxor ferroelectrics.[16] For A0 samples, T_m are at 10°C, 15°C, 15°C and 20°C when the frequencies are at 100 Hz, 1 kHz, 10 kHz and 40 kHz, respectively. At 1 kHz the T_m values of A0 and A1 are taken as 15°C and −5°C respectively. The average values of dielectric maximum (ε_r), the saturated polarization (P_s), the remanent polarization (P_r), coercive field (E), etc. of the samples were summarized in Table 2. The results obtained

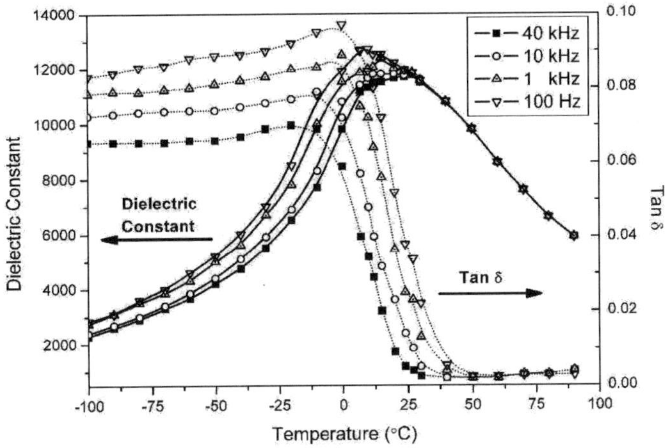

(a)

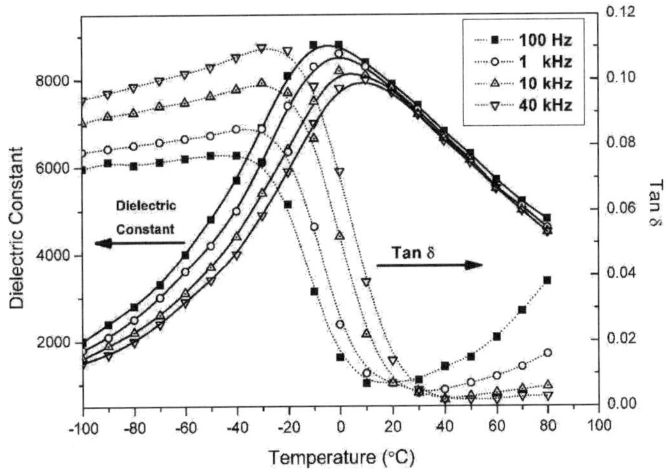

(b)

Fig. 2. Dielectric properties at various frequencies of (a) A0 and (b) A1 sintered at 1000°C.

from other previous works[17–19] were tabulated for comparison. It can be seen that the addition of the oxide additive caused the reduction in both maximum dielectric constant ($\varepsilon_{\max}$) and maximum temperature (T_m). This may be useful in using these materials in a wider range of temperature and applications. From Figs. 2(a) and 2(b), a relaxor-type ferroelectric property could still remain in the sample with addition of Bi_2O_3 and Li_2CO_3 (A1) as the change of frequency affects the dielectric constant of the ceramic which is similar to that found in the non-added sample (A0). Moreover, it can be noticed that the peak of the dielectric constant broadened markedly in the A1 sample as shown in Fig. 2(b).

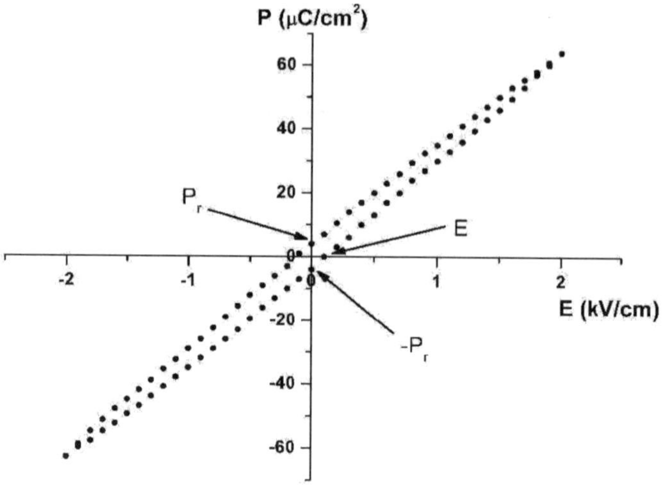

(a)

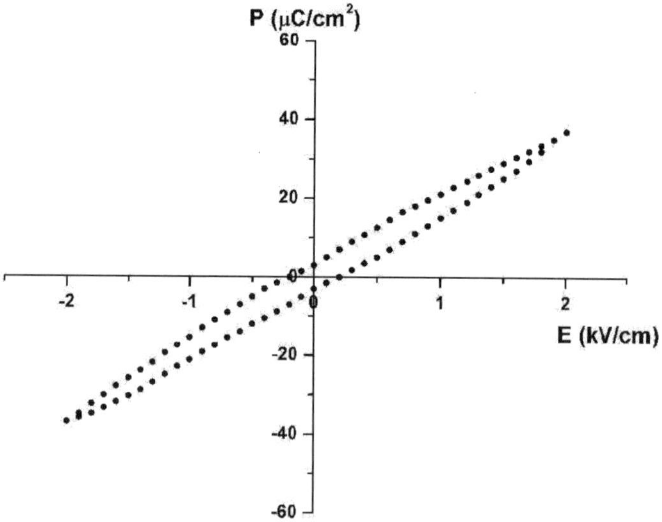

(b)

Fig. 3. Plots of polarization vs. electric fields of selected samples of (a) A0 and (b) A1 sintered at 1000°C.

Figures 3 and 4 illustrate the $P-E$ hysteresis loops and strain vs. polarization relations of selected samples (A0 and A1). The values of P_s and P_r for A0 and A1 samples are closely the same, which confirmed the good ferroelectric property of the oxide-added sample such as the A1 sample. Moreover, these values are comparable to that of Kong *et al.*[17] The study presented above may also indicate that the changes in the B-site cations (mainly ferroelectrically active Nb ions) and uniden-

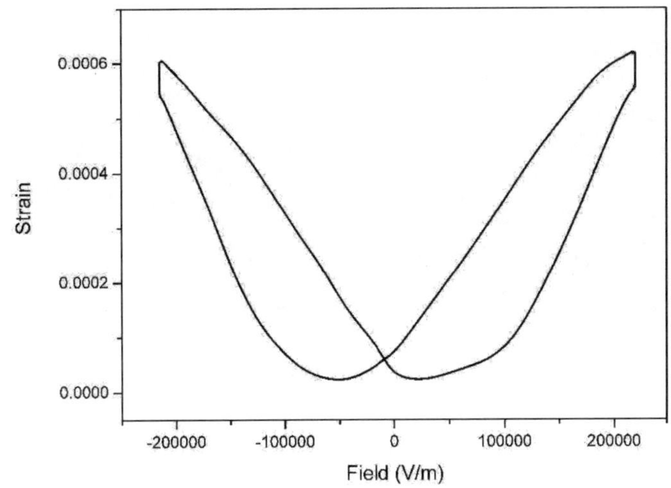

(a)

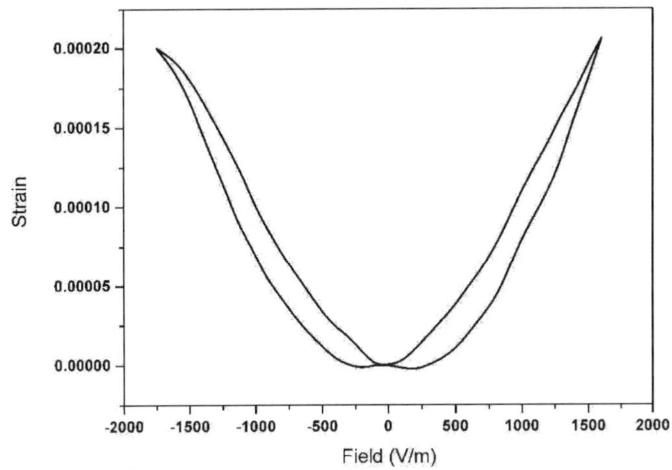

(b)

Fig. 4. Plots of strain vs. electric fields of selected samples of (a) A0 and (b) A1 sintered at 1000°C.

tified phase could influence the dielectric response and electrically decreased strain characteristics at higher oxide contents.

Figure 5 illustrates the SEM pictures of the A0 and A1 samples. It can be seen that the average grain size of the A0 sample (3.0 μm) was smaller than that of A1 (4.3 μm). These samples, however, possessed a good densification which agrees pretty well with the density data. The abnormal grain growth of needle-like grains agglomerated at grain boundaries was also observed in the A1 sample. This may be due to the aid of liquid phase sintering in the oxide additive sample at an elevated

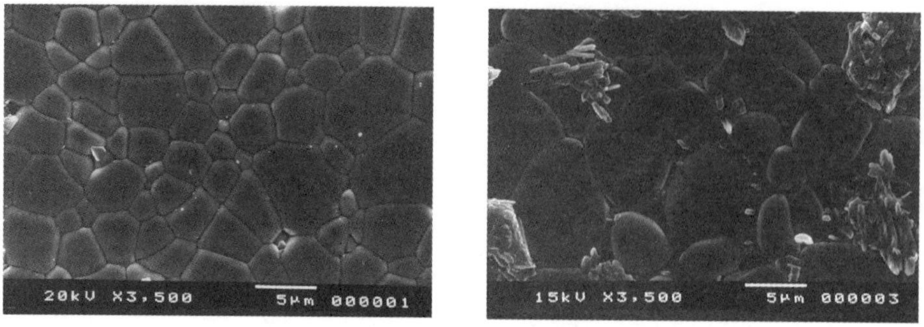

(a) (b)

Fig. 5. SEM pictures of samples (a) A0, (b) A1 sintered at 1000°C.

temperature. The average grain sizes of A1, A2 and A3 were found similar and their values are in the same order as that obtained from Kong *et al.* (6.2 μm),[17] though they prepared the samples through the high energy ball milling process.

4. Conclusion

In this work, it was found that the oxide additives (Bi_2O_3–Li_2CO_3 series) could reduce the sintering temperature of 0.9PMN–0.1PT ceramics considerably. The X-ray diffractometry and scanning electron microscopy showed that only the single perovskite with rhombohedral structure of PMN was detected in all samples and the grain size was enlarged by the oxide additive. The physical and electrical properties such as density, dielectric constant and saturated polarization were comparable to that of non-additive samples. This ensures the potential of developing these materials for electrical and electronic applications.

Acknowledgments

The authors would like to express their sincere thanks to the Thailand Research Fund, the Commission on Higher Education and Graduate School, Chiang Mai University for financial support throughout the project.

References

1. G. H. Haertling, *J. Am. Ceram. Soc.* **82**, 797 (1999).
2. L. E. Cross, *Mater. Chem. Phys.* **43**, 108 (1996).
3. K. Uchino, *Acta. Mater.* **46**, 3745 (1998).
4. H. Takeuchi, H. Mazusawa and C. Nakaya, *IEEE, Ultrason. Symp.* **1**, 697 (1993).
5. J. Chen, A. Shuralnd, J. Perry, B. Ossmann and T. R. Gururaja, *IEEE, Int. Symp. Appl. Ferroelectric.* 2730 (1996).
6. J. Kelly, M. Leonard, C. Tantigate and A. Safari, *J. Am. Ceram. Soc.* **80**, 957 (1997).
7. D. Dong, K. Murakami, S. Kaneko and M. Xiong, *J. Ceram. Soc. of Japan* **101**, 1090 (1993).

8. J. P. Guha, D. J. Hong and H. U. Andersan, *J. Am. Ceram. Soc.* **71**, 152 (1988).
9. D. E. Wittmer and R. C. Buchanan, *J. Am. Ceram. Soc.* **64**, 485 (1981).
10. G. F. Chen and S. L. Fu, *J. Mater. Sci.* **25**, 424 (1990).
11. T. Tunkasiri, *Smart Mater. Struct.* **3**, 243 (1994).
12. Powder Diffraction File, Card No. 27–1199, (Swarthmore, PA, Joint Committee on Powder Diffraction Standards, International Center of Diffraction Data), (2002).
13. Powder Diffraction File, Card No. 6–452, (Swarthmore, PA, Joint Committee on Powder Diffraction Standards, International Center of Diffraction Data), (2002).
14. J. C. Bruno, A. A. Cavalheiro, M. A. Zaghete, M. Cilense and J. A. Varela, *Mater. Chem. and Phys.* **84**, 120 (2004).
15. J. C. Bruno, A. A. Cavalheiro, M. A. Zaghete and J. A. Varela, *Ceram. Inter.* **32**, 189 (2006).
16. A. J. Moulson and J. M. Herbert, *Electroceramics: Materials, Properties and Application*, 2nd edn. (London, 2003), p. 320.
17. L. B. Kong, J. Ma, W. Zhu and O. K. Tan, *J. Mat. Sci. Lett.* **20**, 1241 (2001).
18. L. Wul and Y. C. Liou, *Ferroelectrics*, **168**, 251 (1995).
19. Y. Yang, C. Fang and Y. Yu, *Mater. Lett.* **49**, 345 (2001).

QUANTUM BROWNIAN AND THE CONSTRAINED PATH INTEGRAL

S. BOONCHUI*, V. SA-YAKANIT and P. PALOTAIDAMKERNG†

Department of Physics, Faculty of Science, Kasetsart University, Bangkok 10900
and
Center of Excellence in Forum for Theoretical Science,
Chulalongkorn University, Bangkok 10330

Received 31 July 2008

We show that the environment affects a quantum system in the form of the constrained trajectory. Our result allows one to describe the quantum system in terms of stochastic state vector rather than quantum history. Moreover we can alternatively reduce the time evolution operator. Then the trajectory of system is constrained.

Keywords: Quantum Brownian motion; the spectral density; the constrained path integral.

1. Introduction

The understanding of the dynamics of dissipative quantum system is of fundamental importance both from a practical and a conceptual point of view. The standard quantum Brownian motion (QBM) model[1] is described by a particle with $H(x, p)$, coupled to an environment of harmonic oscillator (P_λ, Q_λ). The effect of the environment on the dynamics of the system can be seen as the interplay of the dissipation and fluctuation phenomena traditionally. For the QBM problem with linear coupling, both dissipation and fluctuation are determined by some specific property of the environment, i.e., "the spectral density", which measures the number of oscillators with a given frequency present in the environment and strength of the interaction between such oscillator and the system,

$$I(\omega) = \sum_i \delta(\omega - \omega_i) \frac{C_i^2}{2m_i\omega_i}. \tag{1}$$

The effect of dissipation and fluctuation in Brownian motion had been thoroughly studied for classical system using semi-phenomenological (Langevin) equation.[2] For a limited extent to QBM, the evolution equation for the reduced density

---

*Author to whom any correspondence should be addressed.
†Author to whom any correspondence should be addressed.

matrix is the master equation. The master equation can be used to examine the properties of the most general QBM.[3] The master equation has been derived with different methods. For example,Unruh and Zurek[4] derive the master equation by using canonical methods. Caldeira and Leggett[5] had derived the master equation for the ohmic environment at high temperature.

The question about the influence of zero temperature environment on the Brownian particle has been discussed in the past years. There have been studies on the temperature-dependent weak localization measurements,[6] reporting residual decoherence in metals at zero temperature, in contradiction to theoretical predictions,[7] and on the zero-point decoherence induced by Coulomb interaction in disordered electron system. We are interested in how the environment affects on the Brownian particle in some averaged way.

In this paper, we show the alternative effect of environment in the zero temperature case, which act on the dynamics of the Brownian particle. The motion of the Brownian particle is restricted to the constraint condition only. In Sec. 1 we show the constrained path integral methods and an example. Next, we reduce the time evolution for the system + environment. Then we obtain the effective time evolution operator for the system and the constraint condition, presented in Sec. 2. This derivation is done by path integral methods. Finally in Sec. 3 we give analytic results.

2. The Constrained Paths Integral

Let us consider the time average of a quantity $\mathbb{F}[x(t)]$ along a path $x(t)$. It is given by the functional

$$\mathrm{F}[x(t)] = T^{-1} \int_{t_0}^{t} \mathbb{F}[x(t)]dt \tag{2}$$

with $T = t - t_0$. The amplitude at which a particle starting in x' at t_0 and ending in x at t has the value of $\mathrm{F}[x(t)]$ exactly equal to f is given by the restricted path integral,

$$K(x, x', t|\ f) = \int D[x]\delta(f - \mathrm{F}[x(t)])e^{iS[x(t)]/\hbar}, \tag{3}$$

where $\delta(x)$ is the Dirac δ function and $S[x(t)]$ is an action of the particle. $S[x(t)]$ is the classical action. Equation (3) is the constrained paths integral with the weight functional as $\delta(x)$ Dirac δ function. It selects only paths which has the value of $\mathrm{F}[x(t)]$ exactly equal to f.

For an example, the restricted paths integral is used to the tunneling time problem. Sokolovski[8] provided the traversal time wave function $\psi(x, t\,|\tau)$ which is the probability amplitude for finding the particle in x to have spent in $\Omega \equiv [a, b]$ prior to time t, a net duration τ. It is obvious that the traversal time wave function $\psi(x, t|\tau)$ is obtained by restricting the system's evolution to particular classes of Feynman paths integral,

$$\psi(x,t|\tau) = \int dx' \int D[x]\delta(\tau - t_{ab}^{cl}[x(t)])e^{iS[x(t)]/\hbar}\Psi_0(x',t_0), \tag{4}$$

where $\Psi_0(x',t_0)$ is the initial wave function.

Here $t_{ab}^{cl}[x(t)] = \int_{t_0}^{t} \Theta_{ab}[x(t)]dt$ is the traversal time, $\Theta_{ab}[x(t)]$ is unity for $a < x < b$ and zero otherwise. We can decompose the final wave function $\Psi_F(x,t)$ into a sum of over all τ

$$\Psi_F(x,t) = \int_{t_0}^{t} d\tau\ \psi(x,t|\tau), \tag{5}$$

when $\Psi_F(x,t) = \mathbf{U}(t)\Psi_0(x,t)$ is the final state wavefunction and $\mathbf{U}(t) = \exp\left[-\frac{i}{\hbar}\int_{t_0}^{t} dt'\mathbf{H}_{sys}\right]$ is the time evolution operator for the quantum system, $\mathbf{H}_{sys}$ is the Hamiltonian of the quantum system.

3. Restricted Motion of the Brownian Particle

The environment is modeled by a set of harmonic oscillators with mass m_λ and natural $\omega_{\boldsymbol{\lambda}}$. The particle is coupled linearly to each oscillator with strength $C_{\boldsymbol{\lambda}}$. The Hamiltonian $\mathbf{H_{total}}$ of the combined system and environment $\mathbf{H_{in+en}}$ is

$$\mathbf{H_{total}} = \mathbf{H}_{sys} - \mathbf{x}\sum_{\lambda} C_{\boldsymbol{\lambda}}\mathbf{Q}_{\boldsymbol{\lambda}} + \sum_{\lambda}\left(\frac{1}{2}m_\lambda\mathbf{P}_\lambda^2 + \frac{1}{2}\omega_{\boldsymbol{\lambda}}^2\mathbf{Q}_{\boldsymbol{\lambda}}^2\right), \tag{6}$$

where $\mathbf{P}_\lambda$ is the momentums of the oscillator, $\mathbf{x}$ and $\mathbf{Q}_{\boldsymbol{\lambda}}$ are the coordinates of the particle and the oscillators, respectively.

We show an analytical approach to the restricted motion of the Brownian particle by a zero temperature environment. Let us consider the partial trace of the evolution operator. We assume that a system A and a system B is composited by coupling interaction. We can find the effective time evolution operator for "system B" by tracing over the degree of freedom of "system A". Usually it contains some effects of "system A" on "system B"

$$\mathbf{U}_{eff}(t) = tr_B(\mathbf{U}_A(t)). \tag{7}$$

This way had been presented by Suratsuji and Iida[9] for considering the Berry phase. We can find the particle trace of the time evolution operator for the Brownian particle by using the coherent state $|a\rangle$ tracing over the environmental degree of freedom α,

$$\mathbf{U}_{eff}(t) = tr_{en}(\mathbf{U}(t,t_0)) = \int d^2a\ \langle a|\,\mathbf{U}(t,t_0)\,|a\rangle\,. \tag{8}$$

The time evolution operator of the Brownian particle interacting with the environment is

$$\mathbf{U}(t,t_0) = \exp\left[\frac{i}{\hbar}\mathbf{H_{total}}(t-t_0)\right]. \tag{9}$$

Then substituting Eq.(9) into Eq.(8), we have

$$\mathbf{U}_{eff}(t) = \int d^2a \ \langle a| \exp\left[\frac{i}{\hbar}\mathbf{H_{total}}(t-t_0)\right] |a\rangle \,. \tag{10}$$

In Eq.(10) one naturally picks up the transition amplitude for the quantum process starting from the initial state $|\alpha\rangle$ and returning to the same state. Now we have the effective propagator as

$$K(x_b, x_a; t-t_0) = \int d^2a \ \langle x_b| \langle\alpha| \exp\left[\frac{i}{\hbar}\mathbf{H_{total}}(t-t_0)\right] |\alpha\rangle |x_a\rangle \,. \tag{11}$$

Then, with the aid of the time-discretization together with the completeness relation holding for x, we get $K(x_b, x_a; t-t_0)$ as

$$\int D[x(t_{N-1})] \ldots \int D[x(t_1)] \langle x_b| \langle\alpha| \mathbf{e}^{\left[\frac{i}{\hbar}\mathbf{H_{total}}\epsilon\right]} |x_{N-1}\rangle \ldots \langle x_1| \mathbf{e}^{\left[\frac{i}{\hbar}\mathbf{H_{total}}\epsilon\right]} |\alpha\rangle |x_a\rangle \tag{12}$$

with $\epsilon = (t-t_0)/N$. Further noting the relation for $\epsilon \approx 0$,

$$\langle x_k| \exp\left[\frac{i}{\hbar}\mathbf{H_{total}}\epsilon\right] |x_{k-1}\rangle \tag{13}$$

$$\simeq \langle x_k| \exp\left[\frac{i}{\hbar}\mathbf{H}_{sys}\epsilon\right] |x_{k-1}\rangle \exp\left[\frac{i}{\hbar}\mathbf{H}_{sys}\epsilon + \mathbf{H}_{in+en}\epsilon\right] \tag{14}$$

$$= \int dp_k \exp\left[\frac{i}{\hbar}\left(p_k\left(x_k - x_{k-1}\right) - \mathbf{H}_{sys}\epsilon\right)\right] \exp\left[\frac{i}{\hbar}\mathbf{H}_{sys}\epsilon + \mathbf{H}_{in+en}\epsilon\right]. \tag{15}$$

So Eq. (11) can be expressed as

$$K(x_b, x_a; t-t_0) = \int D[x(t)] \ \mathbb{T}_a[x(t)] \exp\left[\frac{i}{\hbar}S_0[x(t)]\right] \tag{16}$$

with $S_0[x(t)] = \int_{t_0}^{t} L_{sys}(x, \dot{x}, t)dt'$ is the action of the Brownian particle and $\mathbb{T}_\alpha[x(t)]$ is just the internal transition amplitude,

$$\mathbb{T}_a[x(t)] = \langle a| \exp\left[\frac{i}{\hbar}\int_{t_0}^{t} \mathbf{H}_{in+en}(x(t')) \ dt'\right] |a\rangle \,. \tag{17}$$

Equation (17) is the coherent state propagator for the harmonic oscillator with the external force $x(t)$. It implies that the motion of the Brownian particle is restricted by the internal transition amplitude $\mathbb{T}_\alpha[x(t)]$ as the weight functional in the constrained paths integral. The coherent state propagator for the time-dependent forced harmonic oscillator has been exactly solution[10] as

$$\mathbb{T}_a[x(t)] = e^{\left[\sum_\lambda |a_\lambda|^2 e^{i\omega_\lambda (t-t_0)} + \frac{i}{\hbar}\int_{t_0}^{t} G(\sigma)x(\sigma) \ d\sigma - \int_{t_0}^{t} d\sigma \int_{t_0}^{t} d\beta \ \Theta(\sigma-\beta)x(\sigma)\alpha(\sigma,\beta)x(\beta)\right]} \tag{18}$$

where

$$G(\sigma) = \sum_\lambda \frac{C_\lambda}{\sqrt{2m_\lambda\omega_\lambda}} \left[a_\lambda^* e^{i\omega_\lambda t} + a_\lambda e^{-i\omega_\lambda t}\right] \tag{19}$$

and

$$\alpha(\sigma,\beta) = \sum_\lambda \frac{C_\lambda^2}{2m_\lambda\omega_\lambda} e^{-i\omega_\lambda(\sigma-\beta)}. \tag{20}$$

Here $\alpha(\sigma,\beta)$ is the Fourier transforms of the spectral function of the environmental coupling. Now we have the coherent state propagator for the harmonic oscillator with the external force $x(t)$. Now we define the interference parameter

$$IF = \sum_\lambda |a_\lambda|^2 e^{i\omega_\lambda(t-t_0)}. \tag{21}$$

We assume that environment is modeled by many of harmonic oscillators and has the completely random in the natural frequency ω_λ. Then we can estimate the interference parameter IF to be zero. This approximation is the same as the random phase approximation.[11] Now we can approximate the coherent state propagator to be

$$\mathbb{T}_a[x(t)] = e^{\left[\frac{i}{\hbar}\int_{t_0}^t G(\sigma)x(\sigma)\,d\sigma - \int_{t_0}^t d\sigma \int_{t_0}^t d\beta\;\Theta(\sigma-\beta)x(\sigma)\alpha(\sigma,\beta)x(\beta)\right]}. \tag{22}$$

Then the effective propagator $\langle x_b|\,\mathbf{U}_{eff}(t)\,|x_a\rangle$ for the Brownian system becomes

$$\int D[x(t)] \int d^2a e^{\left[\frac{i}{\hbar}\int_{t_0}^t G(\sigma)x(\sigma)\,d\sigma - \int_{t_0}^t d\sigma \int_{t_0}^t d\beta\;\Theta(\sigma-\beta)x(\sigma)\alpha(\sigma,\beta)x(\beta) + S_0[x(t)]\right]}. \tag{23}$$

Using the complex number integral relation

$$\delta(B)\delta(B^*) = \int d^2A\;\exp\left[AB^* + A^*B\right], \tag{24}$$

Eq. (20) can be rewritten as

$$\langle x_b|\,\mathbf{U}_{eff}(t)\,|x_a\rangle = \int D[x(t)]\;\delta(C)\delta(C^*) e^{\left[\frac{i}{\hbar}S_0[x(t)] - \int_{t_0}^t d\sigma \int_{t_0}^t d\beta\;\Theta(\sigma-\beta)x(\sigma)\alpha(\sigma,\beta)x(\beta)\right]} \tag{25}$$

where $C(t)$ is a correlation function between the stochastic processes $Z(\sigma)$ and the motion of the Brownian particle $x(\sigma)$,

$$C(t) = \int_{t_0}^t Z(\sigma)x(\sigma)d\sigma, \tag{26}$$

where the stochastic processes $Z(\sigma)$ define as

$$Z(\sigma) = \sum_\lambda \frac{C_\lambda}{\sqrt{2m_\lambda\omega_\lambda}}\; e^{i\omega_\lambda\sigma}. \tag{27}$$

Equation (22) implies that the motion of the Brownian particle is restricted by the constraint condition,

$$C(t) = 0, C^*(t) = 0. \tag{28}$$

The result of this is that the state of the Brownian should be moved along a class of paths which satisfy the constraint condition, Eq. (28).

4. Conclusion

We have presented alternatively an approach to consider the QBM problem. This approach provides new physical pictures which describes the effect of the environment on the dynamics of the Brownian motion in the form of the constraint on the Brownian particle. We consider the partial trace of the evolution operator for the Brownian particle coupling with environment. Then we obtain the effective time evolution operator for the Brownian particle and the constraint condition. Therefore any path $x(t)$ satisfying the constraint condition, Eq. (28), is possible while no other path is impossible. Now the effect of the environment on the dynamics of the system can be seen as the interplay of the restricted motion, which act on the dynamics of the Brownian particle.

Acknowledgements

We acknowledge the financial support of Kasetsart University Research and Development Institute (KUDRI). SB wishes to thank the Department of Physics, Faculty of Science, Kasetsart University, for partial support. VK and PP would like to thank Faculty of Science and Center of Excellence in Forum for Theoretical Science Chulalongkorn University for partial support.

References

1. R. P. Feynman and A. R. Vernon, *Ann. Phys. (N.Y)* **24**, 118(1963); A. J. Leggett *et al.*, *Rev. Mod. Phys.* **59**, 1(1987); H. Grabert, P. Schramm and G. L. Ingold, *Phys. Rep.* **168**, 115 (1988).
2. P. Hanggi, in *Noise and Chaos in Nonlinear Dynamical Systems*, Proceedings of the Workshop, Turin, Italy, 1989, eds. F. Moss, L. A. Lugiato, and W. Schleich (CCambridge University Press, Cambridge, England, 1989), Vol. 1: J. M. Sancho and M. San Miguel, *ibid.*; P. Grigolini, *ibid.*
3. G. Lindblad, *Commun. Math. Phys.* **48**, 119 (1973); Ph. Pechukas, in *Large Scale Molecular Systems*, eds. W. Gans *et al.* (Plenum Press, New York, 1991); L. Diosi, *Physica* (Amsterdam) **199A**, 517 (1993); SGnutzmann and F. Haak, *Z. Phys. B* **101**, 263 (1996); A. Rocco and P. Grigolini, *Phys. Lett. A* **60**, 538 (1999).
4. W. G. Unruh and W. H. Zurek, *Phys. Rev. D* **40**, 1071 (1989).
5. A. O. Caldeira and A. J. Leggett, *Physica A* **121**, 587 (1983).
6. P. Mohanty, E.M.Q. Jariwala and R.A. Webb, *Phys. Rev. Lett.* **78**, 3366 (1997).
7. B.L. Altshuler, A.G. Aronov and D.E. Khmelnitsky, *J. Phys. C* **15**, 7367(1982).

8. D. Sokolovski and J.N.L. Connor, *Phys. Rev. A* **44**, 1500 (1991); D. Sokolovski, S. Brouard and J.N.L. Connor, *Phys. Rev. A* **50**, 1240 (1994); D. Sokolovski, *Phys. Rev. A* **57**, 1469 (1998); D. Sokolovski and Y. Liu, *Phys. Lett. A* **281**, 207 (2001).
9. H. Suratsuji and S. Iida, *Prog. Theor. Phys.* **74**, 439 (1985).
10. C. Itzykson and J. B. Zuber, *Quantum Field Theory* (McGraw Hill, New York, 1988).
11. H. Ehrenreich M. H. Cohen, *Phys. Rev.* **115**, 786 (1959).

AUTHOR INDEX